Heterick Memorial Library
Ohio Northern University
Ada, Ohio 45810

GALLIUM ARSENIDE DIGITAL INTEGRATED CIRCUIT DESIGN

McGraw-Hill Series in Electrical Engineering

Consulting Editor

Stephen W. Director, *Carnegie-Mellon University*

Circuits and Systems
Communications and Signal Processing
Control Theory
Electronics and Electronic Circuits
Power and Energy
Electromagnetics
Computer Engineering
Introductory
Radar and Antennas
VLSI

Previous Consulting Editors

Ronald N. Bracewell, Colin Cherry, James F. Gibbons, Willis W. Harman, Hubert Heffner, Edward W. Herold, John G. Linvill, Simon Ramo, Ronald A. Rohrer, Anthony E. Siegman, Charles Susskind, Frederick E. Terman, John G. Truxal, Ernst Weber, and John R. Whinnery

Electronics and Electronic Circuits

Consulting Editor

Stephen W. Director, *Carnegie-Mellon University*

Colclaser and Diehl-Nagle: *Materials and Devices for Electrical Engineers and Physicians*
Franco: *Design with Operational Amplifiers and Analog Integrated Circuits*
Grinich and Jackson: *Introduction to Integrated Circuits*
Hamilton and Howard: *Basic Integrated Circuits Engineering*
Hodges and Jackson: *Analysis and Design of Digital Integrated Circuits*
Long and Butner: *Gallium Arsenide Digital Integrated Circuit Design*
Millman and Grabel: *Microelectronics*
Millman and Halkias: *Integrated Electronics Analog, Digital Circuits, and Systems*
Millman and Taub: *Pulse, Digital, and Switching Waveforms*
Paul: *Analysis of Linear Circuits*
Roulston: *Bipolar Semiconductor Devices*
Schilling and Belove: *Electronic Circuits: Discrete and Integrated*
Smith: *Modern Communication Circuits*
Sze: *VLSI Technology*
Taub: *Digital Circuits and Microprocessors*
Taub and Schilling: *Digital Integrated Electronics*
Wait, Huelsman, and Korn: *Introduction to Operational and Amplifier Theory Applications*
Yang: *Microelectronic Devices*
Zambuto: *Semiconductor Devices*

VLSI

Consulting Editor

Stephen W. Director, *Carnegie-Mellon University*

Elliot: *Microlithography: Process Technology for IC Fabrication*
Fabricius: *Introduction to VLSI Design*
Geiger, Allen, and Strader: *VLSI Design Techniques for Analog and Digital Circuits*
Long and Butner: *Gallium Arsenide Digital Integrated Circuit Design*
Offen: *VLSI Image Processing*
Ruska: *Microelectronic Processing: An Introduction to the Manufacture of Integrated Circuits*
Seraphim: *Principles of Electronic Packaging*
Sze: *VLSI Technology*
Tsividis: *Operation and Modeling of the MOS Transistor*
Walsh: *Choosing and Using CMOS*

GALLIUM ARSENIDE DIGITAL INTEGRATED CIRCUIT DESIGN

Stephen I. Long
Steven E. Butner

Department of Electrical & Computer Engineering
University of California, Santa Barbara

McGraw-Hill Publishing Company

New York St. Louis San Francisco Auckland Bogotá Caracas
Hamburg Lisbon London Madrid Mexico Milan Montreal
New Delhi Oklahoma City Paris San Juan São Paulo
Singapore Sydney Tokyo Toronto

This book was set in Times Roman by Publication Services.
The editors were Alar E. Elken and John M. Morriss;
the production supervisor was Janelle S. Travers.
The cover was designed by John Hite.
Project supervision was done by Santype International Ltd. and Publication Services.
R. R. Donnelley & Sons Company was printer and binder.

GALLIUM ARSENIDE DIGITAL INTEGRATED CIRCUIT DESIGN

Copyright © 1990 by McGraw-Hill, Inc. All rights reserved. Printed in the United States of America. Except as permitted under the United States Copyright Act of 1976, no part of this publication may be reproduced or distributed in any form or by any means, or stored in a data base or retrieval system, without the prior written permission of the publisher.

1 2 3 4 5 6 7 8 9 0   DOC DOC   9 5 4 3 2 1 0

ISBN 0-07-038687-0

Library of Congress Cataloging-in-Publication Data

Long, Stephen I.
    Gallium arsenide digital integrated circuit design / Stephen
I. Long, Steven E. Butner.
        p.   cm. — (McGraw-Hill series in electrical engineering.
Electronics and electronic circuits) (McGraw-Hill series in
electrical engineering. VLSI)
    Includes index.
    ISBN 0-07-038687-0. — ISBN 0-07-038688-9 (solutions manual)
    1. Digital integrated circuits—Design and construction.
2. Gallium arsenide semiconductors—Design and construction.
I. Butner, Steven E.   II. Title.   III. Series.   IV. Series:
McGraw-Hill series in electrical engineering. VLSI, electronics, and
electronic circuits.
TK7874.L66   1990
621.381'5—dc20   89-12671

ABOUT THE AUTHORS

Stephen I. Long is Professor of Electrical and Computer Engineering at the University of California, Santa Barbara. He has often served as a consultant, working with such companies as DELCO, Digital Equipment, Narda Microwave Semiconductor, and Rockwell International. In 1988 he was a research visitor at GEC Hirst Research Centre, UK. Dr. Long was previously manager of GaAs IC design and test at Rockwell International Science Center and a Senior Engineer at Varian Associates. He received his B.S. in Engineering Physics from University of California, Berkeley and his M.S. and Ph.D. in Electrical Engineering from Cornell University. For the past eight years Dr. Long has taught courses at UCSB in integrated circuit fabrication and digital and analog circuit design. He has also been active in short courses in GaAs ICs and picosecond electronics. His current research is in the area of heterostructure devices and the design of GaAs digital and analog ICs. Dr. Long is a member of the American Scientific Affiliation and Tau Beta Pi, and a senior member of the IEEE.

Steven E. Butner is Associate Professor of Electrical and Computer Engineering at the Santa Barbara campus of the University of California. He is also a consultant to the private and public sector, working with such companies as Allen-Bradley, Culler Scientific Systems, and the DELCO Systems Division of General Motors. Dr. Butner has also been a member of the Technical Staff at Bell Northern Research, a Research Associate at the Center for Reliable Computing at Stanford University, and Lead Program Engineer for Honeywell's Utility and Process Automation Products Division. Dr. Butner received his Ph.D. in Electrical Engineering from Stanford University in 1981. Since that time he has taught courses at both the graduate and undergraduate level in computer architecture,

LSI/VLSI design, and integrated circuit testing. He is active in research on the design of high-performance computers, both general- and special-purpose, with a particular emphasis on real-time and fault-tolerant systems. During 1989 he was a visiting researcher at Siemens Research Labs in Munich, Germany. Dr. Butner is a member of ACM, IEEE, and Tau Beta Pi.

To Molly and Lynda
and our families

CONTENTS

Preface — xvii

Introduction — 1

1 Fundamentals of GaAs Devices — 11
1.1 Comparison of GaAs and Silicon for High-Speed Digital Devices — 11
1.2 Schottky Barrier Diodes and Ohmic Contacts — 18
1.3 Principles of GaAs MESFET Operation and Design — 32
   1.3.1 MESFET Drain Current — 33
   1.3.2 Drain Conductance — 37
   1.3.3 Transconductance and Gain-Bandwidth Product — 39
   1.3.4 Source Resistance — 43
   1.3.5 Gate Conduction — 44
1.4 MESFET Second-Order Effects — 45
   1.4.1 Backgating — 46
   1.4.2 Drain Lag Effects — 48
   1.4.3 Subthreshold Current — 50
   1.4.4 Temperature Dependence — 50
1.5 MESFET Fabrication Methods — 53
   1.5.1 Self-aligned MESFET — 55
   1.5.2 Directly-Aligned MESFET — 56
1.6 Heterostructure FETs and Bipolar Transistors — 58
   1.6.1 Heterostructure FETs — 61
   1.6.2 Heterostructure Bipolar Transistors — 69

2 Models for GaAs Devices and Circuits — 79
- 2.1 Schottky Barrier Diode — 80
- 2.2 GaAs MESFET Models — 86
  - 2.2.1 SPICE JFET Model — 88
  - 2.2.2 Modifications to the SPICE JFET Model—Hyperbolic Tangent — 95
  - 2.2.3 Modifications to the SPICE JFET Model—Raytheon Model — 100
  - 2.2.4 Dual-Gate MESFETs — 107
  - 2.2.5 Parasitic Capacitances — 108
- 2.3 Modeling MESFET Second-Order Effects — 110
  - 2.3.1 Backgating Model — 111
  - 2.3.2 Subthreshold Current and Substrate Leakage Current Models — 114
  - 2.3.3 Drain Current Transient (Lag) Effects — 116
  - 2.3.4 Temperature Effects — 120
- 2.4 Heterostructure Bipolar Transistor Model — 120
  - 2.4.1 DC Ebers-Moll Model — 122
  - 2.4.2 Parasitic Resistances and Charge Storage in the HBT — 124
  - 2.4.3 Additional Complexities — 129
  - 2.4.4 Summary of Model Parameters for HBT Model — 133

3 Logic Circuit Design Principles — 140
- 3.1 Noise in Digital Circuits — 142
  - 3.1.1 Sources of Noise — 142
  - 3.1.2 Definitions of Noise Margin — 143
  - 3.1.3 Factors Which Influence the Noise Margin — 148
- 3.2 DC Design of GaAs MESFET/HFET Logic Gates — 149
  - 3.2.1 Graphical DC Design Example — 150
  - 3.2.2 Optimization of DC Design — 157
  - 3.2.3 Design of NOR and NAND Gates — 159
  - 3.2.4 Power Dissipation — 162
  - 3.2.5 Yield of LSI/VLSI Circuits — 164
- 3.3 Transient Design of MESFET/HFET GaAs Logic Circuits — 170
  - 3.3.1 Definitions — 171
  - 3.3.2 Estimation of Propagation Delay — 172
  - 3.3.3 Analysis by CAD Simulation — 183
  - 3.3.4 Dependence of Delay—Width Optimization — 187
  - 3.3.5 Dependence of Delay—Fan-in and Fan-out — 188
  - 3.3.6 Maximum Frequency of Operation — 189
  - 3.3.7 Dynamic Noise Margin — 190

4 Logic Circuit Design Examples — 195
- 4.1 Depletion Mode Logic Circuits — 196
  - 4.1.1 Unbuffered FET Logic — 196
  - 4.1.2 Capacitive Feedforward Logic Circuits — 198
  - 4.1.3 Source Follower Design—Buffered FET Logic — 200
  - 4.1.4 Schottky Diode FET Logic — 206
- 4.2 Enhancement Mode Logic Circuits — 210
  - 4.2.1 Direct-Coupled FET Logic (DCFL) — 211
  - 4.2.2 Source-Coupled FET Logic (SCFL) — 216

		4.2.3 Enhancement/Depletion Buffered FET Logic	223
		4.2.4 Pass Transistor Logic	225
		4.2.5 Complementary JFET Logic	230
	4.3	Dynamic Logic Circuits	231
		4.3.1 Charge Storage Limitations	232
		4.3.2 Single-Phase Dynamic Circuits	233
		4.3.3 Two-Phase Dynamic Circuits	237
		4.3.4 Asynchronous Dynamic Circuits	238

5 Interconnections — 245

5.1	Transmission Line Theory	245
5.2	Analysis of Interconnections	253
5.3	Power Supply and Ground Design	260
	5.3.1 Resistance	261
	5.3.2 Electromigration	262
	5.3.3 Inductance	263
5.4	Delay Estimation	269
	5.4.1 Crossover Capacitance	270
	5.4.2 Effective Source Resistance and Attenuation	271
	5.4.3 Interconnect Delay	280
5.5	Crosstalk	287
	5.5.1 Symmetrical Even and Odd Modes and Their Parameters	289
	5.5.2 Analysis by Superposition of Modes on Symmetric Lines	291
	5.5.3 Case of Purely Backward Coupling	294
	5.5.4 Case of Purely Forward Coupling	295
	5.5.5 Forward and Backward Coupling	297
	5.5.6 RC Approximation	299
5.6	Packaging and Circuit Board Interconnections	301
	5.6.1 Impedance Discontinuities in Point-to-Point Interconnections	303
	5.6.2 Distributed Interconnections	309
	5.6.3 Power Supply and Ground Connections	311

6 Test Methods for Very-High-Speed Digital ICs — 321

6.1	Wafer Probing Techniques	321
6.2	On-chip (Built-in) Test Approaches	325
	6.2.1 Propagation Delay Test Structures	326
	6.2.2 Built-in Test Approaches	330
	6.2.3 Scan-Path Techniques	334
	6.2.4 On-chip Pseudo-Random Testing	337
6.3	Electrooptical Methods	341

7 Using GaAs in Systems — 346

7.1	Architectural Considerations	347
	7.1.1 Architectural Fundamentals	348
	7.1.2 Architectural Problems Unique to GaAs Digital Systems	349
	7.1.3 Synchronous Design	354
	7.1.4 Bus-Type Connections	356
	7.1.5 Interconnection Networks	361

	7.1.6	Point-to-Point Connections	363
	7.1.7	Transaction Flows	364
	7.1.8	Pipelined Systems	364
	7.1.9	RISCs and Pipelining—the CPU/Memory Speed Disparity	366
7.2	I/O and Chip-to-Chip Issues		368
	7.2.1	GaAs-to-GaAs Driver	368
	7.2.2	GaAs-to-Silicon ECL Driver	370
	7.2.3	GaAs-to-GaAs Receiver	371
	7.2.4	Silicon ECL-to-GaAs Receiver	372
	7.2.5	GaAs-to-GaAs 50-Ω Distributed-NOR Bus Driver	372
	7.2.6	GaAs-to-GaAs Bus-Type Receiver	373
	7.2.7	Threshold-Adjustable GaAs Receiver	374
7.3	Asynchronous Techniques		376
	7.3.1	Self-Timed Signaling	376
	7.3.2	Ternary Algebra and Combinational Network Synthesis	378
	7.3.3	Ternary Logic Circuits	379
7.4	Summary		380

8 Physical Design — 384

8.1	Foundry Interface		284
	8.1.1	Scalable Design Style	385
	8.1.2	Caltech Intermediate Format—CIF	386
8.2	Design Rules for the Selective Implant D-Mode Process		388
	8.2.1	Mask Layers	388
	8.2.2	Width and Separation Rules for $\lambda = 0.5$ μm	390
	8.2.3	Active Devices	390
	8.2.4	Contact Structures	393
	8.2.5	Miscellaneous Rules	394
	8.2.6	Electrical Design Rules	395
	8.2.7	CIF Layers and Colors	396
8.3	Design Rules for the Self-Aligned E/D Process		396
	8.3.1	Mask Layers	397
	8.3.2	Width and Separation Rules for $\lambda = 0.4$ μm	397
	8.3.3	Active Devices	398
	8.3.4	Contact and Bonding Pad Structures	398
	8.3.5	Miscellaneous Rules	400
	8.3.6	Electrical Design Rules	401
8.4	Design Examples		401
	8.4.1	D-mode Four-Input NOR	401
	8.4.2	E/D Two-Input NOR	402
	8.4.3	SDFL Three-Input NOR	402
	8.4.4	FET Logic Latch	402
	8.4.5	A Simple GaAs-to-GaAs Pad Driver	403
8.5	Synthesis Using Generic Structures		403
	8.5.1	Programmable Logic Arrays	404
	8.5.2	Finite State Machines	407
8.6	Some Architectural Building Blocks		409
	8.6.1	Steering Logic	409

	8.6.2 Registers		410
	8.6.3 Control and Decode		412

9 Computer-Aided Design Tools and Techniques — 414

- 9.1 Needs of the Designer — 415
- 9.2 The MAGIC VLSI Design System — 418
  - 9.2.1 MAGIC Data Base — 419
  - 9.2.2 MAGIC Display and Plotting Styles — 423
  - 9.2.3 Design Rule Checks — 424
  - 9.2.4 Circuit Extraction — 428
  - 9.2.5 Ext2sim — 432
  - 9.2.6 Foundry Interface — 436
  - 9.2.7 MAGIC's User Interface — 436
- 9.3 Analysis Tools and Techniques — 441
  - 9.3.1 Electrical Rules Check—statG — 441
  - 9.3.2 Switch-Level Simulation — 445
  - 9.3.3 Converting Netlists to SPICE Decks—sim2spice — 453
  - 9.3.4 SPICE — 454
- 9.4 Synthesis Tools and Examples — 455
  - 9.4.1 PLA Generation Tools — 455

Appendixes

- A MAGIC Technology File for Selective-Implant D-Mode Process — 461
- B MAGIC Technology File for GaAs Self-Aligned E/D-Mode Process — 469

Index — 481

PREFACE

This textbook is intended to be a comprehensive guide for the integrated circuit design student who wishes to learn about the opportunities provided by gallium arsenide FETs and bipolar transistors for implementing very high-speed digital circuits. The textbook is focused on communicating the basic principles of very-high-speed digital circuit design with emphasis on providing a logical sequence of material suitable for a 10 or 15 week graduate course. It is expected that the course may include a design project. We believe that the existence of such a book will encourage the offering of GaAs IC design courses at many universities.

A one quarter course on GaAs digital design has been taught at UCSB. The syllabus includes the following material: Chapter 1, Sections 1.0–1.3; Chapter 2, Sections 2.1–2.3; Chapter 3; selected examples from Chapter 4; Chapter 5, Sections 5.1–5.4, 5.6; Chapter 8, the appropriate design rules and examples. A design project is also required. The material presented in Chapter 9 is learned by the students in the VLSI laboratory in conjunction with the laying out and verification of their projects. If a semester were available, it would be desireable to spend more time discussing the theory behind the above selections. Chapters 6 and 7 should be covered in that case, and some discussion of heterojunction transistors would be desireable. Although there is more material provided in the book than can be covered in a one quarter or semester course — especially if the design project is included — much of this material is useful to the reader for future reference and is organized for self teaching, aided by many examples provided in each chapter.

Photographs 8-1 and 8-2 (color insert) are of a 1987 multiproject wafer designed as part of the UCSB GaAs integrated circuits design course and processed by the GaAs MESFET (depletion-mode) foundry of Rockwell International, Thousand Oaks, Calif. The following projects were included on this multi-project run:

Projects Related to the UC-Santa Barbara 1-GHZ Digital IC Tester:

(*a*) Phasing and fanout chip (similar to the structure described in Fig. 7-5)
(*b*) Clock and control chip (a critical synchronizing element in the tester)
(*c*) Thirty-two-bit serial shift register
(*d*) Four-bit parallel/serial shift register

Class-Related Projects:

(*a*) Static RAM
(*b*) Three-bit flash A/D converter
(*c*) Hybrid median FIR filter
(*d*) Dynamic (hot clock) shift register
(*e*) GaAs-to-ECL output driver with temperature compensation

Test Structures:

(*a*) Test strips
(*b*) Schottky metal linewidth test
(*c*) FET addressable matrix
(*d*) Various standard cell test strips
(*e*) Ternary logic test cells

The class-related projects were the result of the third MPC GaAs course taught at UCSB in the winter quarter 1987. Computer-aided design support came from MAGIC (a UC-Berkeley layout editor widely used for VLSI design courses at many universities) and SPICE3 (a circuit simulation tool originally developed by UC-Berkeley). The students were required to verify functionality of their circuit using the dc analysis capability of SPICE3 with the GaAs MESFET model. Emphasis was on optimizing the noise margins of the logic circuit rather than on achieving maximum speed. Layout of the chip was followed by estimation of parasitic capacitances and SPICE simulation of critical timing paths. The layouts were required to be free of design rule violations. A standard pad layout was required so that wafer level testing could utilize a custom high speed (8 GHz) probe card.

Besides serving as a text for university courses the book should also be a reference or self-teaching resource for the industrial users who find themselves confronting an unfamiliar technology. Although far too comprehensive for complete coverage in a short course, the depth of coverage can be selected by the user, as there will be many subsections of the book which can be omitted for a quick introductory class. The book will be invaluable as a reference for students taking such courses.

It is assumed that the reader has completed a standard undergraduate course in analog electronics at the third year (junior) level. Much of the material presented in chapters 2 through 6 requires this knowledge. A similar undergraduate course in digital electronics would be helpful, but is not required. An undergraduate course in semiconductors and devices is necessary to understand chapter 1 and parts of chapter 2. While a course in VLSI design would be helpful for the discussion of layout rules and the CAD tools (chapters 8 and 9), it is not essential as the description contained in these chapters is comprehensive enough to be useful as a text. Finally, familiarity with the SPICE circuit simulation tool is assumed. This is often included as part of an undergraduate electronics or networks course. If the reader is inexperienced with SPICE, the following book is recommended reading: *SPICE. A Guide to Circuit Simulation and Analysis Using PSpice*, by P. Tuinenga, Prentice-Hall, 1988.

The CAD tools used for simulation and layout were selected because of their widespread acceptance and usage in the university VLSI design community and their public domain availability in both university and commercial versions. While SPICE and MAGIC are the major foundations for the design course at UCSB, mention of other tools is also made in Chapter 9. Some of these other tools may be more suitable for less custom or performance-oriented work (gate arrays or standard cells, for example) than is emphasized in this book. If SPICE and MAGIC are selected for use for a GaAs IC design course, the material in this book will make their application to GaAs straightforward by providing detailed information on transistor models and parameters and sample technology files which may be readily modified to support most current foundry processes.

The authors would like to acknowledge the Defense Advanced Research Projects Agency (W. Bandy and J. Toole) and the Jet Propulsion Laboratory (M. Buehler and B. Blaes) for making this book possible through their support of the GaAs IC fabrication costs incurred by the UCSB GaAs IC design course and for their support of research assistantships.

The work of Prof. George Matthaei in the preparation of section 5.5 and for many helpful discussions on the theory of coupled line effects is also gratefully acknowledged. The authors are much indebted to the graduate students who have contributed to the development of many of the ideas and techniques presented in this text and for thoughtful reviews of the manuscript and homework problems. To mention some of these by name: A. Fiedler, D. Fouts, J. Johnson, R. Lewis, K. Macaulay, K. Nary, T. Nguyen, S. Peltan, C.-H. Shu, and L. Yang. Many other valuable comments and suggestions came from: D. Richard Decker, Lehigh University; A. Gopinath, University of Minnesota; Sung Mo Kang, University of Illinois, Urbana-Champaign; T. E. Schlesinger, Carnegie-Mellon University;

Gregory L. Snider, University of California, Santa Barbara; and George J. Valco, Ohio State University. Many hours of help in typing and formatting the manuscript were graciously provided by K. Kramer and A. Sykes. P. Allen and D. Fouts debugged the computer-generated art in Chapters 8 and 9. This work as well as the microphotography of J. Johnson and D. Fouts is greatly appreciated. The help of the GEC Hirst Research Centre library staff in identifying and locating many unusual references for one of us (SL) is also appreciated.

Finally, we would like to thank our families for their encouragement and support while we were preparing this manuscript instead of spending many evenings and weekends in more pleasant activities with them.

Stephen I. Long
Steven E. Butner

GALLIUM ARSENIDE DIGITAL INTEGRATED CIRCUIT DESIGN

INTRODUCTION

Gallium arsenide transistors, both FET and bipolar, are used for digital integrated circuits primarily when the application requires very high speed, and the delay and power requirements of silicon CMOS or bipolar ICs are too high. The high-speed performance of GaAs digital ICs has been excellent. The fast switching capability of these devices is based on their ability to deliver high currents with small changes in the input voltage and on their low internal capacitances.

It is important to understand how the characteristics of the active devices influence the dc levels and the transient response of high-speed logic circuits since the performance is often very sensitive to the design details of the device. Thus, the purpose of this introduction is to set the context for the material to be presented in the chapters which follow, so that the reader will understand why some knowledge of device physics and device equivalent circuit models is essential to have a complete grasp of the tools and techniques available to a digital designer working with very high speed integrated circuits. While the device characteristics and models will be discussed in more depth in Chaps. 1 and 2, some very basic models will first be introduced in this section whose elements can be related to the design of MESFET devices. To relate the device to the circuit response, some particular examples will be presented which are representative of typical logic circuit design and performance considerations. These considerations will be studied much more thoroughly in Chaps. 3 and 4.

DC DESIGN CONSIDERATIONS

A very simple, zeroth-order equivalent circuit model of a switching transistor is illustrated in Fig. 1a. This model uses an ideal switch and a channel resistance, r_{ON},

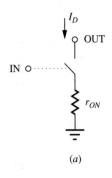

(a)

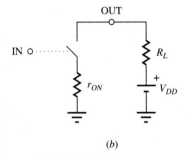

FIGURE 1
(a) Simplified equivalent circuit model of a switching transistor.
(b) Simple inverter is formed by the addition of a load resistor.

to represent the behavior of a transistor in two of its operating regions. In cutoff, the switch will be open and no current will flow through the transistor. When closed, the model represents the low-voltage, high-current region of operation (ohmic region for the FET or saturation region for a bipolar).

An inverter circuit can be constructed by providing a current to the output node of the model with a resistor R_L connected to a power supply V_{DD} as shown in Fig. 1b. When the switch is closed, the output voltage is in the low state V_{OL} which is given by the voltage divider relationship:

$$V_{OL} = \frac{V_{DD} r_{ON}}{r_{ON} + R_L} \tag{1}$$

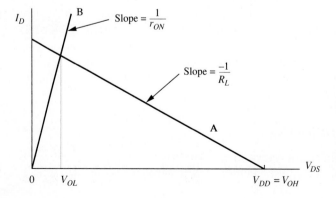

FIGURE 2
The device characteristic and load line for the inverter shown in Fig. 1b. Here, the load line (A) (with slope $-1/R_L$ and intercept at V_{DD}) is shown intersecting the device characteristic (B) (with slope $1/r_{ON}$), at voltage V_{OL}.

When the switch is open, the output voltage is in the high state V_{OH} which will be equal to V_{DD}. This is also shown on the device characteristic and load line in Fig. 2. Here, the load line A with slope $1/R_L$, representing the resistor R_L, is shown intersecting the device characteristic B, with slope $1/r_{ON}$, at voltage V_{OL}. For the open-circuit condition, the resistor voltage pulls up to V_{DD} as shown.

This simple model can be used to compare the output voltages of a MOSFET inverter typical of silicon NMOS and an n-channel MESFET inverter typical of GaAs. The NMOS inverter pair of Fig. 3a is constructed with a resistor load to be analogous to Fig. 1b. The first stage is shown driving an identical second stage. Since the input

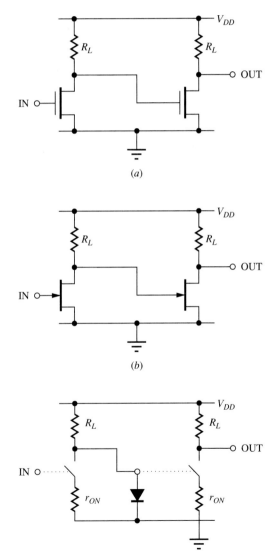

FIGURE 3
(a) MOSFET inverter pair. (b) MESFET inverter pair. (c) FET model modified by addition of a diode to represent forward gate conduction.

to the second stage presents no dc conduction to the output of stage 1, the low and high output states (logic levels) will be the same V_{OL} and V_{OH} as discussed above. However, the MESFET inverter pair shown in Fig. 3b will require a slightly different equivalent circuit model to adequately represent the loading of the second stage. This modified model is shown in Fig. 3c in which a diode is used to represent the forward conduction limit of the gate-source diode of stage 2. This forward bias voltage limit is typically between 0.6 and 0.7 V for a metal gate on n-GaAs.

The load line representation of Fig. 3c is shown in Fig. 4. Here, the diode is represented by the line C which models the diode characteristic with a line of fixed slope $1/r_D$ where r_D is associated with the parasitic series resistance of the diode. Note from Fig. 4 that the high-output logic voltage has been significantly reduced when compared with V_{OH} of the MOSFET shown in Fig. 2. V_{OH} for the MESFET inverter is given by the intersection of the diode characteristic C with the load line A. The supply voltage V_{DD} can no longer be obtained on the output if another stage is being driven.

The forward conduction limit of the MESFET gate/source junction is one of the most significant differences between the MOSFET and MESFET. It reduces the V_{OH} and therefore decreases the ability of the circuit to operate properly in a potentially noisy environment. This ability of the circuit to reject noise can be measured by defining a noise margin for the circuit. The noise margin will be proportional to the difference between the inverter switch threshold (V_{TH}) and the low and high logic voltages (V_{OL} and V_{OH}).

The channel resistance r_{ON} is a function of the channel length of the FET, and it can also be increased by the resistance of the contacts and the series resistance of the channel between the source and drain contacts and the gate. The origin of these resistances can be seen from the MESFET cross-sectional drawing in Fig. 5. For example, r_{ON} would be increased if the contact-to-gate spacing were made larger unnecessarily by poor layout practices. From (1), if r_{ON} is large, then V_{OL} will be increased and less noise margin will be obtained. To regain the noise margin, R_L could be increased to reduce V_{OL}; however, this will degrade the speed (less current

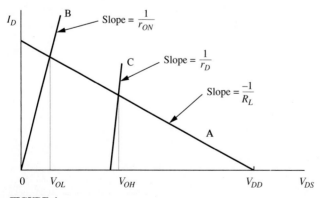

FIGURE 4
The load line representation of Fig. 3c.

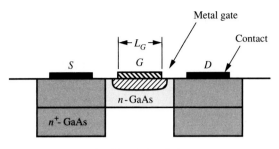

FIGURE 5
MESFET channel cross section.

available to drive the load) and will also shift V_{TH} (which could be seen if a more accurate model were used).

TRANSIENT RESPONSE CONSIDERATIONS

A more complicated model for the FET is required to see how the device parameters influence the speed of switching between the logic states. In most instances, we are concerned about the time required for a signal to propagate through a logic circuit or chain of logic circuits since this constitutes a delay between input and output which we would like to minimize. This delay will be related to the time required to charge or discharge the various load and device capacitances in the circuit.

In lower-speed technologies the delay of the logic gates is not a great concern during the design of an IC. In many instances it is sufficient to design with a simple rule set which guarantees dc functionality but which ignores intrinsic device characteristics and loading effects. Since GaAs logic circuits exist primarily because of their high intrinsic speed, a more comprehensive design strategy is needed to preserve and make the best use of this speed.

Figure 6 is a schematic diagram of an FET equivalent circuit model which can illustrate some important relationships between the delay and the device properties and parameters. C_{IN} represents the input capacitance of the FET caused by the capacitance of the gate to the channel plus any parasitic capacitances due to wiring or metal electrodes on the surface of the semiconductor. It is desirable to minimize this capacitance so that the time constant of the input can be as small as possible. The capacitance can be reduced by decreasing the area of the gate electrode or by

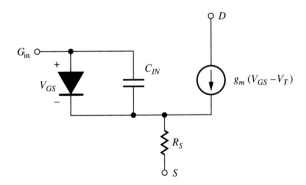

FIGURE 6
Schematic diagram of an FET equivalent circuit model which can illustrate some important relationships between the delay and the device properties and parameters.

reducing the channel doping density and thereby increasing the depth of the junction space-charge region.

The dependent current source, $I_D = g_m(V_{GS} - V_T)$, is responsible for the drain and source currents delivered by the FET. The transconductance g_m is a proportionality factor $(\partial I_D/\partial V_{GS})$ relating the control voltage on the input $(V_{GS} - V_T)$ to the output current I_D. The threshold voltage V_T represents the gate-to-source voltage at which the FET begins to become active. The transconductance factor should be as high as possible in order to obtain the largest current for the smallest change in input voltage. Since the rate at which voltage can change (slew rate) is determined by

$$\frac{dV}{dt} = \frac{I}{C} \qquad (2)$$

devices that can provide large I with small dV and small C can be switched very quickly.

The transconductance will depend inversely on the gate length and on the depth of the gate-to-channel depletion layer. The minimum gate length is limited by the process linewidth design rules. The depletion layer depth can be reduced by increasing channel doping. However, if the doping is increased, the input capacitance will also increase. The tradeoff between g_m and C_{IN} will be discussed in Chap. 3.

The source resistor R_S is a parasitic resistance related to the parasitic contact and source-to-gate spacing parts of the r_{ON} resistance discussed above. Since the source current must flow through this resistor, the control voltage V_{GS} is reduced below the input voltage V_{in} by $V_{GS} = V_{in} - I_D R_S$ thereby reducing the current I_D available from this device.

A real FET is somewhat more complicated and cannot be represented accurately by a current source dependent on a single variable (V_{GS}) with constant transconductance. The characteristics of the drain current with respect to the drain-to-source voltage, V_{DS}, are nonlinear, as can be seen from the plot in Fig. 7. In the figure, V_{GS} is increased in steps starting at V_T, leading to the family of FET characteristics similar to what is seen on a "curve tracer" type of measuring instrument. This nonlinearity is often represented by a convenient function, the hyperbolic tan-

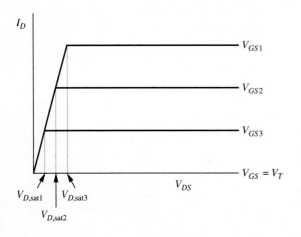

FIGURE 7
The characteristics of the MESFET drain current with respect to the drain-to-source voltage, V_{DS}. In the figure, V_{GS} is increased in steps starting at V_T. $V_{D,\text{sat}}$ is the saturation voltage.

gent, which has a similar shape to the device characteristics for $V_{DS} > 0$. The point at which the drain current saturates is labeled $V_{D,\text{sat}}$ on the figure and is controlled in the model by the parameter α indicated in the following equation:

$$I_D = \beta(V_{GS} - V_T)^2 \tanh(\alpha V_{DS}) \quad \text{for } V_{GS} > V_T \tag{3}$$

This saturation effect comes primarily from the electrons when their velocity can no longer increase in proportion to the electric field in the channel. This velocity saturation effect will therefore be sensitive to the gate length (L_G in Fig. 5) since the average electric field between source and drain will increase if the gate length is made shorter. Thus, a short gate length device will be necessary if the V_{DS} available for device operation is small. The short gatelength is also beneficial as it increases transconductance and reduces input capacitance.

Equation (3) also indicates that the transconductance is a linear function of V_{GS} if V_{DS} is constant. This is only approximately true, and the limitations to this model will be discussed in Chap. 1. Thus

$$g_m = \beta(V_{GS} - V_T) \tag{4}$$

Here, we see that the parameter β is associated with the size of both the transconductance and the drain current.

Now, as an example of how the device parameters affect the circuit speed, the circuit shown in Fig. 8a will be analyzed. This circuit is simply a chain of inverters

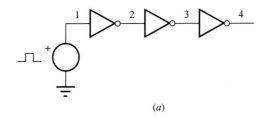

(a)

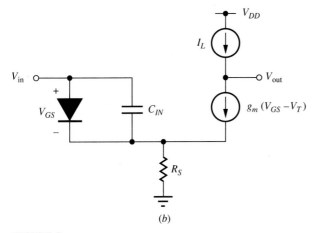

(b)

FIGURE 8
(a) Chain of inverters driven by a pulse. (b) Equivalent circuit used for each inverter.

which will be driven by a pulse with short rise and fall times relative to the speed of the inverters themselves. Each inverter consists of the circuit shown in Fig. 8b, the simplified FET model with an additional current source I_L which will pull the output up toward V_{DD} when the FET is cut off ($V_{GS} < V_T$). The propagation delay of the inverter will depend on the rate that its current sources can charge or discharge the input capacitance of the next stage. Charging current is provided mainly by I_L since the FET will be cut off when the input falls below V_T. The discharge current will be the difference between I_D and I_L.

The inverter chain can be analyzed using SPICE3, a circuit simulation program. The MESFET is represented by the simplified model shown in Fig. 8b. The dependent current source must follow equation (3) which is found in the MESFET device model provided by SPICE3. In order to evaluate the influence of gate electrode length on the switching speed, two different inverter designs are compared, a minimum gate length inverter chain (A) and a twice minimum gate length chain (B). The gate length scaling is taken into account by modifying the device parameters as shown below:

$$C_{IN}(B) = 2C_{IN}(A) \quad \text{and} \quad \beta_B = \frac{\beta_A}{2}$$

Since the gate electrode area is doubled for B, the capacitance can be doubled as well if the influence of fringing capacitances (which do not scale with gate length) are

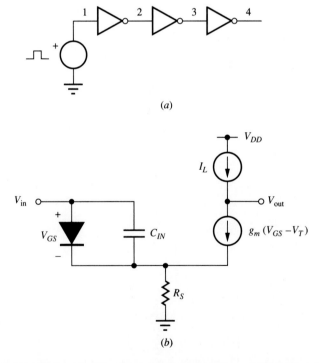

FIGURE 9
Simulated voltages at nodes 2 and 3 of Fig. 8a are plotted. Inverter B was modeled with twice the gate length of inverter A. Measuring the difference in time at the 50 percent voltage point (definition of propagation delay) we find that the delay for B between $v(2)$ and $v(3)$ is nearly a factor of four larger than A.

neglected. While the transconductance parameter scales inversely with gate length for FETs with long gate lengths ($L_G > 5 \ \mu m$), this scaling rule does not strictly apply for short gate FETs since the saturated drain current is influenced primarily by electron velocity saturation which is less sensitive to gate length. Since the drain current available is also reduced in proportion to β, the delay predicted for inverter B will be somewhat pessimistic because of the simplified scaling rules. Finally, in order to keep the same inverter dc characteristics for A and B, it is necessary to scale the load current down by the same factor, $I_L(B) = I_L(A)/2$.

The result of the simulation is plotted in Fig. 9 where a significant difference in delay is noted between the two cases. Voltages at nodes 2 and 3 of Fig. 8a are plotted for each case. Measuring the difference in time at the 50 percent voltage point (definition of propagation delay) we find that the delay for B between $v(2)$ and $v(3)$ is nearly a factor of four larger than A.

We can see from the model that the delay will increase if the capacitance between stages is increased, even if the transistors remain unchanged. A simulation of this loading effect is seen in Fig. 10, where three inverter chains were simulated. The MESFETs were the same in each chain except that the capacitance C_{IN} was set at 50, 100, and 200 fF for A, B, and C respectively. An increase in capacitance between stages could come from long interconnect wiring or large crossover areas

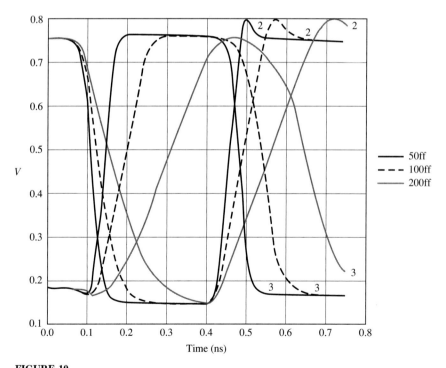

FIGURE 10
Simulated voltages at nodes 2 and 3 of Fig. 8a. Here, inverters A, B, and C have increasing load capacitances of 50, 100, and 200 fF respectively.

between overlapping metal layers. The estimation of the layout capacitance is one of the topics covered in Chap. 5.

The chapters that follow discuss many of the above topics in more depth: device operation and equivalent circuit models (Chaps. 1 and 2), circuit design and design examples (Chaps. 3 and 4), interconnections, packaging, and power supply wiring (Chap. 5), testing (Chap. 6), system design considerations (Chap. 7), layout design rules and examples (Chap. 8), and computer-aided design tools (Chap. 9).

CHAPTER 1

FUNDAMENTALS OF GaAs DEVICES

1.1 COMPARISON OF GaAs AND SILICON FOR HIGH-SPEED DIGITAL DEVICES

Gallium arsenide is a compound semiconductor that has been widely used since the late 1960s for microwave amplification (MESFET; Gunn and IMPATT diodes) and for light emission (LED). Its dominance in the microwave area is still retained with remarkable strides being made in bandwidth, low-noise, and high-power applications, both with the MESFET and with more advanced FETs employing heterostructures as a gate and/or channel barrier. In the optoelectronic area, while GaAs is still used for light-emitting diodes, more recently, rapid developments in the laser area have emphasized other compound semiconductors such as InGaAsP and InP because of the wavelength requirements of fiber optic communication.

The use of the GaAs MESFET for digital applications began in 1974 with some relatively high-power, high-speed SSI divider circuits [1], and has developed over the years into a well-established LSI technology, with some inroads into the VLSI arena. More advanced device technologies employing heterostructures for FETs and bipolar transistors have shown potential for further improvement in performance. The circuits reported have been characterized by higher speeds and lower power levels than are observed in comparable Si MOSFET or ECL circuits.

The structure of a typical GaAs MESFET and silicon MOSFET are compared in Fig. 1-1. The main differences between the devices are: (1) how the channel is formed and (2) how the gate-control electrode is coupled to the channel. The GaAs

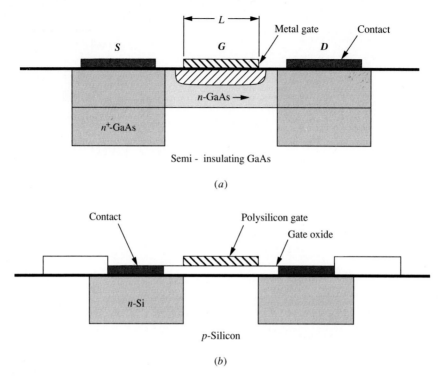

FIGURE 1-1
Schematic cross-sectional representations (not to scale) of (*a*) MESFET and (*b*) MOSFET. V_{DS} is low in both cases.

MESFET uses a thin, doped channel whose thickness is controlled through depletion by a metal/semiconductor junction. The metal gate directly contacts the channel. The MOSFET, on the other hand, forms a channel when the silicon surface is inverted (*p*-type silicon now contains a high density of electrons due to band-bending). The gate electrode is separated from the channel by a thin oxide dielectric layer.

The material and electronic properties of GaAs and silicon are compared in Table 1.1. Each of the properties listed will be briefly reviewed in the following section to indicate its influence on FET performance and in some cases the constraints imposed by the property on circuit design and layout.

The main advantages of GaAs over Si for high-speed FETs can be understood by considering the energy band diagrams and the velocity vs. electric field characteristics shown in Figs. 1-2 and 1-3. GaAs is a *direct-gap* semiconductor, meaning that the minimum energy separation between the conduction band and valence band occurs at the same momentum [Γ(000) in Fig. 1-2] [2]. Silicon, on the other hand, is an *indirect-gap* semiconductor because its conduction band minimum is separated in momentum from the valence band maximum [8]. The direct-gap property, typical of many compound semiconductors and their alloys, is helpful for light emission to occur because an electron must be able to make a transition from the conduction band to the

TABLE 1.1

Property	GaAs	Silicon
Bandgap	Direct; 1.42 eV [2]	Indirect; 1.12 eV [4]
Low-field electron drift mobility	5000 cm^2/(V·s) at $N_D = 10^{17}$/cm^{-3} [2]	800 cm^2/(V·s) at $N_D = 10^{17}$/cm^{-3} [4]
Saturated drift velocity ($\mathscr{E} \gg 10$ kV/cm)	8×10^6 cm/s [2] (100 kV/cm)	6.5×10^6 cm/s [6] (in MOS inversion layer)
Peak electron velocity	1.7×10^7 cm/s [3] at $\mathscr{E} = 3.5$ kV/cm and $N_D = 10^{17}$/cm^{-3}	6.5×10^6 cm/s [6] at $\mathscr{E} \gg 10$ kV/cm
Low-field hole (drift) mobility	250 cm^2/(V·s) [4] at $N_A = 10^{17}$/cm^{-3}	300 cm^2/(V·s) [4] at $N_A = 10^{17}$/cm^{-3}
Substrate resistivity	10^6 to 10^8 Ω·cm	Low
Surface state density	High (10^{12}/cm^{-2} or greater) [5]	Low [7] (10^{10}/cm^{-2})
Native oxide	Several reactive and unstable compounds of Ga and As	SiO$_2$; very stable

valence band without also requiring phonon scattering (a change in momentum) at the same time. The direct gap also leads to some desirable consequences for the electron transport. Note in Fig. 1-2 that in GaAs there is a secondary indirect conduction band valley, separated in energy by 0.29 eV and in momentum [(Γ(000) to $L(\frac{1}{2}\frac{1}{2}\frac{1}{2})$] from the Γ valley. In addition, the curvature of the condition bands differs, with much higher curvature for the Γ valley than for the L valley. Since the effective mass of the electrons depends inversely on the curvature, lower-energy electrons in Γ_6 are expected to have a higher *mobility* (derivative of the electron velocity with respect to the electric field) than the more energetic electrons which have transferred to the upper (L_6) band. This band structure leads to the velocity-electric field characteristic for electrons in Fig. 1-3 calculated by a Monte Carlo simulation at donor concentrations of 0, 10^{15}, and 10^{17} cm^{-3} [3]. At the 10^{17} cm^{-3} doping level typical of the GaAs MESFET channel, the velocity reaches a peak (1.7×10^7 cm/s) at a relatively low electric field of 3.5 kV/cm and then begins to decline at higher fields as more electrons are transferred to the low-mobility band. Thus, at a low electric field, the electron drift mobility is quite high [5000 cm^2/(V·s) at a doping level of 1×10^{17} [2]], providing low resistivity in thin-film layers and high electron velocity at low applied voltage. These properties are desirable for reducing device parasitic resistance and time delay in the control region.

The *saturated drift velocity* is defined as the asymptotic value of the electron velocity at high electric field. In GaAs, this velocity approaches 8×10^6 cm/s for

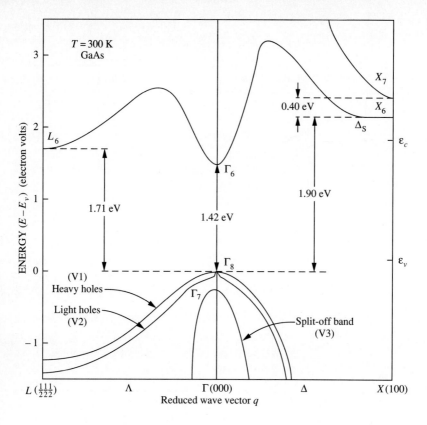
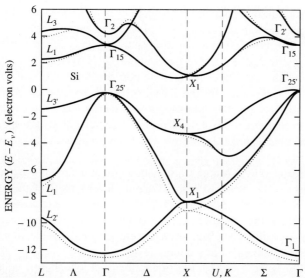

FIGURE 1-2
Energy band structure near the band gap (energy vs. wave vector) for GaAs (*from Blakemore [2]; used by permission*) and silicon (*from Chelikowski and Cohen [8]; used by permission.*)

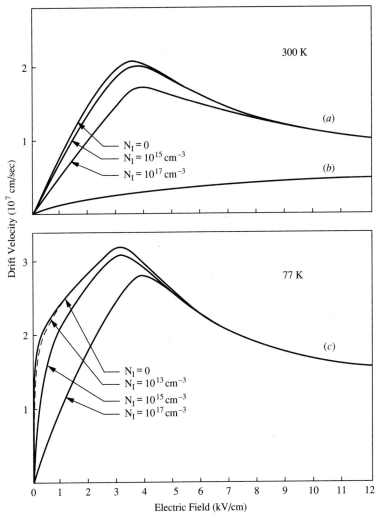

FIGURE 1-3
Electron velocity as a function of electric field for GaAs [(a) and (c) (*from Ruch and Fawcett* [3]; *used by permission*) and silicon (b) (from Sze [4]). The GaAs data is from a Monte Carlo simulation. The influence of impurity scattering for several concentrations, N_I, is shown at (a) 300 K and (c) 77 K. (b) The 300 K silicon data is experimental.

$\mathcal{E}$ = 100 kV/cm in the steady-state condition. In Si, the saturated drift velocity in an inversion layer (required for MOSFET) has been measured as 6.5×10^6 cm/s [6]. The distinction between steady-state and nonequilibrium velocity-field relationships is of significance for GaAs and other III-Vs with similar band structures because a scattering event (phonon) is required for the electron to transfer to the low-mobility, higher-energy conduction band at L. Since the probability of a momentum change is low unless the electron can travel over several mean free paths (the mean free path for a collision is of the order of 15 nm at $\mathscr{E} = 10$ kV/cm [2]), it is widely assumed that very short channel FETs will experience an electron velocity in excess of the steady-state predictions for the given electric field. This effect is often referred to as *velocity overshoot* [9] or in extreme cases as "ballistic" transport. Its effects are not too significant for channels longer than 0.5 µm.

By comparison, the velocity–electric field characteristic of *n*-type silicon [4], shown in Fig. 1-3 curve (*b*), is quite different, having a higher effective mass than GaAs and a slowly increasing velocity with electric field. The electron drift mobility for electrons in Si, doped at levels appropriate for MOS devices, is about 800 cm^2/(V·s) and even lower for bipolar doping levels. While the saturated drift velocity of electrons in Si and GaAs approach each other for very high electric fields ($\mathscr{E} \gg 10$ kV/cm), efforts to indirectly determine v_{sat} from measurement and device models have yielded velocities in excess of 1×10^7 cm/s for GaAs MESFETS [10]. The high velocity at low field of GaAs is what is desired for very high speed FETs or bipolars that also need to operate with small operating voltages to conserve power. The substantial difference between Si and GaAs FETs is displayed when comparing measurements of short channel length (say 1 µm) GaAs MESETs and Si MOSFETs. The rate of increase in current for a small increment in gate voltage (the transconductance) is much higher (typically a factor of three or four) at useful voltage levels (0.6 V above threshold) for the GaAs device than the Si device. The high g_m will help to improve the speed since a smaller voltage swing can be used to obtain the same peak current which is used to charge or discharge load capacitance.

The higher electron velocity at moderate electric field also improves the high-frequency performance of GaAs bipolar transistors since the base and collector transit times are reduced. Access resistance to the base can also be quite small as high doping can be used while still maintaining an acceptable current gain. Bipolars will be discussed further in Sec. 1.6.2.

It is worth while to contrast the advantages of *n*-channel GaAs FETs with the disadvantages of *p*-channel GaAs FETs, since complementary logic circuit designs have been effective and widely used in silicon (CMOS) for reducing static power dissipation. Here, the contrast with silicon is not so favorable; the mobility of holes in *p*-type GaAs is quite low, 250 cm^2/(V·s) being typical at 300 K for doping levels appropriate for FET applications. Also damaging to the usefulness of *p*-channel GaAs FETs is the low barrier height of most metals on *p*-type GaAs (0.45 eV) which renders the leakage currents far too high for most circuit applications (Schottky barrier junctions will be discussed in the next section). The principal device for investigation of *p*-channel GaAs FETs has been the JFET, which overcomes the latter drawback by use of a *pn* junction rather than a Schottky barrier gate [11]. The *p*-channel JFET requires an additional process step, an *n*-type gate implantation. While the hole velocity at high electric fields can be as high as that of silicon, *p*-channel devices with very short gates have not been widely explored because of the barrier height problem. Heterostructure *p*-channel GaAs FETs have demonstrated better performance at low temperatures [12]; however, considerable progress in processing and material growth is needed to make a viable complementary heterostructure FET (HFET or MODFET) process. HFETs will be discussed in more detail in Sec. 1.6.1.

Gallium arsenide is also capable of being grown in a high resistivity form, referred to as *semi-insulating* GaAs. The high resistivity, generally in the range 10^6 to 10^8 $\Omega \cdot$cm, is a consequence of compensation of residual free charge due to nonintentional shallow donors (silicon) or acceptors (carbon) by deep donor (EL2) or acceptor (chromium) levels, not because the material is exceedingly pure (intrinsic GaAs). As might be expected from this picture of delicate balancing of dopants, the material is highly sensitive to growth and annealing conditions, temperature, composition, and defects. Nevertheless, it has proven to be a useful substrate material for GaAs IC fabrication, providing acceptable isolation between devices formed by ion implantation. Contrary to the situation for silicon ICs, no additional processing is necessary to provide this isolation; however, the isolation can be enhanced through some relatively simple and noncritical extra process steps. Charge flowing through the substrate causes interactions between FETs and leads to minimum spacing rules which are less dense than those allowed by silicon MOS devices. Some of the difficulties associated with the substrate will be addressed in Sec. 1.4. An extensive discussion of the physics and characterization of semi-insulating GaAs can be found in the literature [13].

The semi-insulating substrate might be thought to produce very low parasitic capacitance for interconnections on the GaAs surface. While this is true if a metal line is far from its nearest neighbors on the surface compared to the distance to the ground plane (on the bottom of the substrate), for digital ICs this condition almost never exists. It is much more common for lines to be packed closely together in wiring channels for which the width of the conductor and the space between conductors are nearly identical. At this limit, the interconnect capacitance is not much different from that typical of closely packed lines on field oxide, an often-used configuration in silicon ICs [14]. However, the high substrate resistivity does eliminate the occurrence of any slow wave effects that could affect the propagation of high-speed signals over lossy substrates.

A lower parasitic device to substrate capacitance is obtained with semi-insulating substrates, however. In silicon high-speed MOS or bipolar technologies (except for SOS—silicon-on-sapphire), reverse-biased *pn* junctions are required for isolating the channel or collector from the conductive substrate or well. The relatively small depletion depth (< 1 μm) leads to high capacitance per unit, and therefore greater substrate capacitance for the same area device.

A very high surface state density (10^{12} cm^{-2} or higher) is typical at the surface of GaAs [5]. This high density of states within the bandgap tends to pin the Fermi level (energy at which 50 percent of the electron states are occupied) near the middle of the bandgap. Attempts to build MOS devices with deposited dielectrics have not succeeded in producing inverted surfaces. In addition, the threshold voltage has typically exhibited hysteresis. The same is true for Ga and As native oxides which exist in several chemical combinations and are not very stable. Therefore, in order to make a useful FET, the control electrode is limited to either a metal/semiconductor

(MESFET) or *pn* junction (JFET). Thus, the gate voltage is constrained by the forward conduction of these junctions.

In contrast, the silicon surface can have a very low surface state density (10^{10} cm^{-2} or less) when a high-density SiO$_2$ film is thermally grown by oxidation [8]. This property has led to the fabrication and wide usage of well-behaved MOS transistors. The oxide also serves to insulate the gate electrode from the device channel. Therefore, a wide range of gate voltages (subject to reliability constraints) can be used without significant gate conductivity. This property provides a circuit design flexibility which is an advantage of the Si MOSFET.

In the remaining parts of this chapter, the operation and the design of the most significant devices used for GaAs digital ICs will be discussed. This includes the Schottky barrier diode, the GaAs MESFET and JFET, and heterostructure FETs and bipolars. The objective is not to make the reader an expert in all of the nuances of device physics, or to provide a complete background in device design, but to enable the circuit designer to understand how the device design and the important material parameters affect the circuit performance, and to provide a background for the discussion of device models found in Chap. 2. Also, the Schottky barrier diode and GaAs MESFET are often not discussed in undergraduate semiconductor textbooks; therefore a brief description of their operation is appropriate. A brief description of some GaAs IC processes is also included in Sec. 1.5 to give a context to the discussion of design rules to be presented in Chap. 8. Section 1.4 discusses some of the more important MESFET second-order effects. Section 1.6 reviews the operation of heterojunction transistors.

1.2 SCHOTTKY BARRIER DIODES AND OHMIC CONTACTS

The metal/semiconductor junction (Schottky barrier diode) is an integral part of a field-effect transistor, being used as the gate electrode in most cases, and also is often used by itself as a diode for both level-shifting and logic applications. A metal/semiconductor junction is also used to make contact to the channel at the source and drain electrodes of the MESFET or the anode contact of the diode. In this section, a brief overview of the physics of the Schottky diode will be presented, and the distinctions between a Schottky diode and an ohmic contact will be clarified.

Consider the energy band diagram for an *n*-type GaAs/metal junction illustrated in Fig. 1-4*a*. When no external bias voltage is applied across the junction, the Fermi levels in the metal and semiconductor are at the same energy. A potential energy barrier for electrons is formed between the two materials with energy $q\phi_{Bn}$ (barrier height) from metal to semiconductor and with energy $q\phi_{Bn} - (E_C - E_{FS})$ from semiconductor-to-metal (built-in voltage). The net current flow across the junction is zero; on average, equal numbers of electrons pass over the barrier in each direction. Figure 1-4*b* represents the energy band diagram with a forward-bias voltage applied across the junction. The metal Fermi level is shifted down by the positive bias V_F (the energy band diagram plots electron energy by convention), lowering the built-in voltage of the junction by the same amount. Since the conduction band potential

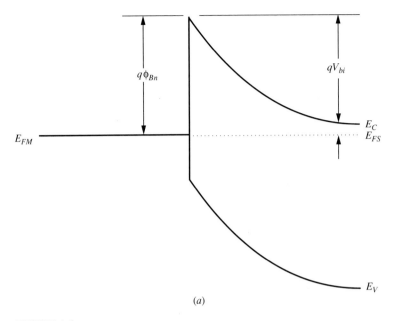

FIGURE 1-4
(a) Energy band diagram for n-type GaAs and a metal. The Fermi levels align when there is no applied voltage.

barrier for electrons is now lower, a larger number of the higher-energy electrons from the conduction band of the GaAs can escape over the barrier. Since the electrons have a Maxwell-Boltzmann thermal energy distribution, this process is known as *thermionic emission*. The number of electrons flowing from the semiconductor to the metal increases exponentially as the barrier is reduced in height. This emission of "hot" electrons over the barrier led to another less frequently used name for the Schottky diode, the *hot carrier diode*. There is also a counter flow of electrons from the metal to the semiconductor, thermionically emitted over the barrier height $q\phi_{Bn}$, which remains independent of the applied voltage. The current density can be described by an equation similar to the equation for a *pn* junction diode

$$J = \left[A^*T^2 \exp\left(-\frac{q\phi_{Bn}}{kT}\right)\right]\left[\exp\left(\frac{qV}{nkT}\right) - 1\right] \quad (1.1)$$

where A^* is called the effective Richardson constant and n is called the ideality factor or emission constant. Here, the first term in brackets is quite different in form from the *pn* junction diode equation in which current flow is by diffusion, but the second term is the same. The thermionic emission model adequately represents the current-voltage characteristics of Schottky diodes in GaAs with doping levels up to mid-10^{17} cm^{-3}. For a more complete derivation and justification of this equation, the reader should consult a textbook on semiconductor device physics [15].

Figure 1-4c shows the energy band diagram under a reverse bias V_R. Note that V_R adds to the built-in voltage, increasing the barrier for electron emission from

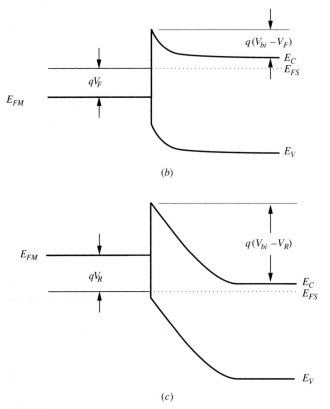

FIGURE 1-4 (*continued*)
(*b*) The energy band diagram for a metal/semiconductor junction with a forward-bias voltage applied across the junction. More electrons will be emitted from semiconductor to metal because the barrier is reduced by V_F. (*c*) The energy band diagram under a reverse bias V_R. The built-in voltage and V_R add, increasing the barrier for electron emission from conduction band to metal.

the conduction band to metal. Therefore, the forward component of the current is suppressed, while the reverse component remains essentially constant (give or take the influence of electrostatic image forces which lower the barrier very slightly), yielding a fixed reverse saturation current when $V_R < 3kT/q$. Therefore, the current-voltage characteristic of the Schottky diode is rectifying and looks very much like that of the *pn* junction. An example of *I-V* and *C-V* characteristics are given in Chap. 2.

In spite of similarities in appearance, the current flow mechanisms for the Schottky diode are quite different from the *pn* junction diode; the majority carriers (electrons) provide the conduction current and reverse saturation current. Therefore, there is no significant storage of minority carriers in the neutral bulk region when the doping of the GaAs is above 10^{16} cm^{-3}. The minority carrier charge is the main factor contributing to the large diffusion capacitance and long recovery times of the *pn* junction, and so the switching time of the Schottky diode from forward bias to reverse bias can be extremely fast. The diode capacitance is simply the space-charge layer or depletion layer capacitance of a one-sided abrupt junction.

The barrier height of the metal/GaAs junction is nearly independent of the work function of the metal. Because of the high surface-state density of GaAs, the Fermi level position is effectively "pinned" within the bandgap. This is in opposition to silicon Schottky barriers where the barrier height depends strongly on the metal work function. Therefore, virtually any metal will form a rectifying junction on a clean GaAs surface. Substantial current flow will occur for forward biases in excess of 0.6 to 0.7 V. Also, the threshold voltage of GaAs MESFETs (to be defined later) will be only weakly affected by the choice of gate metal.

While nearly any metal could be used to form the Schottky barrier on n-type GaAs, the usual choices are limited by compatibility with contact metals or process temperature requirements. If high-temperature processing ($T > 350°C$) does not follow the Schottky metal deposition, a Ti/Pt/Au three-layer sandwich of metals is frequently used because of its moderate chemical stability while in contact with GaAs [16]. It is necessary to prevent the Ga from diffusing out of the GaAs into the Au metallization and the Au into the GaAs. The Ga vacancies which would result from this diffusion cause an undesirable degradation in the electronic transport at the surface and reduction of the barrier height. Lower barrier heights are undesirable for MESFETs or HFETs since the reverse saturation current is increased [see Eq. (1.1)] and the forward turn-on voltage is reduced. If high-temperature processing is required, for annealing of the ion implantations, for example, in a self-aligned process, then a refractory metal silicide or nitride such as WSi_x or WN [17] is generally required to prevent deterioration of the barrier height.

It is worth while to inquire at this point how *ohmic* contacts with *I-V* characteristics that are symmetrical in forward and reverse bias are made. Experimental evidence has shown very little difference in barrier height ϕ_{Bn} between n-type GaAs and any metal deposited on the surface. Figure 1-5a shows the key to understanding the approach. If the width of the barrier is made very narrow through very high concentrations of dopant at the surface, then current can flow through the barrier by tunneling rather than over the barrier by thermionic emission. The current-voltage characteristic can then be nearly symmetrical, since there is a mechanism for large current flow in both directions. Representative *I-V* characteristics for moderately doped and very highly doped n-type GaAs are contrasted in Fig. 1-5b.

The difficulty with this approach lies with the equilibrium solid solubility limitations for dopants in GaAs, typically 10^{19} to 10^{20} cm^{-3} for most dopants, combined with the limited doping efficiency at high doping concentrations. If the space-charge layer width is calculated, one must conclude that the concentrations achievable with standard methods are inadequate for low-resistivity contacts.

Therefore, another approach has been widely employed which utilizes alloyed AuGe/Ni/Au contacts on n^+-type GaAs. The gold and germanium forms an alloy which exhibits a minimum liquidus temperature of 360°C at its eutectic[1] composition. On annealing, the AuGe melts, penetrating below the surface of the GaAs. The formation of Ni_2GeAs regions has been correlated with low contact resistance. This

---

[1] For a detailed discussion of alloys, phase diagrams, and what leads to a eutectic alloy, see C. Kittel and H. Kroemer, *Thermal Physics*, 2d ed., Freeman, 1980.

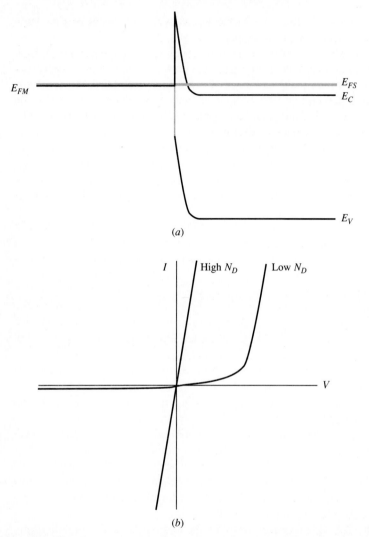

FIGURE 1-5
(*a*) Narrow energy barrier produced by high doping concentration. If the barrier is thin enough, electrons can flow by tunneling. (*b*) *I-V* characteristics for Schottky barrier with low and high doping. Current can flow in both directions by tunneling and tunneling-thermionic emission when doping is high.

may be due to rapid diffusion of Ge from these regions into the GaAs forming heavily n^+ doped material [18]. Thus, tunneling of electrons can occur in these areas through the barrier leading to the ohmic behavior of the contact. Because eutectic alloys freeze in two phases of differing composition, their sheet resistivity is often large compared with a pure metal. The Au overlayer is required to make low-resistivity electrical connection laterally across the contact. Thus, a separate ohmic contact metalization step is generally required for GaAs FET IC processes which requires annealing, which uses a different metal than is required for the Schottky barrier diodes or gates, and which is also distinct from the interconnect metal layers.

The Schottky diodes of interest for GaAs IC applications are planar, i.e., having both contact and barrier metal on the surface of a conductive layer of n-type GaAs which has been laterally defined through masking. The n-type GaAs is therefore a very thin pocket of conductivity (150 to 500 nm deep depending on the design) in an otherwise semi-insulating GaAs substrate. One possible Schottky diode structure is illustrated in both cross section and plan views in Fig. 1-6. Note that the drawing is not to scale; the thickness of the n-type GaAs, b, is much less than the width or length of the ohmic contact metal or the Schottky metal barrier. The n-type doping is produced by ion implantation of Si in most GaAs MESFET processes. There are two implants under the ohmic contact and Schottky barrier in this example of a diode

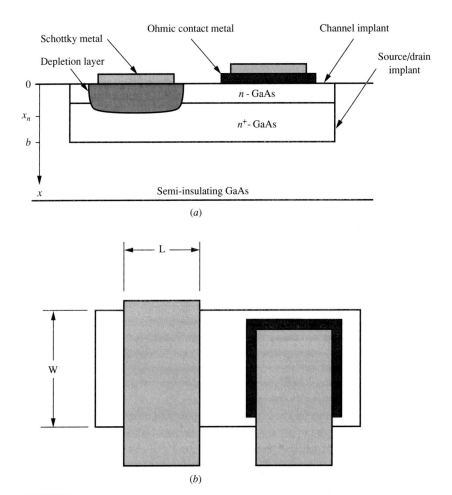

FIGURE 1-6
(a) Cross-sectional view and (b) top view of a Schottky barrier diode (not to scale).

which might be useful for level-shifting applications. The parasitic series resistance of the contact and channel are reduced by the additional implant because of the lower sheet resistance at higher doping. The higher doping also leads to an increase in capacitance (also good for level shifting). Example 1.1 illustrates the advantages of higher doping for level-shift applications.

Example 1.1. Compare the time response of the diode level-shift circuit shown in Fig. 1-7 for diodes with capacitance (measured at zero bias) of 1, 4, and 10 fF/μm^2.

Solution. The time response was determined by simulation with a circuit simulator (SPICE) and is shown in Fig. 1-8. The Spice diode model will be described in Chap. 2 as well as examples of SPICE usage. The diode was driven by $v_{in}(t)$ [curve (*a*)], a falling pulse edge with 10 ps fall time. The output fall time improves with increasing diode capacitance because the displacement current increases correspondingly. The steady-state output voltage is also obtained more quickly for high capacitance. Curve (*b*) with 1 fF/μm^2 shows that the diode is cut off during part of the transition, leading to a very much slower approach to steady state than (*c*) and (*d*) with 4 and 10 fF/μm^2 respectively. This example was chosen to maximize the effect of diode capacitance. A more typical case where the input voltage has a longer fall time will not be as strongly affected by the diode.

The length and width of the Schottky barrier metal must be determined independently for circuit applications. For example, the current density through the diode will not scale inversely with the length of the Schottky metal stripe as the majority of the current enters the stripe within the first 1 μm. It will, however, scale as expected in proportion to the width of the stripe.

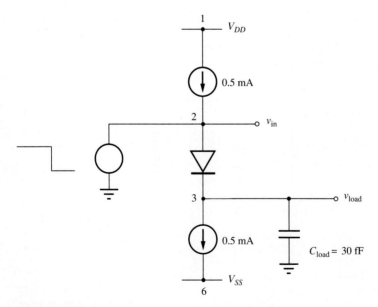

FIGURE 1-7
Level-shifting circuit.

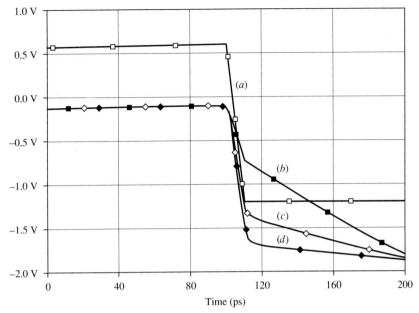

FIGURE 1-8
Time response of the circuit in Fig. 1-7. (a) $v_{in}(t)$. Diode capacitance is varied in (b) 1 fF/μm^2; (c) 4 fF/μm^2; (d) 10 fF/μm^2.

Under forward bias, the electrons will flow through the *n*-type GaAs layer from the contact to the barrier metal. Since the layer is very thin, especially under the Schottky barrier where the semiconductor is depleted of free electrons near the surface, the sheet resistance is relatively high (500 to 1500 Ω/□), and a lateral voltage drop along the length of the channel is expected at higher current densities. As the current flows under the Schottky metal, most of the current will enter the metal very close to the forward edge. From Eq. (1.1), the current depends exponentially on the junction voltage. Therefore, if the voltage decreases significantly due to an IR drop under the metal edge, fewer electrons will be able to overcome the built-in voltage, and the current flow will decrease exponentially with length. This effect is called current crowding, and it must be accounted for in the layout of Schottky diodes intended for use in forward-bias applications.

Typically, in choosing a diode layout, the Schottky metal length, L, will be fixed at a length adequate for reducing series resistance (say 2 μm) and the width, W, will be scaled as needed by the circuit requirements.

Example 1.2. Plot the forward voltage drop across a diode at constant current as the width is changed. Assume that $J_s = 10^{-14}$ A/μm^2, $n = 1.0$, $L = 1$ μm, and $R_s = 500$ $\Omega \cdot \mu$m. $I_D = 1.0$ mA.

Solution. Equation (1.1) describes the *I-V* characteristic of an ideal diode. Real diodes also include series resistance which comes from the ohmic contact, *n*-layer, and the current crowding effect. This series resistance, R_s, scales inversely with the diode width and acts to reduce the voltage seen by the internal (ideal) diode:

$$I_D = J_S L W \left\{ \exp\left[q \frac{[V_D - I_D R_S/W]}{nkT} \right] - 1 \right\} \quad (1.2a)$$

Under forward bias, the exponential term dominates. Solving for voltage drop V_D in terms of W we obtain

$$V_D = \frac{I_D R_S}{W} + \frac{nkT}{q} \ln\left[\frac{I_D}{J_S L W} \right] \quad (1.2b)$$

The result is plotted in Fig. 1-9.

It is useful to find the relationship between the depletion-layer depth, doping concentration, and the applied voltage because of the application of the Schottky barrier as a gate electrode for the MESFET. In this application, the n-type layer doping and thickness are always chosen so that the channel can be totally depleted by the gate so that the drain current of the transistor can be cut off at one particular applied gate-to-source voltage. The applied voltage necessary to accomplish this condition can be calculated from Poisson's equation in one dimension and is often referred to as the *pinch-off* voltage or the *threshold* voltage V_T. In this book, the term threshold voltage will be used to describe this condition in order to avoid confusion with other uses for the term pinch-off and to be consistent with silicon MOS terminology. On the other hand, the doping profile of a level-shift diode is often chosen such that total depletion will not occur for reverse bias of at least V_T. This is because the diode is occasionally used in reverse bias for coupling one logic gate to another (an example can be seen in Chap. 4). In this application, a high capacitance per unit area is needed. Since the

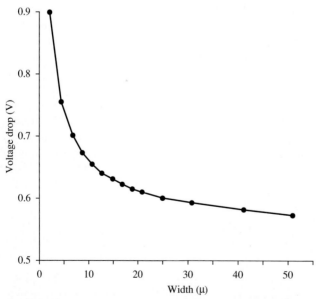

FIGURE 1-9
Forward-biased diode voltage drop as a function of diode width $W(\mu m)$.

capacitance will become very small once the layer has been pinched off, the doping profile is adjusted to prevent this from occurring.

If x is defined as increasing with depth from the surface as shown in Fig. 1-6, then Poisson's equation can be written

$$\frac{d^2 V(x)}{dx^2} = -\frac{\rho(x)}{\epsilon_s} \tag{1.3}$$

where $\epsilon_s = 12.9\epsilon_0 = 1.14 \times 10^{-12}$ F/cm. The net charge density in the semiconductor, $\rho(x)$, is given by

$$\rho(x) = q[p(x) - n(x) + N_D(x) - N_A(x)] \tag{1.4}$$

where $q = 1.6 \times 10^{-19}$ C/charge.

For n-channel MESFETs or Schottky diodes, it is typical for $N_D(x) \gg N_A(x)$ and therefore $p \ll n$. In addition, if it is assumed that the depletion region is free of mobile charge, that fringing effects are neglected, and that the boundary between the depletion layer and the neutral bulk semiconductor is abrupt (i.e., a depletion approximation), then the net charge density reduces down to simply $\rho(x) = qN_D(x)$. The potential $V(x)$ can be found by integration of Eq. (1.3) for an arbitrary doping distribution $N_D(x)$. If the doping concentration were constant (uniformly distributed) so that $N_D(x) = N_D$, then the solution is particularly straightforward, yielding

$$V(x) = \frac{-qN_D}{2\epsilon_s}(x_n - x)^2 + V_{bi} \tag{1.5}$$

where V_{bi} is defined as $\phi_{Bn} - (E_C - E_{FS})/q$. The function $-qV(x)$ is plotted in the energy band diagrams of Fig. 1-4. Letting $x = 0$, Eq. (1.5) can give the depletion depth x_n as a function of applied voltage $V(0) = V_j$. When the applied voltage adds to the potential barrier height, a reverse-bias condition exists, and conversely for forward bias. It can also be seen from Eq. (1.5) that the depletion-layer depth can be varied with applied voltage according to

$$x_n(V_j) = \sqrt{\frac{2\epsilon_s(V_{bi} - V_j)}{qN_D}} \tag{1.6}$$

Thus the height of the MESFET channel shown in Fig. 1-1a can be controlled by V_j. When the channel is pinched off, the depletion depth extends from the surface ($x = 0$) completely through the n layer ($x = b$), and the applied voltage, equal to the threshold voltage V_T, is

$$V_T = \frac{-qN_D b^2}{2\epsilon_s} + V_{bi} \tag{1.7}$$

for the uniformly doped case. Here we assume $\rho = 0$ for $x > b$.

Example 1.3. The channel of a MESFET is to be uniformly doped at a level of $N_D = 3 \times 10^{17}$ cm^{-3}. If the built-in junction voltage of the Ti/Pt/Au Schottky barrier gate electrode/n-GaAs junction is 0.8 V, find the channel thickness required to obtain a threshold voltage of −1.0 volt.

Solution. Since the doping is uniformly distributed throughout the channel depth, then Eq. (1.6) can be used directly to find the channel thickness b. The solution yields $b = 92.5$ nm.

There are two types of MESFETs that are commonly used in digital IC applications. The normally on or *depletion mode* MESFET requires a negative gate voltage relative to the source (V_{GS}) to cut off the drain current under all conditions. From Eq. (1.7), a depletion mode MESFET will be obtained for the case where

$$\frac{qN_D b^2}{2\epsilon_s} > V_{bi} \tag{1.8}$$

It is also possible to produce a MESFET which is normally off or in the *enhancement mode*, where a positive V_{GS} is required to open the channel by selection of N_D or b such that the built-in voltage will totally deplete the channel. In this case, the following condition is true:

$$\frac{qN_D b^2}{2\varepsilon_s} < V_{bi} \tag{1.9}$$

Both depletion and enhancement mode GaAs MESFETs operate by the same mechanism, the depletion of an already existing, doped channel. This is not the case for the Si MOSFET. The enhancement mode MOS transistor functions by inverting the surface of the silicon to produce the channel. The depletion mode device is formed by doping the channel slightly to shift the threshold to a normally on condition.

If the doping concentration in the MESFET channel is not constant with depth, then the charge must be integrated across the region of interest to determine the junction voltage as a function of depletion depth:

$$V(0) = \frac{-q}{\epsilon_s} \int_0^{x_n} x N_D(x) \, dx \tag{1.10}$$

If $x_n = b$ (again assuming for simplicity that $\rho = 0$ for $x_n > b$), Eq. (1.10) can be used to determine the threshold voltage for a given doping profile. The following example illustrates this point for a doping profile formed through ion implantation.

Example 1.4. An ion-implanted MESFET has a channel that is approximated by a gaussian function. As shown in Fig. 1-10, the donor concentration profile which results has a peak of 2×10^{17} cm^{-3} and a standard deviation of 50 nm. The peak in doping occurs 50 nm beneath the surface. Find the threshold voltage for the MESFET if $V_{bi} = 0.8$ V.

Solution. The doping profile shown in Fig. 1-10 can be described by the following equation:

$$N_D(x) = N_p \exp\left[-\frac{(x - R_p)^2}{2\sigma^2}\right] \tag{1.11}$$

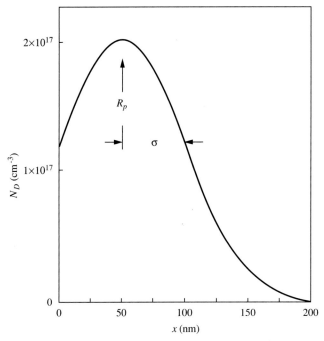

FIGURE 1-10
Ion-implanted doping profile is represented as gaussian.

where N_p is the peak concentration, R_p is the implant projected range (the distance of the peak from the surface), and σ is the standard deviation or straggle. N_p is related to the implant dose Q (atoms/cm²) by

$$N_p = \frac{Q}{\sigma\sqrt{2\pi}} \tag{1.12}$$

Substituting (1.11) into (1.10) and accounting for V_{bi},

$$V_T = V_{bi} - \frac{qN_p}{\epsilon_s} \int_0^b x \exp\left[-\frac{(x-R_p)^2}{2\sigma^2}\right] dx \tag{1.13}$$

Performing the integration required in (1.13) is not straightforward. The solution has been given to be [19]

$$V_T = V_{bi} - \frac{qQR_p}{2\epsilon_s}\left[\mathrm{erf}\left(\frac{b-R_p}{\sqrt{2}\sigma}\right) + \mathrm{erf}\left(\frac{R_p}{\sqrt{2}\sigma}\right)\right]$$

$$+ \frac{qQ\sigma}{\epsilon_s\sqrt{2\pi}}\left\{\exp\left[-\left(\frac{b-R_p}{\sqrt{2}\sigma}\right)^2\right] - \exp\left[-\left(\frac{R_p}{\sqrt{2}\sigma}\right)^2\right]\right\} \tag{1.14}$$

The error function erf(x) is defined as the integral of the gaussian $\exp(-t^2)$:

$$\mathrm{erf}(x) = \frac{2}{\sqrt{\pi}} \int_0^x \exp(-t^2)\, dt \tag{1.15}$$

This is not a simple integral to solve analytically, but tables are readily available for erf(x) in most mathematics handbooks.

The next difficulty is determining b. This is equivalent to defining junction location. Here, we must make an assumption regarding the background concentration. If $N_D(x) = N_A$ at $x = b$, where N_A is the background impurity concentration in the substrate (assume for this example that $N_A = 1 \times 10^{16}$ cm^{-3}), then, from (1.11),

$$b = R_p + \left[2\sigma^2 \ln\left(\frac{N_p}{N_A}\right) \right]^{1/2} \tag{1.16}$$

In this example, $b = 172$ nm. Evaluating (1.13) produces $V_T = 0.8 - 1.86 = -1.06$ V.

An additional characteristic of the Schottky diode which is of interest for digital IC applications is the capacitance of the junction. As is the case for an abrupt, one-sided pn junction, the capacitance will be nonlinear, depending on the applied voltage. This can be seen in Fig. 1-11a in which the charge density ρ for a uniformly doped n-type semiconductor is plotted against position x. The small-signal or incremental capacitance is determined by the amount of additional charge produced for a small voltage change:

$$C_j = \frac{dQ}{dV} \tag{1.17}$$

As illustrated in Fig. 1-11a, an applied bias voltage leads to a depletion depth x_n. When the reverse-bias voltage is increased by some small amount dV, the depletion layer width is also required to move (by dx) in order to expose the positive charge (from ionized donors) needed to maintain charge neutrality. This leads to the shaded area, dQ. Thus, the geometry resembles a parallel plate capacitor of area A in a dielectric with permittivity ϵ_s, and the incremental junction capacitance C_j will vary inversely with x_n according to

$$C_j = \frac{\epsilon_s}{x_n} \quad \mathrm{F/cm}^2 \tag{1.18}$$

Since x_n changes with the dc bias voltage V_j according to Eq. (1.6), the capacitance will also vary with voltage as illustrated in Fig. 1-11b. A linear capacitor, which has constant x_n, is also shown for comparison. If N_D is independent of x, $C(V)$ can be found by substituting (1.6) in (1.18):

$$C(V_j) = \left[\frac{q N_D \epsilon_s}{2(V_{bi} - V_j)} \right]^{1/2} \quad \mathrm{F/cm}^2 \tag{1.19}$$

This can also be written as

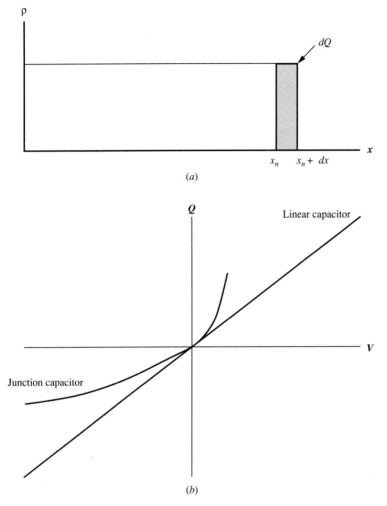

FIGURE 1-11
(a) Charge in the depletion layer vs. depth for a uniformly doped semiconductor. If the applied voltage is increased by dV a small increment of charge, dQ, is exposed by movement of the depletion layer edge by dx. (b) Charge vs. voltage. The slopes of the linear and M/S junction Q-V characteristics are their small-signal capacitances at voltage V.

$$C(V_j) = C_{j0}\left(1 - \frac{V_j}{V_{bi}}\right)^{-1/2} \quad \text{F/cm}^2 \qquad (1.20)$$

where C_{j0} is the junction capacitance per square centimeter at zero applied bias voltage.

If the doping is nonuniform, then an equation for the total charge in the depletion region must be written and Eq. (1.17) applied:

$$Q = \int_0^{X_n} \rho(x)\, dx \qquad \text{C/cm}^2 \qquad (1.21)$$

1.3 PRINCIPLES OF GaAs MESFET OPERATION AND DESIGN

The discussion of the Schottky barrier diode in Sec. 1.2 is directly applicable to the MESFET, since the gate electrode is formed by depositing a metal, refractory metal silicide, or refractory metal nitride film directly on an n-type GaAs channel. In most instances, the channel is formed by ion implantation of silicon or selenium into a semi-insulating GaAs substrate, the same process that was used in forming the Schottky diode. Figure 1-1a presents a schematic (not to scale) representation of a MESFET. The source and drain metallization is typically the Au-Ge/Ni/Au alloyed contact described above. The gate depletion region is drawn parallel to the surface as is the case for the drain-to-source voltage, $V_{DS} = 0$; $V_{GS} = 0$ as indicated. Since the channel is open with $V_{GS} = 0$, the device must be in the depletion mode.

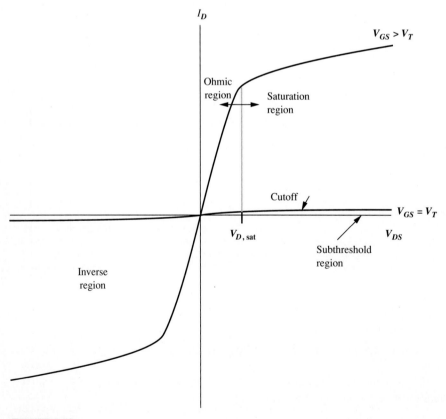

FIGURE 1-12
Regions of operation for an FET.

The operation of any FET is characterized by several regions of operation which are illustrated in Fig. 1-12. One of these regions, *cutoff*, was described in the previous section; the gate depletion region penetrates completely through the channel, blocking the flow of electrons, and therefore cutting off the drain current. It is the purpose of the following section to explain the mechanisms responsible for the ohmic (linear), saturation, and inverse regions. The subthreshold region will be described in Sec. 1.4. The discussion is directed toward the MESFET, as it is the most frequently used device in GaAs ICs; however, the same principles may be applied to the junction FET (JFET) in which a p^+ region between source and drain serves the same role as the metal Schottky barrier gate of the MESFET. A much more complete description of the MESFET physics can be found in Refs. 10 and 20.

1.3.1 MESFET Drain Current

If a small V_{DS} is applied to the FET, Fig. 1-1a illustrates that electrons (arrow) will flow from source to drain. The channel will behave as a resistor, with cross-sectional area $W(b - x_n)$. If V_{GS} is changed, x_n will vary accordingly, leading to a change in resistance as illustrated in Fig. 1-13. This linear voltage–variable resistance behavior

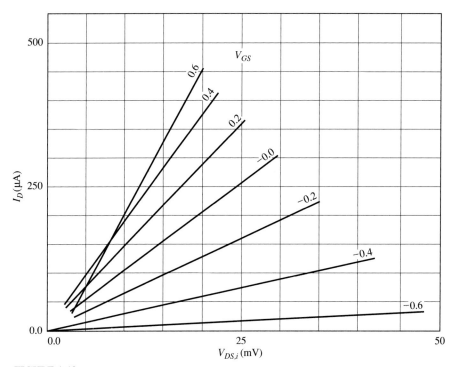

FIGURE 1-13
I_D–V_{DS} characteristic for an FET in the linear (ohmic) region of operation. The incremental conductance g_{ds} is a function of V_{GS} providing a voltage-variable resistance.

leads to the name for this region of operation, the *linear or ohmic* region indicated in Fig. 1-12. The *I-V* characteristic is symmetric about the $V_{DS} = 0$ origin, as expected from the inherent symmetry of the lateral FET structure, leading to an ohmic region in the third quadrant (*inverse* region) as well as in the normal bias polarity (first quadrant). It should be noted, however, that the linearity of the resistance occurs over a rather limited range of V_{DS}. As the drain voltage increases, so does the voltage drop along the channel, leading to an increasing gate-to-channel voltage and a gradually constricting channel cross section as shown in Fig. 1-14; that is, the bias on the gate/channel junction is more reverse biased at the drain end of the channel than the source end. Thus, as V_{DS} is increased, the slope of the *I-V* curve decreases gradually. This slope, the small-signal drain-to-source conductance, is defined by

$$g_{ds} = \left[\frac{\partial I_D}{\partial V_{DS}} \right]_{V_{GS} = \text{constant}} \quad (1.22)$$

When V_{DS} reaches $V_{D,\text{sat}}$, as shown in Fig. 1-12, the drain current flattens out and remains relatively constant for further increases in V_{DS}. A new mechanism governs the operation of this region, called the *saturation* region. There are two common explanations of this phenomenon which describe limiting cases; although the true situation may be somewhere in between, it is always easier to explain and understand the extreme cases than a hybrid case.

The classical explanation, which applies for FETs with gate lengths very much longer than the channel depth, $b - x_n$, attributes the current saturation to the pinching-off of the drain end of the channel [21]. This model is called the *gradual channel model* because the channel height changes slowly with position from source to drain. The channel cross section reduces to zero at the drain end of the gate when $V_{DS} = V_{GS} - V_T$ or, equivalently, $V_{GD} = V_T$. A schematic drawing of this case is provided in Fig. 1-14. This pinched-off condition is strictly true only if the device is purely one dimensional, and the carrier velocity is always related to the electric field by the mobility ($v = \mu \mathscr{E}$). Since the MESFET permits current flow in two dimensions (into substrate and along the channel) and the carrier velocity saturates, this model for FET operation is only good at very low V_{DS}. Since the cross-sectional area of channel at the drain end becomes very small in this model, the electrons must reach infinite velocity in the pinched-off region so that current can be continuous through the channel. In spite of the obvious failure of this model to adequately represent the electron transport in the drain region, it provides good agreement with the *I-V* characteristics of FETs with very long gates, perhaps because this pinched-off region is only a small fraction of the total channel length.

If V_{GS} is made more negative, then the gate depletion region extends further into the channel, and the pinch-off condition at the drain end of the gate can occur at lower values of V_{DS}. Therefore, the saturation voltage $V_{D,\text{sat}}$ is reduced. Also, since the channel cross-sectional area is made smaller as the depletion depth increases, the limiting value of current $I_{D,\text{sat}}$ will also be reduced.

It has been demonstrated that a "square law" drain current-to-gate voltage relationship fits the measured characteristics of long gate length FETs quite well without being strongly affected by the doping profile [20,22]:

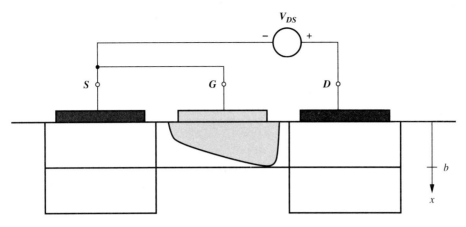

FIGURE 1-14
MESFET channel cross section at a drain voltage corresponding to the onset of the saturation region. As the drain voltage increases, so does the voltage drop along the channel, leading to an increasing gate-to-channel voltage and a gradually constricting channel cross section. The channel is *pinched off* at the drain end of the gate.

$$I_{D,\text{sat}} = \beta(V_{GS} - V_T)^2 \qquad (1.23)$$

This equation is the basis for the SPICE JFET model (Chap. 2). β is a transconductance parameter. Shur has hypothesized that [10]

$$\beta = \frac{2\epsilon_s \mu_n v_{\text{sat}} W}{b(\mu_n V_{po} + 3v_{\text{sat}} L)} \qquad (1.24)$$

which illustrates how the drain current depends on both the drift mobility and saturated drift velocity. V_{po} is the pinch-off voltage ($V_T - V_{bi}$). Thinner channels (b) should also lead to larger current, as will shorter gate length (L).

Equation (1.24) indicates how the MESFET drain current varies with channel width W. As the equation does not specifically account for narrow channel effects, I_D appears to scale directly with W. This is a good assumption for $W \geq 3$ μm. Below this, lateral diffusion effects may cause some error. Also, process linewidth variations may introduce some undesired fluctuation of I_D and so narrow MESFETs should be avoided.

Equation (1.24) also shows that I_D changes nonlinearly with gate length L. For long L, the second term of the denominator is dominant, and I_D varies as $1/L$. For shorter gatelength devices, the current increases less rapidly than $1/L$ as gatelength is reduced. This is because of the effects of velocity saturation. Further increases in I_D (and also g_m) for very short gates will also require a reduction in channel depth b.

The second limiting case attributes the saturation in drain current to the saturation of electron velocity [23] which occurs with increasing electric field as shown in Fig. 1-3. This case will apply best to GaAs MESFETs with short gate lengths ($L < 2$ μm) because the critical electric field ($\mathscr{E}_C = 3.5$ kV/cm) can be obtained in the channel with low gate voltage [$\mathscr{E} \approx (V_{GS} - V_T)/L$]. The channel does not need to pinch off in order for the drain current to saturate [24]. Figure 1-15 illustrates the channel cross section at the onset of velocity saturation. At this condition, the drain current is easily predicted from the device geometry, channel doping, and electron velocity if the electron concentration is assumed to be equal to N_D at the drain end of the gate:

$$I_D = qv_{\text{sat}}N_D W(b - d) \qquad (1.25a)$$

In this equation, d corresponds to the depletion-layer depth just at the point where the velocity is saturated. This will first occur at the drain end of the gate at $V_{GD} \approx \mathscr{E}_c L$. If the doping profile $N_D(x)$ is not a constant with depth, the drain current can be determined by integration [25]:

$$I_D = qv_{\text{sat}}W \int_d^b N_D(x)\, dx \qquad (1.25b)$$

Williams and Shaw [25] have shown that I_D depends critically on the doping profile. This property can be used to optimize the performance of short-channel FETs [26].

Note that the saturation voltage, $V_{D,\text{sat}}$ of Fig. 1-16, for short-channel FETs will be reached at drain voltages well below that required by the classical, gradual-channel pinch-off model V_C, and once V_{DS} is large enough for the velocity to saturate, $V_{D,\text{sat}}$ will be much less dependent on V_{GS}. An additional consequence of the velocity-saturation model is that the drain current in saturation I_{VS} is less than that predicted by the gradual-channel model I_C because the electron velocity approaches a limit. Both of these effects are illustrated by the plot of I_D vs. V_{DS} in Fig. 1-16, in which the predictions of the gradual-channel model are compared with the velocity-saturation model.

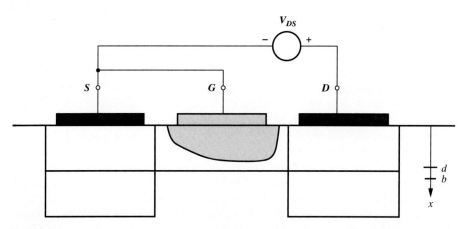

FIGURE 1-15
MESFET channel cross section at the onset of velocity saturation. The electron velocity reaches a limiting value before the channel can pinch off if the gate length is short.

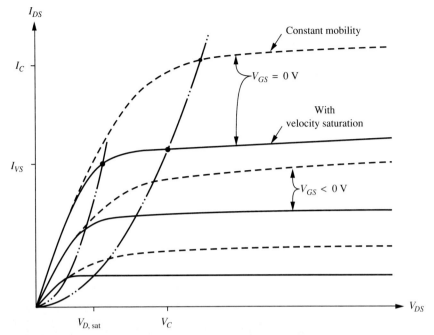

FIGURE 1-16
I_D–V_{DS} characteristic for MESFETs corresponding to a gradual-channel model and a velocity-saturation model. (D. Estreich; used by permission.)

1.3.2 Drain Conductance

Next, consider the behavior of the MESFET while V_{DS} is increased beyond the saturation voltage. While both gradual-channel and velocity-saturation models predict that the drain current remains constant, experiment clearly indicates that it continues to increase, albeit more slowly than in the linear region. Channel length modulation is one explanation for this increase; the pinch-off point or velocity-saturation point moves toward the source as V_{DS} increases [20,27]. If the gradual-channel explanation is applicable, then the current increase is attributed to an increase in carrier velocity due to higher electric field in the region between the source and the pinch-off point. While pinch-off always occurs when the gate-to-channel voltage equals V_T, the higher drain voltage causes greater voltage drop in the channel, so the pinch-off condition occurs closer to the source. Then, since the electric field in this region $\mathscr{E} = (V_{GS} - V_T)/L$, where L is now a function of V_{DS} and V_{GS} and $v = \mu\mathscr{E}$ by assumption, the current must increase as the channel length decreases. This condition is illustrated in Fig. 1-17a.

When the velocity-saturation model is appropriate, channel-length modulation continues to be a source of the finite conductance; however, an alternative interpretation is necessary [27]. Now, as V_{DS} is increased, the electric field in the channel increases, so the velocity-saturation point moves closer to the source. Since velocity

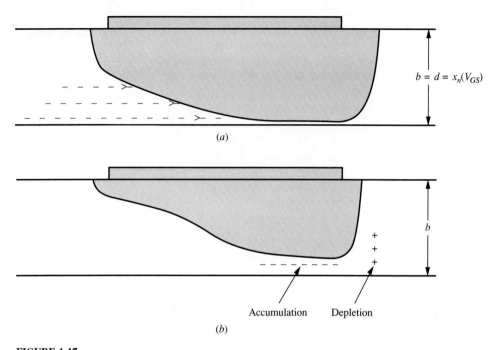

FIGURE 1-17
Further increase in drain voltage leads to a reduction in the channel length and a slow increase in I_D. (*a*) Pinch-off point moves closer to the source for a gradual-channel FET. (*b*) Velocity-saturation point moves closer to the source in the short-channel FET.

saturation always occurs at $\mathscr{E}_C \approx 3.5$ kV/cm, less gate-to-channel voltage is required to produce the critical field and saturate the electrons. Therefore, if the channel voltage is reduced, the depletion depth will also be reduced at this point, increasing the height of the channel $(b - d)$. Since the area for conduction is increased, the drain current also increases. This effect is illustrated in Fig. 1-17*b*.

There is another mechanism that may also add to the conductance, especially for short-channel devices. Since the electric field can become quite high in the drain end of the channel and has a substantial vertical component, electrons may be injected into the substrate [28,29]. Ideally, the electrons are confined to the channel by the built-in potential of the channel/substrate interface which results from the Fermi level difference between the highly doped channel and the intrinsic substrate. This barrier is less than 0.7 V and can be overcome by high electric fields in the channel. That this mechanism contributes to g_{ds} has been demonstrated experimentally by the lower g_{ds} observed when a higher bandgap buffer layer such as AlGaAs or a *p*-type buffer layer is used under the channel [26,30]. The residual g_{ds} for these devices demonstrates that the channel-length modulation effect is also occurring.

The velocity-field relationship for GaAs also leads to accumulation of electrons in the saturated portion of the channel. This accumulation will be necessary to maintain continuity of current, since the velocity of the electrons begins to decrease for $\mathscr{E} >$

$\mathscr{E}_C$, and larger numbers will be needed to retain a continuous current [27]. This is illustrated in Fig. 1-17b by the minus signs under the gate. As the charge spreads out in area from the drain end of the gate to the drain contact region, there is a corresponding net depletion of charge, + signs in Fig. 1-17b below N_D, leading to the formation of a "dipole" region in saturation. The exceptionally low gate-to-drain small-signal capacitance observed in short-channel GaAs MESFETs has been attributed to the formation of this region [31]. Accurate modeling of the drain space-charge extension region requires two-dimensional simulation. The result obtained is very sensitive to the FET geometry.

1.3.3 Transconductance and Gain-Bandwidth Product

Another small-signal model parameter, the transconductance g_m, is of importance for the MESFET. Transconductance is defined by

$$g_m = \left[\frac{\partial I_D}{\partial V_{GS}} \right]_{V_{DS} = \text{constant}} \tag{1.26}$$

The transconductance is significant because it relates the increase in I_D to an increase in V_{GS}, the control voltage of the FET. Thus, it is closely related to the gain of the device. Consider the single-stage amplifier shown in Fig. 1-18a. A small-signal analysis will show that the voltage gain A_v is given by

$$A_v = \frac{v_{\text{out}}}{v_{gs}} = -g_m R_L \tag{1.27}$$

Since resistor loads require large power supply voltages and high resistance in order to obtain high A_v, an active load is more frequently used for IC applications, as illustrated in Fig. 1-18b. The load transistor is a depletion mode MESFET with $V_{GS} = 0$. If the supply voltage is high enough so that the active-load FET is in the saturation region when the logic gate is in transition between states, higher voltage gain can be achieved. This can be seen from a consideration of the MESFET I_D-V_{DS} characteristics in Fig. 1-19. Load lines for a resistor load (a) and an active load (b) are superimposed on the same diagram showing that the active load can lead to much higher gain for the same value of V_{DD} (greater ac output voltage, v_{ds}, for the same ac input voltage, v_{gs}). Simply increasing R_L to a very large value (c) is not satisfactory because the transition region would then occur at very small I_D, where the transconductance is small. A small-signal analysis of Fig. 1-18b yields

$$A_v = -\frac{g_m}{2 g_{ds}} \tag{1.28}$$

if both MESFETs are assumed to have the same channel length and width. This equation also illustrates the necessity of keeping g_{ds} as low as possible, and anticipates some of the difficulties experienced with short channel length MESFETs which have high drain conductance. While high voltage gain per stage is of obvious interest for

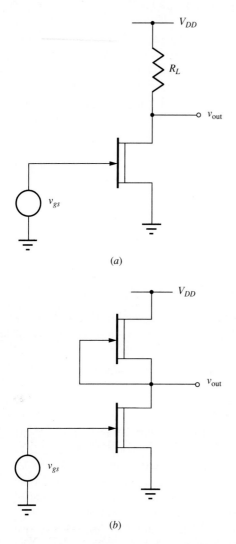

FIGURE 1-18
Single-stage common-source MESFET amplifier with (a) resistor load R_L and (b) active load (MESFET with $V_{GS} = 0$).

analog applications, it is also important for digital circuits since the noise margin (which will be defined in Chap. 3) is closely related to the gain of a logic gate in its transition region.

It is informative to derive an equation for g_m of a MESFET whose drain current is limited by velocity saturation. Although the actual situation is complex, requiring a two-dimensional numerical simulation to explain many observations, a simplified one-dimensional calculation will still illustrate how some of the design choices influence performance. Starting with (1.25) and letting $a = b - d$,

$$g_m = \frac{\partial I_D}{\partial a} \frac{\partial a}{\partial V_{GS}} \qquad (1.29)$$

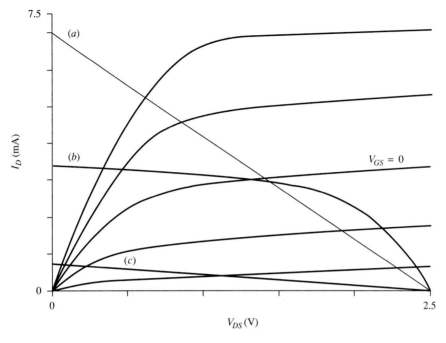

FIGURE 1-19
Load lines for (a) resistor load and (b) active load superimposed on the I_D-V_{GS} characteristic of a MESFET. (c) A large R_L will lead to a small drain current, and the transistor operates in a low transconductance region.

Then, taking the derivative of (1.25a) with respect to a, the first term is obtained. The second term can be found by taking the derivative of (1.6) with respect to the applied voltage, V_{GS}. As a result, g_m is given by

$$g_m = \frac{\epsilon_s v_{sat} W}{d} \tag{1.30}$$

which depends inversely on d. Therefore, to increase g_m, d must be reduced. This may be accomplished by increasing the channel doping N_D according to (1.7). Note that the transconductance is, according to this equation, independent of gate length. The fact that g_m is higher for short gate length MESFETs (below 1 μm) is generally a consequence of larger N_D, which will be explained below.

Example 1.5. Show that the equation for g_m calculated for a nonuniform doping profile given by $N_D(x) = mx$ for $x < d$ is the same equation as for uniform doping (1.29).

Solution. Equation (1.29) will be evaluated for the specified profile. First, Eq. (1.10) will be used to find the relationship between V_{GS} and the depletion depth a:

$$V_{GS} = V_{bi} - \frac{q}{\epsilon_s} \int_0^a mx^2\, dx = -\frac{qma^3}{3\epsilon_s} + V_{bi} \tag{1.31}$$

The derivative needed for (1.29) is now found:

$$\frac{\partial V_{GS}}{\partial a} = -\frac{qma^2}{\epsilon_s} \quad (1.32)$$

Next, Eq. (1.25b) is used to determine the drain current in saturation:

$$I_D = qv_{sat}W \int_{x_n}^{a} mx \, dx = \frac{qv_{sat}\, mW(a^2 - x_n^2)}{2} \quad (1.33)$$

The corresponding derivative can now be found:

$$\frac{\partial I_D}{\partial x_n} = -qv_{sat}\, mWx_n \quad (1.34)$$

Now, by evaluating (1.29) at $a = d$ it is seen that the same equation as (1.30) is obtained. Note, however, that the actual g_m obtained from this profile will be different from that of a uniform profile since the depletion depth d will be different if the same threshold voltages are maintained for the two profiles.

Although higher g_m is desirable for voltage gain, it does not by itself lead to improvement in the high-frequency bandwidth or therefore the switching speed. A better figure of merit is the short-circuit current gain–bandwidth product, f_t, an important measure of the design of the MESFET. The product f_t can be derived from the high-frequency small-signal model for the FET with the output short circuited, and expresses the frequency at which the current gain of a single FET common-source stage falls to unity:

$$f_t = \frac{g_m}{2\pi(C_{gs} + C_{gd})} \quad (1.35)$$

In Eq. (1.35), C_{gs} and C_{gd} are the small-signal gate capacitances with respect to changes in V_{GS} and V_{GD} respectively. The capacitance of a Schottky barrier was defined in (1.18). If we substitute (1.30) and (1.18) into (1.35), and define the gate area $A = WL$, the following equation is obtained:

$$f_t = \frac{v_{sat}}{2\pi L} \quad (1.36)$$

This equation predicts an f_t of approximately 16 GHz for a typical GaAs saturated drift velocity of 1×10^7 cm/s and a 1-μm gate length. This number is in general agreement with f_t values obtained from microwave measurements on 1-μm MESFETs. It is significant to note that while f_t scales inversely with L, this will continue only as long as the gate capacitance continues to scale with L. This is limited by the fringing capacitance, caused by the lateral extension of the depletion layer in the channel to the sides of the gate metal. In other words, the gate capacitance will approach a constant if the gate length is reduced to dimensions comparable with the depletion depth. Therefore, to benefit from reduction in gate length into the submicrometer range provided through the use of more demanding fabrication technology, the channel doping must also be increased to keep the ratio of fringing capacitance to parallel plate capacitance (1.18) reasonably small. This has the added benefit of increasing g_m as shown above.

It can be demonstrated that higher g_m can lead to higher f_t in the presence of parasitic capacitance C_F, which may come from fringing of the gate depletion edge and from interconnections. Using $\omega_t = 2\pi f_t$ and rewriting (1.36) to explicitly include C_F,

$$\omega_t^{-1} = \frac{L}{v_{sat}} + \frac{C_F}{g_m} \qquad (1.37)$$

Even though the intrinsic first term shows no dependence on g_m, Eq. (1.37) shows that ω_t is increased by larger g_m through its effect on the parasitic capacitance. Therefore, for greatest bandwidth or fastest switching, the channel doping should be high. However, the maximum doping level that can be effectively utilized is limited by threshold voltage uniformity considerations. Equation (1.7) illustrates this problem. As N_D increases, b^2 must be reduced to maintain fixed V_T. Assuming that the variation in channel thickness caused by the doping process (ion implantation or epitaxy) is independent of its thickness, then the fractional change in V_T increases as the square root of N_D.

1.3.4 Source Resistance

Perhaps the most important nonideality to be considered in the operation or design of GaAs FETs is the source resistance R_S. This is the access resistance between the source contact and the edge of the gate. As can be seen from Fig. 1-1a, the gate and source are separated by some ungated channel, a doubly implanted region under the source contact, and the contact interface and metal, all of which contribute to the undesired excess resistance. This can be further aggravated by the Fermi level pinning at the surface, causing a surface potential which bends the bands up and depletes part of the surface of free electrons. To minimize this parasitic resistance, efforts are made in the design and layout of the MESFET to increase the doping level in the source and drain regions (self-alignment) [32,33] or to thicken the source and drain regions by recessing the gate in the center of the channel [29,34].

The consequence of the source resistance is to reduce the externally applied gate-to-source voltage, $V_{GS,e}$, by the voltage drop across R_S as illustrated in Fig. 1-20. Thus, the gate-to-source voltage at the edge of the gate is given by

$$V_{GS,i} = V_{GS,e} - I_D R_S \qquad (1.38)$$

A similar effect on g_m is also produced by R_S. If the g_m available outside of the device in the circuit is defined as $g_{m,e}$ and the intrinsic internal g_m as $g_{m,i}$, then it can be shown that (homework problem)

$$g_{m,e} = g_{m,i}\left(\frac{1}{1+g_{m,i}R_S}\right) \qquad (1.39)$$

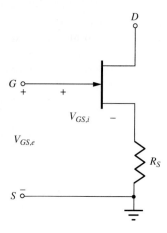

FIGURE 1-20
Source resistance R_S reduces the internal gate-to-source voltage, $V_{GS,i}$, and therefore degrades I_D and g_m.

which demonstrates that not only will the available g_m be reduced by the source resistance but the better the device (the higher its $g_{m,i}$), the more sensitive it is to R_S.

Example 1.6. If $g_{m,i}$ for a mediocre 1-μm MESFET is 100 mS/mm (normalized to an arbitrary 1 mm wide FET), Eq. (1.39) shows that an R_S of 1 Ω·mm will reduce $g_{m,e}$ by only 9 percent. However, for an exceptionally good $g_{m,i}$ of 400 mS/mm (perhaps for an HFET at 77 K), the same R_S causes a reduction in g_m by almost 30 percent. Note, however, that f_t will not be degraded by R_S since it is defined in terms of a current gain, not a voltage gain, so the voltage drop across R_S will not enter into the equation for f_t.

In addition, since the source of resistance acts like a feedback resistor (series-series feedback), it will cause $g_{m,e}$ to vary more linearly with V_{GS} than would otherwise be expected. In Chap. 2, it will be shown that unrealistically high values of R_S can be used to make gradual-channel types of models, which neglect any velocity-saturation effects, emulate the reduced $I_{D,\text{sat}}$ and more constant g_m characteristic of this mechanism.

Finally, the source resistance also affects the apparent value of the incremental drain conductance g_{ds}. It can likewise be demonstrated that (homework problem)

$$g_{ds,e} = g_{ds,i}\left(\frac{1}{1 + g_{m,i}R_S + g_{ds,i}R_S}\right) \tag{1.40}$$

1.3.5 Gate Conduction

One of the principal differences between a GaAs FET (MESFET, JFET, or HFET) and silicon MOS devices is the junction gate and its consequent forward-bias conduction property. While this effect is seldom useful from a circuit design point of view, it can be tolerated when appropriate allowances are made (discussed in Chaps. 3 and 4). If the current flow into the Schottky barrier gate electrode is allowed, the logic high level will be clamped by the gate. If the gate current is allowed to become large, it will also lead to an additional voltage drop across R_S, causing a shift in the I_D-V_{DS} characteristic as illustrated in Fig. 1-21. Here, at low V_{DS} and high V_{GS}, the characteristic curve is shifted to the right by $I_G R_S$. This effect is undesirable when the FETs are used in a logic circuit, because it usually leads to an increase in the logic low-voltage level and a corresponding reduction in the noise margin.

1.4 MESFET SECOND-ORDER EFFECTS

It is unfortunate that the behavior of real MESFET devices does not coincide with the predictions of the first-order explanations. Although the performance of the devices is still quite good, there are several observable effects which should be considered that are associated with the nonidealities of the semi-insulating substrate. These effects include such phenomena as backgating or sidegating, drain current transient lag effects, and frequency dependence. In addition, it is well known that the drain current of an FET does not drop to zero when the gate-to-source voltage is equal to V_T. This residual current is called subthreshold current. Finally, the FET I-V characteristics are temperature-dependent. The origins of these effects will be described in this section. Approaches to the modeling of these effects will be presented in Chap. 2, Sec. 2.3.

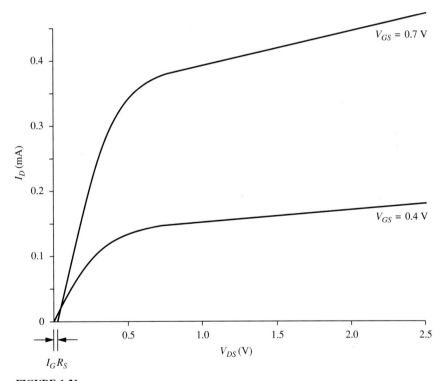

FIGURE 1-21
The gate current, if allowed to become large, will also lead to an additional voltage drop across R_S, causing an offset in V_{DS} near the origin.

1.4.1 Backgating

Backgating or *sidegating* is a widely reported effect [35] in which the drain current of a MESFET or HFET is reduced by the presence of other nearby neighboring FETs which happen to be biased negatively with respect to the source of the first device. Since the neighboring FETs are located to the side of the device under consideration, the applied voltage is sometimes referred to as the sidegate voltage. The possibility for this effect is quite clear; the charge in the MESFET channel is confined to a region beneath the surface by band-bending between the substrate and the channel and the channel and Schottky barrier, as illustrated in Fig. 1-22. At the substrate/channel interface, the Fermi level of the channel aligns with the Fermi level of the substrate because there is no current flowing in a vertical direction. Since the channel is moderately *n*-type doped and the substrate is intrinsic because the excess free-acceptor concentration (carbon) is compensated by deep donors (EL2), there is a built-in voltage as shown, not unlike the situation for a depletion mode MOSFET. If the potential of the substrate were to be biased negatively with respect to the source of the FET, one would expect the height of the potential barrier to be increased, and, as a consequence, the channel would become slightly more depleted of charge from the back side. Therefore, the threshold voltage of the FET will become more positive and the channel current will decrease. For MOSFETs, this phenomena is called the body effect and is a well-known second-order phenomena that is accounted for in all accurate device models. The substrate

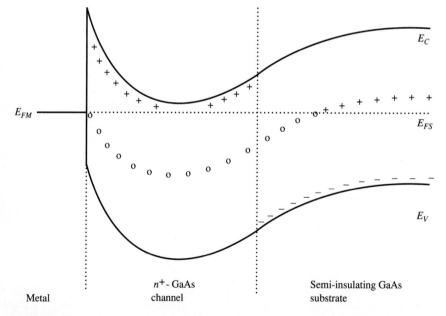

FIGURE 1-22
An energy band diagram drawn under the gate of a MESFET, perpendicular to the surface. Two barriers, the metal/semiconductor junction and the channel/substrate junction, help to confine the free electrons to the FET channel. The minus signs represent ionized acceptors, plus signs represent ionized donors, and dots neutral EL2 deep donors.

potential for the MOSFET is known since the conductivity of the substrate is relatively large. For the MESFET or HFET (nonquantum well), however, due to the very high resistivity of the substrate, the potential is not well known and depends on other FETs in the local environment on the surface of the wafer.

An example of the reduction in current caused by the sidegate voltage is illustrated in Fig. 1-23. In this figure, I_D is plotted for a MESFET with the gate and source shorted ($V_{GS} = 0$) and with the drain voltage fixed at 2.5 V. The upper curve represents the drain current measured when the adjacent MESFET is floating; the lower curves were measured with sidegate voltages of -1 to -4 V. This sample is much worse than usual for room temperature backgating effects, but it illustrates the significant influence that this effect might have on the design of a digital circuit.

It has been observed that the reduction in drain current caused by backgating begins to occur at the same applied (sidegate) voltage at which an abrupt increase in substrate conduction is noted [36]. If a mechanism exists that allows significant charge to be injected into the semi-insulating GaAs substrate and space-charge-limited current to flow, then the backgating effect is not difficult to explain. According to the theory for current conduction in semiconductors with high concentrations of deep levels, the injected charge should be neutralized by the deep centers. Currents will remain at low levels and will be proportional to the applied voltage (ohmic conduction) until the concentration of injected charge exceeds that of the traps. In most samples, however, the problem is that, for the known EL2 and C concentrations, a threshold voltage is predicted which is more than an order of magnitude greater than that observed by experiment [37].

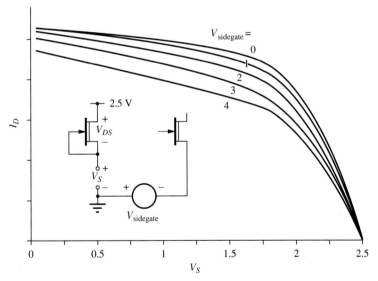

FIGURE 1-23
Reduction in MESFET drain current caused by applying a negative voltage to the source of an adjacent MESFET (sidegate voltage).

A theory explaining this anomaly has been proposed; the deep donor responsible for the semi-insulating behavior of LEC GaAs, EL2, redistributes itself during the high-temperature annealing procedure required to activate the ion-implanted channel. In particular, there is a depletion of EL2 at the surface which would lead to the formation of a lightly doped p-type region due to the ionized carbon acceptors no longer compensated by EL2. Because of the well-known band-bending at the surface of GaAs, this p region will be depleted to a depth of 100 to 200 nm and available for space-charge-limited injection and conduction. Experimental investigation of this phenomena through etching studies of the surface have confirmed this model for the low backgating thresholds sometimes observed [36–38].

It is commonly observed that the degree of backgating varies significantly from substrate to substrate, making the prediction of the backgating threshold unreliable. In order to mitigate these undesirable effects, the isolation between devices can be significantly improved through the use of damage implantations, generally protons, boron, or oxygen, in the field regions between devices [39]. This unannealed implantation produces a high concentration of defects at the surface down to a depth of 300 to 400 nm which act as traps for injected carriers. The backgating threshold voltage is significantly increased through this process step, lending some independence from compositional variations in the substrate, and proton bombardment is for that reason a part of many GaAs IC processes. It is introduced after the ohmic contact alloying step so that no further processing at elevated temperatures (which might anneal some of the damage) will be necessary.

Although the proton implantation is effective in raising the threshold voltage for backgating, it has been observed that this threshold voltage decreases rapidly again when the device temperature is below 10°C [40]. This has been attributed to the reduction in the emission rate of electrons from the deep levels as the temperature is decreased. Therefore, while backgating may be safe to neglect for operation of a circuit with moderate to low power supply voltages at or above room temperature, some modeling is appropriate for "worst-case" conditions which include operation at low temperatures. Modeling will be discussed in Sec. 2.3.1.

Backgating effects are not seen for positively biased sidegate electrodes. If electrons are injected into the substrate from the negatively biased electrode, then the potential under the channel of the negatively biased device will be close to the channel potential, therefore causing little shift in threshold voltage. The device that is positively biased, on the other hand, is under reverse bias with respect to the substrate, and the region under the channel can be shifted in voltage to a much greater extent. The fact that the backgating effect is associated with electron current rather than with the local electrostatics has been demonstrated through experiments with shielding bars located between devices. A reverse-biased Schottky [41] or p^+ [42] shield is effective in reducing both substrate leakage and backgating.

1.4.2 Drain Lag Effects

Another effect, attributable to the semi-insulating substrate, has been referred to as the drain lag effect [43], the tendency for the drain current to overshoot and recover slowly from a step in V_{DS}. Another manifestation of this effect is observed in the frequency domain; the small-signal output conductance in the saturation region, g_{ds}, increases with frequency by as much as a factor of three [44]. This effect occurs between 100 Hz and 1 MHz at room temperature. These effects, unlike backgating, only occur when the device is biased in the saturation region.

Such effects can be explained due to the presence of deep-level traps in the semi-insulating substrate below the channel [43,45]. In saturation, as described in Sec. 1.3, there is an accumulation of electrons beyond the velocity-saturation point as the channel narrows, and the electric field becomes very high at the drain end of the gate. The drain current will therefore be very sensitive to small variations in the height of the channel. If a positive voltage step is applied to the drain, the capacitance through the substrate between the drain electrode and the channel will cause a sudden widening of the channel, leading to an abrupt increase in current. Two-dimensional numerical simulations of the FET have shown that the electric field also has a strong transverse component into the substrate after the drain end of the channel pinches off. Therefore, electrons will also be injected into the substrate at this point. As the number of injected electrons increases, the extra electrons will be trapped, providing additional negative charge in the substrate which will narrow the channel again. This will cause the drain current to slowly recover to its dc value. A negative step has the opposite behavior. Here, a slower recovery would be expected, however, since it takes more time for the traps to release an electron than to capture one. Figure 1-24

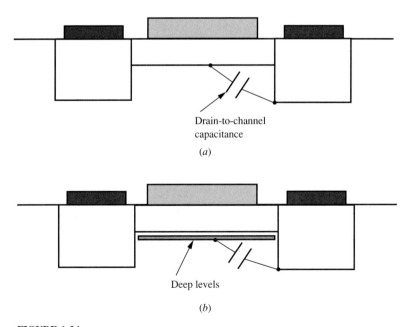

FIGURE 1-24
Schematic cross-sectional view of a MESFET illustrating the drain-to-channel capacitance influencing the interface in (*a*) and shielded by trapped charge in (*b*).

shows a highly schematic cross-sectional view of a MESFET illustrating the drain-to-channel capacitance influencing the interface in (a) and shielded by trapped charge in (b).

This mechanism can also explain the frequency-dependent behavior of g_{ds}. At high frequencies, the traps are too slow to capture and release charge during one cycle of the ac signal. Therefore, they do not counteract the effect of the drain capacitance on the channel/substrate interface, and the drain conductance in saturation is large. At low frequencies, the traps can follow the ac signal and effectively shield the channel from the drain capacitance through the substrate. Therefore, the drain conductance is reduced.

1.4.3 Subthreshold Current

It is well known in MOSFETs [46] and MESFETs [47] that small levels of current continue to flow from source to drain for V_{GS} values below the pinch-off voltage. Pinch-off or cutoff is really a transition between a region of normal conduction dominated by drift of free electrons in the device channel and a region of subthreshold conduction in which electrons are transported by diffusion and drift. Once the channel is totally depleted (by definition) of charge contributed by the donors, there is a potential barrier between the source and drain for electrons attempting to flow parallel to the surface. For low V_{DS}, electrons can be transported by diffusion, since there is a high concentration in the source region and a very low concentration near the drain (positively biased). Therefore, current flow is characterized by an exponential dependence on V_{DS} [46,47], similar to the forward-bias current of a *pn* junction given by

$$J_{ST} = \frac{qDn_0}{L}\left[1 - \exp\left(\frac{-qV_{DS}}{kT}\right)\right] \quad (1.41)$$

where n_0 is the equilibrium electron concentration in the source and L is the gate length.

In addition, as might be expected, the current will also depend exponentially on V_{GS} as this voltage controls the height of the barrier between source and drain. For the short-channel devices of interest, the drain voltage influences the field in the channel and therefore also reduces the height of the barrier. This contribution is similar to a thermionic emission model. The voltage dependences of the subthreshold current can be represented by

$$I_D = I_0\left[1 - \exp\left(-\frac{cqV_{DS}}{kT}\right)\right]\left[\exp\left(\frac{bqV_{DS}}{kT}\right)\right]\left[\exp\left(\frac{aqV_{GS}}{kT}\right)\right] \quad (1.42)$$

where a, b, and c are empirical fitting parameters.

1.4.4 Temperature Dependence

The temperature dependence of the drain current of the MESFET is influenced by two related mechanisms, the variation of the built-in voltage of the channel/substrate interface and the variation in the channel transconductance factor (β) [48]. These two mechanisms can be understood if the channel/substrate interface is modeled as a *pn* junction [19,49]. Referring back to the energy band diagram in Fig. 1-22, there are two built-in voltages of interest which exhibit temperature dependence. Both $\phi_{Bn} - (E_C - E_{FS})/q$, the built-in voltage of the Schottky barrier, and V_{bi}, the channel-to-substrate built-in voltage, will be similarly affected by the temperature dependence of $(E_C - E_{FS})$. $(E_C - E_{FS})$ varies with temperature according to

$$E_C - E_{FS} = kT \ln\left(\frac{n}{N_C}\right) \quad (1.43)$$

where n is the free-electron concentration at the bottom of the conduction band and N_C is the density of states in the conduction band [4]:

$$N_C = 2\left(\frac{2\pi m^* kT}{h^2}\right)^{3/2} \quad (1.44)$$

where m^* is the electron effective mass and h is Planck's constant. The change in these built-in voltages will affect the MESFET threshold voltage, which according to [19] is given by

$$V_T = \phi_{Bn} - \frac{E_C - E_{FS}}{q} - \frac{\overline{Q}}{C_G}\left(1 - \frac{C_G}{C_\sigma}\frac{\overline{Q}}{Q}\right) + \frac{1}{C_G}[2\epsilon_s q N_A(V_{bi} - V_{BS})]^{1/2} \quad (1.45)$$

where the gate capacitance per unit area, C_G, is defined as

$$\frac{1}{C_G} = \frac{R_p}{\epsilon_s} + \frac{\sqrt{8/\pi}}{\epsilon_s}\sigma = \frac{1}{C_R} + \frac{1}{C_\sigma} \quad (1.46)$$

and $\overline{Q}$ is the net implanted charge given by

$$\overline{Q} = \frac{Q}{2}\left[\text{erf}\left(\frac{R_p}{\sqrt{2}\sigma}\right) + 1\right] \quad (1.47)$$

In Eq. (1.46), R_p is the projected range for an assumed LSS implant profile and σ is the corresponding straggle for that distribution. N_A in Eq. (1.45) is the ionized acceptor concentration in the substrate, uncompensated now as the EL2 is neutral while below the Fermi level. The channel-to-substrate voltage V_{BS}, if known, also affects V_T, as discussed in Sec. 1.4.1. Since $(E_C - E_{FS})$ increases with T, the threshold voltage of the MESFET is expected to become more negative as temperature increases. This is verified by measurement of V_T, plotted in Fig. 1-25a, where a significant negative shift with temperature is observed.

The channel transconductance parameter, β, as defined in the square law relationship $I_D = \beta(V_{GS} - V_T)^2$, also varies with temperature. When defined as in Ref. 48, β is given by

$$\beta = \frac{W\epsilon_s \mu_n}{2L\, b^*} \quad (1.48)$$

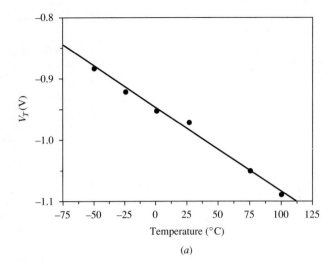

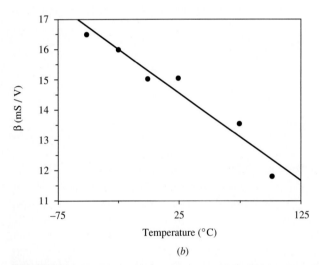

FIGURE 1-25
Threshold voltage (V_T) and the transconductance parameter (β) are both reduced as temperature is increased. (*From S. J. Lee and C. P. Lee* [48]; *used by permission.*)

where μ, the mobility, decreases with T and b^*, the effective channel thickness, increases with T because of the temperature-dependent built-in voltages discussed above. In Eq. (1.48), W is the channel width and L is the gate length. Thus, β decreases as the temperature is increased. Experimental measurements were reported in Ref. 48 and are plotted in Fig. 1-25b.

Because V_T is increasing at the same time that β is decreasing, it is possible for I_D to either increase or decrease, depending on whether V_T or β has the stronger effect. Therefore, the net effect of T on the drain current depends on V_{GS}. This behavior is illustrated by the I_D vs. V_{DS} plot in Fig. 1-26 showing data at 30 and

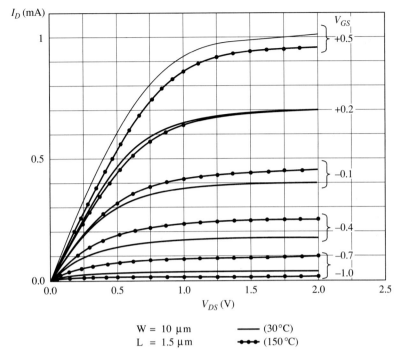

FIGURE 1-26
I_D vs. V_{DS} for a MESFET at 30°C and at 150°C.

150°C. The data, measured on a 1.5×10 μm MESFET, illustrate that V_T shift dominates at V_{GS} values near cutoff, leading to an increasing I_D with T. Conversely, at large V_{GS}, the influence of β overrides that of V_T; therefore, drain current decreases with T. This property can be used for temperature-compensation circuits; e.g., there is a V_{GS} value at which the drain current is independent of temperature. Also, proper selection of V_{GS} can be used to compensate for the negative temperature coefficient of the Schottky barrier diode when combined together in an appropriate circuit.

The Schottky diode (either gate or level shift) also exhibits a complex behavior with temperature. The intrinsic junction will vary with temperature according to Eq. (1.2). The temperature coefficient will depend on the current level. For very low currents (constant current), the diode voltage will decrease by about 2 mV/°C. At higher bias currents, -1.0 to -0.5 mV/°C is observed. The positive temperature coefficient (TC) of the series (or source) resistance, R_S tends to partially compensate for the negative TC of the diode when the diode operates in a series resistance limited extreme.

1.5 MESFET FABRICATION METHODS

In Sec. 1.3 it was shown that the performance of a GaAs MESFET can be improved if the gate length is reduced, if channel doping is increased, and if the source resistance and drain conductance can be minimized. At the same time, it is essential that the FET threshold voltage be exceedingly uniform across a wafer and between wafers so that large numbers of FETs can be interconnected both on-chip and between chips and still function properly as a logic circuit. Fabrication methods for digital MESFET ICs have been developed which address these requirements.

The same basic processing operations are needed to form devices and to interconnect devices in any GaAs MESFET IC fabrication sequence. These operations are outlined below:

1. *Channel definition*. Precisely defined regions on the substrate surface must be doped n type to form FET channels, source and drain regions, and planar Schottky barrier diodes.
2. *Ohmic contacts*. Ohmic contacts must be made to the source and drain electrodes and anodes of Schottky diodes.
3. *Schottky diodes*. Metal must be deposited and patterned to form gates for MESFETs and cathodes for the Schottky diodes.
4. *Interconnections*. At least two levels of metal lines must be provided for wiring. These levels must be insulated from one another by a dielectric. They must be precisely patterned.
5. *Via contacts*. Provision must be made to connect electrically between interconnect metal layers. These contacts are called vias.

In the following section, the above operations will be briefly described. Two representative GaAs IC fabrication methods, one that uses the gate itself to align the source and drain implant regions (*self-aligned*) and one that manually aligns the gate to the channel, will be used as examples of the basic operations and also to provide some context for the discussion of design rules found in Chap. 8. Much more detailed discussions of these process operations can be found in several sources [50, 51].

Ion implantation has been used in the fabrication of GaAs ICs for defining device channels. This technique produces a very precisely controlled doping concentration and depth by bombarding the surface of the GaAs semi-insulating substrate with a beam of ions that has been accelerated to high energies by an electrostatic field. The MESFET channel and source/drain contact areas are often fabricated by ion implantation of n-type dopants. Silicon and selenium are both well-behaved n-type dopants in GaAs, meaning that they do not diffuse at abnormally high rates during the high-temperature annealing process (greater than 800°C). This is important since high-temperature annealing is needed to restore the crystal structure which was damaged by the high-energy ion collisions and to incorporate the dopant atoms substitutionally into lattice sites where they may become electrically active [52].

Beryllium and magnesium have been used effectively as p-type dopants when ion-implanted into GaAs. These might find application in a JFET process in which a p-type gate electrode is required.

An ion beam may be masked by interposing a thin film of a mask material between the beam and the surface of the GaAs substrate. Photoresist can be made thick enough to serve as a mask and can be patterned in selected areas on the surface. Metals would also be effective as an ion mask; however, most metals tend to react chemically with GaAs at the high temperatures required for annealing. This chemical reaction will cause Ga or As vacancies in the channel under the gate or will allow the metal to diffuse rapidly into the sample; either event will degrade the barrier height of a Schottky barrier.

Ohmic contacts are generally formed to n-type GaAs devices using the gold-germanium-nickel alloyed contact described in Sec. 1.2. The thin layers of metal required are most often applied by thermal evaporation onto the GaAs wafer in a vacuum [53]. Prior to the deposition, the areas to receive ohmic contacts are defined by windows through a photoresist layer. After deposition, the undesired contact metal is removed by washing away the underlayer of photoresist with a suitable solvent. This procedure is known as *lift-off* [50].

Schottky metal is deposited on the GaAs wafer either by vacuum evaporation or sputtering [53]. It must be directly in contact with a clean GaAs surface to be useable as a gate electrode for a MESFET or as a Schottky diode. Metals such as Al or Ti/Pt/Au (high electrical conductivity but low temperature stability), or silicides or nitrides of tungsten (low electrical conductivity but high temperature stability), are most frequently used [54]. They are patterned either by lift-off, as described above, or by etching. If the Schottky metal has high conductivity it is often also used as one of the interconnect layers. If not, then a separate deposition and patterning of a high conductivity metal is necessary.

Interconnect metal layers are usually separated by a thin film of a deposited dielectric such as Si_3N_4, SiO_2, or polyimide to allow for crossovers. Contact between metal layers must be provided, where needed, through via contacts. These are windows etched through the dielectric film by plasma etching [53]. Their locations are defined by photoresist. In some cases, the second-layer metal is separated from the first layer by *air bridges* instead of dielectric. These second-layer metal air bridges consist of thick, plated gold bridges suspended over the surface between supporting pillars of gold. The advantage of this approach is the reduced interconnect capacitance which results from the lower relative dielectric constant of air ($\epsilon_r = 1$) instead of $SiO_2 (\epsilon_r \approx 4)$ or $Si_3N_4 (\epsilon_r \approx 8)$. Even without air bridges the capacitance of the second-layer metal is less than the first layer since the effective dielectric constant is reduced by placing the metal on top of the deposited dielectrics with low ϵ_r rather than the GaAs ($\epsilon_r = 12.9$). Many processes are introducing a third-metal interconnect layer which will help to increase wiring density.

1.5.1 Self-aligned MESFET

There have been many approaches reported which seek to reduce the source resistance of a MESFET by using the gate electrode itself as a mask for the source and drain (S/D) n^+ implantation [32,33]. In order to avoid Schottky barrier degradation during annealing, refractory silicides such as WSi_x can be substituted for the gate metal. Such materials are stable in contact with GaAs up to 900°C. The use of the gate itself as a mask eliminates the critical alignment step between the S/D regions and the gate electrode which would otherwise be necessary. Hence, this type of approach is often referred to as a *self-aligned* process. If the gate were inadvertently located on top of this high dose implant in a non-self-aligned process, a large shift in V_T would occur, destroying the functionality of the circuit.

A typical self-aligned process sequence is illustrated in Fig. 1-27. First, the channel is defined by selectively ion-implanting Si through a photoresist (PR) mask. Next, the WSi$_x$ Schottky metal is deposited and etched to form the gate electrode somewhere near the center of the channel implant. This gate can now be used as a mask for the S/D implantation as shown in Fig. 1-27b and c. If a gate length less than 1 μm is desired, then the implant must be set back from the gate edges to avoid the *short-channel effect*, a negative shift in V_T as gate length is reduced. This can be accomplished by a two-step process using a moderate Si implant around the gate, followed by deposition of SiO$_2$. The SiO$_2$ can be etched, leaving a sidewall on the gate as shown in Fig. 1-27c. This sidewall will separate the Si implant away from the gate edge by 100 to 200 nm. Then, the implant can be performed at a very high dose to reduce the contact resistance of the S/D. The implantations can now be annealed to activate the dopants. The S/D ohmic contacts are then deposited, patterned, and the contacts are alloyed (Fig. 1-27d). The first level of interconnection can then be defined, either directly on the GaAs semi-insulating substrate surface, or on a deposited dielectric layer (dielectric 1) which covers the surface. If the dielectric approach is used, then vias must first be etched through the dielectric so that contact can be made between the S/D ohmic metal, the gate material, and the first-layer metal (Fig. 1-27e) as needed. Then another dielectric (dielectric 2) is deposited as an insulator between first and second layers of metal as shown in Fig. 1-27f). Via contacts are again provided where needed by etching dielectric 2 prior to depositing metal 2.

1.5.2 Directly aligned MESFET

A representative directly aligned process, in which the gate is manually aligned between the implanted S/D regions, is illustrated by Fig. 1-28 [55]. The process begins by deposition of an Si$_3$N$_4$ film on the substrate which will be used both for the annealing cap and for protection of the surface. Figure 1-28a shows the formation of the *n*-type channel by selectively implanting Si through windows in a PR mask and through the nitride cap. Next, in Fig. 1-28b, the S/D regions are exposed with PR windows and implanted with a second *n*-type implant which will reduce the source resistance. Sufficient space must be maintained between these implants and the future location of the gate so that some misalignment can be tolerated. Both implants are now annealed. Next, windows must be etched in the nitride for definition of the S/D ohmic contact metal shown in Fig. 1-28c. The contact is patterned by lift-off after deposition and then is alloyed. The gate Schottky metal is defined between the S/D implants by a window etched in the nitride. Then the Schottky metal is deposited and patterned (Fig. 1-28d). This gate metal ordinarily serves a dual role as a Schottky gate and as the first level interconnection metal. In this process, the first layer of interconnect metal is directly on the GaAs substrate and therefore may have somewhat higher

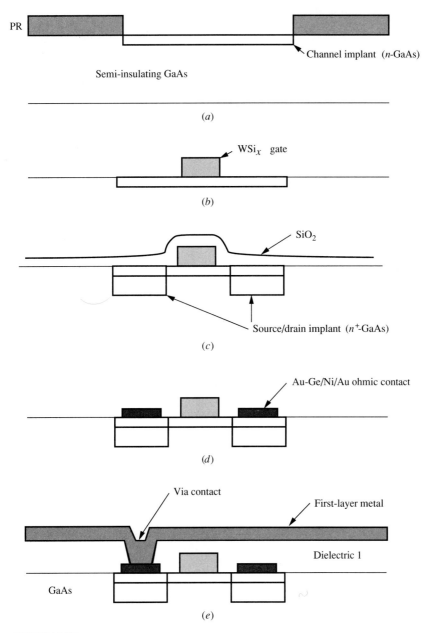

FIGURE 1-27
Fabrication sequence for a self-aligned MESFET.

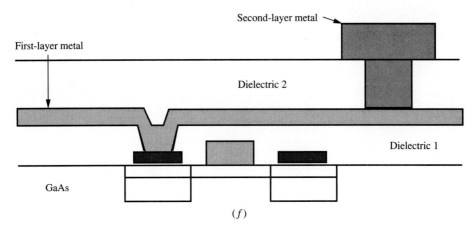

(f)

FIGURE 1-27 (*continued*)

capacitance than the first-layer metal of the self-aligned process described above. Finally, a second dielectric layer is deposited as shown in Fig. 1-28e, and via contacts are etched where needed. Second-level interconnect metal is deposited and patterned as in the self-aligned process.

1.6 HETEROSTRUCTURE FETs AND BIPOLAR TRANSISTORS

The development of molecular beam epitaxial growth (MBE) and metal-organic chemical vapor deposition (MOCVD) techniques since the late 1970s has irreversibly changed the design of new electronic and optoelectronic devices. Very thin layers of compound semiconductor materials can now be grown with great precision in doping and thickness and with good uniformity over 3-in diameter GaAs wafers. The dominant theme for these new devices has been "bandgap engineering" through the use of the heterojunction, an abrupt or intentionally graded junction between two semiconductors of different bandgap [56]. In a more conventional structure with the same semiconductor on both sides of a junction, the electrons and holes experience forces proportional to doping gradients (diffusion) and built-in or applied electric fields (drift). Since the bandgap is constant, the electrostatic forces will be equal and opposite on electrons and holes. Heterojunctions provide an additional degree of freedom in the design of a device; the force acting on electrons and holes can be controlled independently by adjusting both bandgap (composition) and doping.

It is well known that the growth of good quality junctions between binary (III-V) semiconductors and their alloys (III-III-V or III-V-V combinations) is possible if the lattice constants of the two compounds are the same, but this constraint limits the choices of energy gap, conduction band, and valence band offset combinations as illustrated in Fig. 1-29 [57]. This figure plots bandgap vs. lattice constant for the most widely used III-V semiconductors. Combinations which have the same lattice constants, such as GaAs/AlGaAs or InGaAs/InAlAs, align vertically on the chart.

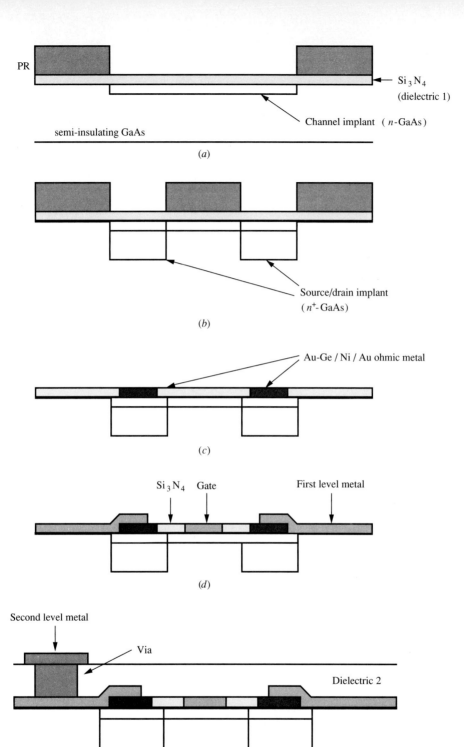

FIGURE 1-28
Fabrication sequence for a directly aligned, selectively implanted MESFET.

It has been recently discovered that non-lattice, matched combinations can also be grown without extensive crystal defects if the desired film is thin enough so that the lattice mismatch can be accommodated by elastic strain rather than by the formation of dislocations (*pseudomorphic growth*) [58]. If the heterojunction is to be formed between semiconductor A and semiconductor B, for example, a sandwich structure (usually called a *quantum* well) A-B-A or B-A-B is required to evenly distribute the stress. This technique opens up a much wider range of heterojunction combinations and has already led to significant improvements in the performance of low-noise microwave FETs through the use of AlGaAs/InGaAs/GaAs pseudomorphic quantum well structures [59].

In the following section, the application of heterojunctions to devices suitable for very-high-speed digital circuits will be discussed—specifically, the operation of the heterostructure FET (HFET or MODFET for modulation-doped FET), also known by many other acronyms (HEMT for high electron mobility transistor, SDHT for selectively doped heterojunction transistor, TEGFET for two-dimensional electron gas FET). Other more recently developed HFET structures (SISFET for semiconductor-insulator-semiconductor FET, HIGFET for heterojunction insulated-gate FET, and complementary HFETs) will be briefly mentioned. The heterostructure bipolar transistor (HBT) will also be described in comparison with the Si bipolar (homojunction) transistor.

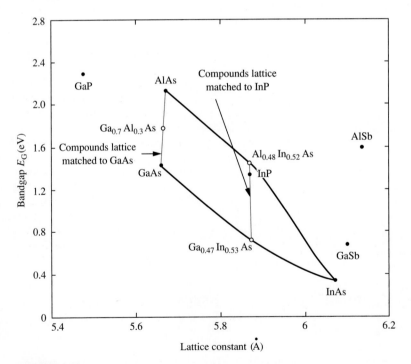

FIGURE 1-29
Bandgap vs. lattice constant for the most widely used III-V semiconductors. (*From C. P. Lee* [57].)

1.6.1 Heterostructure FETs

The original concept for the HFET came from experimental observation of enhanced electron mobility in *modulation-doped* heterostructures at low temperatures [60]. The term modulation doping has led to the name MODFET for this first-generation HFET. By confining the conduction electrons spatially (with a heterojunction) in a GaAs region with few impurities, and separating the conduction electrons from the ionized donors (impurities) that usually are their source, high electron concentrations (a sheet concentration of $n_s > 10^{12}$ cm^{-2}) **and** high mobilities can be obtained in the same sample. This is not the case for bulk semiconductors, as can be seen from Fig. 1-3, where the low field mobility (slope) is greatly reduced as the impurity concentration N_I is increased.

> **Example 1.7. Drawing energy band diagrams at heterojunction interfaces.** Modulation-doped heterostructures are composed of alternating layers of n^+-AlGaAs and non-intentionally doped GaAs(i-GaAs). A construction for the energy band diagram of a single heterojunction of this type is shown in Fig. 1-30. In Fig. 1-30a the two semiconductors are not yet in contact and are drawn such that the conduction band offset ΔE_C and the valence band offset ΔE_V are spaced between the AlGaAs and the GaAs in the experimentally determined ratio: $\Delta E_V = 0.55 x_{Al}$ and $\Delta E_C = 0.75 x_{Al}$, both linear functions of the Al mole fraction [61]. The above ratio applies as long as the AlGaAs continues to have a direct energy gap, up to an Al concentration of about 40 percent. The bandgap of the AlGaAs is determined by the Al concentration, increasing from 1.42 eV for GaAs (direct bandgap) to 2.17 eV for AlAs (indirect bandgap).
>
> The Fermi level in the AlGaAs is shown near the conduction band, because of the n^+ doping level; the Fermi level in the GaAs is mid-gap since the doping is assumed to be low. Next, in Fig. 1-30b, the semiconductors are brought into contact. Since there is no applied voltage specified across the heterojunction, the Fermi levels E_F must align horizontally. Note that the band offsets are not properly accounted for in Fig. 1-30b; the offsets do not depend on doping, only on composition. Their inclusion in the band diagram will produce band bending as shown in Fig. 1-30c and d.
>
> Next, draw the valence bands. ΔE_V must be positioned between the band edges such that the slope in the GaAs and the AlGaAs is the same on both sides of the interface. Poisson's equation requires that the slope of the band edge is proportional to the electric field. Since the permittivity of the materials is the same and the electric field must be continuous across the interface, the slopes must be equal. In the AlGaAs, if there is a uniform donor concentration, the curvature must be parabolic.
>
> The conduction bands must follow their respective valence bands because the bandgap cannot change unless the composition is also changed. The edges should join together with the offset ΔE_C separating them as shown in Fig. 1-30d. The upward curvature produced in the AlGaAs conduction band implies that this region is depleted of electrons. Thus, positive fixed-space charge from ionized donors will be present in this depletion region. The corresponding negative fixed charge required for charge neutrality across the interface must occur in the conduction band notch which serves as a potential well for conduction electrons.

It is the conduction electrons in the potential well that makes this structure interesting for devices. Note that the electrons are separated physically from the

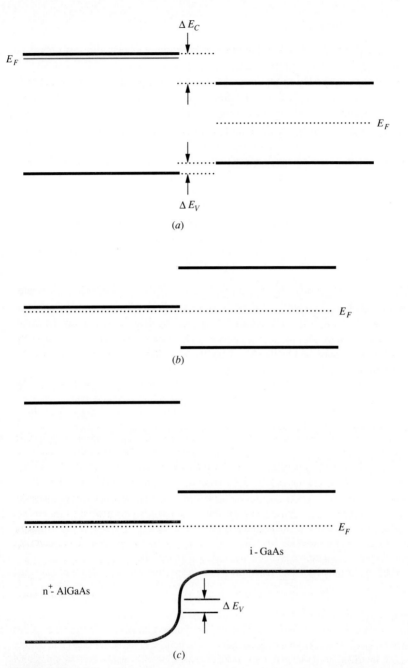

FIGURE 1-30
The construction of an energy band diagram of a single modulation-doped heterojunction: (*a*) before contact, (*b*) Fermi levels are aligned after contact, but offsets are not considered, (*c*) valence band offset included, (*d*) conduction band offset included.

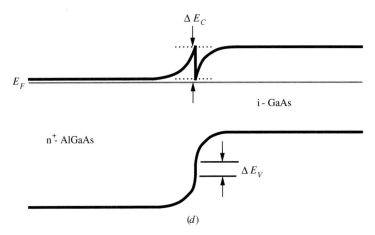

FIGURE 1-30 (*continued*)

donors. Often a very thin, undoped AlGaAs spacer layer (1 to 5 nm thick) is located between the doped AlGaAs and the GaAs well to prevent diffusion of the Si dopant atoms into the well during crystal growth and to avoid electron scattering from the coulomb potential of the ionized donor atoms. Therefore, with this structure, the impurity scattering should be much reduced and higher mobilities at low temperatures are expected and have been observed [as high as 2×10^6 cm^2/(V·s) at 4 K [62]]. Secondly, the electrons are confined to a thin (< 10 nm) well and form a two-dimensional charge sheet parallel to the interface (two-dimensional electron gas). Because of the lack of impurity scattering and the reduced dimensionality of this structure, it has been assumed that the saturated drift velocity of the electrons will also be enhanced. This has never been conclusively proven, however, since most estimates of v_{sat} have depended strongly on the model assumed for transport in the MODFET channel. Finally, high electron sheet concentrations can be induced in the well if the structure is biased so that the Fermi level intersects the well. There will be subbands within the well which are the allowed states for the electrons. The concentration can increase until the electrons begin to escape from the well by thermionic emission from the higher-energy subbands or by hot-electron effects. In the AlGaAs/GaAs system used in the example (and in most devices), the well can accommodate a sheet charge density n_s of about 1×10^{12} cm^{-2} electrons before significant numbers begin to escape into the AlGaAs. Other heterojunction systems may offer higher sheet concentration if the well depth is greater (InAlAs/InGaAs, for example).

This modulation-doped heterostructure lends itself nicely to the formation of a field-effect transistor. As illustrated in Fig. 1-31, a Schottky barrier gate electrode can be placed on the AlGaAs, and ohmic contacts can be made to the two-dimensional electron gas to produce a channel whose electron concentration is controlled by the gate voltage. The doping concentration and thickness of the AlGaAs can be selected so that the depletion layer under the Schottky barrier will reach completely through the AlGaAs if a normally off (enhancement mode) FET is desired. This condition is

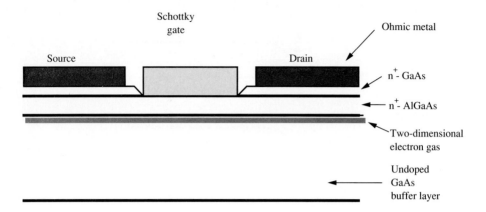

FIGURE 1-31
Schematic cross section of MODFET with Schottky barrier gate.

shown in Fig. 1-32, where the energy band diagram from gate to substrate has been drawn. The well has now been flattened out by the built-in voltage of the Schottky barrier and no electrons can be confined at the interface. Applying a positive bias to the gate will restore the well by offsetting the built-in voltage, a band structure like Fig. 1-30d will be obtained, and charge can flow through the well from source to drain. A normally on (depletion mode) device can also be produced if the AlGaAs doping and thickness is great enough to contain the Schottky barrier built-in voltage without pulling up the energy bands at the AlGaAs/GaAs interface.

The FET produced by this heterostructure (MODFET) is similar to a MOSFET since the channel charge is confined to a thin sheet. The AlGaAs layer serves as a barrier between the charge and the gate electrode, somewhat like the oxide in a MOSFET. However, the AlGaAs/GaAs conduction band offset is much smaller (≈ 0.2 eV) than the SiO_2/Si offset (several volts), so some electrons will escape into the AlGaAs as V_{GS} is made more positive. Some of the electrons injected into the AlGaAs are captured by deep levels (the DX centers); others will drift in the S/D electric field, but at a much lower velocity than in the GaAs. The net effect is an abrupt reduction in transconductance as the gate voltage becomes forward-biased. Figure 1-33 illustrates this property by comparing the measured transconductance of an HFET with a MESFET of similar channel dimensions [61]. The silicon MOSFET would exhibit a smoothly increasing transconductance with gate voltage, but with a slope many times less than either of the GaAs devices.

The MODFET transconductance [see Eq. (1.30)] can be quite high if a thin AlGaAs barrier layer separates the gate and channel charge. Since the dielectric constant is much higher for AlGaAs ($\epsilon_r = 12.9$) than for SiO_2 ($\epsilon_r = 4.0$), even barrier layers of the same thickness should provide three times more g_m for the MODFET. The minimum AlGaAs thickness is limited by the Schottky barrier gate leakage, because thinner barriers will require more doping to produce the same FET

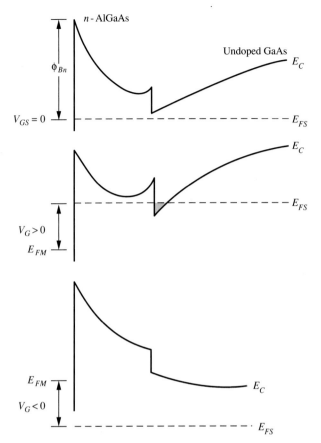

FIGURE 1-32
Energy band diagram drawn vertically through the channel from gate to substrate.

threshold voltage. The leakage current will become excessive if the AlGaAs donor concentration is too high. (Refer to the discussion on contacts in Sec. 1.2.) Nevertheless, g_m of 450 mS/mm has been reported for a MODFET at room temperature and 520 mS/mm at 77 K [63]. Furthermore, as shown in Fig. 1-33, the high carrier mobility and the early saturation of electron velocity produce high g_m at low gate bias above threshold, $V_{GS} - V_T$ [64].

The voltage dependence of the gate (channel) capacitance is similar to that of the MOSFET [61]. Since the MODFET channel charge is formed by inversion in the triangular well at the interface, once the charge is induced the capacitance due to this charge will remain constant; the distance between the gate and the charge is fixed by the AlGaAs layer thickness. The gate-to-source capacitance continues to increase with V_{GS}, however, since at moderate forward bias the depletion layer in the AlGaAs is not necessarily totally depleted. The incremental ionized donor charge in the gate barrier will also be modulated by V_{GS}. Thus, the total capacitance is expected to increase with V_{GS}. The gate-to-drain capacitance will decrease rapidly

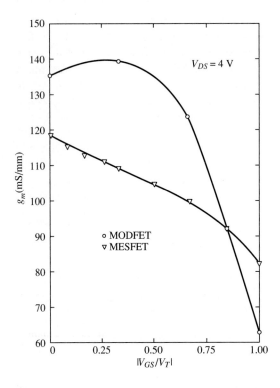

FIGURE 1-33
The transconductance of an HFET and a MESFET of similar channel dimensions are compared. (*From Arnold* et al. [*61*]; ©*1984 IEEE. Used by permission.*)

with increasing V_{GD} as saturation occurs because the well at the drain end of the gate will flatten out as V_{GD} becomes more reverse-biased, resembling the band diagram in Fig. 1-32, Thus, an increase in V_{GD} will have little effect on the amount of charge in the channel.

The MODFET can also be compared to the MESFET. The gate voltage of both devices is limited by conduction of the Schottky barrier. While the barrier height is higher on the wider gap AlGaAs layer, the doping concentration is also higher (1×10^{18} cm^{-3}) in the AlGaAs than is typical for a GaAs MESFET channel. Therefore, the forward conduction limit (typically 0.7 to 0.8 V) is not increased in proportion to the bandgap. Also, the greater barrier height of the AlGaAs makes contacting the two-dimensional electron gas more difficult; generally, an n^+-GaAs cap layer or a self-aligned S/D implantation is required to reduce the source resistance to an acceptably low level [65].

The threshold voltage of the MODFET discussed above depends on several constants and variables which are distinct from the MOSFET or the MESFET that are defined in Fig. 1-30*d* and described in the equation below:

$$V_T = \phi_{Bn} + \frac{\Delta E_C}{q} - \frac{qN_D d^2}{2\varepsilon_s} \tag{1.49}$$

In this equation, N_D is the donor concentration in the AlGaAs layer and d is the thickness of the doped part of the AlGaAs layer. As the equation shows, N_D and d

can be used to set the threshold voltage if the composition of the AlGaAs is fixed. MODFET sensitivity of threshold voltage to these variables is similar to the MESFET. An increase in Al composition will increase both ϕ_B and ΔE_C, but not at the same rate. If the Al concentration is greater than 0.25, the silicon doping in the gate barrier layer leads to a large concentration of deep levels. These deep levels cause a shift in the threshold voltage when the device is exposed to light which, at low temperatures, persists after the light is removed. There is also a significant threshold shift with temperature.

The other second-order effects typical of the MODFET are somewhat less pronounced than the MESFET (described in Sec. 1.4). Backgating tends to be highly dependent on material growth processes, but can be much less than typical in an ion-implated MESFET. Little information on reproducibility is available at this time. The transconductance is higher and more uniform over a wide bias range for the MODFET, as shown in Fig. 1-33. The high-frequency $g_m r_{ds}$ product is also higher. Further improvements are possible with improved device structures with wide-gap semiconductor on both sides of the narrow-gap well.

The first-generation AlGaAs/GaAs MODFET described above is only one of many possibilities. Many variations in design have been reported which address one or more of the shortcomings of this device. For example, HFETs have been demonstrated with undoped AlGaAs gate barrier layers and n^+-GaAs gate electrodes (sometimes called a *SISFET* for semiconductor-insulator-semiconductor FET) [66]. A band diagram for this device is shown in Fig. 1-34a. The threshold voltage for this device is established by the conduction band offset, since ϕ_{Bn} and N_D of Eq. (1.49) are both zero. Since the Fermi level is near the conduction band edge in the degenerately doped GaAs gate layer, the threshold voltage for the SISFET is naturally close to 0 V. Unfortunately, $V_T = 0$ is low for some popular circuit types (discussed in Chaps. 3 and 4), and efforts are being made to modify the structure to provide more positive thresholds. Substituting a Schottky barrier for the n^+ gate is an attractive alternative from a processing point of view since a self-aligned S/D implantation can be used to fabricate this device (called the HIGFET for heterojunction-insulated gate FET) [67]. The threshold voltage is increased by approximately 0.8 V by the inclusion of ϕ_b in Eq. (1.49). Unfortunately, $V_T = 0.8$ V is too high for optimum design of popular GaAs logic circuits, but innovations in device and circuit design may overrule this objection.

The demonstration of *p*-channel AlGaAs/GaAs MODFETs with enhanced hole mobility [12, 67, 68] has encouraged the development of complementary logic circuits with GaAs MODFETs. Figure 1-34b shows the energy band diagram for a *p*-channel MODFET, where now a two-dimensional hole gas is confined in a well at the *p*-AlGaAs/non-intentionally doped GaAs heterojunction. While the room temperature mobility is low, at 77 K, μ_H as high as 6000 cm^2/(V·s) has been reported. The integration of *p*-MODFETs with *n*-MESFETs has been demonstrated [68]. Integration with *n*-MODFETs is significantly more difficult, requiring selective regrowth by MBE or MOCVD. (In other words, after the first growth, *n*-channel device areas are masked and the remaining area etched. A second growth step is then required to produce the *p*-channel MODFET layer structure.)

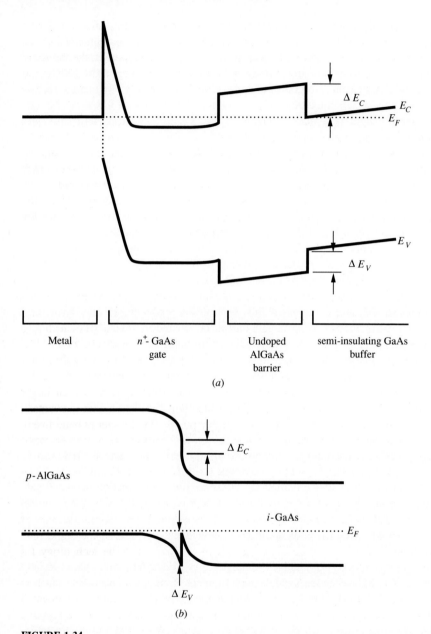

FIGURE 1-34
Energy band diagrams for (a) HFET with i-AlGaAs gate barrier and $n+$ gate electrode (SISFET) and (b) p-channel MODFET.

A p-HIGFET version of this device has also reported. Integration of n- and p-HIGFETs using self-aligned S/D ion implantation on the same epitaxial layered structure has been shown [67]. While low static power has been reported, the threshold voltage is somewhat nonoptimum for complementary logic with n-HIGFETs, and gate leakage due to the small valence band offset limits the logic voltage range prematurely.

Other III-V materials can be substituted for GaAs and AlGaAs which might provide higher n_s due to a deeper well. Higher n_s is desirable because higher drain currents can be obtained for an FET with the same physical dimensions. If silicon-doped AlGaAs can be eliminated from the gate barrier, then the threshold-shift problems associated with the deep levels (DX centers) can also be eliminated.

As there have been many excellent tutorial review articles published on the HFET with extensive bibliographies, the student seeking more detailed information on these promising new devices can refer to one of these [62, 69, 70].

1.6.2 Heterostructure Bipolar Transistors

While the primary subject of this book is the design of high-speed GaAs FET ICs, the developments in GaAs heterostructure bipolar transistors have also been significant. Frequency divider circuits with clock frequencies above 20 GHz and HBTs with $f_{\max}$ of 105 GHz have been reported [71]. Although the technology for fabrication of HBT digital ICs is much less well developed than that for that for HFETs and MESFETs, this situation will change with time, and HBTs may become the preferred device for certain applications which are dominated by load capacitance and for which larger power dissipation can be accommodated. Bipolar digital circuits have the advantage of very uniform threshold voltage, since V_{BE} is determined by the energy band lineup rather than by doping and thickness as for the MESFET or for MODFETs with doped gate barriers. In addition, very high transconductances are obtained with HBTs; over 10 S/mm has been reported as compared with 0.4 S/mm for a very good MESFET or HFET. Thus, the propagation delay does not increase with load capacitance as rapidly for HBT circuits as it does for FET circuits.

The benefits in performance which would be gained by the use of an emitter region with an energy gap wider than the base were recognized at the same time as the original invention of the bipolar transistor [56, 72], but the technology for fabricating such a device did not exist until the 1970s. The benefit of the wide-gap emitter can be understood when the energy band diagram for an npn homojunction bipolar transistor and an Npn heterojunction bipolar transistor (HBT) are compared in Fig. 1-35. From inspection of the valence band edge (A) for the homojunction transistor it can be seen that the potential energy barrier for holes (built-in voltage) between the base and emitter (qV_p) is the same as the barrier for electrons between the emitter and base (qV_n). The collector current consists mainly of the electron current injected from the emitter into the base, I_n, less any bulk recombination in the base (small). The base current consists mainly of the hole current, I_p, injected into the emitter from the base, less any recombination in the base or E/B depletion layer (which should also be small). It is necessary to see how these

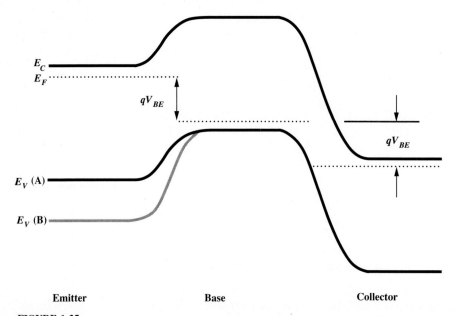

FIGURE 1-35
Energy band diagram for *npn*-GaAs homojunction bipolar (A) and *Npn*-AlGaAs/GaAs heterojunction bipolar transistor (B).

currents are related to the built-in voltages and the doping levels in the base and emitter to understand the HBT.

If recombination currents are neglected (not a good assumption at low current densities; however, bipolar transistors for digital circuits operate at high current densities), first-order theory for the BJT can be applied to compare the magnitude of these two main current components [73]. Their ratio is important because the current gain $\beta = I_C/I_B \gg 10$ for the transistor to be useful in ECL/CML digital or in any analog circuit applications. I_p and I_n are diffusion currents. If the base width between E/B and B/C space-charge layers is W_b, and the emitter width (either physical thickness or diffusion length, whichever is smaller) W_e and nondegenerate doping levels are assumed in both regions, then Boltzmann statistics will provide the minority carrier concentrations; the equations below will result:

$$J_p = \frac{qD_p n_i^2}{W_e N_e} \left[\exp\left(-\frac{qV_A}{kT}\right) - 1 \right] \tag{1.50}$$

$$J_n = \frac{qD_n n_i^2}{W_b N_b} \left[\exp\left(-\frac{qV_A}{kT}\right) - 1 \right] \tag{1.51}$$

In these equations, n_i^2 is the square of the intrinsic concentration for the emitter and base semiconductors, both GaAs or Si for the homojunction BJT. V_A is the applied bias on the E/B junction. The doping concentrations in the *n*-type GaAs emitter is N_e, and in the *p*-type GaAs base, P_b. Taking the ratio of Eqs. (1.50) and (1.51),

$$\frac{I_n}{I_p} = \frac{N_e}{P_b} \frac{D_n}{D_p} \frac{W_e}{W_b} \quad (1.52)$$

The maximum value of β is obtained under the above assumptions. Therefore, if the doping is equal in the emitter and base and the base and emitter widths are equal, then β_{max} will be given by the ratio of the electron to hole diffusivity. This ratio is about 20 for GaAs and about 3 for Si. Therefore, these ratios would be greater than β for the npn homojunction with equal doping levels. So, to obtain useful β in the homojunction device, the emitter doping must typically exceed the base doping by a significant margin.

The valence band edge for the HBT (B) is also shown in Fig. 1-35. The effect of the conduction band offset ΔE_C and the valence band offset ΔE_V is not shown, since it is assumed that the transition between the two materials is graded in composition over a distance sufficient to smooth out the conduction band spike (viz. Fig. 1-30d). Note that the energy-gap difference between the base (say GaAs) and the emitter (perhaps AlGaAs), ΔE_G, is added to the built-in voltage for the holes. The current ratio I_C/I_B for the HBT therefore contains an additional term which reflects the suppression of the hole current by the exponential of the potential energy difference, ΔE_G, which is added to the valence band:

$$\frac{I_n}{I_p} = \frac{N_e}{P_b} \frac{D_n}{D_p} \frac{W_e}{W_b} \exp\left(\frac{\Delta E_G}{kT}\right) \quad (1.53)$$

Since it is possible to obtain tenths of electron volts for ΔE_G in III-V heterojunctions, the maximum current gain can be increased by a very large amount [77]. The large current gain is not very useful (there would be no benefit for it to exceed 100), so it is desirable to increase the base doping–emitter doping ratio. Increasing the base width in the HBT can reduce the parasitic base resistance while still maintaining adequate β, but the base transit time will be increased. Thus, there is a tradeoff between transit time (f_T) and base resistance (f_{max}) in optimizing the high-frequency performance.

To see why the ability to suppress hole injection into the emitter is a big advantage for the HBT over an Si or GaAs homojunction BJT, the factors contributing to fast switching speed in a nonsaturating (ECL or CML) logic circuit should be examined. f_{max}, the maximum oscillation frequency, is an accepted figure of merit with straightforward use for microwave applications, but its applicability for predicting time delay of a bipolar transistor switch in a logic circuit is not clear. A better predictor of switching delay for digital applications which considers the circuit as well as the device has been derived from an analysis of a two-transistor circuit [56,74,75]. The equation for the switching delay, τ_s, is

$$\tau_s = \frac{5}{2} R_b C_c + \frac{R_b}{R_L} \tau_b + (3 C_c + C_L) R_L \quad (1.54)$$

where R_b is the total base resistance, C_c is the base-to-collector capacitance, τ_b is the transit time of electrons through the base, and R_L and C_L are the circuit load resistance and capacitance respectively. This equation illustrates the primary role that base resistance plays in the switching delay, appearing linearly in two terms. This is where the HBT has its greatest advantage, the ability to boost the base doping to solid-solubility limits to reduce the base resistance. If very narrow emitter stripes are also used (1 to 2 μm), the access resistance to the base will then be dominated by the region external to the base/emitter junction. If this area can be made thicker than the internal base layer, doped very heavily by ion implantation or diffusion, and the ohmic contact metal can be located very near to the emitter edge through a self-alignment approach, then the performance of the HBT can be extremely good in a high-speed digital circuit. An example of a device designed with these objectives is illustrated in Fig. 1-36 [76].

Another observation which can be made from Eq. (1.54) is the strong effect that the load resistance has on the switching speed. As pointed out in Ref. 56, τ_s has a minimum with respect to R_L given by

$$R_L = \left(\frac{R_b \tau_b}{3 C_c + C_L} \right)^{1/2} \tag{1.55}$$

which will in turn establish the optimum bias current level for an ECL or CML circuit; i.e., in order to provide a suitable logic swing, ΔV, for driving the next stage, a bias current given by $I = \Delta V/R_L$ is required. Thus, a particular power dissipation is determined which depends on the device base transit time, resistance and collector capacitance, and the circuit load capacitance. For many HBT designs and circuit applications, the optimum value of R_L is rather low, leading to a moderate to high power dissipation per logic gate to obtain the highest speed for the HBT circuits. While non-ECL types of HBT circuits (I^2L, for example) have been demonstrated which do provide low power dissipation, the switching speed is poorer than MESFET circuits and so is not as competitive with high-speed silicon technologies.

The high base doping of the HBT is also an advantage for operation at high current levels; the current density at which high-level injection becomes significant is very large. Therefore, small devices can be operated at high current levels to increase g_m without losing current gain. Base-width modulation effects are also reduced by the high base doping.

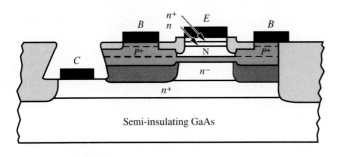

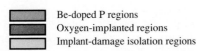

FIGURE 1-36
Schematic structure of HBT with Be implanted external base and self-aligned oxygen implant for reduced parasitic capacitance. (*From Asbeck* et al. [76]; ©*1984 IEEE. Used by permission.*)

Finally, studies of the dependence of f_t on V_{CE} have shown that the highest f_t is obtained at small V_{CE}, a desirable condition for high-speed logic circuit applications [78]. This occurs because the time required for the electrons to pass through the collector space charge layer depends on their velocity. If the potential drop between base and collector is large enough that electrons may transfer into the L conduction band valley, then their velocity will be reduced as the effective mass becomes greater. The collector transit time is a major part of the total transit time for a well-designed HBT.

HOMEWORK PROBLEMS

1.1 Which of the four diodes shown in Fig. P1-1 will make the best level shift diode? Explain why. Rank the others in order of preference.

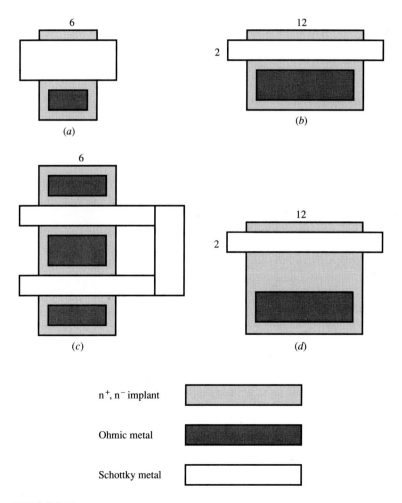

FIGURE P1-1

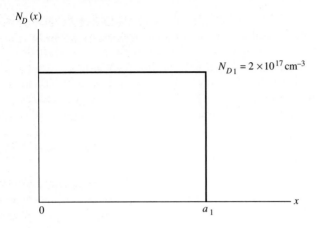
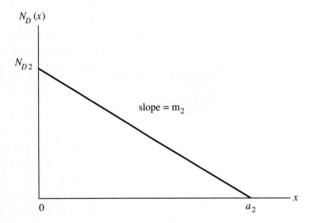
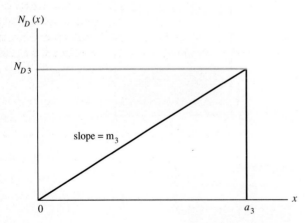

FIGURE P1-3

1.2 Show that the sensitivity of the MESFET threshold voltage to channel thickness increases as the square root of the channel doping if uniform doping is assumed.

1.3 (a) Use the velocity saturation model for drain current to derive equations for the drain current of MESFETs fabricated with the doping profiles shown in Fig. P1-3. Choose the coefficients m_2 and m_3 and depths a_2 and a_3 so that each profile has the same amount of charge and the same pinch-off voltage [79].

(b) Show that g_m for each device is given by the same equation. At any particular V_{GS}, will the g_m be the same for MESFETs made from the three profiles? Explain.

(c) Assume that N_D for the uniformly doped channel is 2×10^{17} cm^{-3} and that the threshold voltage is -1.0 V. Plot several points on the graph of I_D vs. V_{GS} for the three MESFET profiles.

1.4 Derive f_t using the HF small signal FET model.

1.5 Derive Eqs. (1.39) and (1.40) from a small-signal FET model which includes the source resistance R_S.

1.6 Construct the energy band diagram for the n^+-In$_{0.51}$Ga$_{0.49}$P/p-GaAs heterojunction. The conduction band offset is 0.21 and the valence band offset is 0.25 eV [80].

1.7 Construct the energy band diagram for an n-Al$_{0.3}$Ga$_{0.7}$As/p-GaAs/n-GaAs HBT with an abrupt emitter/base interface. Be sure to include the effect of the energy band offsets in the diagram.

REFERENCES

1. Van Tuyl, R., and C. Liechti: "High Speed Integrated Logic with GaAs MESFETs," IEEE *J. Solid State Cir.*, vol. SC-9, pp. 269–276, October 1974.
2. Blakemore, J. S.: "Semiconducting and Other Major Properties of Gallium Arsenide," *J. Appl. Phys.*, vol. 53, pp. R123-R181, October 1982.
3. Ruch, J. G., and W. Fawcett: "Temperature Dependence of the Transport Properties of Gallium Arsenide Determined by a Monte Carlo Method," *J. Appl. Phys.*, vol. 41, pp. 3843–3849, August 1970.
4. Sze, S. M.: *Physics of Semiconductor Devices*, 2d ed., chap. 1, Wiley, 1981.
5. Lile, D. L.: "Metal-Insulator-GaAs Structures," in *Gallium Arsenide* (Eds. M. J. Howes and D. V. Morgan), chap. 7 Wiley, 1985.
6. Fang, F. F., and A. B. Fowler: "Hot Electron Effects and Saturation Velocities in Silicon Inversion Layers," *J. App. Phys.*, vol. 41, pp. 1825–1831, March 1970.
7. Sze, S. M.: *Physics of Semiconductor Devices*, 2d ed., p. 380, Wiley, 1981.
8. Chelikowsky, J. R., and M. L. Cohen: "Nonlocal Pseudopotential Calculations for the Electronic Structure of Eleven Diamond and Zinc-Blende Semiconductors," *Phys. Rev.*, vol. B14, p. 556, 1976.
9. Wada, T., and J. Frey: "Physical Basis of Short Channel MESFET Operation," IEEE *J. Solid State Cir.*, vol. SC-14, pp. 398–411, April 1979.
10. Shur, M.: *GaAs Devices and Circuits*, chap. 7, Plenum, 1987.
11. Zuleeg, R., Notthoff, J. K., and G. L. Troeger: "Double-Implanted GaAs Complementary JFETs," IEEE *Elect. Dev. Lett.*, vol. EDL-5, pp. 21–23, January 1984.
12. Tiwari, S., and W. I. Wang: "p-Channel MODFETs Using GaAsAs/GaAs Two-Dimensional Hole Gas, "IEEE *Elect. Dev. Lett.*, vol. EDL-5, pp. 333–335, August 1984.
13. Lindquist, P. F., and W. M. Ford: "Semi-Insulating GaAs Substrates," in *GaAs FET Principles and Technology* (Eds. J.V. DiLorenzo and D.D. Khandelwal), Artech House, 1982.
14. Yuan, H. T., Lin, Y-T., and S-Y. Chiang: "Properties of Interconnection on Silicon, Sapphire, and Semi-Insulating GaAs," *IEEE Trans. Elect. Dev.*, vol. Ed-29, pp. 639–644, April 1982.
15. Sze, S. M.: *Physics of Semiconductor Devices*, 2d ed., chap. 5, Wiley, 1981.
16. Sinha, A. K., Smith, T. E., Read, M. H., and J. M. Poate: "n-GaAs Schottky Diodes Metallized with Ti and Pt/Ti," *Solid State Elect.*, vol. 19, pp. 489–492, 1976.

17. Zhang, L. C., Cheung, S. K., Liang, C. L., and N. W. Cheung: "Thermal Stability and Barrier Height Enhancement for Refractory Metal Nitride Contracts," *Appl. Phys. Lett.*, vol. 50, pp. 445–447, 1987.
18. Kuan, T. S., Batson, P. E., Jackson, T. N., Rupprecht, H., and E. L. Wilkie: "Electron Microscope Studies of an Alloyed Au/Ni/Au-Ge Ohmic Contact to GaAs," *J. Appl. Phys.*, vol. 54, pp. 6952–6957, December 1983.
19. Taylor, G. W., Darley, H. M., Frye, R. C., and P. K. Chatterjee: "A Device Model for an Ion-Implanted MESFET," *IEEE Trans. Elect. Dev.*, vol. ED-26, pp. 172–182, 1979.
20. Sze, S. M., *Physics of Semiconductor Devices*, 2d ed., chap. 6, Wiley, 1981.
21. Shockley, W.: "A Unipolar Field-Effect Transistor, *Proc. IRE*, vol. 40, p. 1365, 1952.
22. Middlebrook, R., and I. Richter: "Limits on the Power-Law Exponent for FET Characteristics," *Solid State Elect.*, vol. 6, p. 542, 1963.
23. Turner, J. A., and B. L. H. Wilson: "Implications of Carrier Velocity Saturation in a GaAs Field-effect Transistor," *Int. Symp. on GaAs and Related Compounds*, Dallas, Tex., 1968.
24. Kennedy, D. P., and R. R. O'Brien, "Computer Aided Two-Dimensional Analysis of the Junction Field-Effect Transistor," *IBM J. Res. Dev.*, 14: 95-116, Mar. 1970.
25. Williams, R. E., and D. W. Shaw: "Graded Channel FET's: Improved Linearity and Noise Figure," *IEEE Trans. Elect. Dev,*, vol. ED-25, pp. 600–605, June 1978.
26. Tan, T., Stoneham, E. B., Patterson, G., and D. M. Collins., "GaAs FET Channel Structure Investigation Using MBE," *IEEE GaAs IC Symp. Proc.*, Phoenix, Ariz., pp. 38–42, October 1983.
27. Liechti, C. A.: "Microwave Field-Effect Transistors— 1976," *IEEE Trans. Microwave Theory and Tech.*, vol. MTT-24, pp. 279–300, June 1976.
28. Shur, M. S., and L. F. Eastman: "Current-voltage characteristics, small-signal parameters, and switching times of GaAs FETS," *IEEE Trans. Elect. Dev.*, vol. ED-25, p. 606, June 1978.
29. Ladbrooke, P. H.: "GaAs MESFETS and HEMTs," in *Gallium Arsenide for Devices and Integrated Circuits* (Eds. H. Thomas, D. V. Morgan, B. Thomas, J. E. Aubrey, and G. B. Morgan), Peregrinus, London, 1986.
30. Arnold, D., Kopp, W., Fischer, R., Henderson, T., and H. Morkoc: "Microwave Performance of GaAs MESFETs with AlGaAs Buffer Layers—Effect of Heterointerfaces," *IEEE Elect. Dev. Lett.*, vol. EDL-5, pp. 82–84, March 1984.
31. Englemann, R., and C. Liechti: "Bias Dependence of GaAs and InP MESFET Parameters," *IEEE Trans. Elect. Dev.,* vol. ED-24, pp. 1288–1296, November 1977.
32. Yokoyama, N., Ohnishi, T., Odani, K., Onodera, H., and M. Abe: "TiW Silicide Gate Self-Alignment for Ultra-High-Speed MESFET LSI/VLSIs," *IEEE Trans. Elect. Dev.*, vol. ED-29, pp. 1541–1547, October 1982.
33. Ueno, K., Furatsuka, T., Toyoshima, H., Kanamori, M., and A. Higashisaka: "A High Transconductance GaAs MESFET with Reduced Short Channel Effect Characteristics," *IEEE Elect. Device Meeting Tech. Dig.,* pp. 82–85, December 1985.
34. Ohata, K., Itoh, H., Hasegawa, F., and Y. Fujiki: "Super Low-Noise GaAs MESFETs with a Deep-Recess Structure," *IEEE Trans. Elect. Dev.*, vol. ED-27, pp. 1029–1033, June 1980.
35. Kocot, C., and C. A. Stolte: "Backgating in GaAs MESFETs," *IEEE Trans. Elect. Dev.*, vol. ED-29, pp. 1059-1064, July 1982.
36. Chang, M. F., Lee, C. P., Hou, L. D., Vahrenkamp, R. P., and C. G. Kirkpatrick: "Mechanism of Surface Conduction in Semi-Insulating GaAs," *Appl. Phys. Lett.*, vol. 44, pp. 869–871, May 1, 1984.
37. Chang, M. F., *et al.*: "Material Parameters Affecting Surface Leakage in GaAs Integrated Circuits," in *Semi-Insulating III-V Materials* (Eds. D. C. Look and J. S. Blakemore) Kah-nee-ta, Oreg. pp. 378–386, Shiva Publ., 1984.
38. Subramanian, S., Bhattacharya, P. K., Staker, K. J., Ghosh, C. L., and M. H. Badawi: "Geometrical and Light-Induced Effects on Back-Gating in Ion-Implanted GaAs MESFETs, " *IEEE Trans. Elect. Dev.*, vol. ED-32, pp. 28–33, January 1985.
39. D'Avanzo, D.: "Proton Isolation for GaAs Integrated Circuits," *IEEE Trans. Elect. Dev.*, vol. ED-29, pp. 1055–1059, July 1982.
40. Inokuchi, K., Tsunotani, M., Ichioka, T., Sano, Y., and K. Kaminishi: "Suppression of Sidegating Effect for High Performance GaAs ICs," *IEEE GaAs IC Symp. Proc.*, Portland, Oreg. pp. 117–120, October 1987.

41. Lee, C. P., and M. F. Chang: "Shielding of Backgating Effects in GaAs Integrated Circuits," *IEEE Elect. Dev. Lett.*, vol. EDL-6, pp. 169–171, April 1985.
42. Blum, A. S., and L. D. Flesner: "Use of a Surrounding *p*-Type ring to Decrease Backgate biasing in GaAs MESFETs," *IEEE Elect. Dev. Lett.*, vol. EDL-6, pp. 97–99, February 1985.
43. Rocchi, M.: "Status of the Surface and Bulk Parasitic Effects Limiting the Performances of GaAs ICs," *Physica*, vol. 129B, pp. 119–138, 1985.
44. Larson, L. E.: "Gallium Arsenide Switched-Capacitor Circuits for High-Speed Signal Processing," Ph.D. Dissertation, pp. 98–120, UCLA, 1986.
45. Larson, L. E.: "An Improved GaAs MESFET Equivalent Circuit Model for Analog IC Applications," *IEEE J. Solid State Cir.*, vol. SC-22, pp. 567-574, August 1987.
46. Troutman, R. R.: "Subthreshold Design Considerations for Insulated Gate Field-Effect Transistors," *IEEE J. Solid State Cir.*, vol. SC-9, pp. 55–60, April 1974.
47. Lee, S. J., *et al.*: "Ultra-Low Power, High Speed GaAs 256 Bit Static RAM," *IEEE GaAs IC Symp. Proc.*, Phoenix, Ariz., pp. 74–77, October 1983.
48. Lee, S. J., and C. P. Lee: "Temperature Effect on Low Threshold Voltage Ion-Implanted GaAs MESFETs," *Elect. Lett,*, vol. 17, pp. 760–761, Oct. 1, 1981.
49. Houng, Y. M., and G. L. Pearson: "Deep Trapping Effect at the GaAs-GaAs:Cr Interface in GaAs FET Structures," *J. Appl. Phys.*, vol. 49, pp. 3348–3352, 1978.
50. Williams, R. E.: *Gallium Arsenide Processing Techniques*, Artech House, 1984.
51. Howes, M. J., and D.V. Morgan (Eds.): *Gallium Arsenide. Materials, Devices, and Circuits*, Wiley, 1985.
52. Howes, M. J., and D.V. Morgan (Eds.): *Gallium Arsenide. Materials, Devices, and Circuits*, chap. 5, Wiley, 1985.
53. Sze, S. M. (Ed.): *VLSI Technology*, 2d ed., chaps. 5 and 9, McGraw-Hill, 1988.
54. Howes, M. J., and D. V. Morgan (Eds.): *Gallium Arsenide. Materials, Devices, and Circuits*, chap. 6, Wiley, 1985.
55. Welch, B., Shen, Y.-D., Zucca, R., Eden, R. C., and S. I. Long: "LSI Processing Technology for Planar GaAs Integrated Circuits," *IEEE Trans. Elect. Dev.*, vol. ED-27, pp. 1116–1123, June 1980.
56. Kroemer, H.: "Heterostructure Bipolar Transistors and Integrated Circuits," *Proc. IEEE*, vol. 70, pp. 13-25, January 1982.
57. Lee, C. P.: presented at the 1985 IEDM short course on *High Speed Digital Integrated Circuits*.
58. Rosenberg, J. J., Benlamri, M., Kirchner, P. D., Woodall, J. M., and G. D. Pettit: "An $In_{0.15}Ga_{0.85}As$/GaAs Pseudomorphic Single Quantum Well HEMT," *IEEE Elect. Dev. Lett.*, vol. EDL-6, pp. 491–493, October 1985.
59. Chao, P. C., *et al.*: "$0.1\mu m$ Gate-Length Pseudomorphic HEMTs," *IEEE Elect. Dev. Lett.*, vol. EDL-8, pp. 489–491, October 1987.
60. Dingle, R., Störmer, H. L., Gossard, A. C., and W. Wiegmann: "Electron Mobilities in Modulation-Doped Semiconductor Heterojunction Superlattices," *Appl. Phys. Lett.*, vol. 33, p. 665, 1978.
61. Arnold, D. J., Fischer, R., Kopp, W. F., Henderson, T. S., and H. Morkoc: "Microwave Characterization of (Al,Ga)As/GaAs Modulation-Doped FETs: Bias Dependence of Small-Signal Parameters," *IEEE Trans. Elect. Dev.*, vol. ED-31, pp, 1399–1402, October 1984.
62. Dingle, R., Feuer, M. D., and C. W. Tu: "The Selectively Doped Heterostructure Transistor," in *VLSI Electronics, Microstructure Science*, vol. 11, *GaAs Microelectronics* (Eds. N. G. Einspruch and W. R. Wisseman), p. 227, Academic Press, 1985.
63. Dingle, R., Feuer, M. D., and C. W. Tu: "The Selectively Doped Heterostructure Transistor," in *VLSI Electronics, Microstructure Science*, vol. 11, *GaAs Microelectronics* (Eds. N. G. Einspruch and W.R. Wisseman), p. 220, Academic Press, 1985.
64. Drummond, T. J., *et al.*: "Field Dependence of Mobility in AlGaAs/GaAs Heterojunctions at Very Low Fields," *Elect. Lett.*, vol. 17, pp. 545–546, 1981.
65. Feuer, M. D. "Two-Layer Model for Source Resistance in Selectively Doped Heterojunction Transistors," *IEEE Trans. Elect. Dev.*, vol. ED-32, pp. 7–11, January 1985.
66. Solomon, P., Knoedler, C. M., and S. L. Wright, "A GaAs Gate Heterojunction FET," *IEEE Elect. Dev. Lett.*, vol. EDL-5, pp. 379–381, September 1984.
67. Cirillo, N. C., *et al.*: "Complementary Heterostructure Insulated Gate FETs (HIGFETS)," *1985 IEEE Int. Elect. Dev. Mtg., Tech. Dig.*, pp. 317–320, December 1985.

68. Kiehl, R. A., and A. C. Gossard: "Complementary *p*-MODFET and *n*-HB MESFET (Al,Ga)As Transistors," *IEEE Elect. Dev. Lett.*, vol. EDL-5, pp. 521–523, December 1984.
69. Morkoc, H., and P. Solomon: "Modulation-Doped GaAs/AlGaAs Heterojunction FETs (MODFETs), Ultrahigh-Speed Device for Supercomputers," *IEEE Trans. Elect. Dev.*, vol. ED-31, pp. 1015–1027, August 1984.
70. Drummond, T. J., W. T. Masselink and H. Morkoc: "Modulation-Doped GaAs/(Al,Ga)As Heterojunction Field Effect Transistors: MODFETs," *Proc. IEEE*, vol. 74, pp. 773–822, June 1986.
71. Chang, M. F., *et al.*: "Self-aligned AlGaAs/GaAs Heterostructure Bipolar Transistors with Improved High Speed Performance," Paper IVA-1, Device Research Conf., Santa Barbara, Calif., June 1987. For abstract see *IEEE Trans. Elect. Dev.* vol. ED-34, p. 2369, November 1987.
72. Shockley, W.: US Patent number 2,569,347. Filed June 1948.
73. Neudeck, G. W., and R. F. Pierret (Eds.): *Modular Series on Solid State Devices*, vol III. Addison-Wesley, 1983.
74. Ashar, K. G.: "The Method of Estimating Delay in Switching Circuits and the Figure of Merit of a Switching Transistor," *IEEE Trans. Elect. Dev.*, vol. ED-11, pp. 497–506, November 1964.
75. Dumke, W. P., Woodall, J. M., and V.L. Rideout: "GaAs-GaAlAs Heterojunction Transistor for High Frequency Operation," *Solid State Elect.*, vol. 15, pp. 1339–1344, December 1972.
76. Asbeck, P. M., Miller, D. L., Anderson, R. J., and F.H. Eisen: "GaAs/(Ga,Al)As Heterojunction Bipolar Transistors with Buried Oxygen-Implanted Isolation Layers," *IEEE Elect. Dev. Lett.*, vol. EDL-5, pp. 310–312, August 1984.
77. Batey, J., and S. L. Wright: "Energy Band Alignment in GaAs:(Al,Ga)As Heterostructures: The Dependence on Alloy Composition," *J. Appl. Phys.*, vol. 59, pp. 200–209, 1 January 1986.
78. Ishibashi, T., and Y. Yamauchi: "A Novel AlGaAs/GaAs HBT Structure for Near-Ballistic Collection," Paper IVA-6, Device Research Conf., Santa Barbara, Calif., June 1987. For abstract see *IEEE Trans. Elect. Dev.*, vol. ED-34, p. 2371, November 1987.
79. Greiling, P.: example from "Picosecond Electronics", short course class notes, UCSB, 1984.
80. Rao, M. A., *et al.*: "Determination of Valence and Conduction Band Discontinuities at the (Ga,In)P/GaAs Heterojunction by *C-V* Profiling," *J. Appl. Phys.*, vol. 61, pp. 643–649, 1987.

CHAPTER 2

MODELS FOR GaAs DEVICES AND CIRCUITS

The design of high-speed GaAs digital integrated circuits requires the use of accurate models for the active devices and for passive circuit elements. As will be presented in detail in Chap. 3, approximate models (hand analysis) and more detailed models (computer simulation) can be used to evaluate the logic levels, noise margins, power dissipation, and propagation delay of a logic circuit. Since the most widely used circuit simulation tool for IC design is SPICE [1], models that are typically used with this tool to predict current vs. voltage and capacitance vs. voltage will be presented for the Schottky barrier diode, GaAs MESFET, and the heterojunction bipolar transistor. Measurement methods and parameter extraction techniques will be discussed for each model since accurate input parameters* are important if good correspondence between measured data and simulated characteristics are to be obtained. A comprehensive guide

---

*Throughout the book, the SPICE parameters used as inputs to the device models will be normalized to unit area or width. Also, to avoid confusion, all such model input parameters and all SPICE circuit simulation examples will be expressed in lower case characters because of the prevailing lower case usage in UNIX. The symbols and case used in the device equations in the text, however, will follow the standard usage in the electron device literature.

to SPICE has been written by Tuinenga [2] and should be used as a reference by those unfamiliar with this tool.

In addition to the device (active) models, approximate models used for estimating the capacitive loading effects of electrostatic fringing fields from narrow metal contact pads on the devices will also be presented in this chapter. The very important consideration of loading capacitance from metal interconnections between devices and logic gates on chip will be postponed until Chap. 5.

2.1 SCHOTTKY BARRIER DIODE

The Schottky barrier diode is one of the high-performance components available in GaAs IC technology and is extensively used in the design of GaAs circuits. It may find application as a level-shifting element, limited to forward-bias operation, in which case its design should emphasize low series resistance and high junction capacitance. In other circuits, it may also find use as a logic-switching element, in which case both forward and reverse-bias operations are needed, and low series resistance and low capacitance are at a premium.

The physical mechanisms for the operation of the Schottky barrier diode (SBD) were presented in Chap. 1 and have been described in depth in other sources [3, 4]. The SBD of primary interest for application in the GaAs IC consists of a metal layer or perhaps a refractory-silicide layer in intimate contact with n-type GaAs. A variation which might occur in some HFET processes would consist of a metal/n-type AlGaAs junction. Briefly, the main feature that distinguishes the SBD from the pn junction is that the SBD is a majority carrier device. In spite of this major difference, the current-voltage relationship is still accurately described by the ideal diode equation:

$$I_D = I_S\left[\exp\left(\frac{qV_{D,i}}{nkT}\right) - 1\right] \quad (2.1)$$

In this equation, I_S represents the diode saturation current, n is the ideality factor or emission coefficient, and q, k, and T represent charge, Boltzmann's constant, and temperature in kelvin as usual. At room temperature, the thermal voltage kT/q is approximately 26 mV. I_s is generally in the range from 10^{-15} to 10^{-14} A when the diode area is normalized to 1 μm^2 and n ranges between 1.0 and 1.2 for most SBDs on n-type GaAs. $V_{D,i}$ is the intrinsic diode voltage which appears across the junction.

While the above equation adequately represents the junction itself, any practical SBD will also suffer from series resistance caused by the contact and whatever bulk n-type GaAs that might be in series with the junction and the contact. In addition, since the diode is fabricated in a planar structure as described in Chap. 1, there will be a voltage drop laterally under the barrier metal having the highest voltage (and therefore current) at the edge closet to the contact. This effect is called *current crowding*, and the excess resistance is a spreading resistance which will be current-dependent [5]. The total series resistance-induced voltage drop is incorporated into Eq. (2.1) by defining $V_{D,i}$ as

$$V_{D,i} = V_D - I_D R_S \quad (2.2)$$

where R_S is the series resistance due to the contacts, neutral n-GaAs, and current crowding at the edges of the contacts. V_D is the voltage at the diode terminals.

The SPICE diode model assumes a constant R_S. This approximation leads to some inaccuracy at large V_D where the voltage drop across the diode series resistance is a significant fraction of the total. However, this extreme operating point should be avoided for dependable circuit designs anyway; contact resistance and electrode spacings vary significantly from wafer to wafer due to variations in surface preparation and alignment tolerances. A circuit design which would depend on series resistance to obtain a certain voltage drop will be inherently undependable. At smaller current density, the constant R_S approximation will be acceptable in most applications. More complete models which include the spreading resistance variation have been described in the literature [5].

The equivalent circuit used in SPICE for the diode model is shown in Fig. 2-1. The nonlinear dependent current source I_D provides the functional relationship of Eq. (2.1). The resistor models the voltage drop caused by R_S. The voltage variable capacitor C_D models the charge stored in the depletion region of the SBD junction. This capacitance is modeled by the C-V relationship described in Chap. 1, given below:

$$C_D = C_{j0}\left(1 - \frac{V_{D,i}}{V_{bi}}\right)^{-m} \qquad (2.3)$$

where C_{J0} represents the capacitance of the junction at zero bias (typically 1 to 2 fF/μm^2) and V_{bi} is the effective built-in potential (vj in the model parameter set), typically 0.8 V, but which may be higher than the true built-in potential if needed to fit the capacitance-voltage relationship caused by a nonuniform doping profile. The grading coefficient, m, is 0.5 for uniform doping in the GaAs, but can range from 0.33 for linear grading of doping to 1.3 for an ion-implanted gaussian profile.

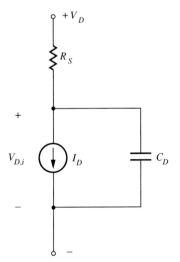

FIGURE 2-1
SPICE diode model equivalent circuit. The dependent current source provides the diode current according to the ideal diode equation (2.1).

It is significant that no provision is needed for diffusion capacitance, the dominant contributor to the forward-bias junction capacitance of the PN junction, because significant numbers of minority carriers are not stored in the neutral GaAs. Therefore, the transit time parameter, τ, of the SPICE diode model is allowed to default to zero when modeling the SBD.

Equation (2.3) would predict infinite capacitance if $V_{D,i} = V_{bi}$, the flat-band condition. This possibility is of little concern, since the series resistance prevents this from happening due to the large current flow that would also occur in a near encounter with the flat band. To be safe, however, the SPICE diode model assumes that the capacitance no longer increases when the internal voltage exceeds 0.5 V. This simplification is unlikely to cause serious error, because the diode conduction current is large in this range, and the charge transported by the conduction current exceeds that of the capacitor by a comfortable margin.

Extracting the important SPICE diode model parameters is, n, rs, $cj0$, vj, and m of Table 2.1 from experimental measurement is straightforward. The dc current-voltage equation (2.1) can be plotted as $\log I_D$ vs. V_D, as shown in Fig. 2-2. A straight line fit is obtained over several decades of current. Outside this region, the current flow mechanism deviates from the semilog relationship at low currents by recombination effects and at high currents because of voltage drop across the internal series resistance. The straight line region will obey the equation

$$\log_{10} I_D = \log_{10} I_S + \frac{qV_D}{2.3nkT} \qquad (2.4)$$

obtained by taking the log of (2.1). Therefore, from Eq. (2.4), the intercept at $V_D = 0$ on the semilog plot provides the saturation current I_S which fits this equation. This becomes the SPICE model parameter is when normalized to unit area. The slope of the line is determined by n and is the inverse of about 60 mV/decade if $n = 1$. It is evident from Fig. 2-2 that at higher forward-current levels the diode current is limited by series resistance. Therefore, R_S can be determined from the plot if the actual device characteristic is compared against the extrapolated ideal straight line. The difference in voltage, $V_{D2} - V_{D1} = \Delta V$, corresponds to the voltage drop

TABLE 2.1
SPICE diode model used for GaAs Schottky diode. Typical parameters are for a diode 1 μm in length and 1 μm in width

Name	Parameter	Units	Default value	Typical value	Area
is	Saturation current	A	1.0e-14	1.0e-14	*
rs	Ohmic resistance	Ω	200	1500	*
n	Ideality factor	—	1.0	1.1	
tt	Transit time	s	0	0	
cj0	Zero-bias junction capacitance	F	0	2.0e-15	*
vj	Built-in potential	V	1	0.8	
m	Grading coefficient	—	0.5	0.5	

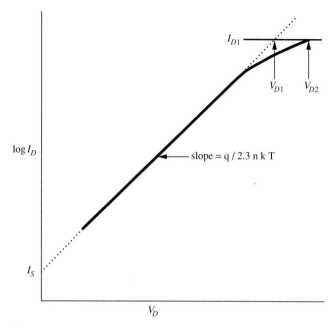

FIGURE 2-2
The dc current-voltage equation (2.1) can be plotted as log I_D vs. V_D in order to facilitate the extraction of the parameters I_S, n, and R_S.

due to series resistance, $I_D R_S$. ΔV and R_S should be determined at several values of I_D and an average taken, because R_S will depend on I_D. Since the model must be most accurate at the current selected for use in level shift diodes, the typical operating current for the given application should guide the choice of I_D for the determination of R_S which when normalized to unit area becomes model parameter rs.

The model parameters required for the capacitance equation (2.3) can also be extracted from measured C-V data. Using a capacitance meters with small-signal excitation (ac voltage < 20 mV or so), first the probe or test fixture capacitance must be nulled out or recorded. Then the zero-bias capacitance can be measured directly. To determine m and V_{bi} guess a reasonable value for m and plot $C_D^{-1/m}$ as a function of V_D in the reverse-bias region, as shown in Fig. 2-3. Forward bias is to be avoided, since most capacitance meters will not measure the correct capacitance when large conductivity is also present. If the correct m value was assumed, the plot will result in a straight line, and the effective built-in voltage can be determined from the intercept point less the Gummel correction kT/q. Fortunately, the equation is not terribly sensitive to the choice of m, and a good fit can be obtained over quite a wide range of values. Of course, the intercept will vary significantly with the choice of m.

A summary of the SPICE diode model parameters used for the GaAs Schottky diode is given in Table 2.1 along with the assumed default values for these parameters. Typical parameter values, normalized to a 1 square micrometer area (1 μm long × 1 μm wide), are also provided to illustrate the need for measurement of the GaAs

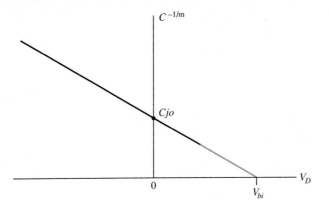

FIGURE 2-3
To determine m and V_{bi} a reasonable value for m is assumed. A plot of $C_D{}^{-1/m}$ as a function of V_D in the reverse-bias region can be used to determine V_{bi} (model parameter pb) from the intercept.

Schottky diode parameters, not simply relying on the default values. The asterisk in the area column indicates that the values will be automatically scaled with the geometric factor specified on the device element line in the SPICE input files.

It is important to keep in mind that the model parameters will depend on the length of the Schottky metal overlap with the active layer in a different way from the way they depend on width. This is because of the current crowding effects, discussed in Chap. 1, which cause most of the forward-bias current to flow into the contact within the first micrometer of the edge. Thus, the model parameters associated with the current flow (is, rs) will scale differently with length than will the parameters (such as n, $cj0$, vj, m) not influenced by resistive loss in the active layer. The diode Schottky metal stripe length (L in Fig. 1-6) is not varied when adjusting the diode area during layout. Diodes should be characterized with a fixed stripe length (1 to 2 μm is typical) and the diode area adjusted by varying the width (W in Fig. 1-6). Therefore, the width should be specified by the geometric factor on the diode line while the length must be determined on the model line by the parameters themselves. An example of an input line describing a diode in a circuit node listing is shown below:

$$d1 \quad 2 \quad 1 \quad lsdiode \quad 4$$

Here, d1 is the device number, 2 and 1 are the positive and negative nodes respectively, lsdiode is the diode model name, defined with a .model statement, and 4 is the area factor, 4 μm^2.

The model parameters scale in the following way with diode area:

is * A
rs / A
cj0 * A

Example 2.1. Measured *I-V* data for a GaAs Schottky barrier diode with area $2 \times 36 \ \mu m^2$ is plotted in Fig. 2-4. (*a*) Estimate *is*, *n*, and *rs* for this diode. (*b*) Use the SPICE diode model with these input parameters to predict *I-V* and plot on Fig. 2-4.

Solution

(*a*) Figure 2-4 represents $\log_{10} I_D$ as a function of the external applied voltage V_D in the forward-bias region. The three parameters needed to describe this forward characteristic are described in Eqs. (2.1) and (2.2). At intermediate current levels (above the region at low currents where surface leakage and recombination may apply and below the high current region where series resistance becomes dominant), the -1 term in (2.1) and the $-I_D R_S$ term in (2.2) can be neglected. A straight line is then predicted for the semilog plot according to (2.4), where $\log_{10} I_S$ is the current intercept at $V_D = 0$. From the extrapolation of the intermediate region in Fig. 2-4, $I_s = 1 \times 10^{-12}$ A. Scaling to a $1 \times 1 \ \mu m^2$ area gives $is = 1.4 \times 10^{-14}$ A.

The slope of this region is 13.9 decades/V or 72 mV/decade. Thus, $n = 72/60 = 1.20$. This parameter is not scaled with area.

The series resistance can be estimated from the behavior of the diode at high current levels where the *I-V* curve is falling away from the extrapolated straight line as shown in Fig. 2-2. From Fig. 2-4, $V_{D2} - V_{D1} = 0.11$ V at $I_D = 3.5 \times 10^{-3}$ A.

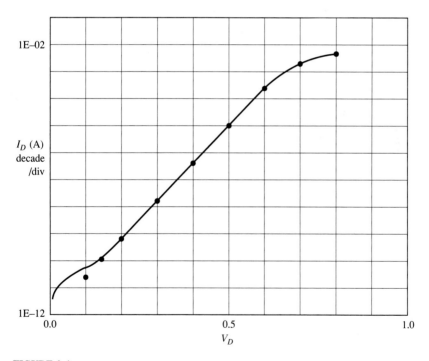

FIGURE 2-4
Measurement of *I-V* characteristic for a GaAs Schottky barrier diode with area $2 \times 36 \ \mu m^2$. Points are the predictions of the SPICE diode model.

```
Example 2.1  sbd simulation
.model sbd d(is=1.4e–14 n=1.2 rs=2230)
v1 1 0 dc
d1 1 0 sbd 72
.dc v1 0.1 1 0.01
.print v(1) i(v1)
.end
```

FIGURE 2-5
SPICE input file used to simulate diode characteristics.

Therefore, $\Delta V/I_D = 31\ \Omega$. Scaling to a $1 \times 1\ \mu m^2$ area, model input parameter $rs = 2230\ \Omega$.

(b) The SPICE input file used to simulate the diode I-V is shown in Fig. 2-5. The points plotted on Fig. 2-4 represent the output of this simulation. Note that the fit is quite good except in the very low current region.

2.2 GaAs MESFET MODELS

Because of the limited voltage swing and low power requirements of dense GaAs logic circuits, an accurate dc model over all operating regions is especially critical. Inaccuracies will lead to incorrect prediction of voltage levels and the possibility of a malfunctioning circuit over normal extremes of process or temperature. A SPICE analysis is obviously only as accurate as the equations used for the circuit elements. These equations must be a compromise between accuracy and excessive computation time. Models must operate efficiently in circuit analysis programs such as SPICE, because there can be many devices included, even in the smaller subcircuits typically evaluated by full analog circuit simulators. Since the model will be called to evaluate the device terminal characteristics many times during a simulation, the time required to execute the equations should be minimized, and the equations themselves must be continuous up to at least the third derivative so that convergence is speedy (the convergence error is estimated from the third derivative of the device I-V relations). Direct current convergence is checked by each device subroutine as it is called for each occurrence of the device in the circuit.

Most MESFET models reported in the literature do not accurately model the dc or transient behavior of real devices over all regions of operation, particularly the region below saturation. This deficiency is perhaps partly due to a historical bias toward microwave applications where the saturation region is most important and the other regions are to be avoided. Also, the drain current in short-gate-length GaAs MESFETs is limited by the drift velocity of the channel electrons which peaks at low electric field (3500 V/cm) and then declines (velocity saturation). Consequently, even low V_{DS} (less than 0.3 V) will show some evidence of saturation. Modes that were designed for use with silicon JFETs which reach saturated velocities at ten times higher electric field and which generally utilize longer channels do not accurately predict all of the detailed structure of the I_D-V_{DS} or I_D-V_{GS} characteristic.

Figure 2-6 presents a schematic diagram of the equivalent circuit commonly used to represent the GaAs MESFET. The controlled current source, $I_D(V_{GS,i}, V_{DS,i})$, at the center is the key element for the dc model. It produces a drain current that

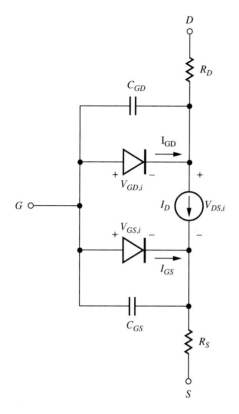

FIGURE 2-6
A schematic diagram of the equivalent circuit commonly used to represent the GaAs MESFET in SPICE.

depends on the internal gate-to-source (GS) and gate-to-drain (GD) voltages, $V_{GS,i}$ and $V_{DS,i}$. These voltages are, in general, different from the terminal (external) voltages, V_{GS} and V_{DS}, because of the parasitic source and drain access resistances, R_S and R_D, discussed in Chap. 1. They are related by the equations below:

$$V_{GS,i} = V_{GS} - I_D R_S \quad \text{and} \quad V_{DS,i} = V_{DS} - I_D(R_S + R_D) \quad (2.5)$$

The GS and GD diodes in Fig. 2-6 represent the gate junction; the forward-bias current flow is modeled with individual Schottky barrier diodes buried within the MESFET model. The accuracy of the gate current model is as important as the drain current, because the forward gate conduction sets the logic high-voltage level in several widely used GaAs logic circuits. Errors in gate current prediction will lead directly to comparable errors in the noise margin predictions for these circuits.

Time-dependent effects are simulated through charge storage elements. The charge stored in the GS and GD depletion layer is represented in Fig. 2-6 by the two voltage-dependent (nonlinear) capacitors, C_{GS} and C_{GD}.

In this section, three popular FET models are reviewed and related to the equivalent circuit of Fig. 2-6, and their predictions are compared with actual device measurements. Modifications are discussed which will improve on the accuracy of the models. Techniques will be introduced in Sec. 2.3 that can be used to model some of the second-order effects presented in Chap. 1. Parameter extraction methods for the models will also be suggested.

2.2.1 SPICE JFET Model

The SPICE JFET model is a popular starting point for the simulation of GaAs MESFETs. Although the existing model is obsolete and has several deficiencies which limit its usefulness for GaAs IC simulation and which will be identified below, the framework of the JFET subroutine is convenient for modification of the model since it is largely self-contained. The model itself may be adequate for noncritical, comparative design simulations if nothing better is available.

The dc drain current predictions of the SPICE JFET are based on the equations of the Shichman-Hodges model [6], a charge-control model that assumes that the drain current is proportional to the total amount of charge in the channel and (in the linear region) to V_{DSi}. It assumes a square law relationship between I_D and V_{GS} as discussed in Sec. 1.2. The total charge is calculated by assuming that all of the charge is confined to a charge sheet at a constant depth. This approximation is most accurate in the case of an impulse or spike doping profile in the channel. The saturation current follows from the maximum of the linear region current. The model does not take into account velocity-saturation effects. The source and drain resistors are treated as separate circuit elements as shown in Fig. 2-6, so the internal node voltages are determined through an iterative process. The equations below describe current in the cutoff, linear, and saturation regions respectively:

$$I_D = 0 \quad \text{if } V_{GSi} < V_T \tag{2.6a}$$

$$I_D = \beta[2(V_{GSi} - V_T)V_{DSi} - V_{DSi}^2](1 + \lambda V_{DSi})$$
$$\text{if } V_{DSi} < V_{GSi} - V_T \quad \text{and} \quad V_{GSi} \geq V_T \tag{2.6b}$$

and $$I_D = \beta(V_{GSi} - V_T)^2(1 + \lambda V_{DSi})$$
$$\text{if } V_{DSi} \geq V_{GSi} - V_T \quad \text{and} \quad V_{GSi} \geq V_T \tag{2.6c}$$

In these equations, the parameter β represents the "k factor" or transconductance parameter and has units of amperes per square volt. For the GaAs FETs, β scales directly with channel width W over a wide range. Scaling with gate length L is less direct, as shown in Chap. 1. It is convenient to evaluate β in the model parameter assignment statement (.model command) in terms of a unit width, generally 1 μm, so that widths in micrometers of each MESFET can be assigned when the transistor input node connections are described in the SPICE input file. β typically ranges from 10^{-6} to 2×10^{-5} A/V^2. V_T is the FET threshold voltage, typically in the range -1.5 to -0.5 V for depletion mode and 0.1 to 0.3 V for enhancement mode devices. Note that the SPICE JFET model connects the linear and saturation regions at the classical pinch-off voltage, $V_{GS} - V_T = V_{DS}$, described in Chap. 1.

The factor $(1 + \lambda V_{DSi})$ is used to empirically account for the increase in drain current in the saturation region as V_{DS} increases. This same factor is used in all of the MESFET models to be described. The channel-length modulation parameter, λ, is equivalent to 1/VA (the Early voltage in bipolar transistors). The parameter VA is often used to describe base width modulation when modeling the bipolar transistor. λ ranges typically from 0.01 to 0.2 V^{-1} for most MESFETs with gate lengths in the 0.8 to 2.5 micrometer range.

The solid curves in Fig. 2-7 show measured *I-V* characteristics for a 1.6 × 50 μm (gate length × channel width), ion-implanted, depletion mode GaAs MESFET. The predictions of the model equations are shown by the solid circles. It can be seen that the fit is adequate to the saturation region, but is in error by as much as 300 mV of V_{DS} in the linear region. This error is largely a consequence of the model not explicitly dealing with velocity-saturation effects. Also, note that the classical pinch-off condition overpredicts the saturation voltage in short-channel MESFETs as described in Chap. 1.

Finally, a word about the convergence of the model. Since the model does not include the subthreshold region, one can see that the drain current will be discontinuous for finite values of λ and V_{DS} when V_{GS} changes from below V_T to above V_T. This abrupt step in current can lead to lack of dc convergence when this range of V_{GS} is involved in the solution. Usually the problem can be avoided, however, by specifying an input voltage or setting initial conditions (.nodeset for dc mode or .ic / <uic> option for the transient mode) which avoid a bias point initially in cutoff. For example, if the input voltage ramps down from the high state (input

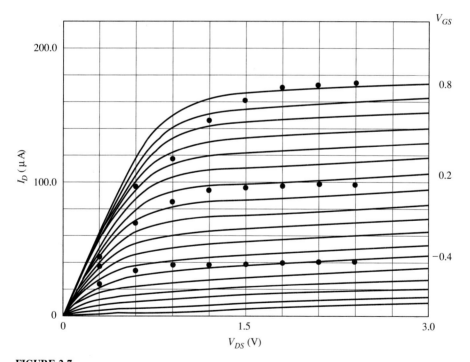

FIGURE 2-7
The solid curves show measured *I-V* characteristics for a 1.6 × 50 μm (gate length × channel width), ion-implanted, depletion mode GaAs MESFET. The predictions of the SPICE JFET model equations are shown by the solid circles.

transistor initially on) to the low state (input transistor in cutoff), the convergence problems will not occur until the simulation has been completed.

An additional convergence problem may also arise if the MESFET model subroutine is modified to improve dc accuracy or to incorporate second-order corrections. Derivatives of I_D and I_G with respect to V_{GS} and V_{GD} are calculated in the model subroutine and will depend on the functional form of I_D. In order to achieve efficient convergence, the iterative solution algorithm depends on calculating slopes and predicting the next point. If the derivatives are in error, then the convergence will be severely hampered. Therefore, any modifications to the equations describing $I_D(V_{GS}, V_{DS})$ must also be reflected in the equations for the partial derivatives of I_D.

Next, a brief description will be given of methods for extraction of the device parameters discussed above. The determination of β, V_T, and R_S and R_D can also be used for the hyperbolic tangent model to be described in the next section. The determination of the dc value for λ applies to all three models to be discussed.

The device parameters β, V_T, λ, R_S and R_D can easily be extracted in a manner that gives a good fit to the saturation region. During this process, however, they lose their correspondence or relationship to the actual measurable physical device parameters V_T, R_S, and R_D because of the shortcomings of the SPICE JFET model. The process is illustrated by Fig. 2-8b which shows a plot of $\sqrt{I_D}$ as a function of $V_{GS,i}$. The drain current of Fig. 2-8b is determined by extrapolating the slope of the I_D-V_{DS} characteristics in the saturation region back to $V_{DS} = 0$ as shown in Fig. 2-8a. The extrapolation removes the influence of the $(1 + \lambda V_{DSi})$ term from Eq. (2.6c). These intercepts on the I_D axis will be referred to as $I_{D,\text{sat}}$, which is a function of V_{GS} and is plotted in Fig. 2-8b. Then, by rearranging (2.6c) and including (2.5), the following equation is obtained:

$$\sqrt{I_{D,\text{sat}}} = \sqrt{\beta}(V_{GS} - I_D R_S - V_T) \qquad (2.7)$$

It can be seen that this equation will produce a straight line if the device obeys the square law saturation current predicted by (2.6c). Even if it deviates from square law, as is normal for short-channel MESFETs, it can be forced to fit (2.7) by selecting R_S such that the data plotted in Fig.2-8b is linear. Under this condition, the intercept of the line gives V_T and the slope is $\sqrt{\beta}$.

This extraction procedure produces an R_S value that is larger than the actual measured source resistance. This large R_S will emulate velocity-saturation effects in the I_D-V_{DS} characteristics such as constant transconductance with gate bias and reduced drain saturation currents, because it is acting as a feedback resistor in the common-source connection typically used for measurement. While the larger R_S provides the desired fit to the saturation region, unfortunately the fit in the linear region is degraded. The only way to minimize the error in the linear region when compared with experimental data is to make R_D very small, typically 0. Figure 2-7 shows the degree of error that is produced by Eq. (2.6b) in the linear region. The nonsymmetric assignment of R_S and R_D also leads to drastic errors in the inverse region, important for the simulation of pass transistors (transmission gates). It also leads to a discontinuity in slope at $V_{DS} = 0$ which disturbs the convergence of circuits employing pass transistors.

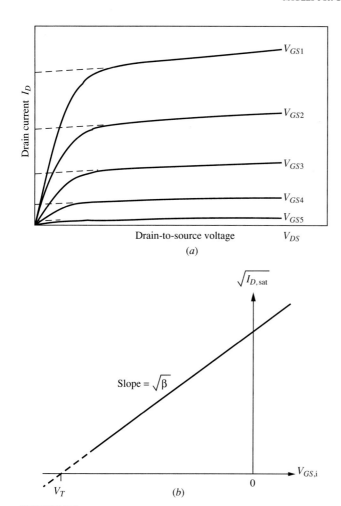

FIGURE 2-8
(a) The saturated drain current of the MESFET is determined by extrapolating the slope of the I_D-V_{DS} characteristics in the saturation region back to $V_{DS} = 0$. *(b)* $\sqrt{I_{D,\text{sat}}}$ is plotted as a function of $V_{GS,i}$.

The dc value for the parameter λ can be evaluated by taking the ratio of two drain currents, I_{D1} and I_{D2}, measured at two values of V_{DS} (V_{DS1} and V_{DS2}) on a constant V_{GS} curve. This will lead to the equation below from which λ can be determined:

$$\frac{I_{D2}}{I_{D1}} = \frac{1 + \lambda V_{DS2}}{1 + \lambda V_{DS1}} \tag{2.8}$$

It should be noted that λ should be extracted from a V_{GS} curve well below the forward-bias gate limit in order to avoid the influence of device self-heating. This thermal effect can produce smaller-than-realistic drain currents at high V_{GS} and V_{DS} when the device is measured on a curve tracer or on a low-frequency parameter analyzer, because the increased channel temperature reduces the drain current through mobility reduction. For a depletion mode MESFET, a V_{GS}-V_T of approximately 0.5 V would be suitable for extraction of λ.

The predictions of gate current utilize the following ideal diode equations for each diode in Fig. 2-1:

$$I_{GS} = I_S \left[\exp\left(\frac{qV_{GS,i}}{kT}\right) - 1 \right] \quad (2.9)$$

$$I_{GD} = I_S \left[\exp\left(\frac{qV_{GD,i}}{kT}\right) - 1 \right] \quad (2.10)$$

In contrast with the SPICE diode model described in Sec. 2.1, this model assumes that the ideality factor is fixed at $n = 1$. However, the actual ideality factor is often close to 1.1 for a gate diode. Therefore, if the diode parameter extraction procedures described in Sec. 2.1 are employed, the prediction of the gate diode voltage drop V_{GS} or V_{GD} at higher current levels (where V is greater than 0.5 V) will be seriously in error. This situation is illustrated in Fig. 2-9 in which log I_G is plotted as a function of V_{GS} for ideality factors of 1.0 and 1.1. If the saturation current I_S is obtained from the extrapolation of the measured data ($n = 1.1$), and Eqs. (2.9) and

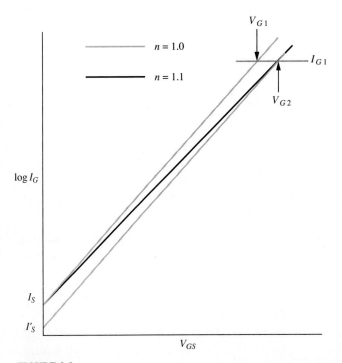

FIGURE 2-9
Log I_G is plotted as a function of V_{GS} for ideality factors of 1.0 and 1.1. I'_S is obtained from a single point fit of Eq. 2.9 with $n = 1$ to a V_{GS}, I_{GS} point selected in the middle of the forward-bias voltage range of interest.

(2.10) ($n = 1.0$) are used to predict the forward gate voltage at a specific current level I_{G1}, then the predicted voltage, V_{G1} in Fig. 2-4, will be incorrect. A nearly correct value of V_{GS} or V_{GD} in the region of importance (0.6 to 0.75 V) can be obtained from these equations, however, if a new (but artificial) saturation current, I'_S, is defined. I'_S is obtained from a single point fit of (2.9) to a V_{GS}, I_{GS} point selected in the middle of the voltage range of concern. As shown by the plot, the gate diode voltage V_{G2} is predicted with adequate accuracy (within 20 mV or so) when I'_S is used with (2.9).

The gate capacitance components C_{GS} and C_{GD} are described in the SPICE JFET model by the depletion-layer capacitance equation (2.3). Unfortunately, only C_{GS0}, C_{GD0}, and V_{bi} (pb) are user-adjustable model parameters. The grading coefficient m is internally set to 0.5, the value for uniformly doped semiconductor layers. The effect of this approximation is illustrated in Fig. 2-10, in which the experimental (measured) total gate capacitance vs. voltage characteristic for an ion-implanted MESFET is plotted on the same scale with the graph of Eq. (2.3) using the measured zero-bias capacitance C_{j0}, fixed $m = 0.5$, and $V_{bi} = 0.8$ V. There are significant discrepancies between the slopes of the two curves. The ion-implanted profile fits well to (2.3) if $m = 1.3$ and $V_{bi} = 1.7$ V. In addition, no provision for the pinch-off of the channel layer is made in the JFET model; rather, the channel doping profile is considered to be of infinite extent. However, Fig. 2-10 shows that the actual measured capacitance drops rapidly after the applied reverse bias falls below the channel pinch-off voltage. Finally, the true gate capacitance is actually a complicated func-

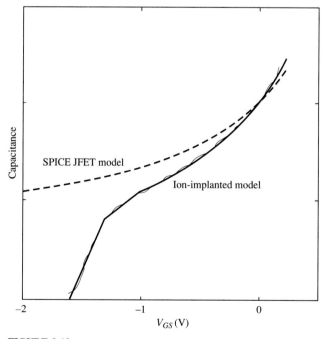

FIGURE 2-10
Measured total gate capacitance vs. voltage characteristic for an ion-implemented MESFET is plotted on the same scale with the graph of Eq. (2.3) using the measured C_{jo}, fixed $m = 0.5$, and $V_{bi} = 0.8$ V.

tion of both V_{GS} and V_{GS}, since the gate depletion region shape depends on both potentials, is two-dimensional in extent, and is also subject to a nonlinear velocity-field characteristic for GaAs. Therefore, the SPICE JFET capacitance model must be modified so that at least pinch-off and the correct slopes are modeled. More detailed capacitance models have been described in several publications [7–9] with sufficient detail to permit modified MESFET capacitance models to be introduced into circuit simulators.

The proper distribution of the total gate capacitance between zero-bias capacitance parameters C_{GS0} and C_{GD0} is unclear with the simple depletion capacitance model. The total gate capacitance C_G is defined as the measured capacitance of the gate when the source and drain are shorted together. While it is common practice for simulating the performance of microwave MESFET circuits to set C_{GD0} at some small fraction of the total gate capacitance, this assumption is only valid if the MESFET is always biased in the saturation region. This assumption is clearly not justified for digital application, where the MESFET typically will be biased in cutoff, linear, and saturation regions during a transition from one logic state to the other. For the digital case, a better assumption is to set $C_{GS0} = C_{GD0}$. Given that these two zero-bias capacitances are equal, it would seem logical to set $C_{GS0} = C_{GD0} = C_G/2$ so that $C_{GS} + C_{GD} = C_G$ when $V_{DS} = 0$. However, in saturation, where C_{GD} drops to nearly zero when the channel is pinched off, $C_{GS} + C_{GD} = C_G/2$. The total capacitance is therefore underpredicted, and propagation delays will be underestimated.

Alternatively, if the assignment $C_{GS0} = C_{GD0} = C_G$ is made, then the model predicts $C_{GS} + C_{GD} = 2C_G$ when $V_{DS} = 0$, an overprediction of total capacitance by a factor of 2 in the linear region. Clearly, the simple single-variable depletion capacitance model does not adequately represent the complex two-dimensional behavior

TABLE 2.2
SPICE JFET model parameters used for GaAs MESFET. Typical parameter values are shown for an ion-implanted, depletion-mode, non-self-aligned MESFET with 1 μm gate length, normalized to 1 μm channel width

Name	Parameter	Units	Default value	Typical value	Area
vto	Threshhold voltage	V	−2.0	−1.0	
beta	Transconductance parameter	A/V^2	1.0e-4	1.0e-4	*
lambda	Channel-length modulation parameter	1/V	0	0.1	
rd	Drain ohmic resistance	Ω	0	0	*
rs	Source ohmic resistance	Ω	0	2500	*
cgs	Zero-bias GS junction capacitance	F	0	1.2e-15	*
cgd	Zero-bias GD junction capacitance	F	0	1.2e-15	*
pb	Gate junction effective built-in potential	V	1	0.8	
is	Gate junction saturation current	A	1.0e-14	3.0e-16	*

of the FET. Nevertheless, the empirically determined partitioning $C_{GS0} = C_{GD0} = 0.75C_G$ yields delay predictions in reasonable agreement with experiment (10 to 20 percent error) on a modified depletion capacitance model which accounts for pinch-off and permits the use of both m and $V_{bi}(pb)$ to optimize the capacitance fitting.

A summary of the SPICE JFET model parameters used with the GaAs MESFET is presented in Table 2.2. In the table, the default parameters are compared against typical measured device parameters for a 1-μm gate length MESFET normalized to a channel width of 1 μm. An asterisk in the area column signifies that this parameter is automatically scaled with the geometric factor specified on the device element line. For example, a 1 × 10 μm MESFET with model name *mes* would be invoked by the following line in the SPICE node listing:

$$j1 \quad 3 \quad 2 \quad 1 \quad mes \quad 10$$

Here, the drain, gate, and source are connected to nodes 3, 2, and 1 respectively.

2.2.2 Modifications to the SPICE JFET Model— Hyperbolic Tangent

A great many dc MESFET models have been reported in the literature which are intended to give better agreement with experimental data than is possible with the simple model described above. Some of these models are physically based, such as the Shockley model [10], which is useful only for very long gate length FETs because it assumes constant mobility in the channel, or one of many two-section velocity-saturation models ([11], for example) in which the electron velocity is held proportional to electric field below a critical field, $\mathscr{E}_c$ (constant mobility), and a constant electron velocity is assumed for $\mathscr{E} > \mathscr{E}_c$. While these more physics-based circuit-simulation models are appealing, they generally do not show good agreement between model I-V curves and the actual devices, especially in the linear region. A more accurate physics-based model has, however, recently been reported [12].

In most cases, better results are obtained with purely algebraic models, which use arbitrary mathematical functions to match all or a portion of measured I-V curves. The saturation drain conductance in the Shichman-Hodges model is an example of this approach. Probably the most widely used of these models is the *hyperbolic tangent (or Curtice) model* [13,14], which uses the tanh function as an empirical, convenient, single-region model, continuous in its derivatives, for fitting all regions of FET operation. The tanh function smoothly approaches unity when the argument becomes large. Calculation of the tanh function and its derivatives may slow SPICE operations to some degree, but the smoothness of the function may help to speed convergence, thus recouping some of the lost time. This function is used in some commercially available versions of SPICE as the MESFET model. The model equation is given below:

$$I_D = \beta(V_{GS,i} - V_T)^2 (1 + \lambda V_{DS,i}) \tanh(\alpha V_{DS,i})$$
$$\text{for} \quad V_{GS} > V_T \quad (2.11a)$$

and $\quad I_D = 0 \quad$ for $\quad V_{GS} \leq V_T \quad (2.11b)$

In these equations, the coefficients β and λ have the same role as in the SPICE JFET model. In fact, the first three terms are the same and include the same square law I_D-V_{GS} dependence as Eq.(2.6c). Therefore I_D approaches the same limiting current in saturation. The model will still overpredict I_D since velocity saturation is not considered explicitly, a larger than physical R_S must be used to linearize transconductance, and a nonsymmetric R_S and R_D may still be required to fit the linear region. The same parameter extraction techniques described in Sec. 2.2.1 apply to Eqs. (2.11).

The main difference between the hyperbolic tangent model and the SPICE2 JFET model is in the transition between the linear and saturation regions. The saturation of current at $\beta(V_{GS,i} - V_T)^2(1 + \lambda V_{DS,i})$ is determined by the fourth term $\tanh(\alpha V_{DS,i})$, dependent only on $V_{DS,i}$. Thus, the saturation will be independent of $V_{GS,i}$ and will occur at a constant voltage, $V_{D,\text{sat}}$. This behavior is more typical of short gate length MESFETs than the gradual increase in $V_{D,\text{sat}}$ produced by $V_{D,\text{sat}} = V_{GS,i} - V_T$ in the SPICE JFET model. The new parameter α affects both the slope in the linear region and the saturation voltage. The effect of α on the slope can be seen from the first derivative of (2.11a) when evaluated at $V_{DS,i} = 0$,

$$\frac{\partial I_D(0)}{\partial V_{DS}} = \alpha\beta(V_{GS,i} - V_T)^2 = \alpha I_{D,\text{sat}} \qquad (2.12)$$

An improved version of this model which includes an additional term for velocity saturation will be described in Sec. 2.2.3. The extra term allows for better fit in the linear region and permits the use of $R_S = R_D$.

Typically, the gate current is modeled by the ideal diode equations, (2.9) and (2.10), although often an ideality factor parameter is added. The simple depletion model for the gate capacitance is also often used with the hyperbolic tangent model. The comments in Sec. 2.2.1 regarding capacitance would still apply.

Extraction of the normalized MESFET model input parameters *beta*, *vt*, *lambda*, and *rs* proceeds in the same manner as discussed in Sec. 2.2.1. *Alpha* is extracted from Eq. (2.12). Since there will be some interaction between the slope of the linear region and the saturation voltage with choice of *alpha*, some optimization of the parameters can be beneficial in improving the fit. This optimization might make use of a global parameter optimization tool such as SUXES [15] or TECAP [16]. As can be seen in Fig. 2-11, a good fit to the data measured on a 1×50 μm MESFET was obtained through optimization, even though R_S was constrained to be equal to R_D. The typical parameters in Table 2.3 are a result of this optimization.

Optimizers are mathematical algorithms that seek to minimize the error between a designated reference (the measured device I_D-V_{DS} data in this application) and the predictions of the designated model equation. The global minimum is sought; however, initial parameter values must be close to the final result in order to avoid convergence into local minima instead. The final optimized parameters will give the best fit with the given model equation, but the parameter values are unlikely to have any useful physical significance.

A summary of the tanh model parameters used with the GaAs MESFET is presented in Table 2.3. In the table, typical extracted device parameters are given

TABLE 2.3
Hyperbolic tangent model parameters used for the GaAs MESFET. Typical parameter values are shown for an ion-implanted, depletion mode, non-self-aligned MESFET with 1 μm gate length, normalized to 1 μm channel width, and were obtained through optimization

Name	Parameter	Units	Typical Value	Area
vto	Threshold voltage	V	−1.0	*
beta	Transconductance parameter	A/V^2	7e-5	
lambda	Channel-length modulation parameter	1/V	0.1	
alpha	Saturation voltage parameter	1/V	3.65	
rd	Drain ohmic resistance	Ω	1650	*
rs	Source ohmic resistance	Ω	1650	*
cgs	Zero-bias GS junction capacitance	F	1.2e-15	*
cgd	Zero-bias GD junction capacitance	F	1.2e-15	*
pb	Gate junction effective built-in potential	V	0.8	
is	Gate junction saturation current	A	3.0e-16	*

for a 1-μm gate length ion-implanted, depletion mode, non-self-aligned MESFET normalized to a channel width of 1 μm. An asterisk in the area column signifies that this parameter is automatically scaled with the geometric factor specified on the device element statement. The model parameters scale in the same way with area as did the SPICE JFET model. Alpha is independent of area.

Example 2.2. A 1 × 50 μm GaAs MESFET I_D-V_{DS} characteristic is shown in Fig. 2-11. Use this data to extract normalized model parameters beta, vto, rs, alpha, and lambda for the hyperbolic tangent model and compare the fit of the model with the measured data.

Solution. Using the approach suggested in Sec. 2.2.1, the first step is to construct a plot of $\sqrt{I_{D,\text{sat}}}$ as a function of V_{GS}. $I_{D,\text{sat}}$ is obtained by extrapolating the I_D-V_{DS} curves in the saturation region back to $V_{DS} = 0$. The result of this process is found in Fig. 2-12. Also shown in Fig. 2-12 is a straight line fit to the low-current region of $\sqrt{I_{D,\text{sat}}}$. This serves to illustrate the departure of the measured data from the ideal square law characteristic. It can also be used to determine R_S. The equation for the straight line is

$$\sqrt{I_{D,\text{sat}}} = (1.76)V_{GS} + 1.98$$

The maximum deviation from ideal is seen at the highest V_{GS}. At this point, $\sqrt{I_{D,\text{sat}}} = 3.21$ mA$^{1/2}$. Thus, the apparent intrinsic V_{GS} is found to be $V_{GS,i} = 0.57$ V from the above equation. Therefore, the model equation $I_D = \beta(V_{GS} - I_D R_S - V_T)^2$ will be the best fit

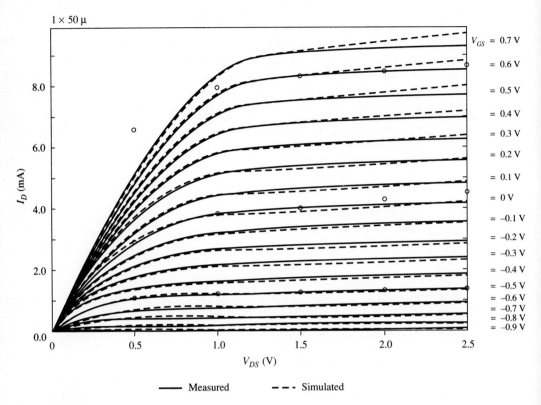

FIGURE 2-11
Comparison of experimental data of a 1×50 μm ion-implanted MESFET with the predictions of Eq. (2.11) using optimized model parameters.

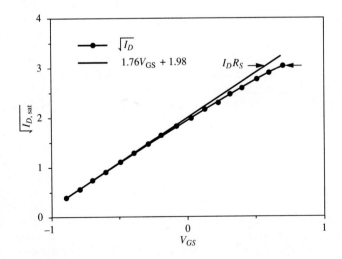

FIGURE 2-12
$\sqrt{I_D}$, sat was determined from Fig. 2-11 and is plotted as a function of V_{GS}.

if $I_D R_S = V_{GS} - V_{GS,i} = 0.13$ V. Since the measured $I_{D,\text{sat}} = 8.86$ mA at $V_{GS} = 0.70$ V, $R_S = 14.7\ \Omega$ for the $1 \times 50\ \mu\text{m}$ MESFET. The model parameter is

$$\text{rs} = 735\ \Omega\cdot\mu\text{m}$$

when normalized to a 1-μm width.

To show that this choice leads to a linear $\sqrt{I_{D,\text{sat}}}$ vs. $V_{GS,i}$ characteristic, the $\sqrt{I_{D,\text{sat}}}$ data is replotted in Fig. 2-13 against $V_{GS} - I_{D,\text{sat}} R_S$. The line can be extrapolated as shown to find parameter vto:

$$\text{vto} = -1.15\ \text{V}$$

The slope of the line gives $\sqrt{\beta}$. So $\beta = 3.1 \times 10^{-3}$ A/V² or

$$\text{beta} = 6.1 \times 10^{-5}\ \text{A/V}^2\cdot\mu\text{m}$$

when normalized as above.

Next, alpha is extracted from Eq. (2.12) with the use of Fig. 2-12 again. Choosing the $V_{GS} = 0$ V curve, the slope at $V_{DS} = 0$ is

$$\frac{\partial I_{D,\text{sat}}(0)}{\partial V_{DS}} = \alpha\beta(V_{GS,i} - V_T)^2 = 7.3 \times 10^{-3}\ \text{S}$$

at $V_{GS} = 0$ V, $V_{GS,i} = -0.056$ V. Using the above value for β yields $\alpha = 3.49$. Therefore,

$$\text{alpha} = 3.49\ \text{V}^{-1}$$

as this parameter is independent of channel width.

Finally, the channel-length modulation parameter lambda can be determined from Eq. (2.8). Using the $V_{GS} = -0.5$ V curve, let $V_{DS2} = 2.5$ V and $V_{DS1} = 0$ V. Then $I_{DS1} = I_{D,\text{sat}} = 1.24$ mA and $I_{DS2} = 1.55$ mA. Rearranging (2.8) to solve for λ,

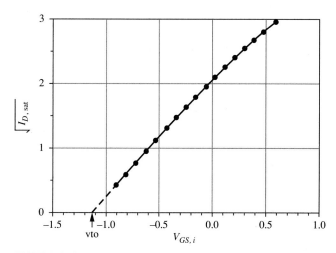

FIGURE 2-13
$\sqrt{I_{D,\text{sat}}}$ replotted as a function of $V_{GS}-I_D R_S = V_{GS,i}$.

$$\text{lambda} = \lambda = \frac{I_{D2} - I_{D1}}{V_{DS2}I_{D1} - V_{DS1}I_{D2}} = 0.10 \text{ V}^{-1}$$

Again, no normalization is required.

The parameters determined above were substituted into a commercial version of SPICE which supports the hyperbolic tangent model [17]. The open circles plotted on Fig. 2-11 show the results for $V_{GS} = -0.5, 0,$ and 0.5 V. Note that the fit is reasonably good at the lower two V_{GS} values and is significantly improved overall by optimization (dashed lines).

2.2.3 Modifications to the SPICE JFET Model—Raytheon Model [7]

Significant improvement in the agreement between dc MESFET measurements and the hyperbolic tangent model can be obtained if the saturation current level is modified to reflect velocity-saturation effects. Since these effects present themselves as a linearization of the transconductance, $\partial I_D/\partial V_{GS}$, an empirical correction can be made by adding a V_{GS}-dependent term to the denominator of Eq. (2.11a). This approach was described in Ref. [7] and has been incorporated into the SPICE3 MESFET model as well as other commercial SPICE products. Since the model was first described by a group of authors at the Raytheon Research Laboratory, we will refer to the model as the Raytheon model in this section.

The drain current is described by the equations below:

$$I_D = \left[\frac{\beta(V_{GS,i} - V_T)^2}{1 + b(V_{GS,i} - V_T)}\right](1 + \lambda V_{DS,i})\left[1 - \left(1 - \alpha\frac{V_{DS,i}}{3}\right)^3\right]$$
$$\text{if} \quad 0 < V_{DS} < 3/\alpha \quad (2.13a)$$

or $\quad I_D = \left[\dfrac{\beta(V_{GS,i} - V_T)^2}{1 + b(V_{GS,i} - V_T)}\right](1 + \lambda V_{DS,i}) \quad \text{if} \quad V_{DS} \geq 3/\alpha \quad (2.13b)$

The denominator of the first term increases in size as V_{GS} increases, thus reducing the current in proportion to the new parameter b as required. I_D begins at low V_{GS} with a quadratic relationship as in the SPICE JFET and hyperbolic tangent models, but then becomes almost linear in $V_{GS} - V_T$ for large V_{GS}. The magnitude of b can be increased as needed to model devices which exhibit stronger velocity-saturation effects. Because of this added independent parameter, the source and drain resistances, R_S and R_D, can now be made equal and are used to represent the parasitic channel resistance instead of being used to emulate velocity saturation. The second term produces a finite slope in the saturation region, proportional to λ, as in Eqs. (2.6) and (2.11). The last term in Eq. (2.13a) is a polynomial fit to the hyperbolic tangent function [7], but executes more efficiently, thereby saving computer time. Figure 2-14 compares experimental data for the 1×50 μm ion-implanted MESFET previously shown in Fig. 2-11 with the predictions of Eqs. (2.13) with optimized model parameters (the optimizer in TECAP [16] was used). The V_{DS} error in the linear region is less than 50 mV for this self-aligned, depletion mode FET. Slightly higher error was found in the linear region of an enhancement mode FET when modeled with Eqs. (2.13).

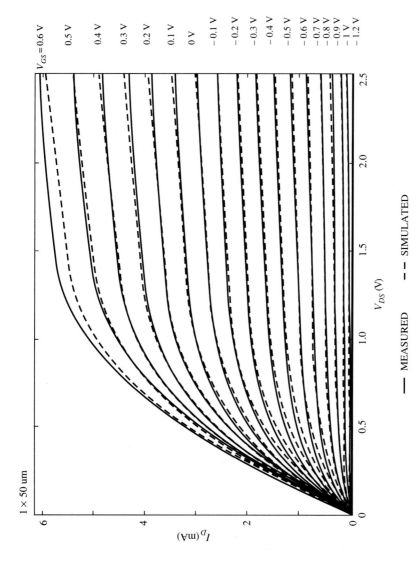

FIGURE 2-14
Comparison of experimental data for a 1×50 μm ion-implanted MESFET with the predictions of Eqs. (2.13) using optimized model parameters.

The parameters for the Raytheon model can be estimated through simple graphically assisted calculations. Referring back to Fig. 2-8b, if $\sqrt{I_{D,\text{sat}}}$ as defined in Sec. 2.2.1, is plotted against $V_{GS,i}$, both $\sqrt{\beta}$ and V_T can be determined. Here, however, a slightly different approach is taken. Rather than finding R_S by linearizing this curve, as was necessary in Secs. 2.2.1 and 2.2.2 to account for velocity saturation, it is desirable to begin with the assumption that $R_S = R_D$ and to determine a value for R_S by measurement. An accurate method for measurement of the source resistance is described in Ref.[18]. Then, if $V_{GS,i}$ is calculated from this value of R_S, any residual curvature in the $\sqrt{I_D}$ curve will be due to velocity saturation. This deviation can be used to estimate the size of b.

Example 2.3. An example illustrating how b is estimated is shown in Fig. 2-15 where $\sqrt{I_{D,\text{sat}}}$ is plotted against $V_{GS,i}$ calculated from the measured R_S. V_T can be evaluated from the intercept on the $V_{GS,i}$ axis. $\sqrt{\beta}$ can be determined from the slope of the straight line portion of the curve, near pinch-off. If a straight line with slope $\sqrt{\beta}$ is extrapolated from this region, the deviation of the actual data from the line will be apparent at larger values of $V_{GS,i}$. In Fig. 2-15, this difference was noted at V_{GS1}, where the extrapolated $\sqrt{I_D}$ yields current I_{D2} and the actual I_D yields I_{D1}. Since the straight line corresponds to Eq. (2.13a) with $b = 0$, an equation representing the actual data (2.13a) can be subtracted from the equation for the straight line which gives

$$I_{D2} - I_{D1} = \Delta I_D = \beta V_{GST}^2 - \frac{\beta V_{GST}^2}{1 + bV_{GST}} \tag{2.14}$$

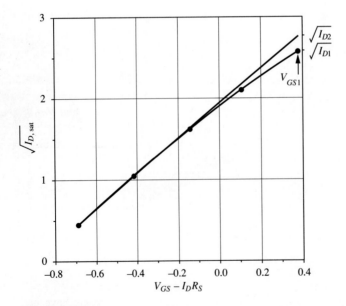

FIGURE 2-15
$\sqrt{I_{D,\text{sat}}}$ is plotted against $V_{GS,i}$. Estimated values of the SPICE3 MESFET model parameters beta, alpha, lambda and b can be extracted from this graph.

where $V_{GST} = V_{GS,i} - V_T$. Solving this equation for b produces the following equation:

$$b = \frac{\Delta I_D/\beta V_{GST}^3}{\left[1 - (\Delta I_D/\beta V_{GST})^2\right]} \tag{2.15}$$

The parameter alpha can be extracted from the slope of the I_D characteristic in the linear region in which V_{DS} is small. Taking the derivative of Eq. (2.13a),

$$\frac{dI_D}{dV_{DS}} = \alpha\left(1 - \frac{\alpha V_{DS}}{3}\right)^2 \left[\frac{\beta(V_{GS,i} - V_T)^2}{1 + b(V_{GS,i} - V_T)}\right](1 + \lambda V_{DS,i}) \tag{2.16}$$

which approaches $\alpha I_{D,\text{sat}}$ as V_{DS} becomes very small. While this approach will give reasonably good fit in the linear region, there may be some applications of the model where the saturation voltage is more important than the slope. In this case, alpha can be obtained from the saturation voltage, since Eq. (2.13a) saturates when $V_{DS} > 3/\alpha$.

As is the case for all of the models discussed, the fit of the model to experimental data can be improved through optimization on a computer. The only constraint that should be imposed on this optimization is to set $R_S = R_D$ so that the optimizer will not destroy the symmetry of the device.

The gate current model included in the MESFET model is identical to the SPICE JFET gate current model described in Sec. 2.2.1. I_S must be specified as a model parameter. The ideality factor n is fixed internally at 1.0 in the SPICE3 implementation, whereas some commercial versions allow for adjustment of n [17].

A new capacitance model has been provided in the Raytheon MESFET routine [17]. The original SPICE JFET model, the Curtice model [13], and most other published modifications (see Refs. 8 and 19, for example) use a straightforward depletion capacitance approach as in the diode model (Secs. 2.1 and 2.2.1). This approach treats each device capacitance C_{GS} or C_{GD} as a function of a single variable, either V_{GS} or V_{GD} respectively. The model described in Ref. 7 has modified the capacitance and charge equations so that they are functions of both V_{GS} and V_{GD}. In addition, the effects of carrier velocity saturation and pinch-off have been included, and a smooth transition into the inverse region is provided so that the device can be completely symmetric. The C-V dependence is best described by Fig. 2-16, which is Fig. 3 of Ref. 7. Here, the behavior of standard diode equations such as (2.3) is illustrated by the dashed lines (Van der Ziel [20]), the velocity-saturated capacitances represent the predictions of Pucel et al. [21], and the one-parameter interpolation represents the functional dependence of Ref. 7. The equations themselves are rather complex, and the behavior of the model is not obvious from inspection of these equations, so they will not be presented in this section. However, the reader is encouraged to read the reference if further information is desired. The accuracy of the capacitance model has not been verified by comparison with the measured device capacitances of short-gate MESFETs since very few such measurements have been published. The behavior of C_{GD} does, however, compare qualitatively with data reported in Ref. 22. The sharp increase in C_{GS} with V_{DS} represents the capacitance

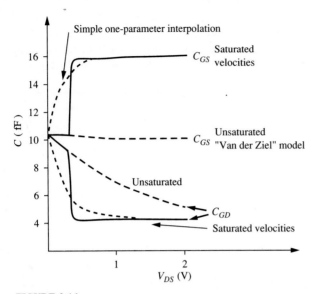

FIGURE 2-16
C_{GS} and C_{GD} are calculated and compared for a 1×20 μm MESFET according to three models, an unsaturated velocity model [20], a saturated velocity model [21], and a one-parameter interpolation. (*From Statz et al.*, [7], ©*IEEE 1987. Used by permission.*)

behavior of a strongly velocity-saturated MESFET (short gate length, highly negative threshold voltage) such as was discussed in Ref. 21. The C_{GS} increase may be due to a strong electron accumulation in the channel. A low threshold voltage ($V_T = -1.2$ V) MESFET can show a flatter, slightly increasing C_{GS} with V_{DS} [23]. Enhancement MESFETs may show even less increase. Unfortunately, the Raytheon capacitance model does not allow for this type of adjustment in the C_{GS} characteristic.

The capacitance parameters in the model listing are named cgs, cgd, and pb. pb is the built-in junction voltage as defined in the diode depletion capacitance equations (2.3), and the grading coefficient is fixed internally at 0.5. However, the definition of cgs and cgd is quite different from that used in the previously described models. Contrary to what is stated in Ref. 7, cgs corresponds to the asymptotic limit of the gate-to-source capacitance for large V_{DS}. Fig. 2-16 shows that C_{GS} saturates under this condition, and the parameter cgs is this limiting capacitance. Likewise, cgd corresponds to the asymptotic gate-to-drain capacitance. While the bias dependences of the gate capacitances can be measured indirectly through microwave S-parameter measurement and used to optimize the fit of the capacitance equations of the model, it is adequate to divide the total measured zero-bias gate capacitance C_G between cgs and cgd by a ratio of 0.85 to 0.15 respectively [24]. pb can be determined from the intercept of C_{GS}^{-2} vs. V_{GS}, as shown in Fig. 2-3, measured under the condition $V_{DS} = 0$ V.

A summary of the SPICE3 MESFET model parameters used with a GaAs

TABLE 2.4
SPICE3 MESFET model parameters (Raytheon model [7]) used for the GaAs MESFET. Typical parameter values are shown for an ion-implanted, depletion mode, non-self-aligned MESFET with 1 μm gate length, normalized to 1 μm channel width

Name	Parameter	Units	Default value	Typical value	Area
vto	Threshold voltage	V	−2.0	−1.1	
beta	Transconductance parameter	A/V^2	1.0e-4	1.1e-4	*
b	Doping tail extending parameter	1/V	0.3	0.38	
lambda	Channel-length modulation parameter	1/V	0	0.05	
alpha	Saturation voltage parameter	1/V	2	2.50	
rd	Drain ohmic resistance	Ω	0	900	*
rs	Source ohmic resistance	Ω	0	900	*
cgs	Zero-bias GS junction capacitance	F	0	1.2e-15	*
cgd	Zero-bias GD junction capacitance	F	0	2.8e-16	*
pb	Gate junction effective built-in potential	V	1	0.8	
is	Gate junction saturation current	A	1.0e-14	5.0e-14	*

MESFET is presented in Table 2.4. In the table, the default parameters are compared against typical extracted and optimized [25] device parameters for a 1-μm gate length, depletion mode, non-self-aligned MESFET normalized to a channel width of 1 μm. An asterisk in the area column signifies that this parameter is automatically scaled with the geometric factor specified on the device element line. For example, a 1 × 10 μm MESFET with the model name mes would be invoked by the following card in the SPICE3 node listing:

$$z1\ 1\ 2\ 3\ \text{mes}\ 10$$

where z1 is the device number, 1, 2, and 3 are the drain, gate, and source nodes respectively, mes is the model name, and 10 is the area factor for a 10 μm channel width. The model parameters scale in the following way with an area factor, A:

$$\text{beta} * A$$
$$\text{rd} / A$$
$$\text{rs} / A$$
$$\text{cgs} * A$$
$$\text{cgd} * A$$
$$\text{is} * A$$

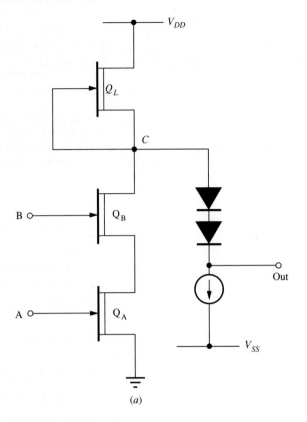

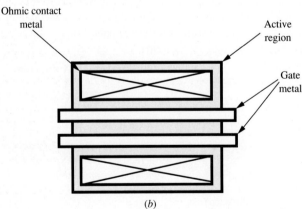

FIGURE 2-17
(a) The circuit diagram of a simple two-input NAND gate. (b) A top view of the layout of the two devices Q_A and Q_B sharing a single channel. Parasitic source resistance and layout area can be both reduced by this dual-gate approach.

2.2.4 Dual-Gate MESFETs

As will be discussed in detail in Chap. 3, many logic gate topologies can be realized by combining FETs in parallel to produce a NOR function or in series to give a NAND function. To see how a logic gate might be implemented with series devices, the circuit diagram of a simple two-input NAND gate is shown in Fig. 2-17a. In this circuit, FETs Q_A and Q_B are being used as switches, and Q_L is connected with $V_{GS} = 0$ as an active load, emulating a current source. A level-shift network is included under the assumption that the FETs are depletion mode devices and will therefore require a negative V_{GS} in order to enter the cutoff region. Level shifting is needed because node C will have a minimum voltage (logic low) determined by the intersection of the load current I_L with the linear region of Q_A and Q_B, and thus will be positive in the range of 0.2 to 0.5 V. This low voltage will be limited by the series resistance of Q_A and Q_B in their on state. This series resistance prevents the use of more than two series transistors in this form of static NAND, because the logic low voltage would be too badly degraded.

In order to reduce the performance degradation caused by the series resistance of Q_A and Q_B, it is common practice to combine the two devices in a single channel as shown in Fig. 2-17b. The gates can then be located closer together and the parasitic channel resistance reduced, especially for self-aligned transistors, because two ohmic contact interfaces have been eliminated by the reconstruction of the series devices. In addition, there will be a savings in layout area which is equally valuable.

From the modeling point of view, this dual-gate FET creates a somewhat different device than two single-gate FETs connected in series. In particular, it is not obvious how the series resistance of the node between Q_A and Q_B should be partitioned. In fact, this presents a serious problem for models that use the source resistance, R_S, as a feedback element to emulate velocity saturation. In general, a good fit of these models to the measured dual-gate characteristics cannot be obtained over a wide range of V_{GS} and V_{DS}. The SPICE3 MESFET model does explicitly contain a correction term for velocity-saturation effects, and it will produce an acceptable fit to a dual-gate MESFET when the model parameters are optimized. The data presented in Fig. 2-18 illustrate the fit of this model to measured data on a 1.5×20 μm ion-implanted MESFET fabricated by the selective implant process. Constraints were imposed on the source and drain resistances. R_D for the upper FET was set equal to R_S of the lower FET. The resistance of the common node was estimated to be equal to R_S, divided equally between Q_A and Q_B. Optimization of the model parameters around this constraint clearly provided a good fit. A higher fractional error in I_D (maximum of 10 percent) was observed for V_{GSA} closer to pinch-off (-0.6 V); however, this bias condition would not be representative of the normal operation of the NAND gate where V_{GSA} and V_{GSB} are either below V_T (shifted logic low) or above 0.4 V (shifted logic high). The error in these two limits is extremely small.

The dual-gate MESFET model could be implemented in SPICE by using two .model statements, one for Q_A and the other for Q_B. Using the model parameters extracted for the 1.5 μm gate length MESFET shown in Fig. 2-18, normalized to a 1-μm channel width, the following model description could be used:

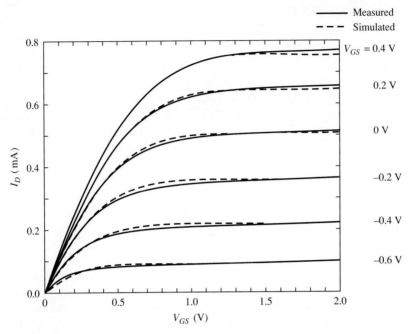

FIGURE 2-18
The data presented illustrate the fit of the SPICE3 MESFET model to experimental data taken on a 1.5 × 20 μm ion-implanted dual-gate MESFET. The gate voltage of Q_A was held fixed at 0 V.

.model mesa nmf (vto = −0.92 alpha = 4.1 beta = 6.45e-5 lambda = 4.3e-2 b = 6.1e-4
+ rs = 2700 rd = 1350 is = 9.1e-16)

.model mesb nmf (vto = −0.92 alpha = 4.1 beta = 6.45e-5 lambda = 4.3e-2 b = 6.1e-4
+ rs = 1350 rd = 2700 is = 9.1e-16)

2.2.5 Parasitic Capacitances

The MESFET models described in Secs. 2.2.1 to 2.2.3 provided estimates of the internal gate depletion-layer capacitances which vary with the internal gate-to-source and gate-to-drain voltages. Additional fixed-capacitance components associated with the MESFET, the fringing capacitance associated with the edges of the depletion layer, and the geometric fringing capacitance originating with coplanar metal strips on the GaAs surface must also be considered for any of the models in order to properly account for the total device capacitance.* The source of these parasitic capacitances is illustrated in Fig. 2-19.

---

*It should be noted that the gate capacitance in the Raytheon MESFET model approaches limiting values at large Vds as shown in Fig. 2-16. The parasitic C_{GS} and C_{GD} may be accounted for by adjusting the model parameters cgs and cgd (which represent the limiting values) to include the parasitic elements. C_{DSP}, however, must be added to all of the models as an additional element.

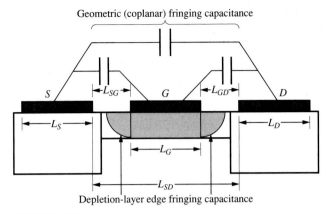

FIGURE 2-19
MESFET parasitic capacitance models. The total device capacitance consists of the depletion capacitance plus the parasitic capacitances.

Note that these fringing capacitances do not include the additional fixed-load capacitance caused by interconnections between FETs or logic gates. If the length of the interconnect wires is long, these parasitic circuit capacitances can be much larger than those associated with the transistors themselves and must also be modeled if accurate predictions of transient delay are to be obtained. The estimation of the interconnect capacitance and other effects associate with interconnections is the subject of Chap. 5.

The *edge fringing capacitance* component arises from the lateral extension of the gate depletion layer beyond the edge of the gate electrode. As the donor concentration in the channel is finite and the potential on the gate is constant, Poisson's equation would dictate that the depletion region must extend beyond the gate edges on both sides. This excess capacitance is above that predicted by the essentially parallel plate calculation of the MESFET models. If the gate length is short, then the excess capacitance can be a significant fraction of the total capacitance and will be very important in the simulation of logic gates with short interconnections.

The depletion-layer edge fringing capacitance will depend to some extent on V_{GS} or V_{GD}, and it is possible to model this effect if one is willing to add equations inside the MESFET model subroutine [8]. From a practical perspective, the edge capacitances change only slowly with voltage, and approximating them as being fixed is reasonable unless the actual FET gate length is very much shorter than 1 μm. This capacitance is assumed to be identical for gate-source (C_{GSE}) and gate-drain (C_{GDE}) edges, and can be approximated from calculations of the capacitance of a biased metal stripe on a plane semi-infinite semiconductor [22,26] whose result is given as

$$C_{GSE} = C_{GDE} = \epsilon_s W \left(1.41 + \frac{0.86 \epsilon_o}{\epsilon_s} \right) = 1.48 \epsilon_s W \qquad (2.17)$$

In this equation ϵ_s is the permittivity of the semiconductor and ϵ_o is the permittivity of free space. For GaAs, $\epsilon_s = 12.9 \, \epsilon_o$; therefore C_{GSE} and C_{GDE} are approximately 0.17 fF per micrometer of channel width. This result is similar to that reported in Ref. 8.

The *"geometric" fringing capacitance*, a second contributor of excess capacitance, is also indicated in Fig. 2-19. This capacitance arises from the fringing capacitance between coplanar, metal electrodes on the surface of a semi-infinite dielectric. This capacitance is independent of any internal device space charge. This excess capacitance to C_{GS}, C_{GD}, and C_{DS} must also be added to the parasitic capacitance. While very accurate calculations of this capacitance have been made for rectangles of arbitrary size using Green's function solution [27], the ratio of the geometric fringing capacitances to the internal depletion-layer capacitance of a typical 1 μm MESFET or to the interconnect capacitances is quite low. Thus, a more approximate solution, such as was developed in Ref. 21, can be successfully employed for typical FET structures without sacrifice in accuracy. The equation for the excess capacitances are reproduced below [21]:

$$C_{GSG}, C_{GDG}, \text{ or } C_{DSG} = (\epsilon_r + 1)\epsilon_0 W \frac{K(1-k^2)^{1/2}}{K(k)} \qquad (2.18a)$$

$$k_{SD} = \left[\frac{(2L_S + L_{SD})L_{SD}}{(L_S + L_{SD})^2}\right]^{1/2} \qquad (2.18b)$$

$$k_{GD} = k_{GS} = \frac{L_{GD}}{L_{GD} + L_G} \qquad (2.18c)$$

The electrode lengths L_S, L_G and L_D and the spacings L_{SD}, L_{SG}, and L_{GD} are defined in Fig. 2-19. It is assumed that $L_{SG} = L_{GD}$ and $L_S = L_D$ on the average and that $L_D \gg L_G$ and $L_D \gg L_{SG}$. $K(k)$ is an elliptical integral of the first kind. [See Chap. 5, where Fig. 5-6 and Eqs. (5.20) and (5.27) can be used to estimate $K(k)$.] ϵ_r is the relative dielectric constant and ϵ_o the permittivity of free space. (Thus, $\epsilon_s = \epsilon_r \epsilon_o$). For typical small GaAs MESFET geometries, the geometric fringing capacitances have been calculated to be approximately 0.11 fF per micrometer of channel width.

Combining both contributions, the total parasitic fringing capacitance becomes

$$C_{DSP} = C_{DSG} = 0.11 \, \text{fF}/\mu\text{m}$$

$$C_{GSP} = C_{GSE} + C_{GSG} = 0.28 \, \text{fF}/\mu\text{m} \qquad (2.19)$$

$$C_{GDP} = C_{GDE} + C_{GDG} = 0.28 \, \text{fF}/\mu\text{m}$$

2.3 MODELING MESFET SECOND-ORDER EFFECTS

The most important second-order phenomena that influence and modify the MESFET device characteristics were described in Chap. 1. Under certain conditions, these effects may be very significant in predicting the performance of a circuit. For example, circuits with high power supply voltages will prove to be more susceptible to backgating, and simulation of this type of circuit without consideration of backgating effects will not adequately predict its performance. Also, low-temperature operation (below 0°C) has been shown to enhance the influence of backgating on MESFETs. Subthreshold current and substrate leakage is another phenomenon which may be very important for modeling of very low power circuits or for dynamic logic circuits. Lag effects reduce the voltage gain of inverters at frequencies above 1 kHz and can cause the threshold voltage of an inverter to depend on its past history; therefore this effect should be considered for accurate simulations. Temperature effects are quite pronounced and must also be investigated to guarantee performance of a design under environmental extremes. Therefore, some effort must be given to incorporating models for these effects into SPICE.

2.3.1 Backgating Model

Backgating or sidegating [28,29] is characterized by a reduction in the observed drain current of a MESFET when a neighboring device is biased negatively with respect to the source of the device under test. The physical basis of this effect was discussed in Sec. 1.4.1. It is called backgating because it results from an increase in the space-charge layer width at the channel/substrate interface of the FET. This increased width causes a positive shift in the threshold voltage of the device. It is not difficult to characterize with simple test patterns, but it is very dependent on the chemistry of the semi-insulating substrate material and on the processing techniques. It is not unusual to observe significantly different degrees of backgating from wafers cut from different ingots or even from one end of an ingot to the other. Therefore, it is difficult to accurately predict for any unknown wafer, and accurate modeling of backgating effects is limited by this uncertainty. Averages over many wafers, if available, may be useful in estimating typical effects and making representative allowances in the design as needed to account for worst-case conditions.

The influence of backgating on a circuit will also depend on the design of the circuit. Circuits requiring large power supply differentials will be more sensitive than those with small voltages. Unannealed proton, oxygen, or boron implantation [30] is also frequently used in GaAs IC processes to reduce the degree of backgating by introducing large trap densities in the field region between devices. The high densities of traps suppress substrate leakage current. In general, substrate quality is improving as better growth techniques are developed, and the magnitude of the backgating effects is much less now with high-quality undoped LEC GaAs than it was when chrome-doped semi-insulating GaAs was still in use.

Since the effect presents itself as a positive shift in threshold voltage, which is dependent on the difference between the source voltage of the MESFET V_S and the most negative supply voltage V_{SS}, it can by modeled as a body voltage term in the equation for the threshold voltage. The relationship between backgate voltage V_{BG} and the MESFET threshold voltage for an ion-implanted gaussian doping profile in a substrate with constant acceptor density has been derived in Ref. 31, and is shown below:

$$V_T = V_{TO} - \frac{1}{C_G}\sqrt{2\epsilon_s q N_A (V_{bi} + |V_{BG}|)} \qquad (2.20)$$

where V_{TO} represents the MESFET threshold voltage in the absence of backgating which is corrected for the backside space-charge layer by the second term, N_A is the assumed substrate acceptor doping concentration, ϵ_s is the semiconductor permittivity, V_{bi} is the built-in junction voltage at the channel/substrate interface, and C_G is the backgate capacitance per unit area given by

$$\frac{1}{C_G} = \frac{R_p}{\epsilon_s} + \frac{\sqrt{8/\pi}}{\epsilon_s}\sigma \qquad (2.21)$$

R_p is the projected range of the implant and σ is the straggle. The backgate bias V_{BG} is given by $V_{BG} = K_{BG}(V_S - V_{SS})$, where K_{BG} is a proportionality constant which will depend on the substrate, distance between devices, orientation of devices, and process [32-34]. It is therefore not predictable but must be measured from suitable test patterns. Since V_{SS} is the largest negative voltage in the circuit, it is assumed that the potential difference between the source of the device under test and V_{SS} will dominate the backgating effect.

It has been shown that the above equations, when used to calculate the threshold voltage in the SPICE JFET model, produce a very accurate correction for the backgating effect [32]. These equations may be introduced into the JFET or MESFET model subroutine to modify the threshold voltage; however, for devices that do not have constant source voltage, V_S, such as an active load transistor with common drain and with gate and source connected together, V_T will depend on V_S. The small-signal derivatives $\partial I_D/\partial V_{GS} = g_m$ and $\partial I_D/\partial V_{DS} = g_{ds}$ will depend on V_T, complicating the equations and slowing down the execution of the model to some degree.

A simpler approach has been proposed [33] which utilizes a linearization of Eq. (2.20) to estimate backgating effects on drain current. The following equation illustrates this simplification, where again V_{TO} is defined as the FET threshold voltage without backgating ($V_{BG} = 0$):

$$V_T = V_{TO} + K_{BG}(V_S - V_{SS}) \qquad (2.22)$$

Obviously, the calculation of the small-signal parameters is greatly simplified, and the parameter extraction is also much easier. The error for this approximation is somewhat greater than when Eq. (2.20) was used, but realistically, since only the average backgating effects are modeled due to the poor predictability of substrate properties, and the backgating effects are rather small in most cases, this more convenient approximation may be quite adequate. In fact, it lends itself to a non-invasive implementation (as a subcircuit rather than a modification to the model) if the user is willing to define the FET model within a subcircuit as shown in Fig. 2-20a. Here, the varying threshold voltage is represented in the SPICE input file by a voltage-dependent voltage source $K_{BG}(V_S - V_{SS})$ in series with the gate connection. The effectiveness of the correction is demonstrated by Fig. 2-20b, an I-V curve for a 1 μm gate length MESFET connected as an active load. Note that as V_{SS} is reduced to -2 V, the drain current is significantly suppressed in this sample, chosen for its pronounced backgating. Most IC processes and substrates available now would not exhibit such large backgating effects. The dashed lines in Fig. 2-20b show the prediction of the SPICE JFET model when used in conjunction with the linearized correction term.

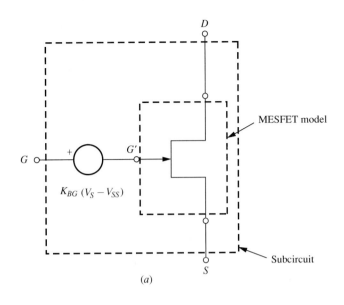

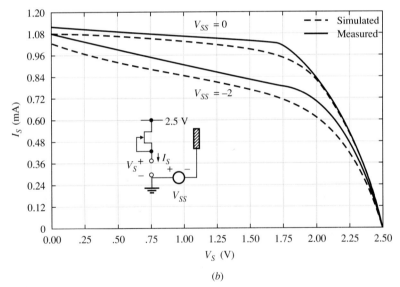

FIGURE 2-20
(a) The threshold voltage variation due to backgating is represented in the SPICE input file by a voltage-dependent voltage source $K_{BG}(V_S - V_{SS})$ in series with the gate connection. (b) I_D vs. V_{DS} for a 1-μm gate length MESFET connected as an active load. The dotted lines compare the prediction of the SPICE JFET model when used with the linearized correction term against the experimental data.

While the correction is not perfect, it does properly predict the increase in g_{ds} and the reduction in I_D produced by the backgating. The residual error in I_D is not significant enough to warrant further correction due to the uncertainties in K_{BG}.

2.3.2 Subthreshold Current and Substrate Leakage Current Models

The mechanisms for subthreshold current and substrate leakage current have been discussed in Sec. 1.3.3. These effects are of little consequence for the design of logic circuits optimized for high speed in which power dissipation is moderate to high ($>$ 1 mW/gate); however, they are significant if very low power operation ($<$ 0.1 mW/gate) is of primary concern, as might be the case in the design of a random access memory. Also, since the leakage currents increase exponentially with temperature, low-power circuits operating at high ambient temperatures may also be affected. Another benefit from consideration of the subthreshold current region of operation is that the dc I_D-V_{GS} characteristic can be made to be continuous for V_{GS} changes from below V_T to above V_T. This is not the case for any of the models described in Sec. 2.2. A smooth transition through V_T will promote convergence and reduce computation time in most cases.

Subthreshold current is the residual leakage current that flows from source to drain when V_{GS} is biased more negatively than V_T. Here, V_T would be defined by the intercept method discussed in Sec. 2.2.1. Careful examination of the measured $\sqrt{I_D}$-V_{GS} characteristic shows that drain current continues to flow beyond this definition of cutoff for several tenths of a volt in V_{GS}. Figure 2-21 illustrates the dependence of log I_D on V_{GS} for a MESFET with $V_T = -0.8$ V at a fixed V_{DS} of 1.0 V. Note also

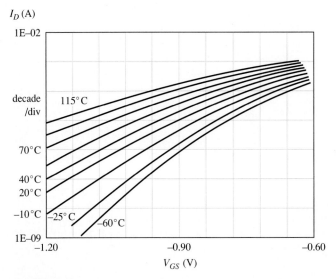

FIGURE 2-21
Plot of subthreshold current as a function of V_{GS}.

in the figure that the subthreshold current increases with temperature. The activation energy depends on whether the dominant current flow mechanism is by thermionic emission or by diffusion and therefore is dependent on the bias point of the FET.

The following equation has been proposed to model the flow of current in the subthreshold region [35,36]:

$$I_D = I_o \left[1 - \exp\left(\frac{cqV_{DS}}{kT}\right) \right] \left[\exp\left(\frac{bqV_{DS}}{kT}\right) \right] \left[\exp\left(\frac{aqV_{GS}}{kT}\right) \right] \qquad (2.23)$$

The parameters a, b, and c are empirical fitting parameters which can be determined from experimental data. The first term of (2.23) arises from the diffusion component of the drain current which can be fit from the I_D-V_{DS} characteristic at low V_{DS}. The second and last terms are due to the thermionic emission component whereby electrons are emitted over the potential barrier between source and drain. These coefficients can be fit from the I_D-V_{DS} and I_D-V_{GS} characteristics measured at $V_{DS} > 0.5$ V and $V_{GS} < V_T$. The coefficient I_O is a saturation current factor which would be fit to the measured I_D at some fixed V_{GS} and V_{DS}.

In addition to the device leakage current beyond pinch-off, there is also leakage current through the substrate which occurs due to injection of charge from a forward-biased contact. Contacts to the substrate could consist of an FET source n^+ implant or just the Schottky interconnect metal in contact with the surface of the GaAs. For small voltage differences below the threshold voltage for backgating effects, current flow is very small and predominantly ohmic through the bulk substrate. This can be modeled quite simply by connecting a large resistor between the circuit nodes of interest. It has been shown that this current is independent of the spacing between the contacts [34], possibly because the thickness of the substrate is quite large compared with this spacing, and current flows into the substrate by spreading from a point source.

If charge is injected from the contact in excess of what can be trapped by the deep levels in the substrate, space-charge-limited current will flow from the forward-biased source to other reverse-biased contacts [29]. When this occurs, the current abruptly increases by two or three orders of magnitude. The applied voltage at which the current step is observed has been correlated with the backgating threshold voltage; therefore, this condition must be avoided if the circuit under consideration is to be well-behaved. Since the space-charge-limited current region is influenced by the trap concentration, unannealed proton implantations have proven to greatly suppress this source of leakage by increasing the threshold for backgating [29,36]. The proton implantations increase the current level in the ohmic region below this threshold, however. The substrate leakage current will also increase strongly with temperature, at the same rate as the intrinsic concentration in the semi-insulating GaAs increases. Additional holes are compensated by the deep donors, leaving free electrons to increase exponentially with temperature with an activation energy of one-half of the bandgap. Thus, if very low power operation is an important factor in the design of a circuit, special care must be taken to characterize these complex effects by designing and evaluating suitable test patterns for characterization of the particular process and materials used to fabricate the circuit.

2.3.3 Drain Current Transient (Lag) Effects

Unfortunately, the saturated drain conductance of the MESFET, g_{ds}, is also influenced by the nonideal semi-insulating substrate and the surface. In the time domain, this effect is manifested as an overshoot in drain current in response to a step in the drain voltage. The magnitude of the overshoot relative to the steady-state drain current will depend on the bias conditions. While the risetime or fall time for a GaAs MESFET logic circuit is quite small (100 to 200 ps), the recovery time of the drain current overshoot can be quite long, often in the 10 to 100 μs range. In the frequency domain, one observes a frequency-dependent incremental drain conductance, g_{ds}, or drain resistance, r_{ds}, in the saturation region. Figure 2-22 illustrates that the drain resistance often drops by a factor of 2 to 3 when the measurement frequency increases from 100 Hz to 1 MHz [37]. There is also a significant phase shift in this range of frequency. Since no effect is observed for MESFETs in the ohmic (linear) region of operation, it is likely that the channel/substrate interface is responsible. Because the frequency of the effect is so low and the recovery time long, the deep levels (traps) in the substrate may also participate in this phenomenon. A theory that attempts to explain this phenomenon was presented in Sec. 1.4.2.

The lag effect therefore results in a lower r_{ds} in saturation for high-frequency signals than would be predicted from curve tracer or parameter analyzer measurement. In addition, the logic gate input threshold voltage (to be defined in Chap. 3) may depend on whether the input is changing from high to low or from low to high.

The easiest, least difficult, and least comprehensive way to model the drain lag effect is simply to increase the λ parameter in the MESFET model from the value

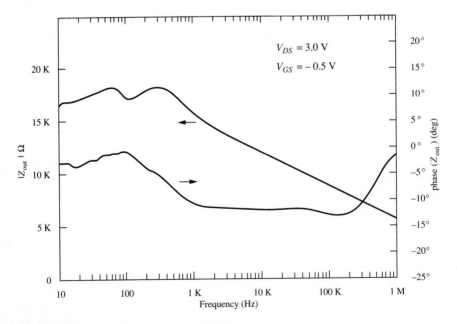

FIGURE 2-22
Measurement of $|r_{ds}|$ and phase (r_{ds}) vs. frequency. (*From L. E. Larson [38]; used by permission.*)

extracted from the low-frequency measurement on the curve tracer or parameter analyzer to its high-frequency value, typically a factor of three larger. This approach correctly models the magnitude variation in g_{ds} (but not the phase) and presents a worst case for the evaluation of the transfer characteristic of a logic circuit composed of such devices (for reasons that will become clear in Chap. 3 when the relationship between the noise margin and voltage gain is presented). Also, the steady-state logic voltages would not be expected to vary significantly due to the lag effect. However, this simple modeling attempt will not represent the long time constant transient behavior of the step response; nor will it adequately represent the device for analog circuit simulation in which the phase response is equally important.

A more comprehensive model of the lag effect is suggested in Fig. 2-23, which will qualitatively predict the drain current transient and the magnitude and phase dependence of r_{ds}. This model utilizes a linear voltage-dependent current source g_2 [where $i = g_2 V(4)$] in parallel with the MESFET to provide the excess transient drain current. This approach is related to a model described in Ref. 38 but is more accurate[1] and executes faster. The controlling node (#4) is driven by a network (R1, C1, and R2) that samples the drain voltage. The controlled source is switched into the circuit when the MESFET $V_{GS} \geq V_T$. A switch model exists in SPICE3. Alternatively, a FET could be used as a switch. The frequency response of the current from g_2 is determined by the network, which provides a zero and a pole at frequencies ω_z and ω_p respectively:

$$\omega_z = \frac{1}{R1C1} \qquad \omega_p = \frac{R1 + R2}{R2C1R1} \qquad (2.24)$$

---

[1] In Ref. 38, a MOSFET (Q2) was used in parallel with the MESFET (Q1). In saturation, $g_{ds} = g_{ds1}$ (MESFET) $+ 2\beta_2 W_2(V_{DS} - V_{TO2})$ which implies a nonphysical g_{ds} dependence on V_{DS}.

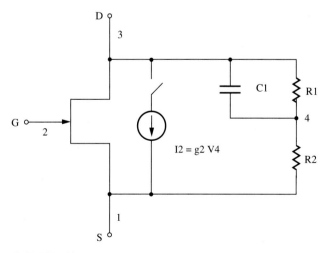

FIGURE 2-23
Subcircuit for simulating the drain lag effect of a GaAs MESFET.

From (2.24), $\omega_z \ll \omega_p$; therefore the control voltage begins to increase with frequency (up 3 dB at ω_z) until it approaches V_{DS} as $\omega \gg \omega_p$. The total drain conductance will increase with the control voltage; thus r_{ds} decreases and then saturates as the frequency increases. The network also provides a phase shift in g_2: first increasing as $\omega \approx \omega_z$, then constant for $10\omega_z < \omega < \omega_p/10$, then decreasing as $\omega \approx \omega_p$. Thus, the phase of r_{ds} will shift in the negative direction as shown in Fig. 2-22. The reduction in magnitude of r_{ds} will be determined by the ratio of g_2, the dependent source transconductance, to $(g_2 + g_{ds1})$, as shown later by Eq. (2.25).

The step response of the circuit in Fig. 2-23 exhibits overshoot and undershoot with a time constant that is controlled by R2C1. The amount of excess current will be dependent on g_2 and can be scaled accordingly. Because the lag effect is correctly predicted, a hysteresis in the transfer characteristic of a MESFET is produced; therefore, the threshold voltage of a logic gate transfer characteristic will exhibit a hysteresis when this model is applied. The overshoot or undershoot phenomena could be described alternatively by a time-dependent threshold voltage for the MESFET.

There are some difficulties in using this model. Firstly, it will significantly increase the simulation time for any given circuit. The components shown in Fig. 2-23 can be implemented in a subcircuit, but the total number of nodes increases as does the number of components. Secondly, a quantitative fit to the experimental data is not possible with a simple pole-zero model. This model predicts that the phase shift will occur mainly within a one-decade frequency range. Measurements indicate a somewhat greater range. Also, other effects (unequal overshoot and undershoot recovery times, for example) may be present that are not modeled by this method. Further complexities can be added, such as multiple time constants or multiple controlled current sources, but the extra components required to accomplish this will detract from the simulation run time even further. Finally, fitting to the model requires accurate analog measurement of r_{ds} and I_D vs. frequency and time, a function not supported by many parametric test systems.

Example 2.4. Determine R1, R2, C1, and g2 of Fig. 2-23 to provide a reasonable fit to the data shown in Fig. 2-22.

Solution. The change in $|r_{ds}|$ ranges from 18 kΩ below 1 kHz to 7 kΩ at 1 MHz. g_2 must be selected to provide the appropriate ratio of low and high frequency $|r_{ds}|$. Since at high frequency $V(4) = V(3)$, and at low frequency $V(4) = [R_2/(R_1 + R_2)] \cdot V(3)$ is much less than $V(3)$,

$$\frac{|r_{ds}(\text{HF})|}{|r_{ds}(\text{LF})|} = \frac{g_{ds1}}{g_{ds1} + g_2} = \frac{7}{18} \qquad (2.25)$$

where $g_{ds1} = \beta W (V_{GS1} - V_{TO})^2 \lambda$ or $|r_{ds}(\text{LF})| = 1/g_{ds1}$. Since $|r_{ds}(\text{LF})| = 18$ kΩ, $g_{ds1} = 5.6 \times 10^{-5}$ S, and from (2.25), $g_2 = 8.8 \times 10^{-5}$ S.

The overall phase response is more difficult to match. Although the phase of $V(4)$ can be matched with the phase of V_{ds} rather closely, $|V(4)|$ does not become significant until $\omega = \omega_p/10$. Thus, most of the phase shift of the model takes place over a one- to two-decade frequency range, not three as shown in Fig. 2-22. Nevertheless, the model is still useful in predicting the qualitative behavior. Also it has been shown that the slope of the plots in Fig. 2-22 depends significantly on the material [37].

To determine R1, R2, and C1, begin by selecting a ratio of ω_z/ω_p:

$$\frac{R_2}{R_1 + R_2} = \frac{\omega_z}{\omega_p} \qquad (2.26)$$

Because of the limitation mentioned above, there is no point in making this ratio less than 0.01. R_2 should be large enough that it does not significantly affect r_{ds}(LF). Thus, $R_2 \gg |r_{ds}(\text{LF})|$. Assuming $R_2 = 1$ MΩ and $\omega_z/\omega_p = 0.01$, then $R_1 = 100$ MΩ. Then, picking $\omega_p/2\pi = 10^5$ Hz (the phase response begins to recover at 100 kHz in Fig. 2-22) will determine C1 = 1.59 pF.

Figure 2-24 shows the result of a SPICE simulation of a 1 × 50 μm MESFET with the parameters given in Table 2.4. The circuit used to determine $|r_{ds}|$ and phase r_{ds} is shown in the inset. The SPICE listing is given in Fig. 2-25. Note that the proper magnitude range is obtained, but the phase response occurs over a smaller frequency range and has a larger angle (–28° as compared with –15°).

Alternatively, if a better match to the phase shift were desired, the maximum phase shift can be determined by the maximum value of θ:

$$\tan\theta = \frac{g_2 \sin\phi}{g_{ds1} + g_2 \cos\phi} \qquad (2.27a)$$

$$\phi = \tan^{-1}(\omega R_1 C_1) - \tan^{-1}\left[\omega R_1 C_1 \left(\frac{R_2}{R_1 + R_2}\right)\right] \qquad (2.27b)$$

θ will reach a maximum for $\omega = \omega_p/2$.

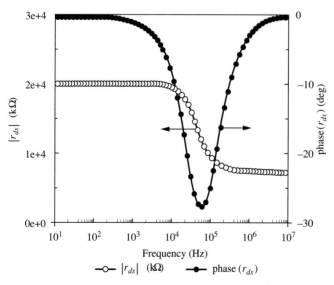

FIGURE 2-24
SPICE simulation of $|r_{ds}|$ and phase (r_{ds}) for 1 × 50 μm MESFET using subcircuit for lag model.

```
fet with rds correction: gdsl V3 + g2 V4
* ac analysis of rds vs freq.
***************
.model mesfet nmf(vto=–1.0 beta=1.1e–4 alpha=2.5 lambda=0.05 rs=885
+ rd=885 b=0.38 is =3.6e–16)
.model swmodel sw(vt=–1.0 roff=1e5)
***************
x1 12 13 0 4 mes50
vgs 13 0 –0.5
vdd 14 0 3
vin 14 12 ac .1
.ac dec 10 10 10meg
***************
.subckt mes50 3 2 1 4
********** D G S control**
z1 3 2 1 mesfet 50
g2 3 1 4 1 8.8e–5
s1 3 6 2 1 swmodel
r1 3 4 100meg
r2 4 1 1meg
c1 3 4 1.59pf
.ends
***************
.end
```

FIGURE 2-25
SPICE listing for simulation in Example 2.4.

2.3.4 Temperature Effects

As explained in Sec. 1.4.4, the MESFET drain current, gate current and threshold voltage are dependent on the channel temperature. While these effects can be predicted, and the device equations can be modified in the SPICE subroutine to account for temperature variation, it is often more convenient and just as effective to simulate the effect of temperature on circuit behavior by preparing separate sets of MESFET and Schottky diode model parameters for the extremes below and above room temperature. Then the circuit can be analyzed in a worst-case manner for extremes in both temperature and process variations and in supply voltages to guarantee that the circuit will meet specification under all environmental, process, and electrical conditions. This approach is essential if a well-behaved and predictable circuit is required.

2.4 HETEROSTRUCTURE BIPOLAR TRANSISTOR MODEL

The heterostructure bipolar transistor (HBT) described in Chap. 1 is being used in very-high-speed digital ICs currently being evaluated in the laboratory. Many of these circuits have demonstrated very high speeds (20 GHz frequency divider, for example [39]). The potential for maintaining that high speed in circuit layouts which are heavily loaded by wiring capacitance[2] (to be discussed in Chap. 5) is excellent, due to the low output impedance of the emitter follower. Commercial availability of the HBT for digital IC fabrication is certain to occur in the near future. Therefore, it is important to review the bipolar junction transistor (BJT) model which has been developed for use in SPICE to determine its applicability for modeling the HBT.

The bipolar transistor model is based on the early work by Ebers and Moll [40] and later work by Gummel and Poon [41]. Getreu [42] has provided an excellent description of how these models can be used to simulate the dc and ac characteristics of the BJT and has described methods for measuring the critical model parameters. In the following discussion, the notation of Getreu and his approach will be employed since any serious attempt to use the BJT SPICE model for simulating an HBT will require a more thorough study than can be provided in this section. Getreu's book is the logical place to continue this study.

The SPICE BJT model may be considered to contain a hierarchy of transistor models progressing from a very simple dc-only representation of the Ebers-Moll model (EM1) which requires only three parameters through levels of additions which include many progressively more complicated effects and more parameters. These additions include:

EM2: parasitic resistances and capacitances.

EM3 and Gummel-Poon: base-width modulation, splitting the base capacitance between internal and external regions, current gain variation with current and voltage, transit time variation with current.

The highest level model, the Gummel-Poon (GP) model, refines the EM3 model and is more closely related to the physics of the BJT, but is more difficult to understand and to determine its proper input parameters. It is only necessary to specify model parameters for the appropriate level of detail required by the application. If not specified, the model parameters default to some harmless, but not necessarily accurate, value. As will be seen below, the typical HBT requires a more complex dc modeling than is typical for most silicon BJTs if the current gain dependence on bias is to be properly predicted.

The SPICE BJT model was designed mainly to be used for large-signal dc and time domain (transient) simulation of digital ICs and small-signal frequency domain (ac) analysis of analog ICs operating at frequencies that are significantly less than the f_t of the transistor. Thus, the EM and GP models provide good dc agreement

---

[2] The *gate array* and *sea of gates* layout style is quite popular because of the simplified layout requirements (only the interconnection mask layers are personalized for a given application). The routing of wiring can be automated by CAD tools. The turnaround time can be very short because of the limited processing required to fabricate gate array wafers. However, the rigid layout constraints (fixed wiring channels) tend to produce ICs with dense wiring that is longer than would be obtained in a full custom layout. Thus, many of the logic gates are heavily loaded by their interconnections, and the propagation delays of these gates suffer from this. This loading-induced speed loss is particularly severe in FET ICs because of the relatively high output impedance of FET drivers.

with measured transistor characteristics and account for internal stored charges. They lead naturally into the popular hybrid-pi model for ac simulation, and because the model elements are calculated from the large-signal device equations, the model can be applied for nonlinear analysis.

The transit time of charge through the neutral base and the collector space-charge region is not explicitly included in the model, however. This extra transit time produces a phase shift in the collector current which becomes significant at frequencies approaching f_t. Therefore, the ac analysis utilizing the hybrid-pi model is somewhat limited; it is most accurate when used for the analysis of microwave bipolar transistors well below their highest frequencies of operation. The common-base T model [43] is more accurate in this application, but is strictly small-signal.

2.4.1 DC Ebers-Moll Model

The basic Ebers-Moll model of the BJT is familiar from nearly all textbooks on semiconductor devices. Its key features can be understood by considering the equivalent circuit shown in Fig. 2-26. The model consists of two dependent current generators with currents:

$$I_{CC} = I_S\left[\exp\left(\frac{qV_{BE}}{kT}\right) - 1\right] \quad (2.28a)$$

$$I_{EC} = I_S\left[\exp\left(\frac{qV_{BC}}{kT}\right) - 1\right] \quad (2.28b)$$

representing the collector current in the *forward active region* (BE junction is forward-biased; BC junction is reverse-biased) and the emitter current in the *inverse region* (BE junction is reverse-biased; BC junction if forward-based) respectively. The control

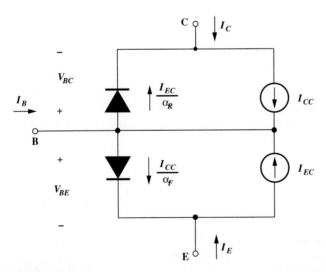

FIGURE 2-26
Ebers-Moll model for the bipolar transistor. (*From Getreu, [42]; ©Tektronix Inc. Used by permission.*)

voltages V_{BE} and V_{BC} are defined as intrinsic (internal) device terminal voltages. The two diodes represent the transistor's BE and BC junctions. The diode currents are related by the associated transport factors α_F and α_R. The advantage of this formulation of the EM model is that a single saturation current parameter, I_S, can be used to specify the generator and diode currents. This model can be used to predict current flow in all four regions of operation.[3] The base current is obtained by adding the currents at the base node:

$$I_B = \left(\frac{1}{\alpha_F} - 1\right) I_{CC} + \left(\frac{1}{\alpha_R} - 1\right) I_{EC} \tag{2.29}$$

The common-emitter forward and reverse current gains β_F and β_R can be defined in terms of the transport factors:

$$\beta_F = \frac{\alpha_F}{1 - \alpha_F} \tag{2.30a}$$

$$\beta_R = \frac{\alpha_R}{1 - \alpha_R} \tag{2.30b}$$

From these definitions, it is equivalent to say that

$$I_B = \frac{I_{CC}}{\beta_F} + \frac{I_{EC}}{\beta_R} \tag{2.31}$$

Following the development by Getreu [42], the EM model can be slightly modified in its topology without changing its functional form in order to obtain a straightforward correspondence with the familiar small-signal hybrid-pi model. This nonlinear hybrid-pi model is shown in Fig. 2-27 where the two dependent current sources in Fig. 2-26 have been merged to provide a single dependent current source. This current source represents the transconductance current generator, $i_c = g_m v_{be}$, in the linear small-signal model:

$$I_{CT} = I_{CC} - I_{EC} \tag{2.32}$$

where I_{CC} and I_{EC} are defined by Eqs. (2.27) and (2.28) as before. In fact, all of the terminal currents are unchanged. The base current I_B is given by (2.31), while I_C and I_E are given below:

$$I_C = (I_{CC} - I_{EC}) - \frac{I_{EC}}{\beta_R} \tag{2.33}$$

$$I_E = -\frac{I_{CC}}{\beta_F} - (I_{CC} - I_{EC}) \tag{2.34}$$

---

[3] The other two regions are the *saturation region* in which both BE and BC junctions are forward-biased and the *cutoff region* where both junctions are reverse-biased. Note the substantially different use of the word saturation in the BJT than for the FET. The saturation region of the BJT corresponds to the linear or ohmic region of the FET.

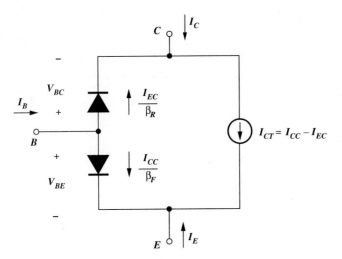

FIGURE 2-27
Nonlinear hybrid-pi (EM1) model. *(From Getreu, [42]; ©Tektronix Inc. Used by permission.)*

Thus, this model requires only three parameters, I_S, β_F, and β_R to describe the basic dc behavior of the BJT. This simplified version of the EM model has been labeled EM1 by Getreu. Some of the small-signal hybrid-pi model parameters (g_m, r_π, and r_μ) have a straightforward correspondence with the circuit of Fig. 2-27. The dependent current source represents g_m. The incremental base-emitter junction resistance r_π is represented by the BE diode. The incremental base-collector junction resistance r_μ is represented by the BC diode.

Unfortunately, this model in its simplicity assumes that β_F and β_R are constants, independent of V_{BE} or V_{BC}. Also, the series resistances connecting the internal transistor nodes to the outside world are not included. Thus, in its simplest form, it cannot even be used to predict the dc behavior of the BJT or HBT.[4]

2.4.2 Parasitic Resistances and Charge Storage in the HBT

The EM1 model can be modified to include parasitic series resistances (linear), and the charge storage in the HBT or BJT is modeled by nonlinear capacitances. These additional elements are shown in Fig. 2-28. The internal box contains the EM1 model elements, while the outer area includes some new elements. The combination of these elements is referred to as the EM2 model [42].

PARASITIC RESISTANCES. $r_{e'}, r_{b'}$, and $r_{c'}$ are the ohmic access resistances between the device contact metallization and the internal device junctions. These resistances

---

[4] In addition, the topology of the Ebers-Moll model does not represent the physics of the bipolar transistor as completely as the common-base T model. Some of its shortcomings for simulation of microwave applications arise from this deficiency.

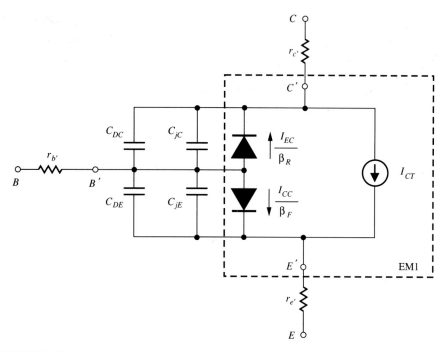

FIGURE 2-28
Parasitic access resistances and stored charges are added (EM2) to improve the EM1 model.

include contact resistance and any series resistance caused by the intermediate bulk semiconductor. Voltage drops occur across these resistors during normal operation which will cause the external and internal node voltages to differ. This effect can be seen from the graph of log I_C and log I_B plotted against V_{BE} in Fig. 2-29. Due to the exponential characteristic of the currents on V_{BE} as shown in Eqs. (2.27) and

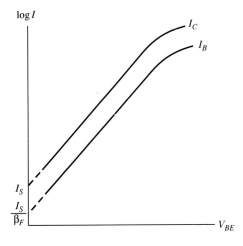

FIGURE 2-29
Semilog plot of the base and collector currents as a function of V_{BE}.

(2.28), log I will follow a straight line over several decades of current. The log I_B and log I_C lines are parallel because the two currents have the same ideality factor of unity in this example. It will be shown later that this is not always the case. At high current levels, a voltage difference between the external device terminals (EBC) and the internal device regions ($E'B'C'$) begins to appear. This can be described by $V_{BE} = V_{B'E'} + I_B r_{b'} + I_E r_{e'}$. Because of this voltage drop, the base and collector currents no longer increase as rapidly as they did at low current levels. Also the collector and base currents may not be exactly parallel at high current levels due to high-level injection effects.

The base resistance consists of three components illustrated in Fig. 2-30. Region 1 is characterized by the specific contact resistance of the base contact material. Region 2 is the series resistance of a thin semiconductor layer constituting the external base. Region 3 is the distributed access resistance into the internal base under the emitter. It is typically approximated by a transmission line model, but in reality is somewhat bias-dependent due to current crowding effects. The total resistance can be approximated by the equation below [44]:

$$r_{b'} = \frac{L_E}{12\sigma_B Z_E W_B} + \frac{S_B}{\sigma_B Z_E W_B} + \rho_c L_B Z_B \tag{2.35}$$

where L and Z represent the length and width of the base or emitter contact metals, S_B is the spacing between the base contact metal and the emitter mesa, σ is the conductivity of the respective semiconductor regions, ρ_c is the specific contact resistance of the base contact, and W_B is the thickness of the neutral or undepleted base (base width).

The emitter and collector parasitic resistances can be estimated by a similar approach. Techniques for measurement of the three resistances are described in Getreu [42].

CHARGE STORAGE. There are also four capacitors added to the model in Fig. 2-26 which represent the internal device charge storage in the space-charge layers

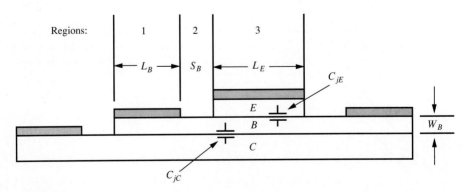

FIGURE 2-30
Cross-sectional view of the HBT emitter, base, and collector regions.

(C_{jE} and C_{jC}) and in the neutral base (diffusion capacitors C_{DE} and C_{DC}). C_{jE} and C_{jC} are also shown in Fig. 2-30. The junction capacitances due to the space-charge layers can be described by the same depletion formula used in the simple Schottky diode and MESFET gate capacitance models:

$$C_{jE}(V_{B'E'}) = \frac{C_{jE0}}{\left(1 - V_{B'E'}/\phi_E\right)^{m_E}} \tag{2.36}$$

$$C_{jC}(V_{B'C'}) = \frac{C_{jC0}}{\left(1 - V_{B'C'}/\phi_C\right)^{m_C}} \tag{2.37}$$

All three parameters are required for each equation to fit the measured data as described in Sec. 2.1. In the HBT, the grading coefficients m_E and m_C are often less than 0.5 since the base dopant beryllium diffuses rapidly at high concentrations. The parameters ϕ_E and ϕ_C are built-in voltages and C_{jE0} and C_{jC0} are the zero-bias capacitances, usually determined empirically from the measured C-V data using the same methods as were described in Sec. 2.1 for the Schottky diode. In order to avoid the problem of Eqs. (2.36) and (2.37) predicting infinite capacitance as $V_{B'E'}$ approaches ϕ_E, the slope of C_{jE} at $V_{B'E} \geq \phi_E/2$ is assumed to remain constant.

The substrate-to-collector capacitance C_{CS} is important for silicon BJTs but not significant for the HBT. This is because the substrate is semi-insulating, and there will be very low capacitance to adjacent surface structures compared with the internal device capacitances or the external wiring capacitance. Thus, C_{CS} can be ignored.

The diffusion capacitances model the additional charge required to add or delete mobile minority carriers in transit through the device. C_{DE} applies to the BE junction in forward bias (forward active mode) and C_{DC} applies to the BC junction when it is forward-biased (in the saturation or inverse mode). These capacitances can be calculated if the total charge in the device is known as a function of BE and BC voltage:

$$Q_{DE} = \tau_F I_{CC} \tag{2.38}$$

$$Q_{DC} = \tau_R I_{EC} \tag{2.39}$$

These charges can be determined from the forward and reverse current components flowing from emitter to base and base to emitter respectively if the time required for transit of the charge is known. For the HBT, the drift/diffusion in the base (τ_B) and the drift in the BC space-charge region ($\tau_{CB_{SCL}}$) are the primary mechanisms leading to the transit delay in the forward direction. The sum of these delays is provided to the model equations by the τ_F model parameter:

$$\tau_F = \tau_B + \tau_{CB_{SCL}} = \frac{W_B^2}{2.4 D_n} + \frac{W_{CB_{SCL}}}{2 v_{sat}} \tag{2.40}$$

Here, the width of the neutral base region is W_B and the width of the collector-base space-charge region is represented by $W_{CB_{SCL}}$. The equation for τ_B applies for the case where the charge is transported across the base purely by diffusion [45] with electron diffusivity D_n ($D_n \approx 25$ cm²/s in heavily doped p-type GaAs). In the reverse direction for the single heterojunction bipolar transistor (BE junction), there will be some additional delay associated with the injection of charge from the base into the collector (a homojunction). The total transit delay for purposes of estimating the diffusion capacitance can be estimated by the following equation:

$$\tau_R = \frac{W_C^2}{2.4D_p} + \frac{W_B^2}{2.4D_n} + \frac{W_{BE\,\text{SCL}}}{2v_{\text{sat}}} \tag{2.41}$$

In this equation, W_C is the width of the neutral collector region and $W_{BE\,\text{SCL}}$ is the width of the base-emitter space-charge region.

The small-signal diffusion capacitances can be calculated from the following definitions:

$$C_{DE} = \left[\frac{\partial Q_{DE}}{\partial V_{B'E'}}\right]_{V_{B'C'}=0} \tag{2.42}$$

$$= \left[\frac{\partial Q_{DE}}{\partial I_{CC}} \frac{\partial I_{CC}}{\partial V_{B'E'}}\right]_{V_{B'C'}=0}$$

$$= \tau_F g_{mF}$$

$$C_{DC} = \left[\frac{\partial Q_{DC}}{\partial V_{B'C'}}\right]_{V_{B'E'}=0} \tag{2.43}$$

$$= \left[\frac{\partial Q_{DC}}{\partial I_{EC}} \frac{\partial I_{EC}}{\partial V_{B'C'}}\right]_{V_{B'E'}=0}$$

$$= \tau_R g_{mR}$$

The forward transit time is most easily determined from measurement of f_t, the extrapolated unity gain frequency intercept of the short-circuit current gain. This is usually determined by calculation of $|h_{21}|$ from the measured S parameters, fitting a line with a 20-dB/decade slope to the data, and finding the frequency where the line intercepts 0 dB. f_t is related to the total emitter-to-collector time constants by

$$\tau_F = \frac{1}{\tau_{ec}} = \frac{1}{(\tau_E + \tau_B + \tau_{CB\,\text{SCL}} + \tau_C)} \tag{2.44}$$

where τ_E and τ_C are time constants associated with the charging of the EB and BC junctions respectively [46]. Ideally, if the HBT exhibits a maximum f_t as a function of collector current, this $f_{t,\,\text{max}}$ corresponds to a minimization of the emitter time constant [42]. Then, if the collector time constant is removed, the transit delays remain:

$$\tau_F = \left(\frac{1}{2\pi f_{t,\,\text{max}}}\right) - C_{jC}(V_{B'C'})r_{c'} \tag{2.45}$$

The V_{BC} dependence of τ_F is modeled inside the SPICE BJT model.

2.4.3 Additional Complexities

While the EM2 level BJT model accounts for most of the important phenomena needed to model the HBT, there are still some significant omissions that must be included in order to obtain an accurate simulation of actual devices. The most important of these can be provided by the EM3 and GP models and are: dependence of β_F on I_C, the partitioning of C_{jC}, and dependence of I_C on V_{BC} (base-width modulation).

The modeling of the variation in β_F is based on the analysis of the semilog plots of I_C and I_B on $V_{B'E'}$ with $V_{B'C'} = 0$, not by the specification of parameter β_F. These plots, shown in Fig. 2-31a, often referred to as *Gummel plots*, clearly show the influence of different mechanisms on the current gain. Thus, there are three distinct regions labeled I, II, and III corresponding to low-current, mid-current, and high-current behavior. The influence of $r_{e'}$ or $r_{b'}$ can be subtracted from the measured data by $V_{B'E'} = V_{BE} - (I_B r_{b'} + I_E r_{e'})$.

CURRENT GAIN. The log of the collector current when plotted against $V_{B'E'}$ should exhibit an ideality factor (n_f) of nearly 1.0 (slope of 59.5 mV/decade of current), characteristic of diffusion under low-level injection. In most HBTs, this is nearly true ($n_f \approx 1.1 - 1.2$). However, at low $V_{B'E'}$ the base current is dominated by recombination (surface and/or bulk) which produces an ideality factor of approximately 2. As shown in Fig. 2-31a, this leads to an intercept point where the two curves cross in region I. Unfortunately, for GaAs/AlGaAs HBTs this crossover current can be relatively high, which necessitates operating the device at high collector currents (large $V_{B'E'}$) in order to obtain acceptable current gains.

REGION I. The high ideality factor, low current region of the base current can be modeled by a second base-emitter diode with ideality factor n_E and saturation current

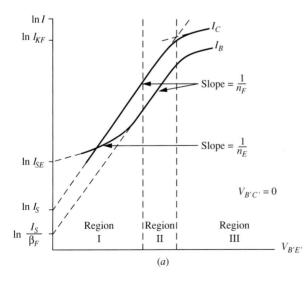

FIGURE 2-31
Plot of $\log_e I_C$ and $\log_e I_B$ as a function of $V_{B'E'}$ with $V_{B'C'} = 0$ for HBT which exhibits three regions of forward active operation. SPICE model parameters are shown as defined in the text. (a) Nonideal device characteristics occuring at low I_C (region I) due to recombination and at high I_C (region III) due to series resistance and high-level injection effects.

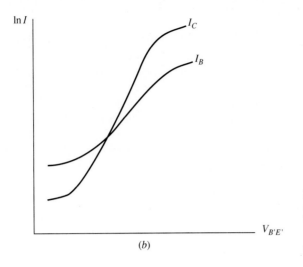

FIGURE 2-31 (*continued*)
(*b*) Gummel plot for HBT with large base recombination. Region II is suppressed by the large base current component.

I_{SE} in parallel with the main $n_F = 1.0$ diode, as illustrated in Fig. 2-32. Now the total base current would be given by

$$I_B = \frac{I_s}{\beta_F}\left[\exp\left(\frac{qV_{B'E'}}{n_F kT}\right) - 1\right] + I_{SE}\left[\exp\left(\frac{qV_{B'E'}}{n_E kT}\right) - 1\right] \quad (2.46)$$

The saturation currents can be selected as illustrated in Fig. 2-31*a*. Since I_S also specifies the collector saturation current and n_F the collector ideality factor, the first term in Eq. (2.46) is not totally independent of I_C but is related by the one parameter β_F. In some HBTs,[5] there is no observable region of $n = 1$ for the base current, as shown in Fig. 2-31*b*. In that case, the low-current behavior extends throughout the operating region of the device until series resistance or high-level injection effects are observed, and the first term of (2.46) should be made small by selecting a large value of β_F (10^4 or 10^5 is sufficient in most cases). In this way it will not interfere with the second term.

Of course, an additional diode could also be added between base and collector if the modeling of the inverse region were important; however, the HBT is not designed to work in this region. Thus, the parameters associated with the second BC diode should not be specified.

---

[5] HBTs with very narrow emitter fingers often have large recombination components to their base current since the edges of the emitter mesa provide a surface with many surface states that facilitate recombination. Since the emitter is narrow, the surface constitutes a large fraction of the total emitter area. This is also true for some planar devices where there are low minority carrier lifetimes in the external base due to residual implant damage.

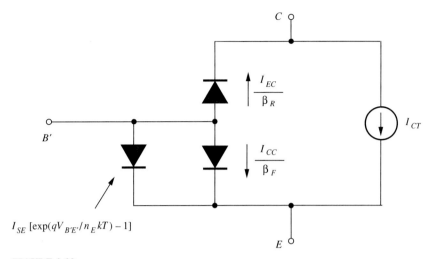

FIGURE 2-32
Schematic diagram of EM1 model modified to include an additional base diode with ideality factor n_E and saturation current I_{SE}.

REGION III. In some cases, the series resistances $r_{e'}$ and $r_{b'}$ are not sufficient to account for the falling off of the collector current from the exponential region II into the high-current region III. In this case, an additional parameter I_{KF} is defined as shown in Fig. 2-31a by the intersection of extrapolated straight lines from region II and region III. At high current levels, the collector current predicted by the GP model approaches the following limit asymptotically:

$$I_C = (I_{KF}I_S)^{1/2}\exp\left(\frac{qV_{B'E'}}{2kT}\right) \qquad (2.47)$$

The saturation current I_S is the same extrapolated intercept of log I_C at $V_{B'E'} \approx V_{BE} = 0$ which was defined above for the EM1 model. The other high-level injection effects typically of concern for the silicon BJT are not as severe for the HBT because of the very high base doping levels ($N_A = 10^{19}$ to 10^{20}) typically used to reduce r_b'. The HBTs tend to fail thermally before β_F or f_t begin to fall significantly.

PARTITIONING OF COLLECTOR CAPACITANCE. The cross-sectional drawing of the HBT shown in Fig. 2-30 makes it clear that the total base-collector capacitance C_{jC} is not completely on the inside of the device (under the emitter region—internal base) but that a significant area (as much as two-thirds) is located in the external base region. Since these regions are separated in the model by $r_{b'}$, as shown by Fig. 2-28, it is helpful to split up C_{jC} into two components, one for the internal base and the other for the external base. This partitioning is illustrated by Fig. 2-33. The model

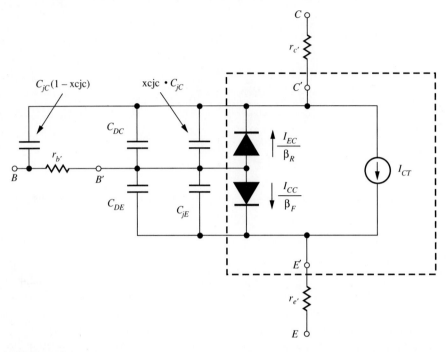

FIGURE 2-33
Schematic diagram of distribution of C_{jC} between internal and external base in the SPICE BJT model.

parameter C_{jC} should be chosen to represent the total BC capacitance, and xcjc is selected according to the area ratio:

$$\text{xcjc} = \frac{A_\text{internal}}{A_\text{external}} \qquad (2.48)$$

BASEWIDTH MODULATION. The width of the base region will vary with V_{BC} due to the BC space-charge layer. This behavior is commonly used in silicon BJTs to explain the positive slope in the collector current characteristic in the forward active region as V_{BC} is increased. While the same mechanism should not be as significant in the HBT due to the high base dopings mentioned above, the devices still may exhibit some finite slope in the I_C-V_{BC} characteristic measured with constant V_{BE}.[6] This slope (if positive) can be modeled with the addition of a new parameter, V_A, which is called the Early voltage. The collector current varies with V_{BC} according to

---

[6] The forward active characteristics have also been observed to have a negative slope for some devices when I_B is held constant.

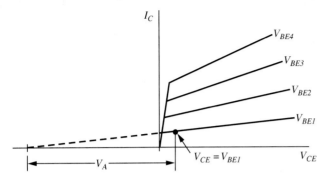

FIGURE 2-34
Collector current as a function of V_{CE} with constant V_{BE}. The parameter V_A is defined on the plot.

the equation below for the forward active region:

$$I_C = \frac{I_S(V_{B'C'} = 0)}{1 + V_{B'C'}/V_A}\left[\exp\left(\frac{qV_{B'E'}}{n_F kT}\right) - 1\right] \quad (2.49)$$

As $V_{B'C'}$ becomes more negative, the collector current increases.

The Early voltage can be defined geometrically as shown in Fig. 2-34 on the I_C-V_{CE} characteristic measured with constant V_{BE}. It is determined by extrapolating the I-V curves in the forward active region back to the negative V_{CE} axis. The voltage between the point where $V_{BC} = 0$ on the curves ($V_{CE} = V_{BE}$) and the intercept with the voltage axis is V_A. If V_A is large, it will be approximately equal to the difference between the intercept point and $V_{CE} = 0$. In the familiar hybrid-pi small-signal model, the base-width modulation effect corresponds to the output resistance or conductance $r_o = 1/g_o$. This resistance would be determined by SPICE from the derivative $\partial I_C/\partial V_{CE}$ with constant V_{BE}, not from the addition of another model element.

2.4.4 Summary of Model Parameters for HBT Model

The model equations and elements discussed above are adequate to simulate the important characteristics of a GaAs/AlGaAs single heterojunction HBT in the forward active, saturation, and cutoff regions. The inverse region is never used as HBTs are not designed to operate effectively under that set of conditions. The model parameters introduced above are summarized in Table 2.5. The names used in Getreu [42] are compared with the names used in the SPICE model. Typical values for a moderately good HBT are compared with the SPICE model default values, and those model parameters that are automatically scaled with area are indicated.

While the SPICE BJT model has many other subtleties for predicting high-current effects, the adaptation of these to the HBT is beyond the scope of this section and is not needed for most applications.

TABLE 2.5
Model parameters needed for simulating HBT with the SPICE BJT model

SPICE name	Getreu name	Parameter	Units	Default value	Typical value	Area
is	$I_S(0)$	Collector current	A	1.0e-16	9e-26	*
bf	β_F	Forward CE current gain	—	100	10^5	
br	β_R	Reverse CE current gain	—	1	1	
re	$r_{e'}$	Emitter resistance	Ω	0	7	*
rb	$r_{b'}$	Zero-bias base resistance	Ω	0	55	*
rc	$r_{c'}$	Collector resistance	Ω	0	27	*
cjc	C_{jC}	BC zero-bias depletion capacitance	fF	0	60	*
vjc	ϕ_C	BC built-in potential	V	0.75	0.6	
mjc	M_C	BC junction exponent	—	0.33	0.33	
cje	C_{jE}	BE zero-bias depletion capacitance	fF	0	32	*
vje	ϕ_E	BE built-in potential	V	0.75	1.25	
mje	M_E	BE junction exponent	—	0.33	0.33	
tf	τ_F	Ideal forward transit time	pS	0	2.5	
nf	N_{CL}	Forward current emission coefficient	—	1.0	1.2	
ise	$C2I_S(0)$	BE leakage saturation current	A	0	2e-21	*
ne	N_{EL}	BE leakage emission coefficient	—	1.5	1.6	
ikf	I_K	Corner for forward beta high current roll-off	A	∞	0.4	*
xcjc	XCJC	Fraction of BC capacitance connected to internal base	—	1	0.33	
vaf	VA	Forward Early voltage	V	∞	35	

HOMEWORK PROBLEMS

2.1. Figure P2-1 is a capacitance-voltage plot for a GaAs Schottky diode.
   (*a*) From the data in Fig. P2-1, determine whether the GaAs is *n* or *p* type and explain why.
   (*b*) Use the data to determine the SPICE model parameters *cjo* and vj. What value of *m* was required to obtain the fit?
   (*c*) The diode area was 5.075×10^{-4} cm^2. Find the depletion depth x_n at $V_D = 0$ and $V_D = -5$ V. Assume the relative dielectric constant of GaAs is 12.9 and that $V_{bi} = 0.8$ V.

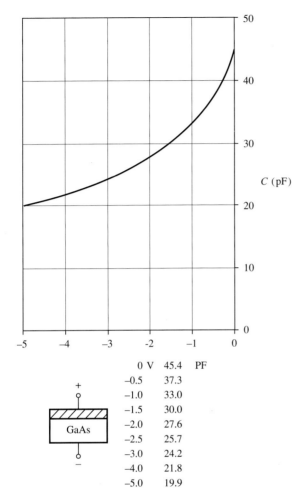

0 V	45.4 PF
−0.5	37.3
−1.0	33.0
−1.5	30.0
−2.0	27.6
−2.5	25.7
−3.0	24.2
−4.0	21.8
−5.0	19.9

FIGURE P2-1
Capacitance-voltage plot for GaAs Schottky diode.

(d) Find the concentration of the dopant at $V_D = 0$ V (assume that the dopant atoms are fully ionized).

2.2. Evaluation of the SPICE3 MESFET model. Measured data for a 1×50 μm MESFET is shown in Fig. P2-2a. A source resistance measurement was performed which found that $R_s = 16.9\ \Omega$. With that information, the intrinsic I-V data was calculated and plotted, as shown in Fig. P2-2b.

Use the intrinsic I-V data in Fig. P2-2b to estimate model parameters alpha, beta, b and lambda for the SPICE3 MESFET model. Scale the parameters properly for a 1×1 μm device. The model equation is shown below:

$$I_D = \frac{\beta(V_{GS} - V_T)^2}{1 + b(V_{GS} - V_T)}\left[1 - \left(1 - \frac{\alpha V_{DS}}{3}\right)^3\right](1 + \lambda V_{DS})$$

2.3. Write and run a test file for SPICE3 using your model parameters which will calculate the I_D-V_{DS} characteristics with the same range of V_{DS} and V_{GS} values used in Fig. P2-2b. (Nested voltage sources can be used in SPICE to calculate families of curves.

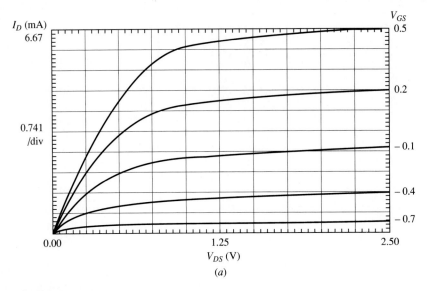

FIGURE P2-2a
Drain current vs. V_{DS} for $1 \times 50~\mu m$ MESFET.

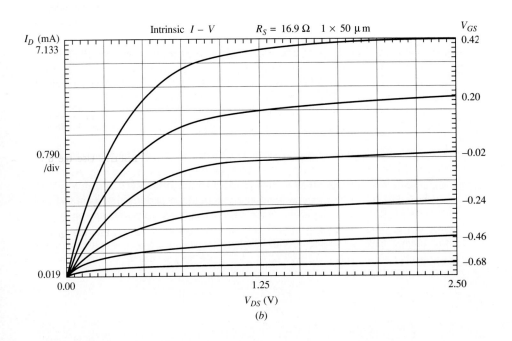

FIGURE P2-2b
Drain current of MESFET from (a) replotted against $V_{DS} - I_D(R_S + R_D) = V_{DS,i}$ where $R_S = R_D = 16.9~\Omega$. $V_{GS,i}$ is the parameter.

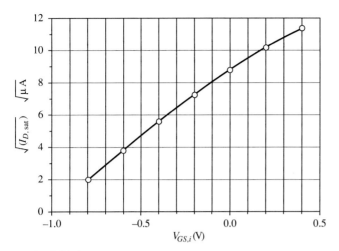

FIGURE P2-6
$\sqrt{I_{D,sat}}$ vs. $V_{GS,i}$ for GaAs MESFET.

For example, the statement:

.dc vds 0 2.5 0.05 vgs 1 − 0.68 0.64 0.22

will perform the dc analysis requested above). Save the output in the form of a table of numbers and plot enough points on Fig. P2-2b for the highest, lowest, and one intermediate V_{GS} value so that the quality of the fit of the model to the intrinsic I-V data can be judged.

2.4. Repeat Prob. 2.3 with the source and drain resistances added into the model. Compare the simulated device characteristics against Fig. P2-2a.

2.5. The SPICE3 MESFET model has a "hidden" parameter which is not described in the user's guide. This is the saturation current of the gate-source and gate-drain diodes, I_S. Perform an experiment with SPICE to determine the default value of I_S. Use the plotting capabilities to make a hardcopy plot of log I_G vs. V_{GS} to find I_S. Also determine the value of the ideality factor, n, which is assumed by the model. (Note that in actual application of the model, I_S can be included in the parameter listing if something other than the default value is required.)

2.6. The data in Fig. P2-6 is plot of $\sqrt{I_{D,sat}}$ vs. $V_{GS,i}$ for a GaAs MESFET. Use the data to determine approximate values for the SPICE3 MESFET model parameter beta, vto, and b.

2.7. A MESFET is connected with gate and source tied together and with the drain attached to V_{DD}. Use the typical model parameters for a 1-μm MESFET in your available SPICE model, and implement the backgating correction discussed in Sec. 2.3.1. Plot the I_D-V_{DS} characteristics for this MESFET when its source voltage is swept from 2.5 to 0 V for V_{SS} = 0 to −5 V in 1 volt steps. Set V_{DD} = 2.5 V and assume that K_{BG} = 0.05.

REFERENCES

1. Nagel, L. W.: "SPICE2: A Computer Program to Simulate Semiconductor Circuits," Electronics Research Laboratory, University of California, Berkeley, paper ERL-M520, 1975; Cohen, E. "Program reference for SPICE2," Electronics Research Laboratory, University of California, Berkeley, paper ERL-M592, 1976.

2. Tuinenga, P. W.: *SPICE. A Guide to Circuit Simulation and Analysis Using PSPICE*, Prentice-Hall, 1988.
3. Sze, S. M.: *Physics of Semiconductor Devices*, 2d ed., chap. 5, Wiley–Interscience, 1981.
4. Rhoderick, E. H.: *Metal–Semiconductor Contacts*, Clarendon, Oxford, 1978.
5. Estreich, D. B.: "A Simulation Model for Schottky Diodes in GaAs Integrated Circuits," *IEEE Trans., Computer-Aided Design*, Vol. CAD-2, pp. 906–111, April 1983.
6. Shichman, H., and Hodges, D. A.: "Modeling and Simulation of Insulated-Gate Field-Effect Transistor Switching Circuits," *IEEE J. Solid State Cir.*, Vol. SC-3, pp. 285–289, September 1968.
7. Statz, H., Newman, P., Smith, I., Pucel, R., and Haus, H.: "GaAs FET Device and Circuit Simulation in SPICE," *IEEE Trans. Elect. Dev.*, Vol. ED-34, pp. 160–169, February 1987.
8. Takada, T., Yokoyoma, K., Ida, M., and Sudo, T.: "A MESFET Variable Capacitance Model for GaAs Integrated Circuit Simulation," *IEEE Trans. Microwave Theory Tech.*, Vol. MTT-30, pp. 719–724, May 1982.
9. Shur, M.: *GaAs Devices and Circuits*, sec. 7.7, Plenum Press, New York, 1987.
10. Shockley, W.: "A Unipolar Field-Effect Transistor," *Proc. IRE*, Vol. 40, pp. 1365–1376, 1952.
11. Turner, J. A., and Wilson, B. L. H.: "Implications of Carrier Velocity Saturation in a GaAs Field-Effect Transistor," *Int. Symp. GaAs and Related Compounds*, Dallas, Tex., 1968.
12. Peltan, S. G., Long, S. I., and Butner, S. E.: "An Accurate DC SPICE Model for the GaAs MESFET, *Proc. IEEE Int. Symp. Circuits and Systems*, Philadelphia, Pa. pp. 6–11, May 1987.
13. Curtice, W. R.: "A MESFET Model for Use in the Design of GaAs Integrated Circuits," *IEEE Trans. Microwave Theory and Tech.*, MTT-28, pp. 448–456, May 1980.
14. Chen, T. H., and Shur, M. S.: "Analytical Models for Ion-Implanted GaAs FETs," *IEEE Trans. Elect. Dev.*, Vol., ED-32, pp. 18–27, January 1985.
15. Doganis, K., and Dutton, R.: "SUXES—Stanford University Extractor of Model Parameters," Stanford Electronics Laboratory Technical Report, November 1982.
16. "Discover TECAP-HP94445A. The Transistor Modeling and Characterization System," HP Product Note HP94445a-1, June 1986.
17. PSPICE version 3.08, Micro-Sim Corp., 1988.
18. Yang, L., and Long, S. I.: "New Method to Measure the Source and Drain Resistance of the GaAs MESFET," *IEEE Elect. Dev. Lett.*, Vol. EDL-7, pp. 75–77, February 1986.
19. Sussman-Fort, S. E., Narasimhan, S., and Mayaram, K.: "A Complete GaAs MESFET Computer Model for SPICE," *IEEE Trans. Elect. Dev.*, Vol. ED-32, pp. 471–473, 1984.
20. Van der Ziel, A.: "Gate Noise in Field Effect Transistors at Moderately High Frequencies," *Proc. IEEE*, Vol. 51, pp. 461–467, 1963.
21. Pucel, R. A., Haus, H. A., and Statz, H.: "Signal and Noise Properties of GaAs Microwave Field Effect Transistors," in *Advances in Electronics and Electron Physics*, (Ed. L. Marton), Vol. 38, pp. 195–265, Academic Press, New York, 1975.
22. Englemann, R., and Liechti, C.: "Bias Dependence of GaAs and InP MESFET Parameters," *IEEE Trans. Elect. Dev.*, Vol. ED-24, pp. 1288–1296, November 1977.
23. Van Tuyl, R.: personal communication.
24. Statz, H.: Raytheon Co., personal communication.
25. Andersson, M., Aberg, M., and Pohjonen, H.: "Simultaneous Extraction of GaAs MESFET Channel and Gate Diode Parameters and Its Application to Circuit Simulation," *1988 IEEE Int. Symp. Cir. and Syst. Conf. Proc.*, pp. 2601–2605, June 1988.
26. Wassertsrom, E., and McKenna, J.: "The Potential Due to a Charged Metallic Strip on a Semiconductor Surface," *Bell Syst. Tech. J.*, Vol. 49, pp. 853–877, May-June 1970.
27. Alexopoulos, N. G., Maupin, J. A., and Greiling, P. T.: "Determination of the Electrode Capacitance Matrix for GaAs FETs," *IEEE Trans. Microwave Theory Tech.*, Vol. MTT-28, pp. 459–466, May 1980.
28. Kocot, C., and Stolte, C. A.: "Backgating in GaAs MESFET's" *IEEE Trans. Elect. Dev.*, Vol. ED-29, pp. 1059–1064, July 1982.
29. Chang, M. F., Lee, C. P., Hou, L. D., Vahrenkamp, R. P., and Kirkpatrick, C. G.: "Mechanism of Surface Conduction in Semi-insulating GaAs," *Appl. Phys. Lett.*, Vol. 44, pp. 869–871, May 1984.

30. D'Avanzo, D.: "Proton Isolation for GaAs Integrated Circuits," *IEEE Trans. Elect. Dev.*, Vol. ED-29, pp. 1055–1059, July 1982.
31. Taylor, G. W., Darley, H. M., Frye, R. C., and Chatterjee, P. K.: "A Device Model for an Ion-Implanted MESFET," *IEEE Trans. Elect. Dev.*, Vol. ED-26, pp. 172–182, 1979.
32. Lee, S. J., Lee, C. P., Shen, E., and Kaelin, G. R.: "Modeling of Backgating Effects on GaAs Digital Integrated Circuits," *IEEE J. Solid State Cir.*, Vol. SC-19, pp. 245–429, April 1984.
33. Kasemset, D,: Rockwell International, personal communication.
34. Subramanian, S., Bhattacharya, P. K., Staker, K. J., Ghosh, C. L., and Badawi, M. H.: "Geometrical and Light-Induced Effects on Back-Gating in Ion-Implanted GaAs MESFET's" *IEEE Trans. Elect. Dev.*, Vol. ED-32, pp. 28–33, January 1985.
35. Troutman, R. R.: "Subthreshold Design Considerations for Insulated Gate Field-Effect Transistors," *IEEE J. Solid State Cir.*, Vol. SC-9, pp. 55–60, April 1974.
36. Lee, S. J., *et al.*: "Ultra-low Power, High Speed GaAs 256 bit Static RAM," *IEEE GaAs IC Symp. Proc.*, Phoenix, Ariz., pp. 74–77, October 1983.
37. Larson, L. E.: "Gallium Arsenide Switched-Capacitor Circuits for High-speed Signal Processing," Ph.D. Dissertation, University of California, Los Angeles, 1986.
38. Rocchi, M.: "Status of the Surface and Bulk Parasitic Effects Limiting the Performance of GaAs IC's," *Physica*, vol. 129B, pp. 119–138, 1985.
39. Wang, K. C., Asbeck, P. M., Chang, M. F., Sullivan, G. J., and Miller D. L.: "A 20 GHz Frequency Divider Implemented with Heterojunction Bipolar Transistors," *IEEE Elect. Dev. Lett.*, Vol. EDL-8, pp. 383–385, September 1987.
40. Ebers, J. J., and Moll, J. L.: "Large-signal Behavior of Junction Transistors," *Proc. IRE*, Vol. 42, pp. 1761–1771, December 1954.
41. Gummel, H. K., and Poon, H. C.: "An Integral Charge Control Model of Bipolar Transistors," *Bell Sys. Tech. J., Vol.* 49, pp. 827–852, May 1970.
42. Getreu, I.: *Modeling the Bipolar Transistor*, Elsevier Scientific Publ., 1978.
43. White, M. H., and Thurston, M. O.: "Characterization of Microwave Transistors," *Solid State Elect.*, Vol. 13, pp. 523–542, 1970.
44. Phillips, A. B.: *Transistor Engineering and Introduction to Integrated Semiconductor Circuits*, Chap. 9, McGraw-Hill, 1962.
45. Phillips, A. B.: *Transistor Engineering and Introduction to Integrated Semiconductor Circuits*, Chap. 8, McGraw-Hill, 1962.
46. Cooke, H. F.: "Microwave Transistors: Theory and Design," *Proc. IEEE*, Vol. 59, pp. 1163–1181, August 1971.

CHAPTER 3

LOGIC CIRCUIT DESIGN PRINCIPLES

The design of GaAs digital logic circuits is a multidimensional problem. The circuit design problem is less constrained than silicon MOS circuit design because there are few standards established for logic levels or supply voltages, and no preferred or obviously superior circuit topology has dominated designs at the present time. In addition, some of the processes available through GaAs foundries provide both enhancement and depletion mode FETs whereas others are currently limited to depletion mode transistors. The threshold voltages of these devices are often different from one foundry to the next, and a relatively wide variance from the mean threshold voltage is allowed. While this problem may disappear in time, the GaAs user must be armed with a more general set of design methodologies than is necessary to complete the design of a standard NMOS or CMOS silicon IC.

The approach taken to optimize the design will depend on what specification of performance is the most important. In general, speed and power can be traded off over about a five to one range for most circuits (with roughly a constant power-delay product) without changing the threshold voltage. With access to the threshold voltage, an even wider range is available. Speed and circuit tolerance (to process variation, supply voltage or ground fluctuations, or temperature variation) are also interchangeable to a degree since a circuit with a low noise margin will exhibit less delay than one that is designed with a larger logic swing and is therefore more robust in a digital system application. In the discussion to follow, the highest priority is assigned to the dc functionality of the circuits over the expected range of process parameters and operating temperatures. Without satisfying this prerequisite, the maximum operating speed of the IC chip has little significance.

The greatest single difference between silicon NMOS design and *n*-channel gallium arsenide MESFET design is imposed by the Schottky barrier junction gate of the MESFET which exhibits a forward voltage conduction limit at about 0.7 V positive with respect to the source. If heterostructure FETs (HFET or MODFET) or JFETs are used, forward bias can be as high as 1 V in some cases. This conduction limit plays a role in most of the GaAs circuit topologies by limiting the logic high-voltage level. For example, Fig. 3-1a shows the circuit schematic for an enhancement/depletion (E/D) NMOS inverter driving another inverter stage. There, when the input is low, the logic high voltage at the output of the first stage will rise to V_{DD}, charging the gate oxide MOS capacitor without conduction limit. In Fig. 3-1b, an analogous E/D MESFET inverter pair is shown. In this case, with the first-stage input low, the output of the first stage will reach a logic high voltage which is clamped to about 0.7 V by the gate-source diode of the second stage, regardless of the V_{DD} power supply voltage. The logic low-output voltage for both circuit types is limited by the voltage drop across the switch FET in its ohmic region, $I_D R_{ON}$, in both cases. Typically, V_{OL} will be in the range 0.1 to 0.2 V. Hence, the logic swing for the NMOS inverter

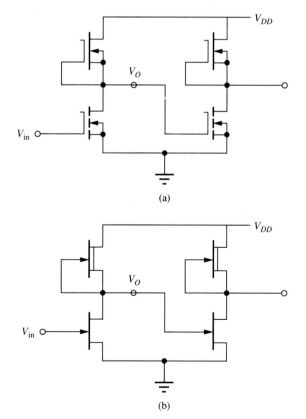

FIGURE 3-1
(*a*) E/D NMOS inverter pair. (*b*) E/D GaAs MESFET or HFET inverter pair.

is V_{DD}-V_{OL}, while for the MESFET inverter it is reduced to 0.7-V_{OL}, a substantial reduction.

Therefore, for GaAs FET logic design, the selection of the circuit topology can be very important. The example above in Fig. 3-1b, while currently popular because of its high circuit density and low power dissipation, places a great deal of emphasis on process uniformity because of the limited logic swing and the constraint that this places on the device threshold control. On the other hand, certain depletion mode circuits, while less attractive due to space and power requirements, require much less stringent threshold control.

In the chapter that follows, a discussion of the sources of noise in digital circuits will be followed by a definition of noise margin and its relationship to the stability of circuits. Then, the optimization of the dc performance through adjustment of device threshold voltage, width ratios, and level shift for maximum noise margin will be presented. The relationship between noise margin and the normally distributed variation of device parameters will be related to the yield of functional circuits. With the functionality of the circuit now established, the transient analysis can proceed so that information on delay as a function of loading, fan-in, fan-out, and device width can be obtained. Representative logic circuit designs using depletion mode and enhancement mode MESFETs will be presented and discussed in Chap. 4.

3.1 NOISE IN DIGITAL CIRCUITS

3.1.1 Sources of Noise

There are numerous internal and external sources of noise that the IC must successfully reject if it is to operate reliably in its intended application. While thermal noise is an ultimate limitation for circuits when voltage and current are scaled down well beyond the limits of today's state-of-the-art fabrication, it is of little concern except for analog applications for realistic devices and circuits at the 1-μm lithographic level. There are much more powerful sources to be controlled, some of which are schematically represented in Fig. 3-2 [1]. Here, external sources such as radio-frequency interference, and internal sources such as mutual inductance and mutual capacitance (crosstalk), inductive and resistive voltage spikes, and drops on power supply and ground lines are identified. These sources are of concern, particularly at the circuit board level of the system hierarchy, and present a noisy supply, ground, and signal environment to the IC. In addition, reflected pulse edges, which arise from unterminated signal transmission lines or lines with nonuniform impedance characteristics, are also very important when the response time of the IC input is short relative to the round-trip propagation time on an interconnecting path. The analysis of the interconnection-related noise sources is discussed in Chap. 5.

Measurements of typical power supply and ground noise amplitudes have been carried out on well-designed multilayer PC boards with packaged ECL and GaAs ICs [2]. Peak-to-peak noise amplitudes of at least 100 mV have been observed in these cases. Much worse environments can be easily created through neglect of important high-speed digital system design principles.

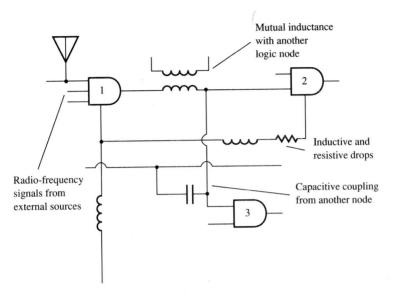

FIGURE 3-2
Sources of noise in digital systems. (*From Hodges and Jackson.* [1] © *McGraw-Hill, used by permission.*)

3.1.2 Definitions of Noise Margin

A function is needed that can measure the ability of a digital logic circuit to operate error-free in a noisy environment. This function is known as the *noise margin,* which can be defined in several ways working from the transfer characteristic[1] of the logic circuit as discussed in Ref. 3. In this section only dc noise voltage will be considered; the transient (dynamic) noise margin is discussed in Sec. 3.3.7.

An example of a typical inverter transfer characteristic, curve $f_1(V_{in})$, is shown in Fig. 3-3a. If the output of inverter 1 is connected to the input of inverter 2 and the transfer characteristics of 1 and 2 are the same, then the inverter 2 characteristic, $f_2(V_{out1})$, can be plotted on the same graph by rotating $f_1(V_{in1})$ and the axes about the 45°, $V_{in} = V_{out}$, line. If a chain of gates is considered, as shown in Fig. 3-3b, every odd stage obeys f_1 and every even stage f_2. Points A and B correspond to the *logic low- and high-output voltages,* V_{OL} *and* V_{OH}, obtained when the inverters are in stable equilibrium. For example, if $V_{in1} = V_{OL}$, then $V_{in2} = V_{OH}$ and $V_{out2} = V_{OL}$, etc. Point C is a point of unstable equilibrium. If $V_{in1} = V_{TH}$ (defined as the *inverter threshold*), then $V_{out1} = V_{in2} = V_{out2} = V_{TH}$ and a small change in V_{TH} or an external noise voltage at one of the inputs will cause inverters 3,4, . . . to snap to stable states A or B.

---

[1] A transfer characteristic can be defined for a logic circuit which relates an output (dependent) variable to an input (independent) variable $[X_{out} = f(X_{in})]$. The derivation of transfer characteristics for particular circuits is the subject of Sec. 3.2. Since the logic circuit being considered has high-input impedance and low-output impedance, the voltage transfer characteristic is the appropriate one to consider [4].

144 GALLIUM ARSENIDE DIGITAL INTEGRATED CIRCUIT DESIGN

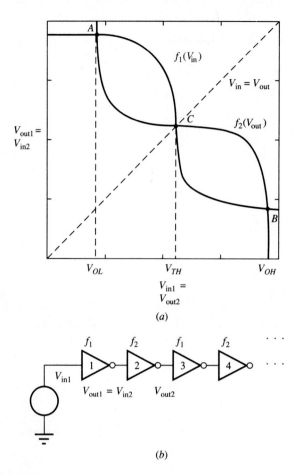

FIGURE 3-3
Transfer characteristic of an inverter $[f_1(V_{in})]$. The characteristic is replotted on the same scale $[f_2(V_{out})]$ by reflecting about the $V_{in} = V_{out}$ line and exchanging the V_{in} and V_{out} axes.

If only one inverter were being considered, and the logic low voltage V_{OL} is applied to the input as shown in Fig. 3-4, then the noise voltage $V_N = V_{INL}$ required to cause the inverter to change state is given by $V_{TH} - V_{OL}$. Similarly, if the input were at V_{OH}, a noise voltage $V_N = V_{INH} = V_{OH} - V_{TH}$ would be required. These critical noise voltages are called the *intrinsic low-noise margin* and *intrinsic high-noise margin* respectively. This single-stage case leads to an unreasonable optimistic

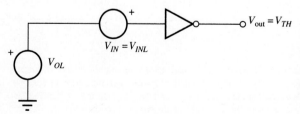

FIGURE 3-4
Definition of intrinsic noise margin for a single inverter. If the noise voltage V_N equals $V_{TH} - V_{OL}$, the inverter output will be at unstable equilibrium (point C in Fig. 3-3).

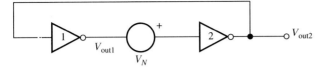

FIGURE 3-5
Definition of noise margin for infinite chain of inverters. A loop of two inverters compose a latch which will be set if the noise voltage V_N at the input to inverter 2 exceeds the noise margin.

noise margin, since the effect of the shifted output voltage on subsequent stages, which can also have noise applied to their inputs, has been ignored.

A more realistic and useful definition is indicated by the connection shown in Fig. 3-5. In this example, a loop of two inverters composes a bistable latch which will be set to its opposite state if the noise voltage V_N at the input to 2 exceeds the noise margin. If the output of 1 is high, then V_N must equal the *noise margin high* V_{NH} to set the latch. If the output of 1 is low, then $V_N = V_{NL}$. These noise margins need not be the same if the transfer characteristics are nonsymmetric. Adverse noise[2] is applied to the input of one inverter only, but because the gates are connected in a loop, an infinite chain of inverters is simulated. As V_N increases V_{in2}, the output V_{out2} decreases V_{in1}; thus both inputs 1 and 2 are moving in opposite directions, as indicated by the arrows in Fig. 3-6. The output will not recover to its original state

---

[2] A noise voltage with polarity which will shift V_{in2} toward V_{TH}.

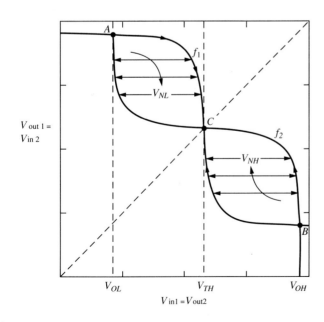

FIGURE 3-6
Transfer characteristic of inverter. The voltages V_{NL} and V_{NH} are noise margins defined by the maximum width of the loops formed by the intersections of $f_1(V_{in})$ and $f_2(V_{out})$.

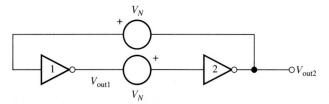

FIGURE 3-7
Latch formed by loop of inverters. Equal and opposite noise voltages, V_N, are inserted within the loop and will upset the latch when equal to V_{MS}, the maximum square noise margin.

when the noise voltage is removed if the noise margin is exceeded. The voltages V_{NL} and V_{NH} are defined in Fig. 3-6 by *maximum width* of the loops formed by the intersections of $f_1(V_{in})$ and $f_2(V_{out})$. The width of the loop represents the noise margin (NM) because the vertical axis plots V_{out} for f_1 and V_{in} for f_2, which are the same in the connection of Fig. 3-5. If the width is exceeded, then the only stable condition available is in the opposite state.

An even more stringent definition of noise margin is obtained when equal and opposite correlated noise sources are attached to each input in the loop, as shown in Fig 3-7. This connection corresponds to the case in which adverse noise is applied between every stage of an infinite chain of inverters. Since the noise V_N is equal and opposite, both V_{in2} and V_{in1} approach V_{TH}. For example, starting from A, $V_{in1} = V_{out2} + V_N = f_2(V_{in2}) + V_N$ and $V_{in2} = V_{out1} - V_N = f_1(V_{in1}) - V_N$. As the arrows in Fig. 3-8 illustrate, the unstable point V_{TH} is approached from two directions forming

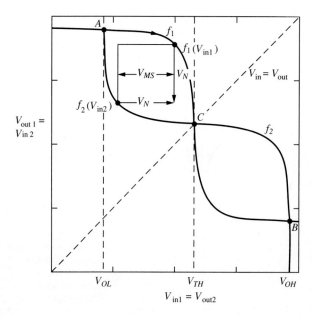

FIGURE 3-8
The noise voltage V_{MS} required to set the latch is given by the width of the largest square that can be drawn inside of the loops of the transfer characteristics. Therefore, this definition of NM, which produce equal low and high NM, is called the maximum square noise margin.

the sides of a square. Hence $V_{MS} < V_{NH}$ or V_{NL}. V_{MS} will be the same whether starting from A or B and represents the noise voltage V_N required to set the latch as defined above. Therefore, this definition of NM is called the *maximum square noise margin* [3]. This definition possibly represents a worst-case condition which is not likely to be exceeded in any real application.

From the above definitions, it is apparent that the same method can be applied to cases in which the logic inverters or gates are not identical simply by drawing the relevant transfer characteristics on the same scales. This might be the case when the gates are unequally loaded or when unequal device characteristics occur due to processing or material variations. Noninverting circuits can also be treated with the same method [3].

Finally, another widely used definition for NM is the slope $= -1$ criterion. As is shown in Fig. 3-9, voltages V_{IL} and V_{IH} are defined at the point on $f_1(V_{in})$ where its slope $= -1$. If the input voltage noise V_N were increased beyond $V_{IL} - V_{OL}$ or $V_{OH} - V_{IH}$, then the logic circuit behaves as an amplifier of noise rather than an attenuator of noise. The noise level will possibly build up in a chain of similar stages leading to an upset event. While this definition is more intuitive than the maximum square definition, Ref. 3 has shown that it can lead to overly optimistic estimates of NM for circuits that have low gain in the transition region. Since this condition occurs often in MESFET logic circuits, the former definitions of the maximum width and maximum square will be used in this text.

Example 3.1. Compare the intrinsic, maximum width, maximum square, and slope $= -1$ noise margins of the inverter whose transfer characteristic is shown in Figs. 3-3 to 3-9. Assume that each axis ranges from 0 to 2 V.

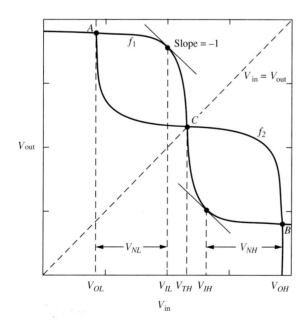

FIGURE 3-9
The slope $= -1$ definition for noise margin.

Solution. The various noise margins can be determined using a ruler to measure the distance from A or B to the appropriate point or the width of the square. The resulting noise margins are given below. Check these against the figure to verify their definition.

$$
\begin{array}{ll}
V_{INL} & 0.71 \text{ V} \\
V_{INH} & 0.79 \text{ V} \\
V_{NL} & 0.64 \text{ V} \\
V_{NH} & 0.69 \text{ V} \\
V_{MS} & 0.45 \text{ V} \\
V_{IL} - V_{OL} & 0.61 \text{ V} \\
V_{OH} - V_{IH} & 0.71 \text{ V}
\end{array}
$$

3.1.3 Factors Which Influence the Noise Margin

Consider an inverter with fixed-logic voltage swing. The noise margins of the inverter will depend on several factors, some of which are accessible to the designer. Figure 3-10 illustrates through transfer characteristics that the NM will be sensitive to (1) voltage gain, and (2) symmetry.

First consider the voltage gain as shown in Fig. 3-10a, where two inverters are compared, one with high voltage gain in the transition region and the other with lower voltage gain. Clearly, the higher gain leads to increased NM since the loops will be wider. The transition region of highest gain occurs when both the switch transistor (pull-down, Q_1) and the load transistor (pull-up, Q_2) are in their saturation regions because the gain is determined by

$$A_V = -g_{m1}(r_{ds1} \| r_{ds2}) \tag{3.1}$$

The incremental output resistance of an FET, r_{ds}, is highest in saturation, as is the incremental transconductance, g_m. Unfortunately, little can be done at the circuit design stage short of adding more devices (cascode, for example) to modify either g_m or r_{ds}. Both quantities scale with channel width so as to keep their product constant.[3] As gate length is increased, g_m will decrease while r_{ds} increases, reaching a maximum product for gate lengths of the order of 3 to 5 μm. Since the f_t of the transistor drops directly as gate length increases, however, increasing gate length is seldom employed for the purpose of increasing voltage gain in digital circuits.

The symmetry of the transfer characteristic will also affect the noise margin, as is shown in Fig. 3-10b. As will be described in the next section, the relative position of the logic gate threshold voltage V_{TH} with respect to V_{OL} and V_{OH} is influenced by the choice of threshold voltage V_T, the ratio of widths of the switch transistor, and the load transistor (width ratio) and also by the level shift network. These can be adjusted during processing or during circuit design to obtain equal low and high noise margins.

---

[3] While r_{ds2} could be increased by reducing the width of the pull-up relative to the pull-down, the dc bias point of Q_1 would be shifted to lower V_{GS} which reduces g_{m1}. The voltage gain will be maximized for some ratio W_1/W_2.

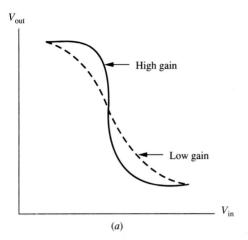

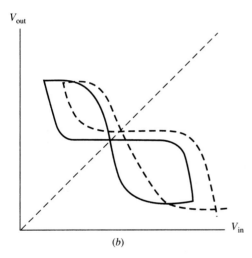

FIGURE 3-10
Transfer characteristics illustrate the influence of (a) voltage gain and (b) symmetry on the noise margins.

3.2 DC DESIGN OF GaAs MESFET/HFET LOGIC GATES

There are several methods that can be used to find V_{OH}, V_{OL}, and the NMs for GaAs MESFET or HFET inverters. While the most fundamental method would be to use the device equations [such as Eqs. (2.6) in Chap. 2] to obtain an analytical solution for the transfer characteristic, the equations become difficult to solve when the source resistance R_S is included. Neglecting the effect of R_S will lead to significant errors in the calculation when applied to GaAs FETs. Therefore, the analytical method is not very useful and will not be presented.

Equivalent circuit methods have been used with some success in which the active devices are modelled as current sources, switches, resistors, and capacitors [5–7]. While these can be useful in gaining insight into the transient response of FET logic circuits and will be employed in Sec. 3.3, their usefulness for dc simulations is limited.

Two other methods will be presented that are sufficiently accurate for dc simulation. Firstly, a graphical (load line) method is described which takes into account the actual device characteristics including R_S. Only the effects of backgating (discussed in Chaps. 1 and 2) are neglected, which is valid except where low-temperature operation is required. The graphical method, while less convenient, provides considerable insight into the relative importance of device parameters, level shifting, and device widths which is lacking in the analytical or computer models. The method then is a necessary supplement to the dc SPICE model which will be used to determine more quantitatively the sensitivity of NM to several device and layout parameters for a depletion mode MESFET circuit example. Both NOR and NAND implementations will be discussed. Finally, the static and dynamic power dissipation will be calculated.

3.2.1 Graphical DC Design Example

The graphical analysis method makes use of actual measured device data, as is shown in Fig. 3-11, which is from a 1×20 μm MESFET. The data shown is a plot of the drain current of a switch or driver FET, I_D, as a function of V_{DS} with V_{GS} as a parameter. From the data it can be seen that this is a depletion mode FET with a threshold voltage of about -0.9 V. This device might correspond to Q_S in the unbuffered FET logic (FL) inverter [8] shown in Fig.3-12. This circuit is selected as being representative of a typical depletion mode logic circuit, and it will be used throughout this chapter as a generic circuit to illustrate design concepts and the use of the analysis and design techniques to be presented. In Fig. 3-12, Q_L and Q_{PD} are connected with $V_{GS} = 0$ as depletion mode active-load devices. Q_L is used to

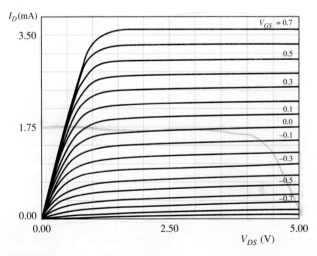

FIGURE 3-11
I_D-V_{DS} characteristic for the switch FET, Q_s in Figure 3-12. The data was measured on a 1×20 μm GaAs MESFET.

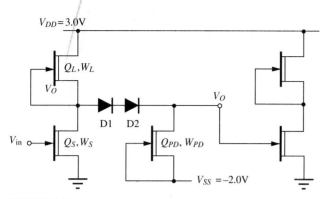

FIGURE 3-12
Unbuffered FET logic inverter (FL) using depletion mode MESFETs.

charge the load capacitance when the input is at V_{OL} and Q_S is in cutoff, while Q_{PD} is used to discharge the load capacitance when the input is at V_{OH}, Q_S is in the linear region, and the diodes may be momentarily out of conduction. Note that current flows at all other times through the forward-biased level-shift diodes from Q_L to Q_{PD}. When the output $V_O = V_{OH}$, then current also flows into the forward-biased gate-source diode of the next stage. This diode clamping at the input will limit V_{OH} to approximately 0.7 V regardless of the width ratios employed by the driving stage. The channel widths of the three MESFETs are W_S, W_L, and W_{PD} respectively.

Example 3.2. Unbuffered FET logic. Construct a transfer characteristic for the FL inverter circuit of Fig. 3-12. Assume that $W_L = W_S = 2W_{PD}$. Determine V_{OL}, V_{OH}, V_{NL}, V_{NH}, V_{MS}. How might W_L, W_S, and W_{PD} be changed to improve noise margin?

Solution. The graphical solution proceeds by assuming that the switch FET has the same width as the device in Fig. 3-11. The output voltage V_O' will be represented by V_{DS} of Fig. 3-11. A static load line can be superimposed on these characteristics as shown in Fig. 3-13 by the heavy solid line beginning at $V_{DS} = V_{DD}$. The load line is the device characteristic for Q_L, which always has $V_{GS} = 0$. If $W_S = W_L$, then the load line can be traced from the $V_{GS} = 0$ curve of Fig. 3-11 on a transparent sheet, flipped upside-down and placed on top of Fig. 3-11. If the widths differ, then the $V_{GS} = 0$ curve must be scaled up or down by the appropriate ratio on a separate plot, then traced and overlayed as shown on Fig. 3-13. The locus of points on the load line in Fig. 3-13 would represent the dc solution for V_O' if there were no current flowing into Q_{PD} or into the next stage.

To account for the extra current required by the level-shift network, the load line must be shifted down by I_{PD} as shown by the load line in Fig. 3-14. This construction assumes that the supply voltage V_{SS} is sufficiently negative that the pull-down device remains in saturation for all values of V_O and therefore provides a current, I_{PD}, which is approximately constant. Thus V_{SS} must be at least $V_{D,\text{sat}}$ ($V_{D,\text{sat}}$ is the V_{DS} voltage at the boundary of the saturation region) more negative than the logic low voltage, V_{OL}. For a short gate length MESFET, $V_{D,\text{sat}}$ is less than $V_{GS} - V_T$ ($= V_T$ for the active-load connection) because of velocity-saturation effects. In this example, V_{OL} will be about -1.3 V and $V_T = -0.9$ V, so a V_{SS} supply voltage of -2.0 V should be sufficient to guarantee saturation.

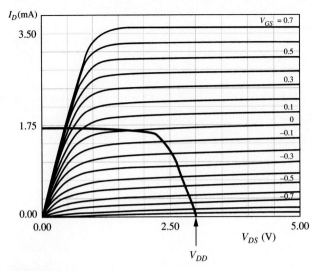

FIGURE 3-13
Device characteristics of Fig. 3-11 with the *I-V* characteristic of Q_L (the load line) superimposed for the case in which $I_{PD} = 0$.

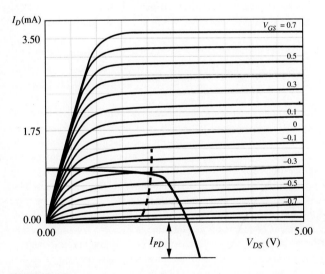

FIGURE 3-14
Device characteristics of Fig. 3-11 with the load line shifted down to account for I_{PD}. Diode characteristic (dashed line) is due to the level-shift diodes and the gate-source diode of the next stage.

Now, the true value of V_O' can be found by the points on the vertically shifted load line that intersect the characteristic curves of Q_S. Note that V_{in} of Fig. 3-12 is equivalent to V_{GS} on Fig. 3-14, and V_O' can now be plotted on a V_O vs. V_{in} graph, as shown in Fig. 3-15. One further adjustment is still required: V_O must be obtained from V_O' by accounting for the level shift. Two forward-biased diodes will provide approximately 1.4 V of level shift. This can be achieved on Fig. 3-15 by shifting the plot of V_O' down by 1.4 V as shown. The dashed line (diode characteristic) on Fig. 3-14 represents the forward conduction limit imposed by the gate source junction of the next stage and the level-shift diodes. This will clamp the logic high level to $V_O' = 2.1$ V maximum or $V_O = 0.70$ V. Thus until the load line is to the left of the diode curve, $V_{DS} = V_O'$ is fixed at 2.1 V. The final result is the complete dc inverter transfer characteristic.

The logic levels and noise margins can be determined from the transfer characteristic in Fig. 3-15 by reflecting it about the $V_{in} = V_{out}$ (45°) line as shown where $V_{out} = V_O$. These quantities are: $V_{OH} = 0.70$ V (limited by forward conduction of the next stage), $V_{OL} = -1.27$ V, $V_{NL} = 0.82$ V, $V_{NH} = 0.89$ V, and $V_{MS} = 0.63$ V. While the low and high NM are nearly equal, V_{NL} could be increased and V_{NH} decreased by a slight increase in W_L/W_S.

Note that the approach in this example provides a considerable amount of insight into the expected influence of changes in design or device parameters. For example, a change in the width ratio W_S/W_L will shift the inverter transfer characteristic to the left or right depending on whether the ratio is increased or decreased respectively. If W_S is increased, less V_{in} is required to obtain the same current I_D; therefore, the characteristic shifts to the left. This effect is illustrated in Fig. 3-16 by the result of a

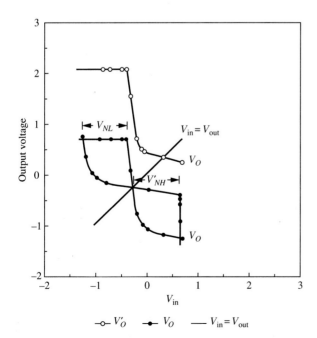

FIGURE 3-15
Transfer characteristic constructed from Fig. 3-14 for the FET logic inverter. The two-diode level-shifting of 1.4 V is accomplished by translating the characteristic down. The shifted characteristic is reflected about the 45° line to determine logic levels and noise margin.

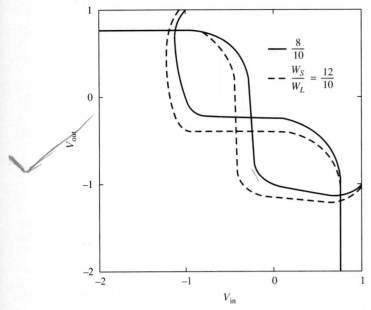

FIGURE 3-16
Transfer characteristic for FL inverters with width ratios W_S/W_L of 12/10 and 8/10.

SPICE simulation in which W_S/W_L was changed from 8/10 to 12/10 while all other parameters and widths were held constant. The V_{TH}, V_{NL}, and V_{NH} for these two cases were determined from these plots and are summarized in Table 3.1.

Changes in the transfer characteristic caused by varying the FET threshold voltage V_T can also be easily understood by graphical analysis. If V_T were made more positive, then the input voltage V_{in} must become correspondingly more positive to begin to activate the switch FET. This will therefore shift the transfer curve toward the right, as shown in Fig. 3-17. This figure shows a SPICE simulation of an FL inverter in which all parameters were held constant except for V_T. The corresponding changes in V_{TH}, V_{NL}, and V_{NH} are given in Table 3.2.

TABLE 3.1
Changes in inverter threshold voltage and the noise margins induced by varying the width ratio W_S/W_L. The FET threshold voltage was held constant at -1.0 V

W_S/W_L	V_{TH} (V)	V_{NL} (V)	V_{NH} (V)
8/10	−0.26	0.78	0.94
12/10	−0.44	0.74	1.15
20/20	−0.3	0.82	0.89

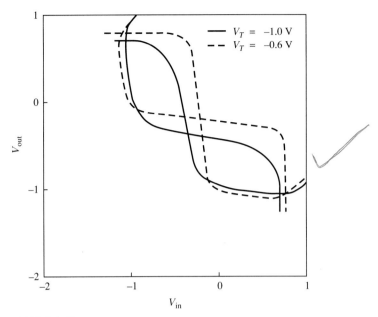

FIGURE 3-17
Transfer characteristic are shown for FL inverters with FET threshold voltages of -1.0 V and -0.6 V.

The number and the width of level-shift diodes will also influence the transfer characteristic in an easily visualized way. A change from two diodes to one diode will reduce the level shift to approximately 0.7 V. The logic low level will be increased by a one-diode drop while logic high level will be unaffected. The threshold voltage of the inverter should not be directly affected. Figure 3-18 presents an example of a SPICE-simulated FL inverter with one- and two-diode drops. The FET threshold voltage was held constant at -0.6 V in both cases. Note that the major change is observed in the logic low level, V_{OL}. Noise margins are also affected, because the width of the loops in the transfer characteristics depends on the logic levels. The observed changes are summarized in Table 3.3.

TABLE 3.2
Changes in inverter threshold voltage and the noise margins induced by varying the FET threshold voltage. The other device parameters were unchanged

V_T(V)	V_{TH}(V)	V_{NL}(V)	V_{NH}(V)
-1.0	-0.35	0.58	0.91
-0.6	-0.21	0.82	0.88

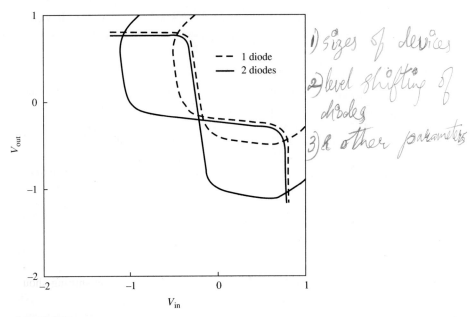

FIGURE 3-18
A SPICE-simulated FL inverter with one- and two-diode drops. The FET threshold voltage was held constant at −0.6 V in both cases.

Fine changes in the logic low level can be made in order to optimize noise margins without affecting the inverter threshold by varying the width of the level-shift diodes. Since the current through the diodes is unaffected by their width, the current density and therefore the voltage drop will be shifted slightly. Care should be taken to avoid diode widths which would result in operating the diodes at very high current densities ($J > 10^4$ A/cm^2). At high current densities, the level-shift diode voltage drop will be very sensitive to the series resistance of the diode, a parameter that can have a larger standard deviation than the forward voltage drop itself when measured at lower current densities.

The large difference between V_{NL} and V_{NH} shown in Table 3.3 for the one-diode level shift indicates that V_{TH} should be changed in the positive direction. This

TABLE 3.3
Changes in inverter threshold voltage, noise margins, and logic low level in an FL inverter induced by reducing the level shift from two to one diode. The FET threshold voltage was held constant at −0.6 V in both cases.

Diodes	V_{TH} (V)	V_{NL} (V)	V_{NH} (V)	V_{OL} (V)
2	−0.21	0.81	0.93	−1.1
1	−0.19	0.23	0.86	−0.43

could be accomplished by adjusting the width ratio (reduce W_S). Ideally, if $V_{NL} = V_{NH}$, the inverter will have the greatest latitude to process-induced variations.

3.2.2 Optimization of DC Design

Circuit simulation tools such as SPICE with appropriate and verified device models and when supplemented by graphical methods can be effective in optimization of the static design. Since our design priority emphasizes stability over speed, the noise margins should be optimized. In this regard, either the loop width definition or the maximum square definition should be useful since the largest square will be obtained in the case when loop widths are equal. In any case, equal low and high NMs, if possible, will provide the best circuit stability and should be the design goal.

In this section, the FL logic gate shown in Fig. 3-12 will be used again as an example; however, the procedure described is generally useful for any circuit topology selected. If the designer is given the option of selecting the FET threshold voltage(s), then this as well as the width ratio and the number and size of level-shift diodes must be optimized. The procedure for optimizing the noise margin consists of evaluating the NM dependence on V_T. Versions with different amounts of level-shifting should be evaluated as well. Then, for the level-shift configuration that provides the best NM in the V_T range of interest, W_S/W_L should be evaluated. The goal of the optimization will be to obtain the largest V_{MS} or $V_{NL} = V_{NH}$ at the mean operating temperature.

Figure 3-19 illustrates the result of a series of SPICE simulations on an FL inverter in which the width ratio was maintained constant while V_T was varied.

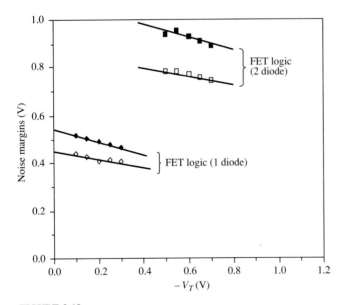

FIGURE 3-19
Noise margin of FL inverter as a function of FET threshold voltage, V_T, for one-diode and two-diode level shifts. All other parameters were held constant. The solid points correspond to the maximum width NM, the open points are for maximum square NM.

Versions of the circuit with one and two level-shift diodes were evaluated. If the FET threshold voltage were more negative than -1 V, a three-diode version should be considered. An example of the SPICE file used to generate the transfer characteristic is shown in Fig. 3-20. Note that the MESFET and diode model information must be determined for each specific IC fabrication process at the appropriate device temperatures; the parameter values shown are by no means to be considered as universal! The transfer curve is reflected around the $V_{in} = V_{out}$ (45°) line and plotted again on the same axes (such as was shown in Fig. 3-16) in order to find noise margins. Both maximum width (solid points) and maximum square (open points) are shown in Fig. 3-19. Then, to maximize NM, the NM as a function of width ratio, W_S/W_L, must be evaluated over a range of V_T. Here, the proper ratio can be more easily found when the transfer characteristics are examined and the width ratio is changed in the correct direction. When the largest, nearly equal, V_{NL} and V_{NH} are identified, then they can be made equal by fine adjustment of the diode width.

Note that W_{PD} must also be optimized. It is determined on the basis of transient performance (to be discussed in Sec. 3.3) instead of static considerations. However, when the optimum ratio W_L/W_{PD} is determined (typically $= 2$ for approximately equal rise/fall times), width W_L can be adjusted to compensate for the change in current I_{PD} to maintain optimum static design.

```
********** SPICE3 MESFET model -- parameters from Table 2.4 **********
.model mesfet nmf(vto=-1 beta=1.1e-4 lambda=.05 alpha=2.5 rs=900 rd=900
+ cgs=1.2fF cgd=0.23fF pb=0.8 is=5e-16)
********** level shift diode model -- parameters from Table 2.1 ***************
.model shdiode d(is=1e-14 rs=1500 cjo=2.0fF vj=0.8 m=.5)
*********** circuit node list *****************
x1 12 4 10 11 inverter
x2 4 20 10 11 inverter
x3 4 21 10 11 inverter
vdd 10 0 2.5
vss 11 0 -2
vin 12 0 dc
************ control statements ****************
.dc vin -2 1 0.05
.nodeset v(4) 0.5
.print dc v(4)
************ subcircuit for NOR gate ***********
.subckt inverter 1 4 10 11
*               in out vdd vss
zsw1 3 1 0 mesfet 10
zpu1 10 3 3 mesfet 12
zpd1 4 11 11 mesfet 6
ds1 3 5 shdiode 6
ds2 5 4 shdiode 6
.ends inverter
.end
```

FIGURE 3-20
SPICE input file for simulation of FL inverter with fanout $= 1$. Note that the model parameters are typical values from Tables 2.1 and 2.4.

It is perhaps more typical that the FET threshold voltage(s) are predetermined by the fabrication facility. In this case, the NM maximization is more constrained and consists of optimization of the width ratios and diode widths.

3.2.3 Design of NOR and NAND Gates

The FL inverter can be modified to perform more complex logic functions. The NOR function can be easily accomplished by placing additional switch FETs in parallel at the input, as shown in Fig. 3-21. Then, if any one FET is in its linear region ($V_{in} = V_{OH}$), the output will be pulled down to the logic low level. If one input is driven while the other inputs are held at V_{SS}, then the transfer characteristic will be identical to that of an inverter if the device widths are the same. For the NOR function, this transfer characteristic will represent the case in which all inputs except for one are low and the last input switches from high to low. However, for inputs changing in the opposite direction, one must allow for the possibility that more than one input will change from low to high at the same time. This case will lead to a new transfer curve shifted to the left as shown in Fig. 3-22a, because the W_S/W_L increases as more inputs are driven in parallel. While the multiple-input case leads to a different set of noise margins than the inverter or NOR with a single input active, it is sufficient that the magnitude of the small-signal voltage gain of the NOR gate with all inputs tied together (slope of V_{out} vs. V_{in} be less than unity (Fig. 3-22b) in the stable logic states at V_{OH} and V_{OL}) so that any common-mode noise voltage on all of the inputs will be attenuated rather than amplified. If this condition is satisfied, then the NOR gate output will be statically stable at V_{OL} when all inputs are high and at V_{OH} when all inputs are low.

Subthreshold current of the switch FETs may also influence the logic high output level for very wide NOR functions. If V_{OL} is not adequately low (at least 0.1 or 0.2 V below V_T) to suppress subthreshold currents, then leakage around the devices may shunt enough load current so that V_{OH} is reduced. This will reduce the low noise margin V_{NL}. Unfortunately, most MESFET and HFET SPICE models do not yet provide a subthreshold region. Therefore, this effect must be evaluated from

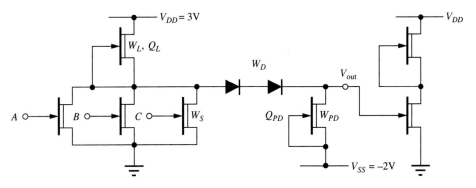

FIGURE 3-21
Schematic diagram of unbuffered FET logic (FL) NOR gate.

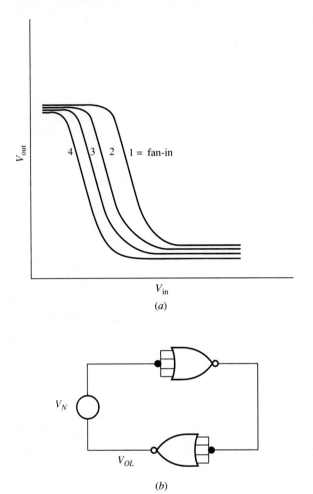

FIGURE 3-22
(a) Transfer characteristic of NOR gate with increasing fan-in. (b) V_N represents a common-mode noise voltage on all of the four inputs.

measured device data and simulated graphically or by the addition of a resistor or current source connecting the drain to the source of each switch FET.

Finally, very wide NOR gates implemented with many parallel FETs (high fan-in) will have slower rise and fall times due to the additional parasitic capacitance on the drain node.

A NAND function can also be generated by placing the switch transistors in series as shown in Fig. 3-23. With this connection, the output will be at logic low only when both inputs are at logic high. The logic low output level obtained from the NAND would not be as low as for the NOR gate, since we now have two devices in series. The total channel resistance and source and drain resistance will be larger than for the single-transistor NOR case. The series transistors can be implemented by constructing a dual-gated FET in which two Schottky metal gate stripes are placed across the same channel implant. Of course, the source-to-drain spacing must be

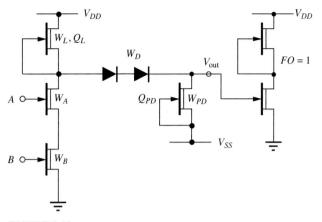

FIGURE 3-23
Schematic diagram of FL two-input NAND gate.

increased to accommodate the modification. (Information on the layout details will be discussed in a later chapter.)

Because we are dealing with a *ratioed* logic[4] with static current flow through the devices, V_{GS} will be different for each input FET due to the voltage drop along the channel. The source voltage for the upper FET, Q_A, is higher than that for the lower FET, Q_B, so

$$V_{GSA} = V_A - V_{DSB} - I_D R_{SA} \qquad (3.2)$$

whereas

$$V_{GSB} = V_B - I_D R_{SB} \qquad (3.3)$$

Therefore, for the same current I_D through both FETs, $V_{GSB} > V_{GSA}$ and $V_{DSB} < V_{DSA}$. As shown in Fig 3-24, V_{DSB} is less than V_{DSA} by quantity ΔV if $W_A = W_B$.

The low and high noise margins will also be different for each input and should be maximized, if possible, by adjustment of the width ratios. The largest NMs should occur when $V_{NL} = V_{NH}$. However, it will not be possible to accomplish this for both inputs simultaneously. Starting from the case where the widths of both FETs are equal, simulation of the transfer characteristic shows that $V_{NLA} > V_{NHA}$ and $V_{NLB} \approx V_{NHB}$. As is the case for the inverter, widening Q_A will cause the current I_D to be reached at a smaller V_{GSA}; thus Q_A can be widened until $V_{NLA} = V_{NHA}$. While Q_A is being widened, ΔV is also decreasing, which reduces V_{OL} and increases V_{NLB}. Looking at Q_B, we now find that $V_{NLB} > V_{NHB}$. Thus, there has been a substantial improvement in V_{NHA} due to the shift in input threshold voltage without adversely affecting V_{NLB} or V_{NHB}. Finally, widening the upper FET will reduce the

---

[4] Logic voltages depend on device widths.

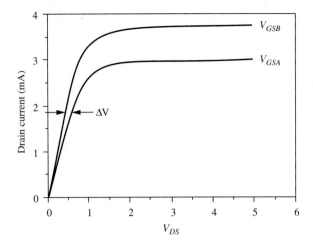

FIGURE 3-24
I-V characteristics of Q_A and Q_B when input *A* and *B* are at V_{OH}. Note the shift in V_{DS} caused by the reduced V_{GS} of Q_A.

offset in the input threshold voltages (the point where $V_{in} = V_{out}$) of the two inputs because Q_A will reach current I_D at a more negative V_{GS} than Q_B. This effect can be significant for logic circuits with a low voltage swing. A shift in the input threshold voltage of 0.1 or 0.2 V could drastically reduce the electrical yield of certain circuits, as described in Sec. 3.2.5.

Because of the shifts in input threshold voltage and in logic low voltage level caused by the addition of a series transistor, the maximum number of inputs that can be safely accommodated for a NAND function in static GaAs logic is two. Three or more inputs will lead to excessive degradation in noise margin which cannot be tolerated if high-yield circuits with large numbers of gates are to be expected. Certain dynamic circuits (Sec. 3.4) can, however, tolerate greater fan-in if no static current flows through the chain of series FETs.

3.2.4 Power Dissipation

The power dissipated by any logic circuit is also an important factor to consider in the overall design optimization. Most of the logic circuit topologies of interest for GaAs FET digital logic are static, in which current is flowing from supply to ground or supply to supply during at least one logic state. However, if the load capacitance is large and the switching frequency high, dynamic power dissipated while charging and discharging the capacitor may also be significant.

Again, the FL inverter will be used as an example for estimation of the power dissipation of a logic circuit. Refer to the schematic drawing shown in Fig. 3-25. Note that the current from the V_{DD} and V_{SS} power supplies is provided through the active load pull-up Q_L and the pull-down Q_{PD}. If it is assumed that the supply voltages are sufficiently large to keep both FETs in their saturation regions, then the input currents I_{DD} and I_{SS} should be roughly constant. The path of the current flow will depend on the logic state, but the total current at each supply will remain to the first order constant. Then, the power can be estimated if the currents are known.

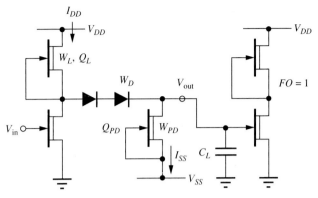

FIGURE 3-25
Schematic diagram of FL inverter.

For 1 μm gate length depletion mode devices with -1 V threshold, I_{DSS} is approximately equal to 80 μA/μm of channel width where I_{DSS} is defined as I_D at $V_{DS} = V_{D,\text{sat}}$ with $V_{GS} = 0$ and $W = 1$ μm. The influence of λ is neglected. Thus,

$$P_{D,\text{static}} = I_{DSS}(V_{DD}W_L + V_{SS}W_{PD}) \qquad (3.4)$$

For the example of Fig. 3-25, if $W_L = 12$ μm, $W_{PD} = 6$ μm, $V_{DD} = 3$ V, and $V_{SS} = -2$ V, the total static power dissipation would be 3.8 mW. I_D per unit width with $V_{GS} = 0$ should scale roughly as V_T^2.

In a more general case, the current will vary as a function of time due to changes in the bias voltages on the devices. If the logic circuit were driven with some periodic waveform, and $I_{DD}(t)$ and $I_{SS}(t)$ were known, the average power dissipation could be calculated by integrating the current over one period as shown below:

$$P_D = \frac{V_{DD}}{T}\int_0^T I_{DD}(t)dt + \frac{V_{SS}}{T}\int_0^T I_{SS}(t)dt \qquad (3.5)$$

The dynamic power dissipation is a consequence of the energy required to charge and discharge the circuit capacitances. In the case of the FL gate, the largest capacitance would most often be the load capacitance C_L, shown in Fig. 3-25. The energy required to charge or discharge this capacitor by the logic voltage ΔV is $C_L \Delta V^2/2$. If the gate is driven by a periodic input signal with frequency f, the capacitor will be charged and discharged once every cycle. Therefore, the dynamic power dissipation is given by

$$P_D(\text{dynamic}) = fC_L \Delta V^2 \qquad (3.6)$$

For example, for the FL gate with $V_T = -1$ V, the logic swing will be approximately 1.7 V. If the load capacitance were 100 fF (high for an unbuffered gate) and $f = 1$ GHz, the dynamic power dissipation would be 0.29 mW, less than one-tenth of the static power. However, there can be instances where long lines are driven (clock lines, for example) where the dynamic power may be of significance. These would generally be a small number of cases on a chip.

The total power dissipated in the logic circuit will be the sum of the static and dynamic power.

3.2.5 Yield of LSI/VLSI Circuits

Uniformity of the logic gate input threshold voltage is a necessary requirement for electrically functional ICs. Process and material nonuniformity cause nonuniform diode and transistor characteristics which, in turn, cause logic gate threshold variations. These variations influence the electrical yield of ICs in VLSI. This electrical yield requirement must be satisfied in addition to the processing yield.

The level of uniformity required of the device characteristics is determined by the circuit configuration (topology) and the design choices (width ratios, V_T, etc.). The circuit configuration fixes the noise margin, the tolerance window available for the high and low logic levels. It is the ratio of threshold variation to noise margin that is critical for determining the IC electrical yield. For example, silicon CMOS circuits have quite large noise margins due to the high power supply voltage (5 V), so uniformity of device parameters seldom limits the functional yield. Speed, however, is strongly influenced by certain device parameters. Silicon bipolar circuits are inherently uniform due to the bandgap-derived input threshold voltage, so noise margin seldom limits yield in this case either.

Material and process variations limit the logic gate threshold uniformity in both GaAs FET and silicon MOS integrated circuits. However, when combined with the small noise margins in GaAs FET ICs due to low supply voltages and forward gate conduction clamping in MESFETs or HFETs, the GaAs circuits are more sensitive to parameter variation, and poor yield of completely functional ICs will occur when the number of gates becomes sufficiently large. In fact, in this case, parametric nonuniformity may constrain VLSI yield before the power dissipation (thermal) or process limitations (defects) take effect. Therefore, the application of statistical models to predict electrical yield is appropriate and should provide guidance on circuit design choices and the required level of device parametric uniformity for a specific number of gates.

Electrical yield can be predicted from statistical models of device uniformity. Several approaches [9,10] have been reported, which have been applied to GaAs and Si. Device parameters are assumed to vary randomly with a normal distribution. The method used to predict electrical yield in this chapter is similar to the used to evaluate CMOS inverters in Ref. 10. First, the circuit under consideration is analyzed to determine the noise margin. Model parameters are measured from actual e-MESFET and d-MESFET devices typical of those which would be encountered in a GaAs foundry process, and SPICE is used to simulate the logic gate transfer characteristic. The circuit design is optimized by adjusting width ratios of the MESFETs to maximize low and high noise margins. This optimized noise margin becomes the selected design target, M^*. Next, the circuit is analyzed for the sensitivities of noise margin to changes in device parameters. Each critical device model parameter, x_i, is varied around its target value, in turn, and the sensitivity of the noise margin to each change is determined, again through simulation. The model parameters to be considered would include: V_T (the FET threshold voltage), β (the k factor for the FET), λ or g_{ds} (the saturation region slope parameter), and R_S (the parasitic source resistance). For many circuits, the most sensitive parameter is found to be V_T, but in cases with very low target noise margin, M^*, the other sensitivities may also be significant. The parameters are presumed to vary independently of each other. Therefore, the mean square variance (standard deviation) of the noise margin, M_σ, can be written as

$$M_\sigma^2 = \sum_{i=1}^{n} (R_i \, \sigma_i)^2 \tag{3.7}$$

where R_i represents the sensitivity of the NM (or slope $\partial M^*/\partial x_i$) to each device parameter x_i and σ_i is the associated variance.

Finally, the available noise margin, ΔM, is computed:

$$\Delta M = M^* - M_{min} \tag{3.8}$$

where M_{min} is the minimum acceptable noise margin, arbitrarily assumed to be 0.1 V, which provides a design safety factor for other "noise" sources. The additional sources include on-chip noise generation, off-chip power supply and ground noise, and crosstalk which will exist in any real digital application and will also reduce the available noise margins.

The ratio of ΔM to the total variance, M_σ, defined as X_1, can be related to the electrical yield by the following probability distribution function [10]:

$$P(-X_1 < X < X_1) = \frac{1}{\sqrt{2\pi}} \int_{-X_1}^{+X_1} \exp\left(-\frac{X^2}{2}\right) dX \tag{3.9}$$

$$X = \frac{M}{M_\sigma} \tag{3.9a}$$

$$X_1 = \frac{\Delta M}{M_\sigma} \tag{3.9b}$$

This equation describes the probability of finding a single inverter within the window of functionality, $M^* \pm \Delta M$, if the mean NM of the inverters sampled, M_μ, is identical to the target NM, M^*. The noise margin M is the independent variable in (3.9).

If N inverters or gates are considered in one circuit, the probability is given by

$$P(N) = P(X)^N \tag{3.10}$$

which is plotted in Fig. 3-26. This figure describes the probability that an ensemble of N gates or inverters will all fall within the NM window centered at M^* with width $2\Delta M$. If $M_\mu = M^*$, then the probability is a function of X_1 as shown in Fig. 3-26.

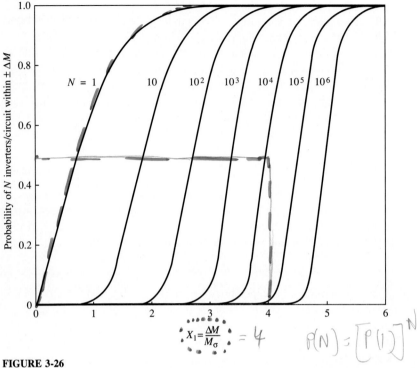

FIGURE 3-26
Probability that N inverters will have a noise margin in a window of width $2\Delta M$, centered at M^*. (*From Buehler and Griswold* [10]. © *Electrochemical Society. Used by permission.*)

If the mean NM of the inverter sample is not identical to M^*, then the probability that a particular inverter in the sample will function properly is further reduced. Equation (3.9) is modified as shown below:

$$P(X) = \frac{1}{\sqrt{2\pi}} \int_{X_2 - X_1}^{X_2 + X_1} \exp\left(-\frac{X^2}{2}\right) dX \tag{3.11}$$

$$X_2 = \frac{M^* - M_\mu}{M_\sigma} \tag{3.11a}$$

Figure 3-26 is still usable for finding the probability; the curves are shifted to the right by X_2, and the new, reduced, probability is shown by the curve corresponding to N at X_1.

In addition, allowance must be provided for the effect of temperature variations on the devices. This effect can be evaluated by extracting device model parameters at the extremes of the required operating temperature range. The new M_μ can then be calculated using SPICE with the respective model parameters, applying the statistical method described above, just as the method would be used for a parameter shift caused by process variation.

Example 3.3. Analysis of a DCFL inverter with fan-out = 2.
(a) Estimate the threshold voltage variance, σ_{VT}, required to obtain a yield of 50 percent for a 10,000-gate DCFL circuit. Assume that the e-mode and d-mode device threshold voltages track one another when varied, and that σ_β, σ_λ, and σ_{RS} are each 10 percent of their mean values.
(b) Estimate the yield if the mean threshold voltage shifts by 0.1 V. What value of σ_{VT} would be required to maintain the 50 percent yield?
(c) Repeat the analysis of part (a), assuming now that the e-mode and d-mode threshold voltages are independent.

Solution
(a) The circuit diagram of a direct-coupled FET logic (DCFL) inverter with a fan-out loading of 2 is shown in Fig. 3-27. This circuit, to be discussed in more detail in the next chapter, requires both enhancement mode and depletion mode GaAs FETs in its implementation. The e-mode devices serve as switches, while the d-mode devices provide active loads.

The analysis of this circuit proceeds by simulating the transfer characteristics using SPICE with a typical set of device parameters (representing the average for devices from the foundry). Width ratios can then be optimized for equal low and high noise margins. The transfer characteristic shown in Fig. 3-28 was simulated for this optimized circuit with MESFET threshold voltages $V_{TE} = 0.2$ V and $V_{TD} = -0.9$ V. The pronounced increase in V_{out} for $V_{in} > 0.7$ V is due to the voltage drop across the source resistor of the e-mode switch because of gate current.

Sensitivities R_i of the NM to model parameters V_{TE}, V_{TD}, β, λ, and R_S can be found by varying the parameters in small increments around their average values. Table 3.4 presents a summary of these sensitivities. Both the enhancement and depletion threshold voltages are assumed to vary together, as might happen in the case of thresholds controlled by ion implantation. Independent variation of V_{TE} and V_{TD} could also be easily simulated.

The results of the DCFL parametric yield analysis are presented in Table 3.5 for the maximum-square NM (MS) and the maximum-width NM (MW). The maximum-width noise margin, $M^* = 0.22$ V, was simulated for $V_{TE} = 0.2$ V and $V_{TD} = -0.9$ V. Assuming that $M_{min} = 0.1$ V and that $\Delta M/M_\sigma = 4$, so that 10,000

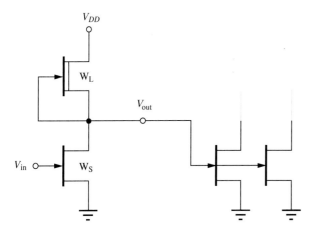

FIGURE 3-27
Schematic diagram of direct coupled FET logic (DCFL) gate with fan-out of 2.

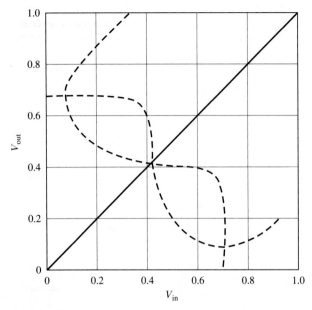

FIGURE 3-28
Simulated DCFL transfer characteristic. $W_S/W_L = 10$. $V_{TE} = -0.2$ V; $V_{TD} = -0.9$ V.

gate circuits are expected to have a 50 percent yield according to Fig. 3-26, $M_\sigma = 30$ mV is required. σ_{VT}, the threshold voltage variance, can therefore be calculated from M_σ (assuming that the variances of β, λ, and R_S were constant with typical values of 10 percent). The result, $\sigma_{VT} = 53$ mV, can be seen in Table 3.5. This result shows that a 10,000-gate DCFL chip is possible with values of σ_{VT} that have been obtained in practice [11].

The accuracy of the prediction is very dependent on the choice of $M_{\min}$. The proper value can be determined for a particular process if a large experimental data base of circuit results is available. Otherwise, the designer must estimate a value that would seem reasonable.

(b) Now consider another case in which the mean threshold voltage is off by only 0.1 V ($V_{TE} = 0.3$ V, $V_{TD} = -0.8$ V). Simulation of the inverter NM for this threshold voltage pair yields $M_\mu = 0.16$ V. For an M_σ of 30 mV, $X_2 = (0.22 - 0.16)/0.03 = 2$. Thus, the new probability of a functioning 10,000-gate circuit is seen to be zero on

TABLE 3.4
Sensitivities of the maximum-width (MW) and maximum-square (MS) noise margin to model parameters V_T (e- and d-mode threshold voltages shift together), β, λ, V_{TE}, V_{TD}, and R_S for a DCFL inverter

NM	R_{VT}	R_β (V/%)	R_λ (V/%)	$R_{V_{TE}}$	$R_{V_{TD}}$	R_{RS} (V/%)
MW	0.56	4.85e-4	1.6e-4	0.64	0.16	3.2e-4
MS	0.22	2.42e-4	0.8e-4	0.28	0.08	2.4e-4

TABLE 3.5
Projected number of functional gates (# gates) at the 50 percent yield level for DCFL inverters. The projections are based on the *maximum-width* NM and the *maximum-square* NM

NM	V_{TE} (V)	V_{TD} (V)	NM (V)	σ_{VT} (mV)	$\Delta M/M_\sigma$	# Gates
MW	0.2	−0.9	0.22	53	4	10K
MW	0.3	−0.8	0.16	53	2	10–100
MW	0.3	−0.8	0.16	25	4	10K
MS	0.2	−0.9	0.15	52	4	10K
MS	0.3	−0.8	0.12	52	1.7	10–100
MS	0.3	−0.8	0.12	23	4	10K

Fig. 3-26 when the curves are shifted to the right by X_2. In this condition, only tens to hundreds of gates/chip are expected to work properly.

To restore the former yield of 50 percent for 10,000 gates, the new $\Delta M = M_\mu - M_{\min} = 0.06$ V applies. $\Delta M/M_\sigma = 4$ as before; therefore, M_σ must be reduced from 30 to 15 mV. Applying the same sensitivity factors from Table 3.4, σ_{VT} of only 25 mV is allowed, as shown in Table 3.5. The data based on the maximum-square NM shows similar trends. It is interesting to note that essentially similar threshold voltage tolerances were obtained for both NM definitions. The smaller maximum-square NM had correspondingly smaller sensitivities, so that the σ_{VT} permitted came out almost equal for both NM definitions. This reinforces the prediction, making it independent of the NM definition.

(c) Table 3.6 summarizes threshold tolerances and permitted yields for the DCFL case, assuming that the threshold voltages for the enhancement mode switch and depletion mode load vary independently (as might happen in the case of threshold voltages controlled by the recessed gate process), the variances being assumed identical to those above. The sensitivities were recalculated to reflect the independent variation. The trends are the same as in the previous case, the only difference being the marginally lower threshold variance permitted in the latter case.

TABLE 3.6
Projected number of functional gates at the 50 percent yield level for DCFL, assuming the e- and d-mode thresholds change independently, but by the same amount

NM	V_{TE} (V)	V_{TD} (V)	NM (V)	σ_{VT} (mV)	$\Delta M/M_\sigma$	# Gates
MW	0.2	−0.9	0.22	45	4	10K
MW	0.3	−0.8	0.16	45	2	10–100
MW	0.3	−0.8	0.16	22	4	10K
MS	0.2	−0.9	0.15	39	4	10K
MS	0.3	−0.8	0.12	39	1.7	10–100
MS	0.3	−0.8	0.12	12	4	10K

The above example illustrates the point that wafer-to-wafer (or global) uniformity of the device threshold is as critical for DCFL VLSI as the uniformity obtained on wafer. Global uniformity is more difficult to control or predict due to the variability of substrate material and the consequent variability of implant activation. Consequently, fine-tuning of V_T by wet chemical etching is unacceptable because it destroys on-chip uniformity, and use of reactive Schottky-barrier materials to adjust V_T causes a built-in reliability problem [12].

3.3 TRANSIENT DESIGN OF MESFET/ HFET GaAs LOGIC CIRCUITS

Once the design of the logic circuit has been optimized for dc performance, it will have the best probability of functioning properly over a range of process parameters typically encountered during fabrication. However, no consideration has yet been given to evaluating the time-domain properties of the logic circuit. Since low propagation delay is the primary reason for using GaAs digital circuits, clearly this element of the design must also be carefully evaluated, but, as mentioned above, not at the expense of static functionality or potential circuit yield. As is true for any technology, the effort required in design increases in a nonlinear manner as the upper limits for speed are approached; sizing of transistors, buffering, and minimization of parasitic capacitances are all essential if the high-speed intrinsic performance of the GaAs FET is to be retained.

While it would be nice if circuits could be designed that contribute no delay, in reality, there are several contributors to the delay experienced in switching between logic states. First, and usually most important, are the parasitic wiring capacitances which will be discussed in Chap. 5. These capacitances are dominated by coupling between interconnections on the surface. This can take the form of parallel interconnect runs where there is coupling from line to line or by intersecting wires on different levels that couple in their overcrossings. It is relatively easy to estimate the effect of the wiring on the delay if capacitances can be determined, since the load capacitance due to wiring does not vary with voltage. Only in extremely simple, and usually not very useful, circuits such as minimum geometry ring oscillators can the wiring capacitances be reduced to very low levels, and the performance is then dominated by the intrinsic device capacitances.

Another source of load capacitance comes from fan-out. Each driven gate presents some input capacitance, due to active devices and parasitics, to the output of the driving gate. This capacitance generally varies with voltage and so the delay caused by this loading is more difficult to estimate. A method for determining an equivalent, fixed, large-signal capacitance, which accounts correctly for total charge, will be described in this section. This equivalent capacitance will facilitate the modeling of fan-out loading as well as the internal loading of the logic gate itself, simplifying it to the point where hand calculations can be used to estimate the step response.

A third effect comes from the drive capability of the device itself. As was described in Chap. 1, there is an intrinsic delay associated with the device itself, caused by the transit time of free electrons through the channel under the gate. This transit time τ is modeled by the intrinsic f_T; in fact, $\tau = (2\pi f_T)^{-1}$. τ can be reduced through improved device design, i.e., by employing materials or structures with higher electron drift velocities or by reducing the gate length. The HFETs appear to be an example of this approach, especially at reduced operating temperatures, with doubly confined channels (single quantum well), and when using channel materials such as GaInAs. Increasing this intrinsic f_T is only part of the solution, however. Equally as important is the ability to realize the high g_m of the intrinsic device in an actual circuit. Minimization of the parasitic source resistance R_S was shown (Chap. 1) to be the key to this problem. The higher the intrinsic g_m, the more important it is to reduce R_S. Finally, g_m can be increased by reducing the spacing between the gate electrode and the channel charge. While the intrinsic $f_T = g_m / C_G$ remains constant during this exercise because the gate capacitance C_G also increases proportionally, the influence of the loading capacitance on the circuit speed is reduced. This occurs because the loaded f_T, f'_T is controlled by both device and circuit capacitances, as shown below:

$$(2\pi f'_T)^{-1} = \frac{C_G}{g_m} + \frac{C_L}{g_m} \qquad (3.12)$$

In the subsections that follow, the influence of these three factors on circuit speed will be made evident and more quantitative. The unbuffered FET logic circuit [8] will once again be used as an example to illustrate techniques for analysis and design of high-speed GaAs logic gates. Optimization of the design for speed must naturally also consider power dissipation as speed and power trade off over a wide range. Loading effects of fan-in (number of inputs) and fan-out (number of outputs) are also important and must be considered with other design decisions. Finally, another noise margin definition will be introduced, dynamic noise margin, which will be used in estimating the response of circuits to fast noise spikes. While the selection of the circuit topology also has a profound effect on the speed and power tradeoffs, this subject will be reserved for Chap. 4.

3.3.1 Definitions

The time delay required for a signal to propagate through a logic gate or for an output to change in response to an input is expressed in relation to the input and output voltages for GaAs FET logic circuits. Figure 3-29 illustrates the standard definitions for propagation delay, rise time, and fall time relative to V_{OL} and V_{OH} [1]. Referring to Fig. 3-29, the propagation delay is defined as the time interval between the input 50 percent point and the output 50 percent point. In general, the high-to-low, t_{PHL}, and low-to-high, t_{PLH}, transitions will have different delays. Rise time t_r and fall time t_f are defined as the 10 to 90 percent transition times measured on the input waveform, whereas t_{LH} and t_{HL} are the equivalent on the output voltage. The cycle time t_{cyc} is the period of the clock frequency. These definitions would be convenient for use on measured or simulated voltages for a logic gate.

When simplified (hand) analysis is to be employed for quick estimates or for developing insight into the critical paths, the input waveform is more conveniently

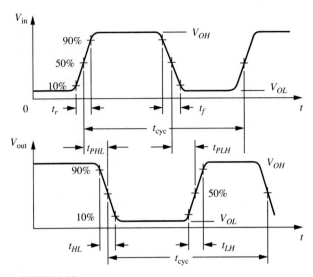

FIGURE 3-29
Standard definitions of propagation delay, rise and fall times. (*From Hodges and Jackson [13]. © McGraw-Hill. Used by permission.*)

represented as a step or pulse function. The delays predicted by this method will differ from those determined with a realistic input as shown above. However, as stated in Ref. 13, it is possible to estimate the response of the gate to an input transition with known rise or fall times by correcting the step response with a root mean square summing function of the form shown below:

$$t_{PHL}(\text{actual}) = \sqrt{[t_{PHL}(\text{step input})]^2 + [t_r/2]^2} \qquad (3.13)$$

3.3.2 Estimation of Propagation Delay

The analysis of the delay of a circuit under consideration can proceed on two levels. The most accurate approach, within the limitations of the device models, is to use a circuit simulation program, such as SPICE, which computes the voltage-varying currents and capacitances at discrete timesteps and solves for the node voltages. While circuit simulators provide accurate estimates or answers to the questions posed, they do not necessarily help the designer find the proper questions to ask. Since transient simulations require more computer time than dc analysis, even more time can be wasted answering the wrong questions than was the case for the dc analysis in Sec. 3.2. Therefore, a more direct hand analysis method is preferred when the objective is to develop insight into the technology or the particular circuit under consideration. Once the critical circuit elements are determined and the tradeoffs understood, the estimates obtained can then be refined by simulation methods.

The fundamental equation relating the current available to charge or discharge a capacitor to the rate of change of voltage with time (slew rate) is shown below:

$$I = C\frac{dV}{dt} \tag{3.14}$$

It is possible to use this equation to estimate the delay with hand analysis if a simplified model for the circuit and device is used. Figure 3-30a illustrates a simple inverter circuit, showing the fixed and voltage-variable capacitances associated with the devices and layout. The origin of these capacitors has been described in Chaps. 1 and 2. It can be seen that there are as many as five capacitors to be considered for each transistor, two variable and three fixed. In addition, the current of the transistors is not constant over the full logic swing. In order to keep I and C constant so that the above equation can be easily solved, some approximations are necessary.

Firstly, the device currects can be assumed to be constant if the change in output voltage is sufficiently small and if the respective devices are in the saturation region. Since t_{PHL} and t_{PLH} are defined over only 50 percent of the logic swing,

$$\Delta V = \frac{V_{OH} - V_{OL}}{2} \tag{3.15}$$

then this range of voltage, ΔV, will be used for the approximation. Figure 3-30b illustrates an approximate equivalent circuit for the low-to-high transition for which Q_S is in cutoff and Q_{PU} is in saturation. The saturated device is approximated as a

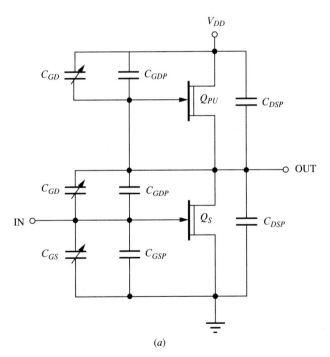

FIGURE 3-30
(a) Schematic diagram of simple GaAs FET inverter showing fixed (parasitic) and variable (depletion-layer) device capacitances.

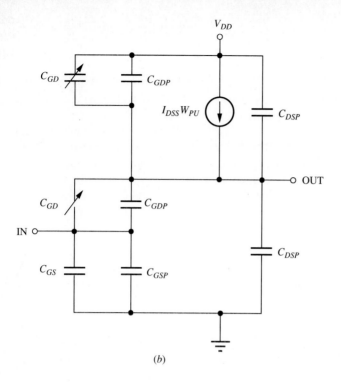

(b)

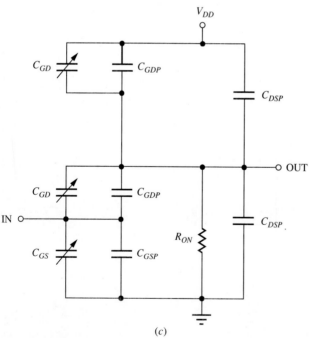

(c)

FIGURE 3-30 (*continued*)
(*b*) Simplified circuit for determination of t_{PLH} assuming that the pull-up FET is in saturation. (*c*) Simplified circuit for determination of t_{PHL} assuming that the switch FET is in the linear region.

current source, and the cutoff device is represented by an open circuit. The solution to Eq. (3.14) will be a linear ramp function in this approximation if the capacitance is constant.

On the other hand, if a device is in the linear (ohmic) region during most of the voltage swing, then an approximation as a resistor, with resistance of R_{ON}, may be more accurate. Figure 3-30c shows the equivalent circuit to be used in this instance. Now, the solution to Eq. (3.14) will be exponential.

In addition, the simplest voltage-variable, small-signal capacitance model for the gate-source and gate-drain depletion regions predicts that

$$C(V) = C_{j0}\left(1 - \frac{V}{V_{bi}}\right)^{-m} \tag{3.16}$$

for $V > V_T$ and $C(V) = 0$ for $V < V_T$. C_{j0} is the zero-bias capacitance and V_{bi} is the effective built-in potential. In order to simplify the evaluation of (3.14), it is helpful to linearize $C(V)$. A large-signal linear equivalent capacitance, C_{EQ}, can be defined by

$$C_{EQ} = \frac{\Delta Q}{\Delta V} = \frac{Q(V_2) - Q(V_1)}{V_2 - V_1} = K_{EQ}C_{j0} \tag{3.17}$$

where V_2 represents the gate-to-source or gate-to-drain voltage in the final state and V_1 the initial state. Q_2 and Q_1 are the corresponding total gate charges at each gate voltage as determined by integration of the capacitance equation (3.16). K_{EQ} is defined as a scaling parameter which relates the equivalent capacitance to the zero-bias capacitance C_{j0}. This definition allows the voltage-variable (nonlinear) capacitors to be replaced with equivalent large-signal linear capacitors, greatly simplifying the estimation of propagation delay.

The equivalent circuit for calculation of t_{PLH} can be redrawn as in Fig. 3-30d. In this figure, C_1 is the total equivalent node capacitance to ground on the drain node

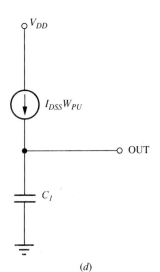

FIGURE 3-30 (continued)
(d) Equivalent large-signal capacitance and current source used to estimate t_{PLH}.

of the inverter calculated by assuming that both V_{DD} and the input are at ac ground potential:

$$C_1 = (K_{EQ1}C_{GD0} + C_{DSP} + K_M C_{GDP})W_S + (C_{GDP} + C_{DSP})W_{PU} \quad (3.18)$$

In this equation, C_{GD0} represents the zero-bias depletion capacitance of the gate/drain junction, normalized to a unit area (usually $1 \times 1\ \mu m$), and the C_{DSP} and C_{GDP} terms are the parasitic capacitances per unit width (analogous to the sidewall capacitance in MOS devices) associated with the respective device terminals (as defined in Chap. 2).[5] K_M is a scaling factor which is introduced to account for the increase in charge on C_{GDP} caused by the Miller effect.

To evaluate the equivalent capacitances, first the total charge must be determined by integrating the capacitance equation (3.16) over the proper voltage range. The result of the integration yields the equation below:

$$K_{EQ} = \frac{-V_{bi}}{(V_2 - V_1)(1-m)} \left[\left(1 - \frac{V_2}{V_{bi}}\right)^{1-m} - \left(1 - \frac{V_1}{V_{bi}}\right)^{1-m} \right] \quad (3.19)$$

If the input (GS voltage of Q_S) is assumed to change as a step function from V_{OH} to V_{OL} at $t = 0$, the GD voltage range to be considered for Eq. (3.19)(to calculate K_{EQ1}) is

$$V_2 = V_{OL} - \frac{V'_{OH} + V'_{OL}}{2} = V_{OL} - V'_{50\%}$$

and $$V_1 = V_{OH} - V'_{OL} \quad (3.20)$$

The symbols V_{OL} and V_{OH} refer, as was the case in Sec. 3.2.1, to the logic low and high at the level-shifted output of the gate, whereas V'_{OL} and V'_{OH} refer to these voltages at the drain node of the switch FET. $V'_{50\%}$ is defined above as the midpoint of the logic swing at the drain node. Recall that the propagation delay is defined with respect to the midpoint. To determine K_{EQ2}, the gate-to-drain voltage range on Q_{PU} must be used in evaluating Eq. (3.19). In this case,

$$V_2 = V_{DD} - V'_{50\%} \quad \text{and} \quad V_1 = V_{DD} - V'_{OL}$$

At this point, it is worth noting that the voltage difference $V_2 - V_1$ for the GD junction of the switch FET is over twice that of its GS junction. This is a consequence of the voltage gain of the inverter, and, as a result, the influence of GD capacitance is magnified (Miller capacitance). Since an equivalent capacitance between the output node and ground (C_1) is desired, the effective capacitance to ground of C_{GDP} must also be determined. It can be seen that $\Delta Q_{GDP} = C_{GDP}(V_2 - V_1)$. In order to account for this extra charge, an effective value can be determined by scaling C_{GDP} by the ratio of the actual voltage difference across the GD junction to the voltage difference across the GS junction of Q_S. This is the scaling factor K_M in Eq. (3.18).

---

[5] As discussed in Chap. 1, the depletion capacitances will generally vary as (doping)$^{1/2}$. The fringing capacitance due to the depletion-layer edge should scale at this rate as well. However, the geometric fringing capacitance will be independent of doping.

The total equivalent capacitance C_1 is now calculated from Eq. (3.18). Then, the propagation delay can be determined through Eq. (3.14) as shown:

$$t_{PLH} = C_1 \frac{V_{OH} - V_{OL}}{2I_{DSS} W_{PU}} \quad (3.21)$$

A similar procedure can also be followed to estimate the high-to-low propagation delay time, t_{PHL}. There two possibilities for the analysis of t_{PHL}, as mentioned above. Figure 3.31a illustrates the equivalent circuit for the case where Q_S is in saturation during the V_{OH} to $V_{50\%}$ transition and Q_{PU} is in the linear region for most of the transition. This case might apply if the positive power supply voltage were only marginally greater than V_{OH}. Then, Q_{PU} is represented by a resistor, R_{ON}.

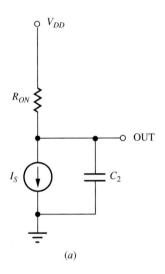

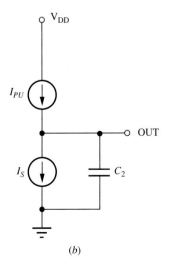

FIGURE 3-31
Equivalent large-signal capacitances are used to estimate t_{PHL}. (a) Pull-up FET is in linear region. (b) Pull-up FET is in saturation region.

178 GALLIUM ARSENIDE DIGITAL INTEGRATED CIRCUIT DESIGN

The other possibility is shown by Fig. 3-31b, in which Q_{PU} is in saturation and can therefore be approximated by a current source. C_2 is the total large-signal equivalent capacitance calculated with the approach indicated above.

In most cases of interest, the situation is slightly more complicated. Fan-out loading on the output of a logic gate is to be expected in all cases of interest. Therefore, an equivalent input capacitance for the loading gate(s) should also be included. Wiring capacitance is also often an important consideration. In fact, if wiring capacitance is significantly larger than the internal equivalent capacitances, a satisfactory estimate might be obtained without estimating these C_{EQ} values.

In addition, for depletion mode logic circuits, some level shifting is always required between input and output, so additional circuit elements must also be considered. An example will be presented to illustrate the technique.

Example 3.4. Estimation of the propagation delay of an unbuffered FET (FL) logic gate. Figure 3-32 illustrates a circuit diagram of an FL inverter (Q_1, Q_2 and Q_5) driving a fan-out of one consisting of the input to an identical inverter (Q_3 and Q_4) connected to the output. The device and loading capacitances of importance are shown in the diagram. Device channel widths are also indicated. The approach to be followed consists of developing equivalent circuits with constant, equivalent capacitances, current sources or resistors, and applying Eq. (3.14) in order to estimate t_{PHL} and t_{PLH}. For purposes of this example, let $V_{DD} = 3.0$ V, $V_{SS} = -2.0$ V, and $V_T = -1.0$ V.

(a) *Estimate t_{PHL}.* Assume that the input is a step function, which changes from V_{OL} to V_{OH} at $t = 0$. If it is also assumed that the shunt capacitance of the level-shift diodes is small compared with the sum of the load (wiring and capacitance to ground of the level-shift diodes) capacitance C_L and the input capacitance C_{in} of the fan-out load, then the diodes will be cut off before these load capacitances are discharged. Therefore, the discharging of the level-shifted output (to the right of the diodes) is dominated by the pull-down transistor, Q_5. The equivalent circuit for this condition is shown in Fig. 3-33a, which we hope to simplify to the circuit shown in Fig. 3-33b by representing Q_5 as a current source, calculating large-signal equivalent

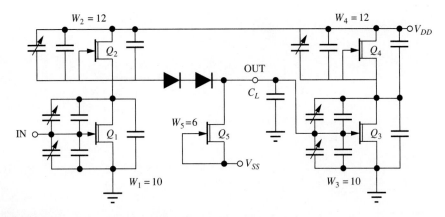

FIGURE 3-32
Schematic diagram of unbuffered FET logic (FL) inverter showing all of the device and circuit capacitances that must be considered in the estimation of propagation delay.

LOGIC CIRCUIT DESIGN PRINCIPLES   179

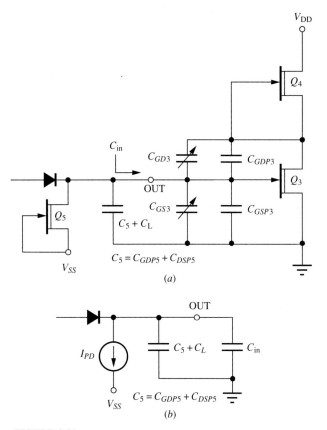

FIGURE 3-33
Estimation of t_{PHL} for the FL gate. The simplified equivalent circuit in (b) is derived from the complete circuit in (a).

capacitances for the voltage-variable elements, and referencing all load capacitances to ground.

In this example, the output (level-shifted) has an initial voltage of V_{OH} and a final voltage of $V_{50\%}$. Thus, Q_5 is in saturation over the full range of output voltage for reasonable values of V_{SS} and can be modeled as a current source. C_{GD5} will always be zero, since the FET channel is assumed to be in pinch-off. Q_4 will also be in saturation, so $C_{GD4} = 0$. Q_3 will be traversing from the linear to the saturation region, so both C_{GS3} and C_{GD3} must be evaluated:

$$C_{GS3} = C_{GS0} W_3 K_{EQ1}$$

K_{EQ1} can be determined by evaluating Eq. (3.19) at the appropriate final and initial voltage limits. Here, the gate-to-source voltages are $V_2 = V_{50\%}$ and $V_1 = V_{OH}$ respectively. The Eq. (3.19) can also be used to evaluate C_{GD3} where the respective gate-to-drain voltages are $V_2 = V_{50\%} - V'_{50\%}$ and $V_1 = V_{OH} - V'_{OL}$. Then, $C_{GD3} = C_{GD0} W_3 K_{EQ2}$.

From the dc analysis carried out in Sec. 3.2.1, it can be shown that $V_{OL} = -1.1$ V, $V_{OH} = 0.7$ V, and, therefore, $V_{50\%} = -0.2$ V. The level shifting will raise

V'_{OL}, V'_{OH}, and $V'_{50\%}$ by approximately 1.5 V. Assuming that the capacitance model parameters of Sec. 2.2.1 apply to this example, let $C_{GS0} = C_{GD0} = 1.2$ fF, $m = 1.3$, and $V_{bi} = 1.7$ V. Also, let $C_{GSP} = C_{GDP} = 0.28$ fF/μm and $C_{DSP} = 0.11$ fF/μm. The calculated values for C_{GS3} and C_{GD3} are shown in Table 3.7. Note that $C_{GD3} = 0$ if $V_{GD} - V_T \leq -1.0$ V. Therefore, the upper integration limit, V_2, should be -1.0 V for calculation of K_{EQ2}.

Since an equivalent capacitance between the output node and ground (C_{in}) is desired, an effective value for C_{GDP3} is needed which accounts for the voltage gain of the switch FET and the consequent increase (Miller effect) in the GD voltage range. It can be seen that $\Delta Q_{GDP3} = C_{GDP3}(V_2 - V_1) = C_{GDP3}(-1.9 \text{ V})$. In order to account for this extra charge, an effective value can be determined by scaling C_{GDP3} by the ratio of the actual voltage difference across C_{GDP3} to the voltage difference across the GS junction, -0.9 V. Therefore, $C_{GDP3(eff)} = 2.1 C_{GDP3} = 2.1 C_{GDP} W_3 = 5.9$ fF in this example.

Finally, all of the other components of C_{in} can now be calculated and are summarized in Table 3.7. While C_{in} was calculated for a single inverter load (fan-out $= 1$), higher fan-outs can also be evaluated by multiplying C_{in} by the fan-out factor. The wiring capacitance, $C_L = 5$ fF, is representative of a short interconnection between the driving inverter and its load. C_5 is the sum $C_{GDP5} + C_{DSP5}$.

The propagation delay t_{PHL} can now be evaluated. I_{PD} is assumed to be constant over the half logic swing being considered, with a value of $I_{PD} = I_{DSS} W_5$. For the example, let I_{DSS} be 80 μA/μm. Then Eq. (3.14) can be evaluated as follows

$$t_{PHL} = (C_{in} + C_L) \frac{V_{OH} - V_{OL}}{2 I_{PD}}$$

Therefore, $t_{PHL} = 73$ ps for fan-out $= 1$

$t_{PHL} = 131$ ps for fan-out $= 2$

This result is the estimate of the step response of the FL inverter. If the response to a more realistic input is needed, Eq. (3.13) can be used to correct for the finite rise or fall time of the actual input. For example, if the input rise time were 150 ps,

TABLE 3.7
Capacitances contributing to the effective large-signal input capacitance C_{in} of the load inverter in example 3.4, part (a). MESFET Q_3 has width 10 μm and Q_5 has width 6 μm

Capacitor	Capacitance (fF)
C_{GS3}	15.1
C_{GD3}	7.5
C_{GSP3}	2.8
C_{GDP3}	5.9
C_{GDP5}	1.7
C_{DSP5}	0.67

then the corrected estimate for t_{PHL} would be 105 ps for fan-out = 1 and 151 ps for fan-out = 2.

Before proceeding to t_{PLH}, another possibility should be considered. If the shunt capacitance of the level-shift diodes (C_D) is *large* compared with the total load capacitance, then the discharge of the load capacitance will be assisted by the displacement current through C_D which is flowing to ground through the switch, Q_1. Q_1 is also in saturation. This approach will permit the use of a much smaller pull-down FET, Q_5, since now both Q_1 and Q_5 are responsible for discharging $C_{in} + C_L$ [14,15]. Current from Q_2 must also be accounted for in Q_1. A similar analysis can be applied in this case. (Homework problem 3.3.)

(b) *Estimate t_{PLH}*. Assume a step input that begins at V_{OH} and ends at V_{OL} at $t = 0$. In this case, the drain node capacitance of the inverter, C_{out}, as well as C_L and C_{in} are all charged by the current from the pull-up FET Q_2. For purposes of determining the propagation delay, the output will swing from an initial state at V_{OL} to a final state at $V_{50\%}$. Q_2 will be in saturation over this full range. Figure 3-34a illustrates the circuit representing this case. It can be seen that some of the current from Q_2 is

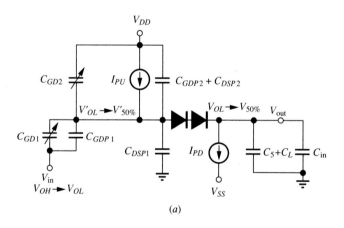

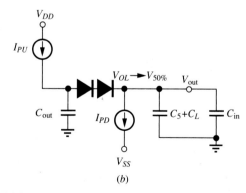

FIGURE 3-34
Estimation of t_{PLH} for the FL gate. The simplified equivalent circuit in (b) is derived from the complete circuit in (a).

consumed by the pull-down, Q_5, which is always in saturation for V_{SS} sufficiently negative (-2.0 V is sufficient). Therefore, the net charging current will be

$$I_{CHG} = I_{DSS}W_2 - I_{DSS}W_5$$

The objective will once again be to calculate equivalent large-signal capacitances, referenced to ground, for C_{out} and again for C_{in} under its new voltage range as illustrated in Fig. 3-34b.

To determine C_{GD1}, Eq. (3.19) must be evaluated at V_2 and V_1. For the GD junction of Q_1,

$$V_2 = V_{OL} - V'_{50\%} = -1.1 - (1.3) = -2.4 \text{ V}$$

and

$$V_1 = V_{OH} - V'_{OL} = 0.7 - (0.3) = 0.4 \text{ V}$$

Since $C_{GD1} = 0$ when $V_{DG1} - V_T = -1.0$ V, then the limits of integration are the same as those used to find K_{EQ2} in part (a). Therefore,

$$C_{GD1} = C_{GD0}W_1 K_{EQ2} = 7.5 \text{ fF}$$

The Miller capacitance for C_{GDP1} must also be determined. $V_2 - V_1$ for the GD junction of Q_1 is -2.8 V, whereas $V_2 - V_1$ for the GS junction is -1.8 V. Therefore, $K_M = 1.56$, and

$$C_{GDP1(eff)} = C_{GDP}W_1 K_M = 4.4 \text{ fF}$$

Since all other capacitances contributing to C_{out} are referenced to ground and $C_{GD2} = 0$, C_{out} can be determined:

$$C_{out} = C_{GD1} + C_{GDP1(eff)} + C_{DSP}(W_1 + W_2) + C_{GDP}W_2 = 17.7 \text{ fF} \quad (3.23)$$

Both C_{GS3} and C_{GD3} must be recomputed, since the voltage limits under consideration are different from those in part (a). To obtain C_{GS3}, evaluate Eq. (3.19) at the limits

$$V_2 = V_{50\%} \quad \text{and} \quad V_1 = V_{OL}$$

which yields

$$C_{GS3} = C_{GS0}W_3 K_{EQ}(V_2, V_1) = 8.3 \text{ fF}$$

It can be seen that Q_3 is always in saturation in this case, so $C_{GD3} = 0$. However, K_M must be found for C_{GDP3}. For

$$V_2 = V_{50\%} - V'_{50\%} = -1.5 \text{ V} \quad \text{and} \quad V_1 = V_{OL} - V'_{OH} = -3.4 \text{ V}$$

$V_2 - V_1$ for the GD junction is 1.9 V whereas $V_2 - V_1$ for the GS junction is 0.9 V. Therefore, $K_M = 2.1$ and

$$C_{GDP3} = 2.1 C_{GD0}W_3 = 5.9 \text{ fF}$$

Again, Q_4 does not contribute to C_{in}. Therefore,

$$C_{in} = C_{GS3} + C_{GDP}(2.1 W_3) = 14.2 \text{ fF}$$

C_5 is found to be 2.3 fF as in part (a) and C_L is again 5 fF. The propagation delay t_{PLH} can now be calculated from Eq. (3.14) as shown:

$$t_{PLH} = (C_{in} + C_5 + C_L + C_{out})\frac{V_{OH} - V_{OL}}{2 I_{CHG}} = 74 \text{ ps} \quad \text{for fan-out} = 1$$

For fan-out = 2, 102 ps is found. Correcting these with Eq. (3.13) for an input fall time of 150 ps, t_{PLH} = 105 ps for fan-out = 1 and 126 ps for fan-out = 2.

(c) The propagation delay t_{PD} can now be determined by averaging the results of parts (a) and (b):

$$t_{PD} = \frac{t_{PLH} + t_{PHL}}{2} \quad (3.24)$$

From this equation, t_{PD} = 105 ps for fan-out = 1 and 139 ps for fan-out = 2.

3.3.3 Analysis by CAD Simulation

The propagation delay can also be estimated with somewhat greater accuracy with the use of circuit simulation software such as SPICE [16] which accounts for the voltage variation of capacitances and currents. Numerical methods are used to solve sets of coupled nonlinear differential equations so that the time dependence of the node voltages can be obtained. A nodal formulation of the circuit equations is solved iteratively to obtain voltages and currents in the circuit. SPICE makes initial guesses of the node voltages, calculates branch currents, and then calculates linearized conductances valid at the assumed voltages. The new node voltage vector is obtained by the Newton-Raphson method, and the process then repeats until convergence is obtained. Time dependence of node voltages is determined by the same process, with the coefficient matrix being reevaluated at every time step and the node voltage determined through iteration. Capacitor currents are evaluated at each time step through the equation below:

$$I(n+1) = \frac{Q(n+1) - Q(n)}{t(n+1) - t(n)} \quad (3.25)$$

Since the charge Q is a function of the node voltage, I/V can be obtained for the conductance matrix. Although the simulated waveforms at all circuit nodes can be obtained with great precision if desired, it is important to remember that the simulated results are only as accurate as the device models.

Also, because the time-domain solution of the node equations requires large amounts of computation, SPICE is suited for detailed analysis and design of smaller circuit blocks for relatively short time intervals. The complete simulation of larger circuits is generally not possible due to the large amounts of computer time required. Logic simulators using gate-level macromodels are more appropriate for this purpose. Used sensibly, SPICE can be a powerful tool for optimizing the design of logic gates and circuits. The tradeoffs between device widths, power dissipation, and delay can be determined. Also, the dependence of delay on fan-in, fan-out, V_{DD}, V_{SS}, temperature, and worst-case process variations can be obtained in a straightforward manner, leading to a well-characterized and (hopefully) dependable circuit.

When used to evaluate the design tradeoffs for a logic gate or circuit, it is helpful to include a realistic input waveform for driving the circuit under investigation and to provide realistic loading for the circuit as shown in Fig. 3-35. The realistic input waveform can be accomplished by preprocessing the input source voltage, usually

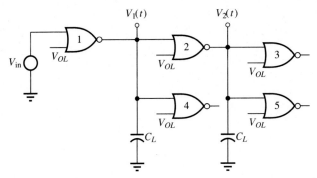

FIGURE 3-35
Schematic diagram of a chain of three NOR gates (1–3) to be used for simulation of propagation delay. Loading is provided by gates 4 and 5 (fan-out) and C_L (capacitive loading). The simulated delay is determined by the time difference between $V_2(t)$ and $V_1(t)$.

obtained with a pulse or piecewise linear function and an independent voltage source, with a typical logic gate of the family being considered (gate 1 loaded by gate 4). The delay of gate 2 is to be determined. Realistic loading is provided by the use of fixed-load capacitors (C_L) to represent wiring capacitance and by fan-out, provided by gates 4 and 5 in this example. FET parasitic capacitances C_{GSP}, C_{GDP}, and C_{DSP} are usually not included in the FET models and must be added as fixed capacitors in the input node list. The delay between $V_1(t)$ and $V_2(t)$ is the desired propagation delay. The subcircuit function in SPICE is useful for simulating structures that must be repeated several times (such as the NOR gates in the above example).

It should be noted that use of the input preprocessing approach is not a good idea if the dc transfer function of the gate is being evaluated. Since gate 1 should provide significant voltage gain in the transition region, its input step size will be amplified, and the input voltage to gate 2 will step through the region of interest with few points. Then, due to the improper shape of the transfer characteristic, accurate determination of the noise margin will not be possible. The use of realistic fan-out loading for dc simulations is often quite important.

Example 3.5. Estimate the propagation delay of a FL inverter using SPICE. Implement the circuit diagram of Figs. 3-32 and 3-35 as a SPICE input file. Assume that $C_L = 5$ fF. Simulate t_{PHL} and t_{PLH} for the fan-out $=1$ and 2 cases for $V_{DD} = 2.5$ V and $V_{SS} = -2.0$ V, and compare the results with the estimates of Example 3.4.

Solution. Figure 3-36 is a listing of the SPICE file used to simulate the response of the circuit. In order to be able to make meaningful comparisons with the results of Example 3.4, the SPICE JFET model was used with gate-source and gate-drain depletion capacitance equations (3.16) modified to include pinch-off and with $m = -1.3$. pb $= 1.7$ was selected in order to best fit the ion-implanted capacitance profile as discussed in Sec. 2.2.1.

First, the inverter and load devices can be set up as subcircuits for convenience in interconnections. The inverter subcircuit includes Q_1, Q_2, Q_5 and the level-shift diodes.

```
simulate fet logic inverter chain. CL=5ff spice jfet model fanout=2
.model mes njf(vto=-1.0 beta=1.0e-4 lambda=0.1
+ rs=2500 rd=0 is=3e-16 cgs=1.2ff cgd=1.2ff pb=1.7)
.model dio d(is=1e-14 rs=1500 n=1.1 cjo=2fF vj=.8 m=.5 eg=1.4)
.options gmin=1e-9
*
x1 12 4 10 11 inv
x2 4 5 10 11 inv
x21 4 51 10 11 inv
x3 5 6 10 11 inv
x31 5 61 10 11 inv
c11 4 0 5fF
c12 5 0 5fF
vdd 10 0 2.5
vss 11 0 -2
vin 12 0 pulse(-1.1 0.7 0p 100p 100p 500p 1.3n)
.tran 5ps 1.3ns
.nodeset v(4)=0.6 v(5)=-1.2
.print tran v(4)
.print tran v(5)
* in1 out vdd vss
.subckt inv 1 4 10 11
z1 3 1 0 mes 10
z2 10 3 3 mes 12
z5 4 11 11 mes 6
ds1 3 5 dio 6
ds2 5 4 dio 6
cgdp2 3 10 3.3ff
cdsp2 3 10 1.3ff
cgdp1 3 1 2.8ff
cgsp1 1 0 2.8ff
cdsp1 3 0 1.1ff
cgdp5 4 11 1.7ff
cdsp5 4 11 0.66ff
.ends inv
.end
```

FIGURE 3-36
SPICE input file for simulation of a chain of three FL inverters as described in Figs. 3-32 and 3-35. Fan-out loading of 2 and capacitive loading of 5 fF is included.

The parasitic device fringing capacitances for each MESFET, described in Sec. 2.2.5, were also included in the subcircuit. A separate load subcircuit, including only Q_3 and Q_4 and their parasitic capacitances, was also constructed for convenience.

A chain of three such inverters with loading was connected in the node list and driven from a piecewise linear source. The output waveforms $V(4)$, the input to the second inverter, and $V(5)$, the output of the second inverter, are plotted in Fig. 3-37 for the fan-out = 2 case described by Fig. 3-36. The propagation delays t_{PHL} and t_{PLH} can be measured from the 50 percent points of the waveforms. An additional simulation for the fan-out = 1 case is also required. Note that the rise and fall times are not equal and are not the same as the 150 ps estimated in Example 3.4.

A summary and comparison of the results is presented in Table 3.8. It can be seen that the accuracy of the average capacitance method used in the hand analysis is surprisingly good, even when an average rise and fall time of 150 ps is assumed. The accuracy would not be expected to be much better than this; the simulation output plotted

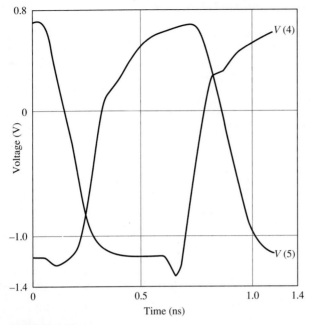

FIGURE 3-37
Plot of SPICE simulation output showing $V(4)$, the output of the first inverter, and $V(5)$, the output of the second inverter. The propagation delay is found from the time difference at the 50 percent points.

in Fig. 3-37 shows that the assumption of constant current is not strictly valid, especially as the output approaches the logic high level. Here, the pull-up FET is beginning to fall out of saturation at the same time as the gate is beginning to require input current due to the forward bias. The output voltage charges up more exponentially than linearly in this region.

TABLE 3.8
Delay estimates for an FL inverter chain. The SPICE2 simulation results are compared with the simplified effective capacitance model (Example 3.4)

Fan-out	t_{PHL} (ps)	t_{PLH} (ps)	t_r (ps)	t_f (ps)
1	89	84	165	125
2	137	115	247	190

Hand analysis (Example 3.4) with assumed t_r, t_f				
Fan-out	t_{PHL} (ps)	t_{PLH} (ps)	t_r (ps)	t_f (ps)
1	105	105	150	150
2	151	126	150	150

3.3.4 Dependence of Delay—Width Optimization

The material presented in the previous section illustrates that the delay of a logic gate is very dependent on device and circuit capacitances and on the currents available to charge and discharge these capacitors. Therefore, one would expect a reduction in delay in a circuit path if the widths of the FETs in the driver logic gate were increased, because the current available to drive the load capacitances, C_L and C_{in}, would increase proportionally. However, an increase in current also results in a proportional increase in power dissipation, as discussed in Sec. 3.2.4. Therefore, there is a direct tradeoff between speed and power when device width is varied in this manner.

If all of the gates in the entire circuit were scaled in width by the same factor, there would still be some asymptotic improvement in propagation delay, but the extent of the improvement would be quite dependent on layout [17]. This can be seen by considering the load capacitance in a particular path of the circuit. Referring to Fig. 3-34b, for example,

$$C_{load} = C_{out} + C_5 + C_L + C_{in} \quad (3.26)$$

Of the four components of C_{load}, clearly C_{out}, C_5 and C_{in} will scale directly with the width of the FETs if all of the FETs are scaled by the same factor. Only the wiring capacitance C_L does not scale directly. However, even the length of the interconnect wiring will also depend to some extent on the device widths due to the area required to lay out a gate. The current available to drive the capacitances scales directly with width. As demonstrated in Ref. 17, propagation delay improves very slowly with an increase in device width when uniform scaling is applied, whereas the power dissipation increases directly with width. For example,

$$t_{PLH} = \frac{[(C_{out} + C_5 + C_{in})W + C_L]\Delta V}{I_{DSS}W} \quad (3.27)$$

and, from Sec. 3.2.4, Eq. (3.4),

$$P_D = I_{DSS}(W_2 V_{DD} - W_5 V_{SS}) \quad (3.28)$$

Therefore, for small widths, the *power-delay product* (PDP) approaches a constant. Propagation delay will increase directly as the device widths are reduced because the indirectly scaled wiring capacitance (C_L) does not decrease as rapidly as do the device intrinsic and parasitic capacitors (C_{in} and C_{out}). Thus, power conservation by uniform reduction in device widths can be utilized, but careful modeling is needed to determine how much speed is being forfeited. For large widths, the power-delay product increases directly with width. Reduction in delay approaches an asymptote, so again careful modeling is needed to identify the point of diminishing returns. This relationship is illustrated in Fig. 3-38.

Determining the minimum useful or acceptable FET widths for a particular logic family and fabrication technology is an important part of the optimization of design. In general, this optimization is very sensitive to the layout design rules. One must establish the size of the basic building block before proceeding with the details of chip layout.

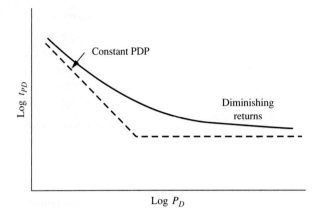

FIGURE 3-38 Propagation delay as a function of FET width (and power dissipation). For large device widths, the delay is reduced asymptotically; for small widths, a constant power-delay product is approached.

Although the technique of uniform width scaling is of limited usefulness for performance optimization, width scaling can be effective in reducing the delay of a particular path in a circuit design. When used for this purpose, Eq. (3.27) shows that if C_L and C_{in} remain fixed, while current and C_{out} are increased, t_{PD} will be reduced. The intersection point of the two asymptotes in Fig. 3-38 moves toward the right. However, the use of a buffer, as described in Chap. 4, may be more efficient in many instances.

3.3.5 Dependence of Delay—Fan-in and Fan-out

The influence of fan-in, the number of inputs to a logic circuit, on the propagation delay will naturally depend on the circuit topology. Specific examples will be presented in Chap. 4. In general, the fan-in will increase the effective internal node capacitance C_{out} of Fig. 3-34b and will therefore become a factor in the calculation of the total capacitance. This can be illustrated through reconsidering Eq. (3.23) from Example 3.4. In this FL example, the terms associated with Q_1 are multiplied by a fan-in factor, FI, to account for additional input FETs. Note that this affects only t_{PLH} for this circuit because of the assumptions regarding the small shunt capacitance of the level-shifting diodes:

$$C_{out} = [C_{GD1} + C_{GDP1(eff)} + C_{DSP}W_1]FI + (C_{DSP} + C_{GDP})W_2 \quad (3.29)$$

Fan-out, the number of identically sized gates loading the output of the driver logic gate, will also increase the propagation delay. Its influence is also greatly dependent on the topology of the logic gate being considered. Specific examples will be given in Chap. 4. For the FL gate, which has been our standard example, fan-out has a corresponding influence on C_{in}, the effective input capacitance. Referring again to Example 3.4, Eq. (3.21) is modified to include a fan-out factor FO:

$$t_{PHL} = (FOC_{in} + C_L)\frac{V_{OH} - V_{OL}}{2I_{PD}} \quad (3.30)$$

Also, the equation for t_{PLH} is modified accordingly to show the fan-out effect:

$$t_{PLH} = (FOC_{in} + C_L + C_{out}) \frac{V_{OH} - V_{OL}}{2 I_{CHG}} \quad (3.31)$$

3.3.6 Maximum Frequency of Operation

Many design specifications will include a maximum clock frequency at which the circuit is required to function without error. It is useful to consider how the requirements for propagation delay of the logic gates in the critical paths of the circuit derive from the maximum clock frequency specification so that the design tradeoffs presented in Sec. 3.3.4 can be made in a more quantitative manner. The objective is to determine the highest toggle frequency for which the amplitude of the output waveform remains at its maximum extreme [18].

A useful design guideline can be derived by using the simplified model of the input and output waveforms shown in Fig. 3-39. The linear rising and falling edges would occur in the limiting case in which: (1) charging and discharging currents are constant, as in the case where the active-load transistors always remain in the saturated region, (2) the capacitance is not voltage-dependent, which is almost true when the fixed-load capacitance exceeds the device capacitances by a significant margin, and (3) the slew rate of the input and output rising and falling waveforms are the same. Under these conditions, the output waveform, $V_2(t)$, begins to change when the input, $V_1(t)$, exceeds the logic gate input switching threshold, V_{TH} (assumed here to be at $V_{50\%}$, the middle of the logic voltage swing). To maintain the full logic voltage swing for $V_2(t)$, the input may reach its maximum value, V_{OH}, just when the output will have dropped to $V_{50\%}$. At this time, the input may begin to approach V_{OL} again, and will not cross $V_{50\%}$ until the output has just reached V_{OL}. From this model it can be seen that the maximum clock frequency will be limited to

$$f_{C,\max} = \frac{1}{4} t_{PD} \quad (3.32)$$

if the above assumptions are justified and if V_2 is not degraded.

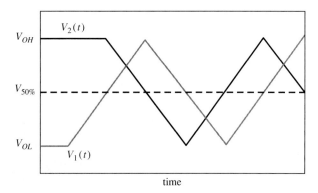

FIGURE 3-39
Model for determining the maximum clock frequency of a logic gate.

While this simple model provides a useful guideline, in a specific case for which the highest performance is needed it will still be necessary to verify this condition by actual computer simulation of the waveforms for the worst-case loading temperature extremes, and parameter spreads. Also, the assumptions of the model may not be totally justified, in which case exponential waveforms or unequal rise and fall times might be observed. This model does, however, provide a good starting point for this analysis, and for less critical design requirements may prove to be adequate without further simulation.

3.3.7 Dynamic Noise Margin

The noise margin of a logic gate was defined in Sec. 3.1.2 in terms of a dc noise voltage applied in series with the input terminals for the input at V_{OL} or V_{OH}. While this definition is sufficient for the dc design of a logic circuit, it does not account for the fact that the output voltage will be delayed with respect to the input if a pulse rather than a dc signal is applied. This delay in the gate response, which can be attributed to the charging of device and interconnect capacitances described by Eq. (3.14), means that a higher noise voltage can be accommodated on the input of the gate without upsetting the output so long as its time duration is comparable with the propagation delay [4]. This increased noise margin for short pulses is called the dynamic noise margin (DNM) and will vary with the width of the pulse for a given gate design. Naturally, the DNM will also vary with the design of the gate and the loading of the gate; for a fixed pulse width, a greater NM will be observed for circuits with slower response. As a consequence, the DNM is quite difficult to predict by any means other than computer simulation and is not characterized by a single number for all conditions, as was the static case. The DNM approaches the static NM if the pulse width is wide because the circuit can charge to the full-voltage range if not constrained in time.

Figure 3-40 illustrates this trend with a plot of the amplitude of the applied pulse as a function of the duration of the pulse as defined in the figure. It can be seen that the DNM exceeds the static noise margin (SNM) by a significant amount as the duration approaches the propagation delay (in this case, about 100 ps). It can also be noted from the figure that variation in one of the power supplies (V_{SS} in this example) causes a shift in the curve corresponding to the change in SNM. The NM is decreased when V_{SS} is reduced in this case because the pull-down FET drops out of saturation at low V_{DS}.

Figure 3-40 was generated through simulation of a cross-coupled set-reset NOR flip-flop, as indicated in Fig. 3-41. An input pulse train, $V_1(t)$, of fixed duration but gradually increasing amplitude was simulated as a voltage source using the piecewise-linear function of SPICE. The latch was initialized by resetting with the initial conditions. Then the input pulse train is applied until the output, $V_2(t)$, indicates that a logic upset has occurred. The latch will hold this new state until reset. Therefore, a simple flip-flop circuit also makes a useful noise margin sensing device for experimental verification of the DNM.

LOGIC CIRCUIT DESIGN PRINCIPLES   **191**

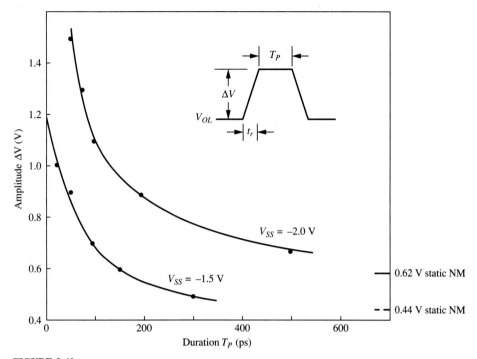

FIGURE 3-40
Contours of dynamic noise margin. The input pulse amplitude for upset of a latch is plotted against the duration of the pulse.

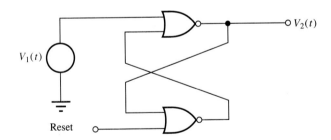

FIGURE 3-41
Set-reset flip-flop used for simulation or measurement of dynamic NM.

An alternative method for the simulation of dynamic transfer characteristics was presented in Ref. 7. In this case, the transient response of the active-load and switch transistors in an inverter chain was simulated using a circuit analysis program. The dynamic current and voltage relationships were used to construct a dynamic load line, which can be used to determine the total charge required to swing the output. In addition, a dynamic transfer characteristic can be constructed from this simulation, which indicates the increased noise margins as expected.

The DNM simulation finds application in evaluation of the response of a logic gate to noise signals of short duration, such as might be generated from crosstalk or reflections between or on transmission lines. This subject is of importance when considering the design of packages and interrconnections and will be analyzed in detail in Chap. 5. It is not necessary to consider the DNM in the design optimization of a logic gate. If the design is optimized for the largest SNMs, and the DNM will always be larger than the SNM, then the design will only be more robust for short input pulses. Also, since the DNM is highly sensitive to loading conditions, it will vary with the application on the IC chip, and no single DNM evaluation would be truly representative of all instances of the gate.

HOMEWORK PROBLEMS

3.1. Design a NAND gate using the unbuffered FET logic (FL) circuit shown in Fig. 3-23. Use SPICE to optimize the widths of the series switch transistors, Q_A and Q_B, so that V_{NL} and V_{NH} are equal for input B. A model for the dual-gate MESFET such as described in Chap. 2 should be used. Show the final transfer characteristic for both inputs.

3.2. A chain of two-input FL NOR gates, such as shown in Fig. 3-35, is to be evaluated for tradeoff between device widths, power dissipation, and propagation delay. First verify that your dc transfer characteristic is optimum, then for fixed $C_L = 50$ fF and fan-out $= 2$, use SPICE to model the t_{PD} as a function of device width. Scale the widths of FETs and diodes. Determine the limiting asymptotes of performance.

3.3. A diagram of a capacitor-diode FET logic (CDFL) gate is shown in Fig. P3-3.

    (*a*) Perform a hand analysis of t_{PHL} and t_{PLH} for the CDFL circuit. Assume that the diode and MESFET parameters in Tables 2.1 and 2.2 apply to this circuit. Let $m = 0.5$.

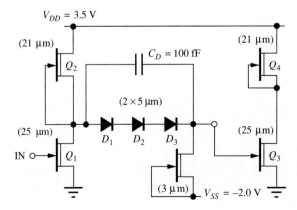

FIGURE P3-3
Circuit diagram of capacitor-diode FET logic (CDFL) inverter.

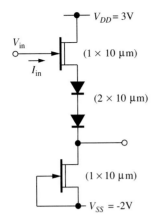

FIGURE P3-4
Source follower.

(b) Estimate power dissipation per stage.
(c) Verify (a) and (b) by simulation using the SPICE JFET model.

3.4. (a) Estimate the input capacitance of the source follower circuit shown in Fig. P3-4 using hand analysis. Assume that the diode and MESFET parameters in Tables 2.1 and 2.2 apply to this circuit.

(b) Verify your estimate using SPICE.

Hint: A current source dependent on I_{in} can be used to charge up a known capacitor. The voltage across the capacitor can be used to determine the total input charge, $Q = CV$.

3.5. Evaluate the dynamic noise margin of an unbuffered FET logic NOR gate (FL) with fan-out = 3. Plot ΔV as a function of T_{pulse} as in Fig. 3-40.

REFERENCES

1. Hodges, D. A., and Jackson, H.: *Analysis and Design of Digital Integrated Circuits*, pp. 23–24, McGraw-Hill, New York, 1983.
2. Gilbert, B. K.: "Exploitation of GaAs Digital ICs in Wideband Signal Processing Environments," *IEEE GaAs IC Symp. Proc.*, Phoenix, Ariz., paper Ch1876-2, 58-61, October 1983.
3. Hill, C. F.,: "Noise Margin and Noise Immunity in Logic Circuits," *Microelectronics*, vol. 1, pp. 16–22, April 1968.
4. Lohstroh, J.: "Static and Dynamic Noise Margins of Logic Circuits," *IEEE J. Solid State Cir.*, vol. SC-14, pp. 591–598, June 1979.
5. Eden, R. C., Welch, B. M., Zucca, R., and Long, S. I.: "The Prospects for Ultrahigh-Speed VLSI GaAs Digital Logic," *IEEE Trans. Elect. Dev.*, vol. ED-26, pp. 299–317, April 1979.
6. Helix, M. J., Jamison, S. A., Chao, C., and Shur, M. S.: "Fan-out and Speed of GaAs SDFL Logic," *IEEE J. Solid State Cir.*, vol. SC-17, pp. 1226–1231, December 1982.
7. Simmons, J. G., and Taylor, G. W.: "An Analytical Treatment of the Performance of Submicrometer FET Logic," *IEEE J. Solid State Cir.*, vol. SC-20, pp. 1242–1251, December 1985.
8. Barna, A., and Liechti, C.: "Optimization of GaAs MESFET Logic Gates with Subpicosecond Propagation Delays," *IEEE J. Solid State Cir.*, vol. SC-14, pp. 708–715, August 1979.
9. Vogelsang, C. H., Hogin, J. L., and Notthoff, J. K.: "Yield Analysis Methods for GaAs ICs," *IEEE GaAs IC Symp. Proc.*, Phoenix, Ariz., pp. 149–152, October 1983.
10. Buehler, M. G., and Griswold, T. W.: "The Statistical Characterization of CMOS Inverters Using Noise Margins," *J. Electrochem. Soc. Extended Abstracts*, vol. 83-1, pp. 391–392, May 1983. This paper was originally presented at the Spring 1983 Meeting held in San Francisco, California.

11. Yokoyama, N., Ohnishi, T., Odani, K., Onodera, H., and Abe, M.: "TiW Silicide Gate Self-Alignment Technology for Ultra-High-Speed GaAs MESFET LSI/VLSI's," *IEEE Trans. Elect. Dev.,* vol. ED-29, pp. 1541–1547, October 1982.
12. Toyoda, N., Mochizuki, M., Mizoguchi, T., Nii, R., and Hojo, A.: "An Application of Pt-GaAs Solid Phase Reaction to GaAs IC," *Proc. Int. Conf. on GaAs and Related Compounds,* Institute of Physics Conference Series 63, p. 521, 1982.
13. Hodges, D. A., and Jackson, H.: *Analysis and Design of Digital Integrated Circuits,* Chap. 3, McGraw-Hill, New York, 1983.
14. Eden, R. C.: "Capacitor Diode FET Logic (CDFL) Circuit Approach for GaAs d-MESFET ICs," *Gallium Arsenide IC Symp. Proc.,* Boston, Mass., October 1984.
15. Namordi, M. R., and White, W. A.: "A Novel Low-Power Static GaAs MESFET Logic Gate," *IEEE Elect. Dev. Lett.,* vol. EDL-3, pp. 264-267, September 1982.
16. Nagel, L.: "SPICE2: A Computer Program to Simulate Semiconductor Circuits," Electronics Research Laboratory, University of California, Berkeley, memo ERL-M592, 1976.
17. Van Tuyl, R., Liechti, C.A., Lee, R.E., and Gowen E.: "GaAs MESFET Logic with 4-GHz Clock Rate," *IEEE J. Solid State Cir.,* vol. SC-12, October 1977.
18. McCormack, Gary D.: TriQuint Semiconductor, unpublished work.

CHAPTER 4

LOGIC CIRCUIT DESIGN EXAMPLES

The design techniques and principles that were described in the last chapter can be applied to the design of GaAs logic circuits with depletion mode, enhancement mode, or mixed E/D MESFETs, JFETs, or MODFETs. In review, the main property that distinguishes the design of high-speed GaAs FET circuits from NMOS or CMOS circuits is the forward-bias gate conduction resulting from the use of a Schottky barrier (metal/semiconductor junction) on top of the n-type channel. This gate conduction may clamp the logic voltage swing in the positive direction if a V_{GS} of 0.6 to 0.7 V is present (MESFET or HFET), or 1.0 V (JFET). Another significant distinction between Si and GaAs FETs is the much greater transconductance of n-channel GaAs FETs which results from the higher mobility and electron drift velocity as explained in Chap. 1. Factors of five higher g_m and f_t are not uncommon for GaAs FETs with equivalent gate width and length to silicon FETs. This provides higher drive currents at smaller voltage swings, leading to higher speed and lower power. Finally, p-channel GaAs FETs are not widely used because of the low mobility and high electric field required to obtain saturated velocity for holes in GaAs. However, p-channel JFETs have been effective in implementing complementary JFET logic and memory when very low static power is essential.

Another distinction can be made between the design of circuits in which speed is of primary concern and the design of lower-speed circuits where reduction in delay is not a high priority. The low-speed ICs can be constructed by following very simplified rules such as might be used in the layout of a gate array. Wiring of gate arrays can be done automatically in some CAD scenarios, without reference to the characteristics or properties of the active devices or without much concern for loading. Although successful functionality may be virtually guaranteed, when speed specifications fail to be met, or timing problems arise, the designer may be ill equipped to analyze and correct the problem.

On the other hand, the nature of high-speed circuits requires an understanding of design at the device level; a knowledgeable digital designer must also be a capable analog designer. The device and interconnect parasitics will often limit performance, so proper optimization of device widths, the use of buffers when needed, and minimization of interconnect capacitance become essential if the best performance is expected.

In this chapter, representative examples of circuit topologies for high-speed GaAs logic will be presented to illustrate a few of the many possibilities available to the designer. This will by no means be an exhaustive survey of every reported circuit idea. The benefits gained and compromises incurred by each of these example circuits will be discussed. Some of these circuits are available to be assigned to the reader as homework for more detailed analysis.

4.1 DEPLETION MODE LOGIC CIRCUITS

Circuits designed with depletion mode (normally-on) GaAs FETs are characterized by unequal input and output voltage levels; V_{GS} must be negative in order to cut off the FET while V_{DS} must be positive at all times. Consequently, level-shifting networks, typically composed of forward-biased Schottky diodes, are necessary. In addition, since both positive and negative signals are required, two power supply voltages, a positive V_{DD} and a negative V_{SS}, are needed. Therefore, additional power is dissipated in the circuit, and the higher supply differentials make depletion mode (d-mode) logic more susceptible to backgating effects (Secs. 1.4 and 2.3).

On the other hand, larger logic voltage swings can be obtained using d-mode FETs because the gate electrode often varies between a negative voltage close to V_T and a positive voltage near 0.7 V. The voltage range is determined by the choice of threshold voltage, V_T, and the amount of level shifting required. This is not the case with enhancement mode (E-mode) MESFET/HFET circuits where V_{GS} is always positive. The higher logic voltage swing of the d-mode circuit leads to larger drive currents $[I_D = \beta W(V_{GS} - V_T)^2]$, but not necessarily to lower delays. It also increases noise margins, giving higher yields. Finally, the power dissipation is often higher because of the dual power supplies and larger current levels.

4.1.1 Unbuffered FET Logic

The unbuffered FET logic circuit (FL) was studied extensively in Chap. 3 as an example illustrating the dc design methods and transient (propagation delay) analysis techniques. An analytical study, which neglects the effect of source resistance, has been published [1] which provides a detailed comparison of this circuit and the buffered version to be described in Sec. 4.1.3. The circuit diagram for an FL inverter can be found in Fig. 3-12. Plots of the simulated transfer characteristics of FL inverters implemented with 1 μm directly aligned, selectively implanted MESFETs were shown in Figs. 3-15 through 3-18. The influence of various design conditions (width ratios, threshold voltage, and level-shift diodes) was evaluated. The result of the transient simulation of a chain of three FL inverters was presented in Fig. 3-37 and Table 3.8.

The circuit has a number of attractive features: high noise margins and a high logic voltage swing provide good parametric yield for any reasonable threshold voltage variation, the power dissipation is moderately low, and the propagation delay is low if the output is lightly loaded.

The dc characteristics of optimized ($V_{NL} = V_{NH}$) FL inverters as simulated by SPICE are summarized in Table 4.1. The results for MESFET threshold voltages of -0.6 and -1.0 V are both recorded in the table. The channel width of the 1-μm gate length MESFETs used for the simulation depended on the FET threshold voltage: 7 μm for Q_1, 10 μm for Q_2, and 5 μm for Q_5 ($V_T = -1.0$ V) and 17/10/5 respectively for $V_T = -0.6$ V. Note that the logic high-voltage level, V_{OH}, is clamped to approximately 0.7 V (for the $V_T = -0.6$ V case) by the fan-out loading of subsequent stages. These variables were evaluated with power supply voltages of $+2.5$ and -2.0 V which are adequate from a noise margin point of view for either of the V_T cases. It can be seen from the reduction in V_{OH} for $V_T = -1.0$ V that the pull-up or load MESFET, Q_2, is no longer in saturation. There is not enough current for V_{OH} to reach the forward gate conduction limit of ≈ 0.7 V.

The deficiency of pull-up current will have a detrimental effect on the low-to-high propagation delay, t_{PLH}. As described in Sec. 3.3, the current available to charge the load capacitance is governed by the difference between the drain currents of Q_2 and Q_5. If Q_2 has not yet reached saturation, then its current will be significantly less than optimum, and the delay will suffer. V_{DD} then should be selected so that it just saturates Q_2, that is, $V_{DD} - (V_{OH} + 2V_D) = V_{D,\text{sat}}$ in order to minimize t_{PLH}. Here, V_D is the forward voltage drop across the level-shift diode. $V_{D,\text{sat}}$, the drain-to-source voltage required to saturate the drain current, depends on the threshold voltage

TABLE 4.1
Typical dc characteristics of the unbuffered FET logic gate of Fig. 3-12 when simulated with 1-μm GaAs d-mode MESFETs

Variable	$V_T = -0.6$ V		$V_T = -1.0$ V	
V_{OH}	0.7	V	0.6	V
V_{OL}	-1.2	V	-1.1	V
V_{NL}	0.83	V	0.6	V
V_{NH}	0.83	V	0.6	V
V_{MS}	0.64	V	0.34	V
V_{DD}	2.5	V	2.5	V
V_{SS}	-2.0	V	-2.0	V
Power dissipation	0.9	mW	2.5	mW
W_S	17	μm	7	μm
W_L	10	μm	10	μm
W_{PD}	5	μm	5	μm

and gate length of the FET; it will decrease as V_T becomes more positive and as gate length L becomes shorter, as described in Sec. 1.3.

The high-to-low propagation delay, t_{PHL}, may be reduced for the same reason by decreasing the V_{SS} supply voltage (more negative) to the point where V_{DS} of the pull-down FET, Q_5, is just in the saturation region. The effect of the switch FET, Q_1, on t_{PHL} is assumed to be not significant, as the level-shift diode will typically be cut off by Q_1 before it can participate in discharging the load capacitance if the diode capacitance is much smaller than the load capacitance.

The optimization of delay is not without penalty in power dissipation which will be increased for larger V_{DD} or more negative V_{SS}. Until saturation is reached, the drain current is also significantly increasing with an increase in supply voltage leading to P_D proportional to V_{DD}^2 or V_{SS}^2. Beyond the point where the active-load FETs are saturated, power increases approximately linearly with the respective supply voltage. The large increase in power dissipation with more negative threshold voltage indicated in Table 4.1 is predicted from the (assumed)[1] square law current-voltage relationship for the FET in saturation, $I_D = \beta W(V_{GS} - V_T)^2$.

4.1.2 Capacitive Feedforward Logic Circuits

One of the deficiencies in the design of the unbuffered FET logic circuit is that the current available for charging the load capacitance is fixed by the difference between pull-up (Q_L of Fig. 3-12) and pull-down (Q_{PD}) active-load currents. Since the pull-down size is fixed by the current required for discharging the load capacitance, there is a significant amount of power dissipated just in the static biasing of the level-shift network. If this current could be reduced somehow without degrading the propagation delay, then the overall power-delay product of the gate might be greatly improved. Fortunately, this type of modification is possible and has been demonstrated to be effective [2–4].

Consider the schematic diagram of the modified FL gate shown in Fig. 4-1. The analysis of t_{PHL} on the FL gate showed that the level-shift diodes cut off during the discharge of the load capacitance, and thus the current-sinking capability of the switch transistor (Q_1) for discharging this load is lost. The width of this FET (W_1) is dictated by noise margin requirements. From the example data for a two-level-shift FL gate in the previous section, W_1 is often substantial and would significantly help in reducing t_{PHL} if it could be coupled to the load. This is accomplished in Fig. 4-1 by increasing the capacitance of the level-shifting network [3,4]. A large-area reverse-biased diode (D_4), usually in parallel with a metal-insulator-metal overlap capacitor (C_D), is placed in parallel with the level-shift diodes D_1 to D_3. This additional shunt capacitance provides a low impedance (*feedforward*) path for the high-frequency signal components which comprise the rise or fall of the output waveform. In other words,

---

[1] This approximation is most accurate for V_{GS} near threshold. Since I_D is set by the active-load FETs, $V_{GS} = 0$. If $V_T < -0.6$ V, the I_D-V_T relationship will generally become more linear due to source resistance and velocity-saturation effects. Thus, the square law estimate is only an approximation.

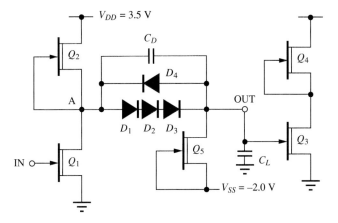

FIGURE 4-1
Schematic diagram of the unbuffered FET logic gate modified to include capacitive feedforward coupling (CDFL).

Q_1 can now sink current from the discharge of C_L, the load capacitance, due to the displacement current through D_4 and C_D. This approach has been called *capacitor-diode FET logic*, or CDFL [4]. The capacitance of D_4 and C_D must be moderately large in comparison with $C_L + C_{in}$ (where C_{in} is the effective input capacitance of Q_3) in order for this technique to be effective, since the high-frequency path sees a capacitive voltage divider, and the pulse will be reduced in amplitude by the capacitive divider ratio:

$$V_{out} = V_{DS}\left(\frac{C_D + C_{D4}}{C_D + C_{D4} + C_L + C_{in}}\right) \qquad (4.1)$$

While this requirement will present some limitations on the maximum-load capacitance for a given level-shift capacitance, it is possible to extend the technique by partitioning C_L and placing part of it at node A in Fig. 4-1. This could be accomplished by locating the pull-down and diodes at the input of the load (Q_3) so that the interconnect capacitance would be directly coupled to the drain of Q_1.

Three diodes in series are used for level shifting for two reasons: first, the current through the diodes is now greatly reduced since W_5 can be made much smaller, leading to less voltage drop under forward bias, and, second, the noise margin for a $V_T = -1.0$ V process can be increased by using a larger level shift. W_2 can also be reduced in comparison with the FL circuit since the static current wasted by the level-shift diodes is now much less.

The power supply voltages for this circuit will be determined by the need to saturate Q_2 and Q_5. The propagation delay t_{PHL} may be reduced by increasing V_{DD} up to the point where V_{OH}, the logic high-voltage level, becomes clamped; beyond this point, the drive current of Q_1 will no longer increase. Therefore, one would not expect the delay to be reduced by further increases in V_{DD}. In fact, if the transconductance of the FET begins to drop at V_{GS} levels near the forward conduction limit, as is typical if the source resistance is not minimized or for first-generation HFETs (Sec. 1.5), I_D

of Q_1 will not increase in proportion to the logic swing, and t_{PHL} will increase as a consequence of the following equation:

$$I = C\frac{dV}{dt} \qquad (4.2)$$

The simulation of the dc and transient characteristics of the CDFL circuit is left to the reader as a homework exercise.

4.1.3 Source Follower Design— Buffered FET Logic

It is normal in the design and layout of logic circuits to find that the load capacitance for gates and inverters can vary over a wide range depending on the fan-out requirements and the length of the interconnection lines between the output and the load. Since the propagation delay of the circuit is predicted by Eq. (4.2), it will be necessary to increase the current available to charge and discharge the load capacitance when heavy loading is present. There are two ways that this can be done. In the cases considered above, the FL gate was modified with the addition of a coupling capacitor (CDFL gate) in order for the switch FET (Q_1) to contribute its current to the discharging and to make more current from Q_2 available for charging C_L. In this approach, a modification of the circuit is required. This scheme increased the peak-load current–static current ratio, which means that the circuit is now a more efficient driver of the load capacitance. The propagation delay will be reduced even though the power dissipation of the CDFL circuit is less than that of the FL circuit; thus, there is a decrease in the power-delay product.

It is also possible, of course, to proportionally increase the widths of all transistors driving the large load so that more current will be available. This possibility was discussed in Sec. 3.3.4. If the widths were severely undersized for the application, this scaling operation will keep the power-delay product constant until the device capacitance becomes comparable with the load capacitance. At this point, further width increases are very costly in terms of power dissipation.

Circuit modifications that fall into the first category are preferred and should be further considered if loading is large. A very popular circuit which can be used to reduce propagation delay is the source follower shown in Fig. 4-2. The source follower or common drain circuit is characterized by very high current gain but low voltage gain, always less than unity. The operation of the circuit and the low voltage gain that results can be understood by considering the I_D-V_{DS} device characteristics for a depletion mode MESFET, Q_{SF}, shown in Fig. 4-3. Here, V_{GS} is a parameter held constant on each curve plotted in the figure. If the current source I_{PD} were ideal ($r_{pd} = \infty$), then the load line (a) will be horizontal. In this example, I_{PD} was selected so that V_{GS} of the source follower FET is close to 0 V. The output voltage,

$$V_{\text{out}} = V_{DD} - V_{DS} \qquad (4.3)$$

will be given by the intersection of load line (a) and the I_D; thus very little change in V_{GS} will be necessary to change V_{out} over its full range. If V_{GS} remains nearly constant, then V_{out} follows V_{in}—hence the name of the circuit.

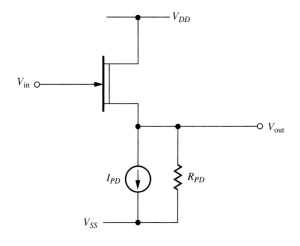

FIGURE 4-2
Source follower. The pull-down current source often uses an active-load FET ($V_{GS} = 0$ V).

Figure 4-3 also illustrates some factors that would produce a voltage gain less than unity. For example, the load line (*b*) represents the more typical case for a finite r_{pd}, provided by a depletion mode MESFET connected as an active load with $V_{GS} = 0$ V. Because of the slope of load line (*b*), a greater change in V_{GS} is now needed to change V_{DS} over its full range. A similar effect will occur if the output conductance of Q_{SF} were increased, making the slope of the device characteristics larger in the saturation region. Finally, if V_{DS} of Q_{SF} or Q_{PD} were to fall below $V_{D,\text{sat}}$, typically 0.5 to 1.0 V depending on the threshold voltage, then the output voltage will change very little with large changes in V_{GS} (clipping). This condition can be avoided by the proper selection of V_{DD} and V_{SS}.

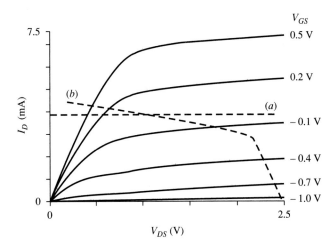

FIGURE 4-3
Load-line construction for source follower: (*a*) Ideal current source and (*b*) nonideal current source with shunt resistance r_{pd}.

Example 4.1. Calculate the ac small-signal voltage gain of the source follower circuit shown in Fig. 4-2. Include the effect of incremental output resistances r_{pd} of the pull-down and r_{ds} of Q_{SF}. Comment on the accuracy of the analysis.

Solution. Redraw the circuit including the small-signal model for the MESFET as shown in Fig. 4-4. The ac small-signal model is enclosed in the dotted lines. From the diagram it can be seen that

$$v_{in} = v_{gs} + v_{out} \tag{4.4}$$

where the lower case v implies ac (sinusoidal steady-state) voltage. Also,

$$v_{out} = g_m v_{gs}(r_{pd} \| r_{ds}) \tag{4.5}$$

Substituting Eq. (4.5) into Eq. (4.4) and solving for v_{out}/v_{in}, the voltage gain is obtained:

$$A_v = \frac{v_{out}}{v_{in}} = \frac{1}{1 + 1/[g_m(r_{pd} \| r_{ds})]} \tag{4.6}$$

If both r_{ds} and r_{pd} were infinite, then the voltage gain is equal to unity.

The accuracy of this analysis depends on g_m, r_{ds}, and r_{pd} remaining constant for all v_{in} values of interest. Since V_{GS} tends to remain constant, g_m should not vary greatly. If the SF and PD FETs remain in saturation, the drain resistances should also be relatively constant. The result will break down completely if either FET drops out of the saturation region. In that case, a large-signal analysis using the device equations will be needed to evaluate A_V.

The input impedance of a source follower is very high because V_{GS} remains nearly constant. C_{GD} is very small since Q_{SF} is always in the saturation region, and the voltage across C_{GS} is V_{GS}. The effective small-signal input capacitance is $C_G = C_{GD} + C_{GS}(1 - A_v)$, where A_v is the voltage gain. The real part of the

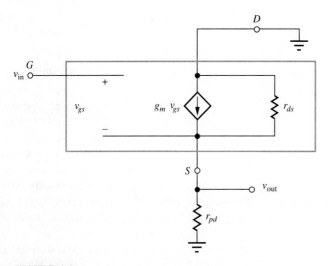

FIGURE 4-4
Small-signal model for source follower stage.

input current will be very small if V_{GS} always remains below the forward conduction voltage. This low input capacitance is beneficial since the loading on the internal logic gate circuitry can be minimal if the source follower is used to buffer the output.

The source follower is also characterized by a low output impedance. Small-signal analysis shows that the midfrequency output impedance approaches $1/g_m$. This is much lower than the output impedance of a common-source stage (r_{ds}), since a change in output voltage of the source follower leads to a change in v_{gs} if v_{in} is held constant. This change is amplified by the dependent current source to produce a large change in current. The low-impedance property is useful for driving large loads.

Next, the transient behavior of the source follower must be considered. In most applications, large capacitive loads are being driven, and equal rise and fall times are desired. The voltage across the load capacitor will change relatively slowly compared with V_{in}. The current needed to charge or discharge the capacitor will be the difference between I_{SF} and I_{PD}. Charging current will result when the V_{GS} of Q_{SF} exceeds its steady-state value. From Fig. 4-3 it can be seen that the drain current of Q_{SF} can approximately double when V_{GS} increases from -0.1 to 0.5 V. If the load causes the output voltage to resist change while V_{in} increases, then V_{GS} will quickly increase and charging will begin. It will end when $I_{SF} = I_{PD}$. The capacitor will be discharged by I_{PD} because Q_{SF} will cut off when V_{in} rapidly decreases. Thus, to obtain approximately equal rise and fall times when driving a constant load capacitance, the following condition must be satisfied:

$$I_{SF}(V_{GS} = 0.7 \text{ V}) - I_{SF}(V_{GSQ}) = I_{PD} \tag{4.7}$$

This condition can be satisfied by selection of the proper steady-state V_{GSQ} for the source follower FET which is determined by the width ratio of Q_{SF} to Q_{PD}. Therefore, the ratio of the peak load current to the static current of the source follower is equal to unity.

Example 4.2. Compare the peak load current–static current ratio of the FL inverter to the same ratio for a source follower.

Solution. Consider the FL inverter circuit shown in Fig. 3-32. In accordance with the design examples in Sec. 3.3, equal rise and fall times for a fixed capacitive load will occur when $W_5 = W_2/2$. The peak-load current for charging will be $I_{DSS}(W_2 - W_5)$ and for discharging, $I_{DSS}W_5$. The average static I_{DD} will be approximately $I_{DSS}W_2$ since this will either flow through Q_1 (output low) or through Q_5 and the gate of Q_3 (output high). Therefore, the peak load current–static current ratio for the FL gate is $W_5/W_2 = 0.5$, so the source follower (ratio = 1) is twice as efficient in driving a capacitive load for the same static I_{DD}. As will be seen below, the propagation delay is lower for the SF than the FL driver for the same power and loading conditions.

The excellent drive capability of the source follower has led to its incorporation into a logic gate [5]. A schematic of a three-input NOR gate, implemented with the SF as an integral part of each logic gate, is illustrated in Fig. 4-5. This circuit approach is called *buffered FET* logic (BFL) because of the noninverting internal driver with low output impedance (buffer). Note that the level shifting required for depletion mode logic is accomplished by including forward-biased diodes between the SF FET (Q_5)

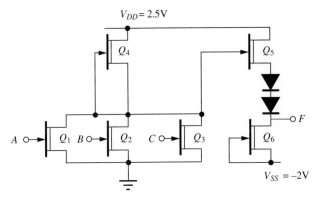

FIGURE 4-5
Schematic of three-input NOR gate implemented with buffered FET logic (BFL).

and the pull-down FET (Q_6). The internal high-impedance node (D) is buffered from the output load capacitance by Q_{SF} as discussed above. Power supply voltages should be selected so that Q_{SF} and Q_{PD} are in the saturation region at all times if a full logic swing on the output is desired. The width ratio of Q_4 to Q_1, Q_2, or Q_3 will be dictated by noise margin considerations, where the objective is to obtain equal low and high noise margins while driving a single input. The other inputs are held at V_{OL}. The dc characteristics of a BFL gate are summarized in Table 4.2.

A simulated comparison between the FL NOR gate and a BFL NOR gate with the same power dissipation (5 mW/gate) is summarized in Fig. 4-6 where load capacitance is the independent variable. For the fan-out = 1 case shown, the FL gate is actually faster than BFL when C_L is less than 15 fF, a length of interconnect less than 100 μm if on the lowest (first) level. The reduced slope for the BFL gate illustrates its superior drive capability which would result in significantly reduced propagation delays if long interconnect lines were being driven. At fan-out = 2 and above, the BFL circuit is always faster.

The FETs that create the logic function are well buffered from the load capac-

TABLE 4.2
Typical dc characteristics of the buffered FET logic gate of Fig. 4-5 when simulated with 1-μm GaAs d-mode MESFETs

Variable	$V_T = -0.6$ V	$V_T = -1.0$ V
V_{OH}	0.66 V	0.58 V
V_{OL}	−1.2 V	−1.1 V
V_{NL}	0.98 V	0.74 V
V_{NH}	0.80 V	0.74 V
V_{MS}	0.65 V	0.50 V
V_{DD}	2.5 V	2.5 V
V_{SS}	−2.0 V	−2.0 V
Power dissipation	1.5 mW	4.1 mW

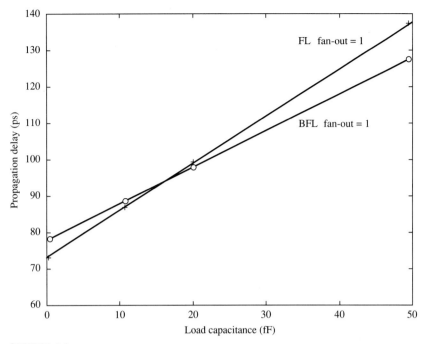

FIGURE 4-6
Propagation delay (SPICE simulation) as a function of load capacitance is compared for FL and BFL circuits. $V_{DD} = 2.5$ V, $V_{SS} = -2.0$ V, and power dissipation = 5.0 mW for both circuit types.

itance by the source follower. Therefore, it is possible to use narrower FET widths for Q_1 to Q_4 of Fig. 4-5 than for Q_{SF} and Q_{PD}. The optimum width ratio of Q_1 to Q_{SF} for minimum propagation delay will be a function of the load capacitance.

In the same manner as FL (Sec. 3.2.3), a two-input NAND function can also be obtained with BFL by using series transistors on the input. More than two inputs on a NAND may result in problems with noise margins. A series/parallel combination sharing a common pull-up active-load transistor can be used to perform an AND/OR/INVERT function.

The efficiency of the buffer can be increased even further through the use of either a dynamic pull-down circuit or a quasi-complementary buffer (superbuffer). The first of these circuits, illustrated in Fig. 4-7, is more suitable for use with depletion mode FETs; the superbuffer will prove to be useful with e/d circuits and will be discussed later in this chapter. The dynamic pull-down uses capacitor coupling to temporarily modify the gate voltage on the pull-down FET to improve on the peak load current–static current ratio. At steady state, the circuit will have the same biasing conditions and thus the same power dissipation as an equivalent BFL gate. During the transition, V_{GS} of the PD device is either increased or decreased dynamically to make more current available in the discharging or charging of the load respectively. The RC time constant is selected so that the dynamic BFL gate will recover to its steady-state condition before the next transition occurs. R can be implemented with either an

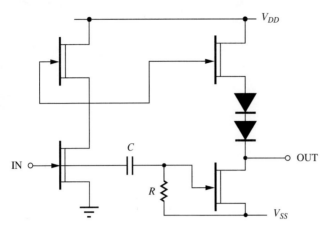

FIGURE 4-7
Schematic of BFL gate with dynamic pull-down.

implant resistor or a weak pull-down with minimum channel width and longer gate length (> 1 μm) if necessary.

4.1.4 Schottky Diode FET Logic

In the circuits discussed above, the logic function has always been performed by FETs, in parallel for the NOR function, in series for a NAND, or series/parallel combinations for more complex functions. The role of the diodes has been strictly for level-shifting purposes. However, the diodes are also well suited for use as switching elements: they have low capacitance, low series resistance, and there are no minority carrier charge storage problems typical of *pn* junction diodes. The area of a diode can be made smaller than a FET; therefore if logic is performed by diodes, some savings in area may be possible.

The Schottky diode FET logic (SDFL) circuit was designed so that diodes could be used to perform logic with a net savings in area and power dissipation [6]. A circuit diagram for a three-input SDFL NOR gate is shown in Fig. 4-8. In this circuit, diodes D_1 to D_3 are logic diodes (described in Chap. 1) which provide both level shifting and the logical OR function on the input. If any one input (A, B, or C) is pulled high, then V_{GS} of Q_1 is high and the output, F, will be low. Diode D_4 is strictly for level-shifting. These diodes can be made quite small in area since only small currents are needed for forward-biasing; a 1×2 μm overlap area has been successfully used for logic diodes in minimum-size SDFL gates [6]. FETs Q_1 and Q_2 serve as an inverter to restore the logic swing after the diode logic is performed.

The width ratio W_2/W_1 will set the threshold voltage for the transfer characteristic, V_{TH}, as defined in Chap. 3. Ideally, W_3 can be selected strictly on dynamic grounds; the pull-down current, I_3, is responsible for discharging the gate-to-source capacitance of Q_1 when the input voltage changes from high to low. I_3 is also needed to forward-bias the diodes. This means that the output of an SDFL gate must provide static current to its fan-out loading from Q_2. Therefore, the fraction of I_2 remaining

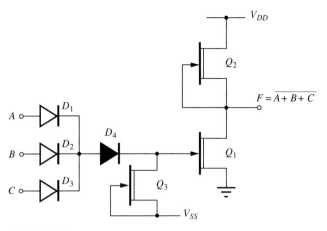

FIGURE 4-8
Schematic of three-input SDFL NOR gate.

to pulled low by Q_1 depends on the fan-out (FO) and, as a consequence, the logic gate threshold voltage and noise margins also depend on fan-out. This limitation sets a practical ratio of $W_2/W_3 > 4$ in order to maintain an adequate noise margin at FO = 3. The minimum width of a MESFET is usually constrained by the implant design rule to 2.5 or 3 μm. Therefore, if 1-μm gate length FETs are used throughout, the minimum W_2 will be at least 10 or 12 μm, and the lowest power dissipation per stage will be constrained by these widths. This problem can be avoided by increasing the gate length L of Q_3 from 1 μm to 1.5 or 2 μm, thereby reducing I_{DSS} of Q_3 by a factor of 1/L. An analytical model that relates the design parameters to the dc logic levels, noise margins, and speed has been reported [7].

The dc characteristics of an SDFL logic gate are summarized in Table 4.3. Here, a three-input NOR gate with $W_1 = 15\ \mu\text{m}$, $W_2 = 17\ \mu\text{m}$, and $W_3 = 3\ \mu\text{m}$

TABLE 4.3
Typical dc characteristics of the Schottky diode FET logic gate of Fig. 4-8 when simulated with 1-μm GaAs d-mode MESFETs. The MESFET threshold voltage is −1 V

Variable	FO = 1	FO = 2	FO = 3
V_{OH}	2.3 V	2.2 V	2.1 V
V_{OL}	0.35 V	0.30 V	0.25 V
V_{NL}	0.82 V	0.78 V	0.66 V
V_{NH}	0.76 V	0.78 V	0.80 V
V_{MS}	0.46 V	0.45 V	0.42 V
V_{TH}	1.37 V	1.25 V	1.12 V
V_{DD}	2.5 V	2.5 V	2.5 V
V_{SS}	−1.5 V	−1.5 V	−1.5 V
Power dissipation	2.6 mW	2.6 mW	2.6 mW

was simulated using the SPICE3 MESFET model. The widths were selected as a compromise so that acceptable noise margins could be obtained at both FO = 1 and FO = 3. Although the noise margins are relatively high, the shift in V_{TH} with FO will erode some of it in normal circuit applications.

Assuming that the SDFL gate (X_1) is driving identical gates as its load, if the fan-out exceeds 3, the total width of pull-downs being driven (FO × W_3) will approach the width of W_2. Therefore, V_{OH} will drop and V_{NL} will become too small for good functional yield. This condition requires either that the widths of the FETs

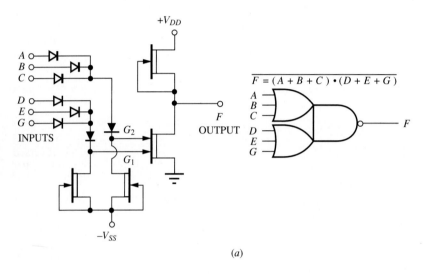

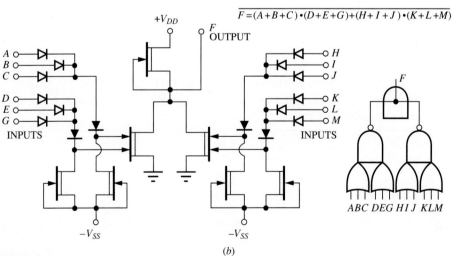

FIGURE 4-9
Multilevel SDFL logic gates: (*a*) OR/NAND, (*b*) OR/NAND/ wired AND. (*From Eden et al.* [8]. © *IEEE. Used by permission.*)

in X_1 be scaled up, or a buffer such as in Fig. 4-2 must be added between the gate output and the load. The disadvantage of the former solution is that W_3 must be increased proportionally with W_1 and W_2; therefore any SDFL gates driving X_1 may be in danger of the same error (and so on back down the line!). Therefore, any customizing of widths to accommodate fan-out or to reduce propagation delay may induce a cascade of design changes. The latter solution is safer, except that the buffer approximately doubles the power dissipation per gate, bringing SDFL into the same power range as BFL.

More complex logic functions are also possible by combining diodes with series transistors as in the OR/NAND gate illustrated in Fig. 4-9a or by sharing a pull-up active-load FET as in the OR/NAND/wired AND gate shown in Fig. 4-9b [8]. These multilevel logic approaches are good at saving power and area since the added diodes are small, and only the added pull-down FETs, usually minimum width, dissipate power.

Two-level diode logic has also been proposed as a means of power and area savings [9], but can be shown to be much less effective. Figure 4-10 contains a schematic of an AND/OR/INVERT logic gate using logic diodes with minimum width (W_3) pull-downs for the OR and logic diodes with a twice minimum width pull-up ($W_4 = 2W_3$) for the AND. The order of logic operations could also be reversed (OR/AND/INVERT). At least a 2 to 1 current ratio must be maintained between W_4 and W_3 so that the voltage at node E will approach V_{OH} when A*B = 1 and to equalize rise and fall times of the input diode logic. The larger width of W_4 creates an even more severe fan-out limitation on the SDFL output switch transistor, now endangering V_{OL} if FO × $W_4 \geq W_1$. In addition, the power dissipation of the gate is increased significantly by the added pull-ups, which are now each half the width of

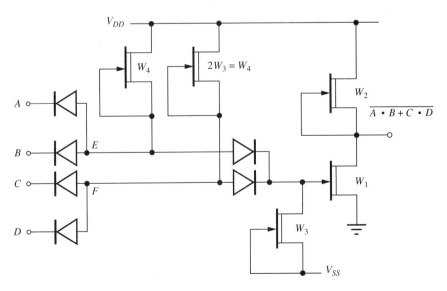

FIGURE 4-10
Schematic of an AND/OR/INVERT logic gate using two levels of diode logic.

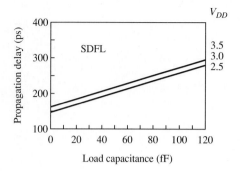

FIGURE 4-11
The average propagation delay measured on nine-stage SDFL ring oscillators with fan-out = 2 as a function of load capacitance. (*From Ekroot and Long [10]. © IEEE. Used by permission.*)

W_2 (if $W_2/W_3 = 4$). Finally, the propagation delay and the power-delay product of the two-level diode logic gate is inferior to normal SDFL or the multilevel SDFL of Fig. 4-8 [9].

The transient behavior of SDFL ring oscillators with fan-out = 2 and fixed capacitive loading has been reported [10], and the experimental propagation delay as a function of load capacitance and V_{DD} is shown in Fig. 4-11. The ring oscillators were composed of SDFL gates with 1-μm gate length MESFETS with widths of $W_1 = 15$ μm, $W_2 = 12$ μm, and $W_3 = 3$ μm, a design best suited for FO = 1. Power dissipation per gate was 2.1, 2.7, and 3.3 mW at $V_{DD} = 2.5$, 3.0, and 3.5 V respectively. As V_{DD} increased from 2.5 to 3.0 V, the propagation delay also increased, indicating that the logic voltage swing was increasing while the charging current was relatively constant. No further increase in delay was observed for V_{DD} over 3.0 V; the input logic diodes and gate diode have clamped V_{OH} at this point. A fan-out loading factor was also measured; the delay increases by 38 ps when fan-out is increased from 1 to 2.

4.2 ENHANCEMENT MODE LOGIC CIRCUITS

Logic circuits can also be implemented with enhancement mode MESFETs or MODFETs that have $V_T \geq 0$ V. In this section, V_{TE} and V_{TD} will be used to distinguish between the threshold voltage of e-mode and d-mode FETs as they are often both used in the same circuit. Since both V_{GS} and V_{DS} are always positive, no level shifting from output to input is needed, and a single power supply voltage, V_{DD}, is usually sufficient. Also, the saturation voltage $V_{D,sat}$ of an enhancement mode FET is less than that of a depletion mode FET. Thus, power supply voltages can be lower for e-mode circuits than their d-mode equivalent. The above properties can save power and layout area.

The gate-to-source voltage will be constrained (without the use of level-shifting or dynamic circuit techniques) to a minimum of 0 V and to a maximum of 0.7 V by forward conduction. The small logic swing will provide high speed if the device transconductance is sufficiently high at small V_{GS}. This requirement emphasizes optimum design of the e-mode FET (as discussed in Chap. 1) to minimize source resistance by self-alignment or recessed-gate fabrication techniques and to minimize the depth of the gate depletion layer to increase the intrinsic g_m.

The small range of V_{GS} also requires uniformity of threshold voltage. Both the mean V_T and the standard deviation on a single wafer and wafer to wafer must be very tightly controlled so that the parametric yield of large circuits will be finite. In this area, the circuit design technique can play an important role in reducing sensitivity of circuits to parameter variations.

An additional problem is also encountered in the design of e-mode GaAs circuits. An active load can be provided with an e-mode MESFET by connecting gate and drain together. While this can be used as a substitute for a resistor, it is not suitable for use in this form for high-speed circuits, since the charging current diminishes as the voltage across the load capacitor increases. Also, the resistor or diode-like load line produces low voltage gain which will reduce the slope in the transition region of the V_{out} vs. V_{in} transfer characteristic as illustrated by Fig. 3-10a, leading in turn to small noise margins. In the GaAs MESFET or MODFET wired with $V_{GD} = 0$ V, the gate conduction also limits the maximum V_{DS}. Therefore, d-mode GaAs FETs are also required for active-load current sources in most e-mode circuits.

4.2.1 Direct-Coupled FET Logic (DCFL)

The first, simplest, and most widely used e-mode GaAs logic circuit is the direct-coupled FET logic circuit (DCFL). A schematic diagram for this circuit is shown in Fig. 4-12. In this drawing, the symbol for the d-mode MESFET is distinguished from the e-mode by a double bar on the channel. Note the superficial similarity of this circuit to E/D NMOS circuits (see Fig. 3-1 for a direct comparison).

Example 4.3. Determine $V_{OH}, V_{OL}, V_{NL}, V_{NH}, V_{MS}$ from the transfer characteristic of the DCFL NOR gate of Fig. 4-12. Use the device characteristics for the e-mode MESFET in Fig. 4-13 to determine the transfer characteristic graphically.

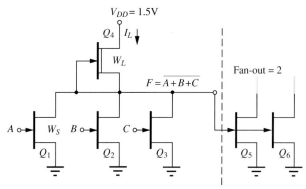

FIGURE 4-12
Schematic of direct-coupled FET logic DCFL 3-input NOR gate. FO = 2.

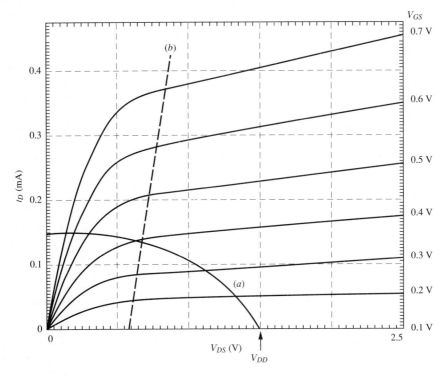

FIGURE 4-13
Load line for a d-mode active load (a) and gate-source diode (b) constructed on I_D-V_{DS} characteristics for e-mode MESFET.

Solution. The voltage swing of DCFL can be easily determined from a load-line construction on a set of I_D-V_{DS} characteristics as shown in Fig. 4-13. The approach is similar to that used in Chap. 3 to graphically derive the transfer characteristic for the FL gate. The e-mode MESFET in this example can be seen to have a threshold voltage of about 0.15 V. The $V_{GS} = 0$ V curve can be obtained from the corresponding d-mode device characteristics and plotted on Fig. 4-13 [line (a)]. A V_{DD} of 1.5 V is assumed. The unused inputs are connected to ground.

The input characteristic of the fan-out loading must also be plotted on these characteristics. For this example, the input characteristic will be the I_G vs. V_{GS} of two forward-biased gate diodes in parallel. For convenience, a linear approximation can be used since the diode current will be limited by the R_S of the load FETs after the current level is high enough to make the diode incremental resistance small. Therefore, a line with slope FO/R_S is drawn [dashed line (b) on Fig. 4-13] starting at $I_D = 0$ from the turn-on voltage (0.6 V for a typical Schottky barrier on n-type GaAs at a low current level). While the value of R_S should be taken from measured data, for this example the typical value from Table 2.4, 900 Ω, will be assumed.

From this construction, it can be seen that $V_{OH} = 0.67$ V, which occurs at the point of intersection between lines (a) and (b). V_{OH} will be clamped by the gate/source junctions of the fan-out (Q_5 and Q_6 in Fig. 4-12) load. The logic low voltage ($V_{OL} \approx 0.15$ V) can be determined by the intersection of load line (*a*) and the e-mode FET characteristic for $V_{GS} = V_{OH}$. For this example, V_{TH} can be estimated from the characteristics to be $V_{in} = V_{GS} \approx 0.45$ V.

The load-line construction represents a particular width ratio which was selected when the d-mode load line was plotted. For that ratio, the transfer characteristic can be derived by finding the intersection points at the available values of $V_{GS}(V_{in})$. A plot of these points is shown in Fig. 4-14. From this figure, $V_{OL} = 0.17$ V. V_{TH} is 0.45 V, as surmised above. The accuracy of the noise margins will be affected by the small number of data points plotted, but a useful estimate can still be made from measuring the maximum width of loops and the maximum square. From Fig. 4-14, $V_{NL} = 0.2$ V, $V_{NH} = 0.17$ V, and $V_{MS} = 0.10$ V.

If desired, V_{OL} could be made quite low by using a very large width ratio W_S/W_L. The logic gate threshold voltage V_{TH} would approach V_T at this limit. This limit is also approached in a high fan-in NOR gate if all of the inputs are switched in tandem. V_{TH} could then be increased by making the threshold voltage of the e-mode FET, V_{TE}, more positive. Then, larger V_{GS} is needed to turn on the switch and so the V_{TH} and V_{NL} are both increased. The combination of large W_S/W_L and high V_{TE} (0.25 to 0.30 V for MESFETs) will improve the noise margin of the DCFL gate, but at some cost in speed. The high V_{TE} will reduce $V_{GS} - V_T$ for the switch MESFETs (Q_1 to Q_3), thus reducing I_D available per unit width and increasing t_{PHL}. Also, the small W_L will reduce the current available to charge the load capacitance, thus raising t_{PLH}.

From this example, it can be seen that the noise margin of DCFL gates is rather small, limited on the high side by the forward gate conduction of the next stage and

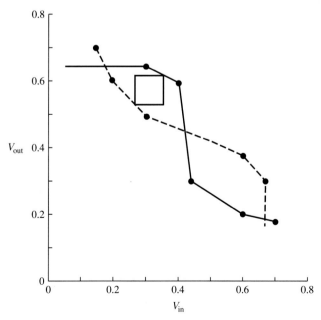

FIGURE 4-14
Plot of the DCFL gate transfer characteristic as determined by the intersection points in Fig. 4-13.

on the low side by the saturation voltage of the e-mode FET. Because of the low noise margin, the NAND function obtained by series connections of e-mode MESFETs is not recommended. If higher V_{OH} can be obtained through increased barrier height (MODFET, WN gate, JFET), then the NAND would be feasible.

V_{TE} and the width ratio plays an important role in the sensitivity of the noise margin to parameter variations. If W_S/W_L is high and V_{TE} is about 0.25 V, then the design will be less sensitive to variation in threshold voltages, but the current drive capability of Q_S is diminished by the relatively small voltage range above threshold ($V_{GS} - V_{TE} = 0.4$ V). This reduced current drive will increase t_{PHL}. On the other hand, if a low V_{TE} (0.1 V) is selected to obtain a lower propagation delay, the width ratio must be made smaller to center V_{TH} in the voltage swing, and the sensitivity of noise margin to FET threshold voltage changes is quite high. A summary of the simulated dc characteristics of a typical DCFL gate is given in Table 4.4.

The fan-out and capacitive load sensitivity for the DCFL gate is moderately high, but less so than the d-mode FL circuit, since the peak load current–average current ratio is unity; i.e., when the input is low, Q_S is cut off and the active-load current I_L is available to charge the load capacitor. If the gate is designed for equal rise and fall times, then the switch FET current at $V_{GS} = V_{OH}$ should be twice I_L. Thus the net discharge current is also I_L under this condition. Very low propagation delays (< 10 ps/gate) have been reported for DCFL ring oscillators with minimal loading [11]. With realistic fan-out and loading, the measured propagation delay increases significantly to the 100 ps and higher range depending on the loading.

The peak load current–static current ratio can be improved significantly through the use of a quasi-complementary output driver (superbuffer), as illustrated in Fig. 4-15. FETs Q_1 and Q_2 are a DCFL inverter; Q_3 is connected as a source follower driven from the inverter output. Q_4 is a pull down, active when the input is high, but is cut off for a low input. Dynamically, the operation is slightly more complicated. The L to H transition on the output follows the inverter output. After t_{PLH} of the inverter has elapsed, Q_4 is already cut off, and the load capacitance will receive the

TABLE 4.4
Typical dc characteristics of the direct coupled FET logic gate of Fig. 4-12 when simulated with 1-μm GaAs e-mode MESFETs. The d-mode threshold voltage is $V_{TD} = -0.7$ V. Fan-out = 2. $R = W_S/W_L$

Variable	R = 4	R = 10
V_{TE}	0.2 V	0.2 V
V_{OH}	0.62 V	0.58 V
V_{OL}	0.18 V	0.08 V
V_{NL}	0.21 V	0.24 V
V_{NH}	0.11 V	0.18 V
V_{MS}	0.07 V	0.13 V
V_{DD}	1.5 V	1.5 V
Power dissipation	0.3 mW	0.2 mW

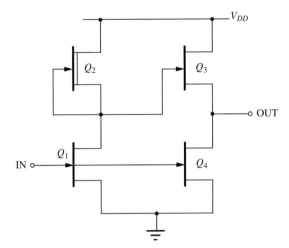

FIGURE 4-15
Schematic of quasi-complementary inverter (superbuffer).

full current of Q_3. However, on the H to L transition, Q_4 begins to discharge the load before the inverter has cut off Q_3. Therefore, on this transition, there will be a momentary current spike on V_{DD} and ground during which both Q_3 and Q_4 are in conduction. As a consequence, circuits that use superbuffers must be designed with ample power bus width so that a momentary voltage drop during the H to L transition does not dynamically upset other logic elements. Nevertheless, after the propagation delay of the inverter has passed, the load receives the full current of either Q_3 or Q_4. The static power of the superbuffer driver is the same as that of the inverter.

The superbuffer concept can also be expanded beyond an inverter into a NOR gate called superbuffer FET logic (SBFL), as shown in Fig. 4-16 [12]. The exclusive

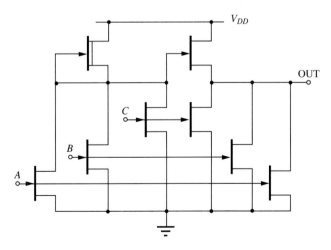

FIGURE 4-16
Three-input NOR gate implemented with SBFL (superbuffer FET logic).

use of superbuffer gates on a chip will result in slightly higher noise margins than DCFL because of the reduced V_{OL} of SBFL. Since Q_4 is never fighting against a static pull-up, V_{OL} can be close to 0 V. Then, V_{TE} and the width ratio can be reduced for higher speed. While standard DCFL gates could also be mixed into the circuit (input and output levels are compatible), the noise margin advantage will be lost because of the higher V_{OL} of the DCFL gates. Since the effective input FET width is increased by using the SBFL design, the input capacitance and likewise the load capacitance per fan-out will be higher. However, the better current drive capability still yields lower delay for SBFL. In a comparison between SBFL and DCFL ring oscillators [12] with 1-μm gate length MESFETs, for a fan-out = 3 and 0.1 mm of interconnection length, the SBFL circuit produced 178 ps/gate while the DCFL produced 276 ps/gate. A 40-ps propagation delay was measured with a power dissipation of 0.35 mW/gate and a 27 ps/fan-out loading factor in a rather extreme design (from the perspective of noise margin) with $V_{TE} \approx 0$ V [13]. Multiplexers and demultiplexers with a 2-GHz clock rate were also reported in this work.

The low (200 mV or less) noise margins which are obtained with standard MESFET DCFL circuits places much emphasis on uniformity of material and process parameters. As discussed in Sec. 3.2, small variations in the FET threshold voltages can reduce noise margins even further. As a result, there have been efforts reported to increase the logic voltage swing by increasing the barrier height of the gate Schottky or by use of JFETs rather than MESFETs. As discussed in Chap. 1, the MODFET gate forward conduction limit is slightly higher than that of a MESFET as well. In addition, a number of circuit design alternatives have been reported which may alleviate the strict process uniformity required by DCFL. A few of these designs will be presented in the following sections.

4.2.2 Source-Coupled FET Logic

Differential amplifier circuit structures have been successfully used for logic applications for years. Bipolar ECL or CML circuits are the most widely recognized examples. This same technique has also been employed in GaAs MESFET logic circuits. A source-coupled FET differential amplifier circuit, shown in Fig. 4-17, serves as the basis for expansion into a source-coupled FET logic (SCFL) [14,15] and will be discussed below.

There are several potential advantages of the differential circuit over more conventional logic circuits which switch between the cutoff and ohmic regions of the FET. Firstly, if current sources with high incremental output resistance are available, differential circuits provide the benefit of good common mode rejection. A common mode input signal is one that drives both inputs of a differential amplifier in the same direction, rather than in opposite directions as for a differential mode signal. These signal voltages are defined in Fig. 4-17 as V_{DM} and V_{CM}. The ideal differential amplifier should produce no output for a common mode input. In a well-designed op-amp, the ratio of differential mode gain (A_{DM}) to common mode gain (A_{CM}) can be greater than 10^4. This property is an advantage for logic applications because any wafer-to-wafer variation in e-mode FET threshold voltage is a common mode voltage and will not strongly affect the switching threshold of the gate.

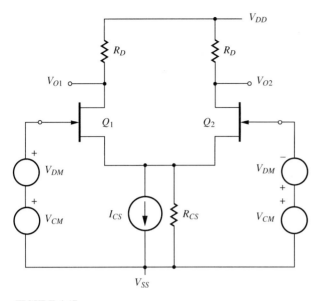

FIGURE 4-17
Schematic of source-coupled FET differential amplifier. The differential input voltage sources are V_{DM} and the common mode input sources are labeled V_{CM}.

It is instructive to see that it is not simple to achieve near-ideal properties in a MESFET differential amplifier. Referring to the diagram in Fig. 4-17, the small-signal A_{DM} and A_{CM} are given by

$$A_{DM} = \frac{V_{O1}}{V_{DM}} = -g_{m1}(R_D \| r_{ds1}) \qquad (4.8a)$$

$$A_{CM} = \frac{V_{O1}}{V_{CM}} = \frac{-g_{m1}R_D}{1 + 2g_{m1}R_{CS}} \qquad (4.8b)$$

where g_{m1} and r_{ds1} are the transconductance and drain resistance in saturation of the e-mode FET, Q_1. From Eq. (4.8b), it is seen that R_{CS} must be very large in order to diminish the CM gain. Assume that d-mode active-load FETs with channel widths W_L and W_{CS} are used to implement R_D and I_{CS} respectively. Then, the incremental drain resistance of the pull-up FET in the saturation region is r_{ds}/W_L, where W_L is the width of the pull-up FET and r_{ds} is the incremental drain resistance normalized to 1 μm of width, and the pull-down current source I_{CS} will have a drain resistance of r_{ds}/W_{CS}. From Eq. (4.8b), assuming that $g_m R_{CS} \gg 1$, the CM gain reduces to

$$A_{CM} = -\frac{W_{CS}}{2W_L} \qquad (4.9)$$

Therefore, A_{CM} is approximately unity.

The differential gain, therefore, should be made as high as possible in order to obtain a good CMRR *(common mode rejection ratio)*, equal to 20 log $|A_{CM}/A_{DM}|$

dB. The differential gain (4.8a) can be rewritten in terms of the FET widths, and with g_m and r_{ds} normalized to 1 μm:

$$A_{DM} = -(g_m r_{ds})\left(\frac{W_1}{W_1 + W_L}\right) \qquad (4.10)$$

Typical values for 1×1 μm MESFETs at high frequency are $g_m = 0.15$ to 0.2 mS and $r_{ds} = 100$ kΩ, which gives a $g_m r_{ds}$ product of about 10. Since the width ratio in Eq. (4.10) will be in the range of 0.5 to 0.9, differential gains between 5 and 10 are to be expected. Likewise, the CMRR is only about 10 (20 dB). To obtain higher differential gain and lower common mode gain requires better design techniques which incorporate cascodes, bootstrapped current sources, and common mode feedback loops into the stage [16,17]. Unfortunately, these additions require larger power supply voltages than are desirable for low power-delay product logic circuits, and so are generally not used. Even so, the limited performance of the simple differential amplifier logic with MESFETs provides a significant improvement in tolerance to FET threshold voltage variations when compared with DCFL, in which the differential mode and common mode input voltages are the same.

Another advantage of the differential circuits for logic applications is the low C_{GD} and high g_m of Q_1 and Q_2 which results from both devices always being biased in the saturation regions. This condition leads to higher cutoff frequencies for the SCFL circuit than are obtained in a conventional common-source connection; thus, the switching speed of source-coupled logic should be better than DCFL.

Figure 4-18 shows a circuit diagram of a three-input OR/NOR SCFL gate implemented with all e-mode FETs. To benefit from the CMRR, both inputs and outputs are fully differential, resulting in a larger number of interconnections than single-ended logic circuits. However, in many cases, fewer gates are required to realize common logic functions (EOR, XOR, D flip-flop) if both data and its complement are available, so there will be a net savings in power and area in these cases. Source followers are included on the outputs to minimize loading of the SCFL circuit and to improve drive capability as discussed in Sec. 4.1.3; however, these can be omitted if loading is light. Level shifting is required between stages to avoid driving the differential amplifier transistors into the ohmic region. Also, when the series transistor approach is used as in Fig. 4-18, level shifting is needed so that inputs at the B or C level of the next stage can be driven from outputs (OR_B, OR_C, NOR_B, NOR_C) which will maintain proper biasing of their respective FETs. The single-diode drop level shift per FET implies that the saturation voltage, $V_{D,\text{sat}}$, of the e-mode FETs must be less than 0.7 V if these devices are to remain in saturation during normal operation. Fortunately, this is generally the case.

Resistors (R_D) are used on the drain node in the circuit of Fig. 4-18 rather than FET active loads in order to simplify the biasing of the SCFL gate. When d-mode active-load FETs are used for pull-ups, it often proves difficult to maintain all FETs in the saturation region with reasonable power supply voltages, a necessary condition if the differential gain is to be adequate and power per gate low. However, active loads have been successfully used in analog GaAs differential amplifier circuits which operate at higher power supply voltages [17].

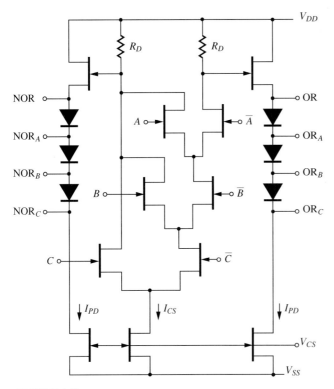

FIGURE 4-18
Source-coupled FET logic (SCFL) three-level OR/NOR gate.

Since the current source and pull-down MESFETs are also e-mode, V_{CS} must exceed $V_{SS} + V_{TE}$. V_{CS} can be available off-chip as an adjustable bias voltage although it could be generated on-chip. A current mirror connection for V_{CS} is also helpful in reducing sensitivity of NM to changes in V_{TE} or V_{TD}.

A typical V_{DD} voltage for the SCFL circuits has been selected on the basis of compatibility with ECL or TTL or CMOS. A V_{DD} of 0 V and V_{SS} of -4.5 or -5.2 V are normally the design target.

The SCFL circuit can be designed for a specific V_{OH} and V_{OL} according to the requirement of the application by selection of FET widths, the source current I_{CS}, and the load resistor R_D. The noise margins will vary along with the logic swing according to the design. There are two conditions that must be satisfied by the design. Firstly, the logic swing, ΔV should not exceed $0.7 - V_T$. Otherwise, the FETs will be switching between the ohmic region and cutoff, which defeats the SCFL design goal of keeping the FETs in saturation at all times. Secondly, there must be enough level shifting between input and output so that the drain-to-source voltage V_{DS} is large enough to avoid the ohmic region when the input is at its maximum voltage. The relationship between these variables can be seen by performing an approximate analysis using the SPICE JFET model equations described in Chap. 2.

Example 4.4. Design the SCFL circuit in Fig. 4-19 given the following specifications:

1. Logic high, $V_{OH} = 1.3$ V
2. Logic low, $V_{OL} = -1.7$ V
3. $V_{DD} = 0$ V; $V_{SS} = -5.0$ V
4. $P_D = 5$ mW

Solution. The design variables R_D, I_{CS}, and W need to be determined. Assume that the transconductance parameter, $\beta = 1 \times 10^{-4}$ A/(V²·μm), the parasitic source resistance, $R_S = 1500\ \Omega\cdot\mu$m, and the FET threshold voltage, $V_T = 0.1$ V, are known.

Firstly, I_{CS} can be determined from the power dissipation specification. Assume (arbitrarily) that the power is equally divided between the two source followers and the logic circuit. The proper partitioning for highest speed depends on the load capacitance and can be determined by simulation, but this exercise will not be carried out in this example. If $V_{DD} - V_{SS}$ is 5 V and the power per gate is 5 mW, then $I_S = I_{DD} = 1$ mA. One-third of this is allocated to I_{CS}, so $I_{CS} = 0.33$ mA. Then, with I_{CS} determined, R_D can be found from the middle of the output voltage range (average of the logic voltage swing) measured at the drains of Q_1 and Q_2.

However, the amount of level shifting must first be established so that the drain voltages of Q_1 and Q_2, V_{D1} and V_{D2}, can be found. If the output is taken from B in Fig. 4-19, then the 0.7-V diode drop and the voltage shift across the source follower are added. The source follower will have at least a drop of V_T, but in order to obtain reasonable g_m (proportional to $V_{GS} - V_T$) and to maintain reasonable width for Q_3 and Q_7, some positive V_{GS} must be allowed. In this example, the midpoint between V_T and 0.7 V, 0.4 V, will be selected for V_{GSQ} (the dc quiescent point). Then, the total level shift, V_{LS}, from V_{D2} to V_B is 1.1 V. So, from the specification for the logic voltages,

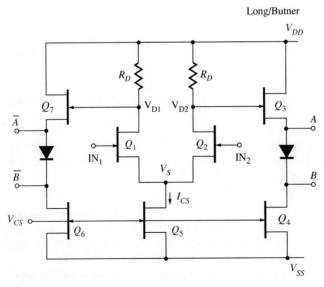

FIGURE 4-19
Source-coupled OR/NOR gate for Example 4.5.

V_{D1} and V_{D2} range from -0.2 to -0.6 V. R_D can be determined from

$$\frac{-0.2 + (-0.6)}{2} = \frac{I_{CS} R_D}{2} \tag{4.11}$$

This equation comes from the fact that the source current I_{CS} is split equally between Q_1 and Q_2 when the outputs are equal (at the midpoint of the logic swing). From this, R_D is found to be 2400 Ω.

The width of Q_1 and Q_2 (W) can be derived from the device drain current vs. V_{GS} equation. If the source resistance is included and the FETs are assumed to be always in the saturation region, then Eq. (2.6) from Chap. 2 can be used to estimate W. β and R_S are normalized to a 1-μm channel width:

$$I_D = \beta W \left[V_{GS} - V_T - \frac{I_D R_S}{W} \right]^2 \tag{4.12}$$

If the same V_{GSQ} is used for Q_1 and Q_2 as was selected for Q_3 and Q_7 (the midpoint between threshold and 0.7 V is reasonable so that the margin against cutoff or ohmic regions is provided and gain is large), then the maximum V_{GS} will be $V_{GSQ} + (V_{OH} - V_{OL})/2 = 0.6$ V. When V_{GS} is maximum, then the output voltage is at V_{OL} and $V_{D1} = -0.6$ V. Therefore, the maximum I_{D1} is

$$I_{D,\max} = \frac{V_{OL} + V_{LS}}{R_D} = \frac{V_{D1}}{R_D} = 0.25 \text{ mA} \tag{4.13}$$

Substituting $V_{GS,\max}$ and $I_{D,\max}$ into Eq. (4.12) and solving for W, a channel width of 11.5 μm is obtained.

Finally, it is important to check the biasing of Q_1 and Q_2 to make sure that $V_{DS} > V_{D,\text{sat}}$ when the input is at its extreme voltages. $V_{GSQ} = 0.4$ V was selected for this example; thus the source voltage V_S will remain at a constant voltage 0.4 V below the midpoint of the input logic swing:

$$V_S = \frac{V_{OL} + V_{OH}}{2} - V_{GSQ} = -1.9 \text{ V} \tag{4.14}$$

Since the minimum V_{D1} or V_{D2} is -0.6 V, $V_{DS,\min} = 1.3$ V, more than adequate to avoid the ohmic region. In fact, it may be possible to use the output A to drive the next stage, since $V_{DS} = 0.7$ V would still be available.

The current source active load Q_5 and pull-downs Q_4 and Q_6 are implemented in Fig. 4-19 with e-mode MESFETs. This is possible since a separate gate voltage $V_{CS} > V_{SS}$ is provided for control of the current, biasing V_{GS} above threshold. A possible advantage of the use of e-mode active loads is that only one type of FET needs to be included in the process. A disadvantage would be that another power supply voltage (and package pin) will be required for the implementation shown. A current mirror could also be used to provide V_{GS} by the addition of one more transistor. This would eliminate the need for V_{CS} and also will provide some compensation for temperature and threshold variations. The width of Q_5 can be determined from I_{CS} and the design value of V_{CS} using Eq. (4.12).

Finally, the design of the source follower stage can be established using the same techniques as described in Sec. 4.1.3. FET widths would be determined by the power budget for the source follower.

Once an approximate set of design parameters has been determined by analysis, the circuit should be further optimized by simulation. This process could include both common mode and differential mode input voltage sweeps, the common mode to verify that the selected Q point provides adequate V_{DS} and V_{GS} range for Q_1 and Q_2 and the differential mode to determine voltage swing and noise margins. Note that $V_{\text{out}} = V_B - V_{\bar{B}}$ and $V_{\text{in}} = \text{IN}_1 - \text{IN}_2$ for the purpose of finding the differential mode noise margin.

The simulated transfer characteristic will often show signs of limiting since the differential voltage gain often exceeds unity. Thus, in reality, Q_1 and Q_2 may be driven into cutoff or the ohmic region. In some cases, clamp diodes or series limiting diodes are used to create a limited voltage swing while maintaining Q_1 and Q_2 in the saturation region.

The D flip-flop is a good example of the high circuit complexity per logic gate that is possible with SCFL (or ECL) [18]. Figure 4-20 shows a diagram of a master-slave D flip-flop implemented with SCFL two-level OR/NOR gates. This circuit is directly analogous to ECL flip-flops. The clock inputs are applied on the lower level and the data inputs on the upper level. Only two SCFL gates are required to implement this circuit. If standard DCFL NOR gates were to be used, at least eight gates would

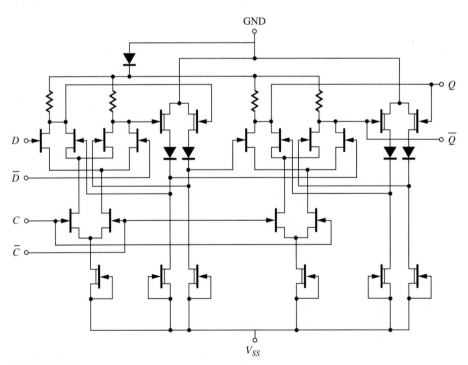

FIGURE 4-20
Master-slave D flip-flop implemented with two SCFL gates. (*From Tamura et al.* [18]. *Used by permission.*)

be needed to achieve the same function. Therefore, the functional equivalence of each SCFL OR/NOR gate in this application is 4. Power dissipation of the SCFL gates is ordinarily higher than DCFL, even with the high functional equivalence considered, since a 5 V instead of 1.5 V power supply is used. Nevertheless, the reduced sensitivity to threshold voltage variation, higher speed, high functional equivalence, and, in some cases, reduced layout area are strong advantages of SCFL when compared with many of the other circuit possibilities.

4.2.3 Enhancement/Depletion Buffered FET Logic

In Sec. 4.2.1, the DCFL gate was shown to have low noise margin and relatively high sensitivity to fan-out and capacitive loading. Efforts to improve on these shortcomings, particularly important for large-scale integration where uniformity is critical and gate array applications where loading is large, have resulted in an enhancement/depletion (e/d) buffered FET logic (BFL) circuit in which a source follower output driver is used to reduce delay [19]. A diagram of a double two-input NOR gate e/d BFL circuit cell is shown in Fig. 4-21.

In order to understand the operation of this circuit, first consider the two-input NOR gate in the left half of Fig. 4-21. The sources of the input FETs Q_1 and Q_2 are biased positively by one diode drop above V_{SS}. Note that the diode is intended to be biased globally rather than by individual gates and so its voltage drop (0.6 to 0.7 V) acts as a reference voltage. This reference voltage can also be used to forward-bias the gate of the d-mode pull-down FETs (Q_5, Q_7) if desired, so that narrower widths can be used for some savings in area. In the analysis to follow, $V_{GS} = 0$ on Q_5 and Q_7. If a typical $V_{TE} = 0.2$ V is assumed, then the source follower will level-shift down by some amount exceeding this voltage because it must be biased in the saturation

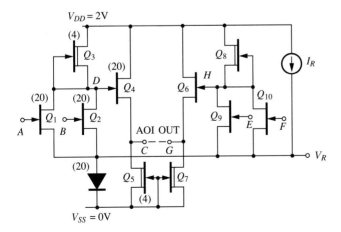

FIGURE 4-21
Enhancement/depletion (e/d) buffered FET logic circuit. A source follower output driver is used to reduce delay [19].

region. If a V_{GS} of 0.4 V were assumed for the source follower and $V_{DS} = 0.2$ V for Q_1 or Q_2, then the logic low voltage level for the output (C) would be about 0.5 V, which actually reverse-biases the input to the next stage slightly, thereby increasing the V_{NL}. The logic high level will be limited by the forward conduction of the next stage to under 1.4 V. As a result, the voltage swing has increased from about 0.5 to 0.6 V for DCFL to about 0.9 V for the e/d BFL circuit. Noise margins will be 0.3 to 0.4 V, a substantial improvement.

Transfer characteristics for the e/d BFL circuit consisting of half of Fig. 4-21 are shown in Fig. 4-22 for V_{TE} of 0.1 and 0.2 V. The corresponding V_{TD} was −0.5 and −0.7 V respectively. The device widths are also in the figure enclosed in parentheses. The input voltage is applied at A, B is kept at 0 V, and the output voltage taken from C. The logic input threshold V_{TH} is too high for symmetric noise margins for the 0.2 V case. Because of the 0.7 V diode shift on the source of Q_1 and Q_2, V_{in} must be larger than 0.9 V before these FETs are even out of the cutoff region. Thus, even with very large ratios of width between Q_1, Q_2, and Q_3 (undesirable for speed),

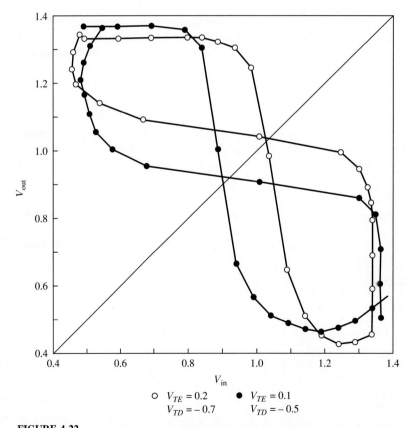

FIGURE 4-22

Transfer characteristic simulated for e/d BFL gate with two threshold voltage combinations: $V_{TE} = 0.1$ V; $V_{TD} = -0.5$ V and $V_{TE} = 0.2$ V; $V_{TD} = -0.7$ V. Device widths are as shown in Fig. 4-21.

V_{TH} cannot be reduced sufficiently to produce equal low and high noise margins. However, by reducing V_{TE} from 0.2 to 0.1 V, this problem can be solved, as shown in Fig. 4-22. Here, nearly equal low and high NM are obtained, with $V_{NL} = 0.37$ V and $V_{NH} = 0.43$ V.

Increased functional complexity per gate can be achieved by the use of a source follower OR connection, as illustrated by the dashed line connecting nodes C and G in Fig. 4-21. This combines two two-input NOR gates to provide an AND-OR-INVERT (AOI) function. The SF OR will provide a high output if either half of the multilevel gate is high. The other SF, if at logic low, will be reverse-biased and therefore inactive under this condition. Some loss in noise margin will be inevitable with the SF OR. This is because the width ratio of SF (Q_4 and/or Q_6) to pull-down ($Q_5 + Q_7$) depends on the logic state. If $W_4/W_5 = W_6/W_7 = R$, then the ratios shown in Table 4.5 will apply if a single pull-down is connected. In this table, D and H refer to the inputs to the SF OR, labeled accordingly in Fig. 4-21. The symbol 0 refers to a logic low level and 1 to the high level. In the 0,0 state, the normal ratio is obtained and V_{OL} should be at the same level as the single two input NOR. In the 0,1 and 1,0 states, less than the normal V_{OH} for a single NOR gate will be obtained. For the 1,1 state, the ratio will increase, and V_{OH} should be at its normal level again.

The power dissipation for the e/d BFL circuit was reported to be 2.5 mW [19] for the two-gate configurable cell shown in Fig. 4-21. The NOR gate cell had an unloaded propagation delay of 120 ps, with added delay factors of 19.5 ps/fan-out and 0.59 ps/fF of load capacitance.

4.2.4 Pass Transistor Logic

Pass transistor logic (also known as transmission gate logic) is a circuit design technique that is widely used in NMOS and CMOS to save power and layout area. It can find application for both static and dynamic logic circuits. In the section that follows, only the static applications will be discussed; dynamic logic is the subject of Sec. 4.3. The pass transistor logic approach is illustrated by a simple example in Fig. 4-23. The output of two inverters (1 and 2) are selected by the pass transistors Q_1 and Q_2 according to control inputs A and $\overline{A}$ so that one of the two inverters (but not both!) drives the input to inverter 3. This two-to-one multiplexer circuit could

TABLE 4.5
Width ratio of two-input source follower OR logic (see Fig. 4-22). D and H are the logic states of each input to the SF OR. $W_4/W_5 = W_6/W_7 = R$

D	H	Ratio
0	0	R
0	1	R/2
1	0	R/2
1	1	R

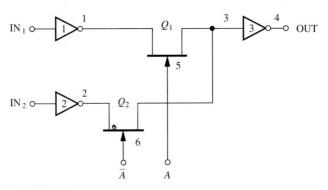

FIGURE 4-23
Two-to-one multiplexer with pass transistor logic.

also be realized with three two-input NOR gates, but there is a significant saving in power and layout area if the function can be accomplished instead by two FETs.

Although the advantages of this approach are quite evident, it is important to understand the problems that must be addressed before pass transistor logic can be used reliably. The logic circuits which have been discussed so far have been *restoring* in function; i.e., they regenerate the full logic swing after each stage, thus helping to recover noise margin and attenuate noise. The pass transistor logic, as will be described below, is *non-restoring*; there is a loss in logic swing which must be regenerated after one or two stages of pass transistors by a standard static restoring logic circuit.

Also, generic to any application of pass transistor logic is the potential for timing problems and unintended data paths. Control signs A and $\overline{A}$ in Fig. 4-23 must be properly synchronized with the outputs from inverters 1 and 2 so that data glitches or the wrong logic value does not propagate to inverter 3. If A and not A are both high at the same time due to clock skew, for example, then the output of 1 and 2 will be connected together, an unintended connection. The resulting width ratio will not be proper, and the logic voltages may not be valid.

An additional problem is also prevalent when only *n*-channel FETs are available, as in most GaAs circuits. If the pass transistor is e-mode, then its output voltage (source voltage) can be no higher than $V_{GS} - V_{TE}$, even if the drain voltage is greater than this voltage. In other words, a minimum voltage drop between gate and source equal to the threshold voltage of the pass transistor is required to open the channel. If the drain voltage is higher, the pass transistor will enter the saturation region. If d-mode pass transistors are used, then the threshold drop problem can be avoided. However, a gate voltage of at least $V_{TD} + V_{OL}$ will be required to cut off the pass transistor. This more negative voltage is often inconvenient to obtain without adding an extra power supply and introducing more level shifting.

The application of pass transistor logic to GaAs FETs is further complicated by the forward gate conduction problem. The gate conduction can degrade both V_{OH} and V_{OL} because this conduction will occur on the gate of the pass transistor itself, forcing current into the output of the driving gate (inverter 1 or 2 of Fig. 4-23). The input to the driven gate (inverter 3) will also clamp the logic swing when the output of the pass transistor is high. The current required by this clamping will cause an added voltage drop across the pass transistor, further reducing V_{OH}. The use of enhancement mode pass transistors is also constrained by the threshold drop problem discussed above; the control voltage on the gate of the pass transistor must be at least $V_{OH} + V_{TE}$ to avoid further reduction in V_{OH} at the output.

Since the voltage swing will be reduced by the pass transistor logic, and noise margins are already on the verge of being too low for circuits such as DCFL, only circuits with higher logic swing such as SBFL or e/d BFL should be considered for use with pass transistors. An example using two e-mode pass transistors with e/d BFL gates will be presented to illustrate the design process.

Example 4.5. Pass transistor logic. Design a multiplexer circuit that uses pass transistors instead of NOR gates to accomplish a two-to-one input selection. Use the e/d BFL gate described in Sec. 4.2.3 as the restoring logic for the data path. $V_{DD} = 2.0$ V.

Solution. A two-to-one multiplexer can be implemented by pass transistor logic as shown in Fig. 4-23. The challenge in this design is to select the logic voltage levels and threshold voltages so as to maximize the noise margin at the input to inverter 3. The e/d BFL circuit is chosen for this application because it has a much larger logic swing ($\Delta V = 0.9$ V) and higher noise margins than DCFL or SBFL, and some degradation of both is expected when pass transistors are inserted into the data path. From simulations of e/d BFL transfer characteristics (Fig. 4-22), it is clear that the optimum NM is obtained with V_{TE} of 0.1 V and V_{TD} of -0.5 V, so this will be the starting point for the design. The data shown in Fig. 4-22 was simulated for an e/d BFL inverter with the following FET widths: e-mode, 20 μm; d-mode, 4 μm. Referring to Fig. 4-21, the d-mode pull-down (Q_5) is connected with $V_{GS} = 0$ V, and the source follower OR was not connected. The circuit will also be evaluated at $V_{TE} = 0.2$ V and $V_{TD} = 0.7$ V in order to estimate the sensitivity to variation in threshold voltages.

Next, a decision must be made as to whether to use e-mode or d-mode FETs for the pass transistors, and an estimation of control voltage levels will be required. First consider the ON state of the pass transistor with the output of inverter 1 at V_{OH}. Although this could rise to $V_{DD} - V_{TE}$ if unloaded, the logic high state will be clamped by gate conduction to about 1.4 V for the e/d BFL with one fan-out load. If one or more pass transistors are inserted into the data path, there will be a voltage drop due to the input current required by inverter 3, and the FET channel resistance will degrade V_{OH}. Figure 4-22 shows that V_{NH} is the smallest of the two noise margins, especially for $V_{TE} = 0.2$ V, so it may be difficult to obtain adequate V_{NH} for an e-mode pass transistor, since the channel resistance in the ohmic region will be larger than for an equivalent d-mode FET.

Therefore, the d-mode MESFET should be used. The d-mode MESFET can achieve higher gate voltage above threshold before gate conduction sets in; therefore larger current can flow for a given V_{DS}. With the drain at V_{OH}, the control voltage for a d-mode MESFET must be a minimum of $V_{OH} + V_{TD}$ (0.9 V for $V_{TD} = -0.5$ V). A higher voltage (1.4 V) will be needed if V_{OH} is to be kept as high as possible at the source. Note that even with 1.4 V on the gate, an e-mode pass transistor would just barely be in conduction, and the source voltage will drop to 1.2 V, too low for a good noise margin.

Now consider the ON state when the output of inverter 1 is at V_{OL} (about 0.5 V). If the gate of the pass transistor is maintained at 1.4 V, the gate conduction of this FET will pull V_{OL} up to about 0.8 V, resulting in a significant reduction in V_{NL}.

Since we are reluctant to give up any NM unnecessarily, it will be desirable to modify the control-line drive circuit so that the control voltage will drop under dc loading. An approach to accomplish this will be shown after the OFF state requirements are considered.

The control stage necessary to guarantee that the pass transistor is off is a maximum of $V_{OL} + V_{TD}$ (0 V for $V_{TD} = -0.5$ V and less than that for more negative thresholds). The maximum OFF state voltage also does not provide any allowance for subthreshold leakage or threshold variations. Therefore, to effectively use the d-mode pass transistor in this application, a negative OFF state control voltage will be needed. While this is a nuisance since another power supply will be required, remember that two supplies are routinely used in the d-mode logic circuits anyway (Sec. 4.1) and that it is more important to design a dependable circuit than one that is attractive, but on the borderline of disaster.

The design of a quasi-complementary (superbuffer) control-line driver which will satisfy the above requirements is shown in Fig. 4-24. The superbuffer topology is desirable for power savings and drive efficiency; Q_3 and Q_4 do not draw static current, and because V_{OL} will be very close to V_{SS}, a smaller V_{OL} will be possible. A V_{SS} of –0.5 V would be sufficient to guarantee that the pass transistor will be off. The circuit is slightly modified relative to that shown in Sec. 4.2.1 by the addition of D_1. This level-shift diode causes V_{OH} to drop with dc loading on the output, as is needed to prevent degradation of V_{NL} by gate conduction, as discussed above. With no static loading, the output voltage will rise to nearly 1.5 V. A possible drawback may be that logic high state will charge more slowly once the diode current begins to drop. The input to the circuit is shifted by two diode drops to make the logic level compatible with e/d BFL. If C can be made large compared with the gate capacitance, then the pull-down current through Q_5 can be made very small, as was true for the depletion mode CDFL circuit discussed in Sec. 4.1.2.

The transfer functions for the e/d BFL multiplexer circuit shown in Fig. 4-23 were simulated for the $A = 1$ and $\overline{A} = 0$ states. Figure 4-25 plots the input to Q_1,

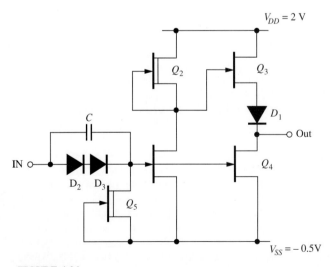

FIGURE 4-24
Level-shifted superbuffer driver for the control inputs (A and $\overline{A}$) of the pass transistor multiplexer.

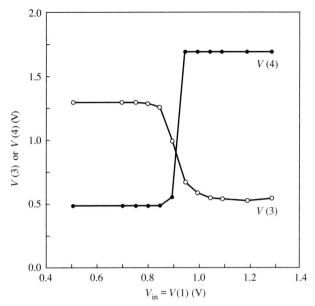

FIGURE 4-25
Simulated logic voltages at nodes 3 and 4 of the two-to-one multiplexer of Fig. 4-23.

$V(1)$ and outputs $V(3)$ and $V(4)$ voltages, when IN_1 is ramped from 0.5 to 1.5 V. The important design variables obtained from the simulation are summarized in Table 4.6. It can be seen from the table that a V_{NH} of 0.36 V is predicted in the best case, while in the worst case it falls to 0.23 V. V_{NL} is always higher. This worst case will be sufficient for many applications, even with a high level of integration, if good control of threshold voltage is provided by the process.

TABLE 4.6
Low and high noise margins and logic levels at the input to inverter 3 of Fig. 4-23 were determined by simulation with the d-mode pass transistor in the ON state. Two typical e- and d-mode MESFET threshold voltages were used in the simulation along with a V_{DD} of 2.0 V to evaluate the design latitude for threshold variation

Variable	Voltage (V)			
V_{TE}	0.1	0.1	0.2	0.2
V_{TD}	−0.5	−0.7	−0.5	−0.7
V_{OH}	1.29	1.31	1.28	1.29
V_{OL}	0.52	0.52	0.45	0.46
V_{NL}	0.36	0.40	0.51	0.54
V_{NH}	0.36	0.33	0.27	0.23
V_{MS}	0.24	0.23	0.22	0.19

The example above illustrated the balance between gate conduction, threshold drops, and subthreshold conduction which must be established over a range of device parameters if pass transistor logic is to be used successfully with GaAs MESFETs. If the forward conduction limit is higher, as is the case for JFETs or some types of heterojunction FETs, then the problem becomes somewhat less constrained.

4.2.5 Complementary JFET Logic

Although the barrier height of metal Schottky barriers on *p*-type GaAs is too low to be useful for MESFET gate electrodes, the *pn* junction can be used to control the channel in a junction FET. The JFET has the advantage of tolerating relatively high forward voltage (1.0 V) prior to the onset of gate conduction. This property will improve noise margin over that of similar MESFET circuits by roughly 0.1 to 0.2 V. Both *p*- and *n*-type devices can be made which opens up the door for complementary logic.

Unlike silicon CMOS, where the ratio of electron mobility to hole mobility is a factor of 3, *p*- and *n*-type GaAs JFETs exhibit a ratio $\mu_n/\mu_p = 10$. The transconductance factors (β) have about the same ratio; therefore, a complementary JFET logic circuit will require a width ratio $W_P/W_N = 10$ to provide equal rise and fall times. The large additional channel width of the *p*-JFET will load the circuit with extra capacitance and reduce the overall speed [20]. Little has been published on the propagation delay properties of this type of circuit.

Nevertheless, there may be certain applications (such as memory circuits) that require the ultra-low static power which the complementary circuit can provide. A circuit diagram of a six-transistor static RAM cell is shown in Fig. 4-26. This uses

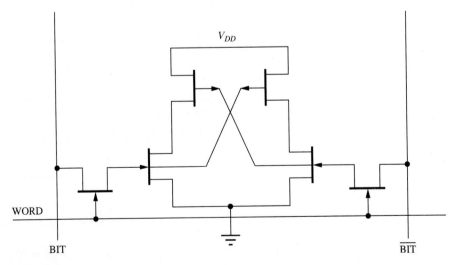

FIGURE 4-26
Circuit diagram of six-transistor static RAM cell using complementary JFET logic.

p-JFETs to pull up the outputs to V_{DD}. The bit lines are accessed through *n*-JFET pass transistors. These cells have been used in 4K × 1 SRAMs, which provided access times of about 10 ns with chip power dissipation of 200 mW. The standby power per cell was very low, about 4 μW/bit [21]. Complementary JFET logic has also been used in the design of a 32-bit RISC architecture microprocessor [22].

4.3 DYNAMIC LOGIC CIRCUITS

Much of the motivation for using E/D logic circuits with GaAs MESFETs, JFETs, or MODFETs has been to obtain low power per gate and high circuit density so that LSI and VLSI circuits can be achieved within a manageable thermal budget and chip size. It is also important to maintain the high-speed operation that is the main reason for using GaAs logic rather than CMOS. Thus, the power-delay product is an important factor which can be reduced by circuit design tradeoffs and topologies.

Several *static* logic circuit approaches, in which direct current flows at all times from the power supply, were described in Secs. 4.1 and 4.2. Static logic is usually dependent on width ratios to control logic levels and noise margins, and hence these circuits are often referred to as *ratioed logic*. For a given logic circuit, speed and power are inversely exchanged over a wide range as device widths are scaled (see Sec. 3.3.4). Changes in FET threshold voltages also strongly influence both speed and power. Power reduction in static logic also requires circuits and devices which operate efficiently with low logic voltage swings so that power supply voltages can be reduced. Logic swing reduction also implies using FETs with more positive threshold voltage so that the full on-to-off ratio of current can be achieved. Finally the device widths are also reduced so that supply currents can be made small. While all of these factors contribute to lower dc power dissipation, the lower voltage swing reduces noise margin as well, requiring more stringent process control to achieve acceptable circuit yield as discussed in Chap. 3. For VLSI circuits, high yield is not just desirable but is essential for successful commercial production. Finally, the smaller current levels often lead to increased propagation delay.

While there have been many demonstrations of LSI circuits with some of the above techniques, the higher complexity examples have demanded very strict control of process parameters and uniformity of device parameters. Hence there is an interest in alternative circuit techniques that offer power reduction in combination with an acceptable noise margin.

Dynamic logic circuits have been widely used in silicon MOSFET ICs for power conservation, increasing functional complexity per stage and increasing the overall circuit density. A dynamic circuit performs logic by the storage and evaluation of charge on circuit nodes which can be isolated temporarily from the rest of the circuit. Since the charge cannot be stored indefinitely, there is a minimum clock frequency for operation below which the logic voltage levels may no longer be valid. Logic levels are not established through width ratios, but by clock voltages and device thresholds; hence dynamic circuits are often described as being *nonratioed*. Therefore, FET widths can be selected to optimize speed. Circuit nodes are precharged to establish their reference level and to further reduce delay. Thus, there are more degrees of freedom in designing a dynamic circuit, which makes the design more difficult but improves the overall performance. The CMOS domino circuit is a good example of a dynamic logic circuit that has enjoyed popularity because it substantially reduces the number of *p*-channel MOSFETs required to achieve a given circuit function, it is fast, and it conveniently works with a single-phase clock [23].

Dynamic logic has been relatively unexploited with GaAs FET technology, possibly because of the concern for operation over a wide range of clock frequencies. The earlier applications were mainly oriented toward very-high-speed SSI circuits, such as frequency dividers [24,25] and shift registers for prescalers and digital RF memories [26]. The results have been very impressive, with 26 GHz clock frequencies [25] being observed for dynamic divide-by-two circuits. Power dissipation has been relatively high for these applications. More recently, low-power GaAs dynamic logic has also been a focus of interest [27], with projections indicating that high functional complexity and low power-delay product can be obtained.

4.3.1 Charge Storage Limitations

Store of charge on an isolated node is the central design principle for dynamic circuits. The length of time that the charge can be held without loss of data is proportional to the noise margin and inversely proportional to the leakage currents affecting the isolated mode. In silicon MOS circuits, the principal sources of leakage current are the reverse-biased junctions isolating the MOSFET sources and drains from their wells and the subthreshold current of the transistors. The gate itself is well isolated by the gate oxide. Also, both NMOS and CMOS circuits have large V_{NH}, providing a wide tolerance for leakage.

In GaAs dynamic circuits, however, the voltage swing is, as usual, constrained by the gate-source diode. To prevent unacceptably high leakage current, most of the forward-bias region must be avoided. Therefore, depletion mode FETs appear more attractive than enhancement mode ones if the gate is to be connected to the storage node. For the same reason, noise margins are lower for GaAs MESFETs than for Si MOSFETs. Sources of leakage current include the gate junction, subthreshold current in cutoff FETs, and substrate conduction, all of which are strongly temperature-activated. Thus, for the reasons above, GaAs FET circuits will exhibit a shorter holding time for the stored charge than their silicon counterparts, and they will require more frequent refreshing or precharging in order to provide error-free operation.

The more severe limitation on minimum operating frequency is not as onerous a restriction for application of dynamic logic as might be at first assumed. There are a large class of applications that require clocking (synchronous circuits) anyway; since the reason for the use of GaAs ICs is to operate at their maximum clock frequency rather than their minimum frequency, the storage time constraint does not restrict the application of dynamic logic to these applications.

A very wide range of minimum clock frequencies for dynamic GaAs circuits has been reported. The highest frequency divider circuits, designed for minimum delay, operated properly only above 7 GHz [25]! However, other GaAs dynamic circuits designed for low power and a high noise margin operated successfully at 100 kHz

[27]. Thus, the storage limitation is quite subject to circuit design choices and to some degree process technology.

4.3.2 Single-Phase Dynamic Circuits

Most dynamic circuit approaches require at least two clock phases to evaluate inputs and propagate data from stage to stage in order to avoid race conditions or having the data propagate through several stages asynchronously. However, a single-phase circuit, the CMOS domino [23], has found widespread application due to the many performance advantages provided. A circuit that operates with a single-phase clock is desirable since there are no clock skew or overlap requirements which must be fulfilled.

The domino circuit has been adapted for implementation with GaAs MESFETs [27], as shown in Fig. 4-27. There are three stages in a domino gate: input, which

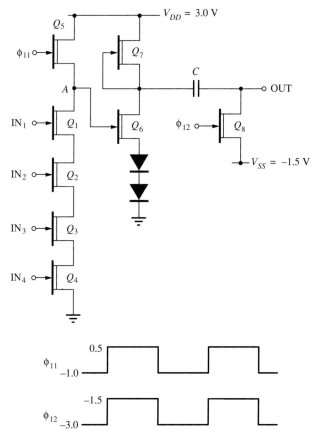

FIGURE 4-27
GaAs domino four-input AND gate.

computes the combinational logic function, the inverter stage, which responds to the charge stored on the gate capacitor, and the level-shift stage, which predischarges the output and precharges the level-shift capacitor. The basic circuit operations are as follows. When the clock is high (precharge phase), node A is precharged and the output is predischarged to V_{SS}. Therefore, during the precharge clock phase, all of the inputs and outputs are low, so there is never any static current flowing through the input combinational network. In this way, the domino logic is nonratioed, very complex combinational functions can be accomplished in a single logic gate structure, and the widths can be selected to optimize the transient performance. Speed of the gate is enhanced because the full pull-up current of the precharge device is available for charging the gate node capacitance. When the clock is low (evaluation phase), Q_5 is cut off. The charge of the internal node A may be discharged through the FET chain in the input stage, depending on the outputs of the previous stages. The full pull-down current of the combinational logic tree is available during the evaluation phase for discharging this node. Then the information propagates, rippling from stage to stage as in a chain of dominos. Note that the only change possible is a low-to-high transition. Glitches that might be generated by high-to-low spurious transitions are prevented.

As shown in Fig. 4-27, a static inverter is included as part of a domino gate; therefore the domino circuit is not totally dynamic. Static power will be required during the precharge phase and during the evaluation phase until node A is discharged. However, the nonratioed input structure makes it possible to realize a much more complex logic function in a single stage than would be feasible with standard static ratioed logic gates. For example, Fig. 4-27 represents a four-input AND gate using series transistors. No more than two transistors in series could be used in ratioed logic without incurring a noise margin problem due to threshold shifts as described in Sec. 3.2.3. Therefore, the static power tends to be averaged over a larger equivalent gate count, so there is a net improvement in power-delay product.

Dynamic power will also be dissipated by these circuits. Since operation is of interest at high clock frequencies, it might be assumed that the dynamic power could be quite large. Here, the GaAs MESFET domino circuit has a significant advantage over CMOS because the logic voltage swing is in the 1 to 1.5 V range as compared with 5 V. Since the dynamic power is proportional to ΔV^2, the power will be a factor of 11 to 25 less for GaAs if all other factors are equal. Testing of the four-input AND circuit in Fig. 4-27 showed only 0.55 mW per gate at a 110-MHz clock frequency for the total (static + dynamic) power [27]. Propagation delay was approximately 200 ps per gate.

Two level-shift diodes are used between the source of the switch device and ground for biasing because the threshold voltage of the d-mode FET is negative. Node A is the node that must be precharged to logic high so that the output voltage of the inverter stage is always at V_{OL} during this phase. Here, the negative threshold voltage of the FET can be an advantage to help reduce gate leakage, because V_{GS} of Q_6 in Fig. 4-27 can be set to around zero volts. Even at 0 V, there is still enough current through Q_6 to quickly pull the output of the inverter low. By avoiding substantial forward bias on the gate, the leakage current through the gate of Q_6 is minimized.

Thus, the use of a d-mode FET is beneficial for increasing the refresh time. If the input circuit discharges A to ground, then Q_6 will be cut off, and the output will be at logic high.

The level shifting required between stages is done by the coupling capacitor to reduce the delay and power consumption of this stage. During the evaluation phase the capacitor is precharged to about 1.5 V by the output of the inverter and FET Q_8. The value of capacitance needed depends on the output voltage swing desired, the load capacitance due to interconnect lines, and the number of the fan-out. The charge stored in the capacitor will be shared by the output (load) capacitance when switching. The coupling capacitance used for single fan-out is approximately the same as the gate capacitance of the next stage. If large load capacitances must be driven, either the coupling capacitance must be increased in size to remain comparable to the load or a static buffer provided. In cases with very long lines or high fan-out, the second option is more attractive.

Two clock inputs, Φ_{11} and Φ_{12}, are shown in Fig. 4-27. These are not two separate phases, but are the same phase; they differ in dc level only. Their voltages as required for a $V_{TD} = -1.0$ V process are also shown in this figure.

Besides the gate current of Q_6 which is reduced by the $V_{GS} = 0$ bias point, other leakage currents in this circuit are the subthreshold current of Q_1 to Q_4 and the substrate leakage current. The dominant leakage path is through the subthreshold current of Q_1 to Q_4 if the isolation between the devices is good. The isolation can be enhanced by proton bombardment, keeping the substrate leakage current quite low for the potential differences used in these circuits ($<$ 10 nA). Subthreshold current can also be further suppressed by reducing the logic low voltage level, V_{OL}, which is easily accomplished by making V_{SS} more negative. Therefore, operation of the GaAs domino circuit at clock frequencies as low as 100 kHz has been demonstrated.

The noise margin can be about 0.5 V or larger, depending on the circuit design. The logic high level will depend on the ratio of the coupling capacitance to the load capacitance and on the choice of V_{DD}. The logic low level is established by V_{SS}. Thus, the voltage swing and also the noise margin can be arbitrarily selected by exchanging power dissipation for voltage swing.

Since the logic voltage swing of the domino circuit ($-V_{SS}$ to approximately 0 V) is about the same as that of buffered FET logic (BFL), the yield of functional logic circuits in an environment in which device parameters vary randomly across or between wafers is expected to be higher than that of a low noise margin design such as direct-coupled FET logic (DCFL). The power dissipation per gate is more similar to DCFL than BFL; therefore circuit complexity in the 4000 gate level would be feasible with a 2-W budget for logic circuit power. As mentioned above, the functional gate complexity can be significantly greater than unity for the domino gates, so equivalent gate counts in the 10,000 gate range should be possible.

Charge sharing is a concern for dynamic circuits that utilize a combinational logic tree such as the domino circuit. This phenomenon can be understood by considering Fig. 4-28 where four inputs are arranged in a tree structure. C_B and C_C represent the parasitic capacitance of nodes B and C which can be greater than that of node A

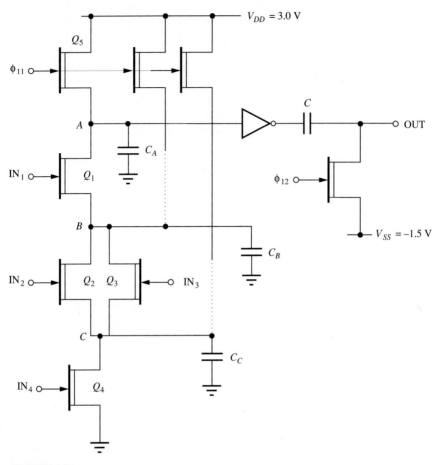

FIGURE 4-28
Domino gate with charge-sharing problem.

depending on the layout and numbers of devices at each node. Assume that (worst case) during the previous evaluation phase, IN_2, IN_3, and IN_4 became high, but IN_1 remained low. Now, nodes B and C have been discharged to ground, and remain there during the next precharge phase since all inputs are discharged to V_{OL}. Node A is precharged to approximately 1.5 V as determined by Φ_{11}. Next, suppose that during the evaluation phase, IN_1, IN_2, and IN_3 go high, while IN_4 remains low and Q_4 is cut off. Now, the charge stored on C_A must be shared with C_B and C_C while still maintaining adequate logic high voltage on node A. If node A drops below the logic threshold voltage, then the output of the gate will make a false transition from low to high. This condition can be avoided if nodes B and C are also precharged by additional precharging FETs, shown in Fig. 4-28 connected by dotted lines. These extra FETs are an added expense in area, but they may only be necessary in some of the more extreme cases.

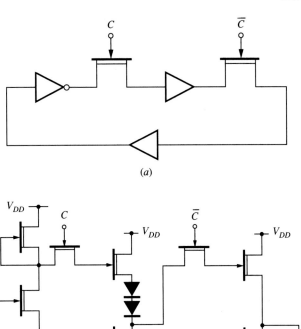

FIGURE 4-29
Dynamic divide-by-two circuit using buffered FET logic and pass transistors. (*a*) Logic diagram. (*b*) Schematic diagram.

4.3.3 Two-Phase Dynamic Circuits

Dynamic logic circuits using two-phase clocks in concert with pass transistor logic have been used for high-speed frequency divider applications [24,25]. A typical circuit of this type, shown in Fig. 4-29, resembles a two-stage ring oscillator (one inverting, one noninverting) with pass transistors to gate the flow of data from stage to stage on alternating clock phases. The complementary clock inputs, C and $\overline{C}$, must be nonoverlapping in order to avoid a free-running oscillation condition. The inverter stage might consist of a BFL or DCFL inverter. The noninverting buffer would generally be implemented by another source follower. Note in Fig. 4-29*b* that the pass transistor is inserted between the inverter and the source follower of the BFL stage. The source follower (SF) input has very low dc input conductance, and the input capacitance is also low [$C_G = C_{GD} + C_{GS}(1 - A_v)$], as discussed in Sec. 4.1.3. Therefore, the input to the SF can be switched very quickly and will not discharge as quickly as a common source input which would be in forward bias with a logic high at the input. The clock voltages on the pass transistors would be determined by the considerations outlined in Sec. 4.2.4.

The maximum clock frequency of a dynamic divider circuit will be in the range

$$\frac{1}{2t_{PD}} < f_{\max} < \frac{1}{t_{PD}} \tag{4.15}$$

since the delay of a source follower stage and of pass transistors is less than t_{PD} for an inverter stage. Operating frequencies of 10.2 GHz have been reported for 1-μm BFL implementations, with an associated power dissipation of 43 mW per gate [24]. MESFETs with 0.2 μm gate length have provided 26-GHz divide-by-two operation with 150 mW per gate [25].

4.3.4. Asynchronous Dynamic Circuits

A dynamic form of DCFL has been suggested [28] which takes advantage of both the conductance and capacitance of a forward-biased Schottky diode for interstage coupling. As shown in Fig. 4-30, the coupling diode will be forward-biased when the output of the first stage is at the logic high voltage, $V_{OH} = 1.4$ V, twice that of DCFL. The depletion-layer capacitance is at a maximum (minimum width) under this condition and is thus fully precharged by the forward conduction. When the output changes from high to low, the diode will fall out of conduction, but some of the charge will remain in the depletion layer. V_{OL} will be approximately 0.1 to 0.15 V. The displacement current through the diode will discharge the gate capacitor even though there is no dc path to ground. If $C_D \geq C_G$, the voltage at the gate node (A) will be driven below ground, reaching a minimum of $V_{OL} - V_D$ when $C_D \gg C_G$. The low noise margin will be dynamically enhanced by this effect.

Eventually, V_{GS} will discharge to a steady-state voltage above ground due to leakage currents through the diode and through the gate/source and gate/drain junctions of the FET. Therefore, maintaining the dynamically enhanced noise margin

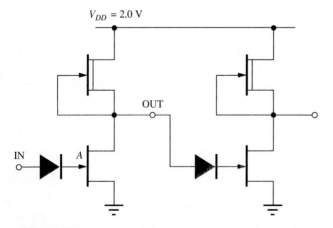

FIGURE 4-30
Two DCFL inverter stages using a coupling diode between stages to dynamically enhance noise margin.

requires a minimum operating frequency of the circuit to keep the diode precharged. Placing a weak pull-down device between A and ground will ensure static functionality regardless of the ratio in area between the diode and the gate.

An application of the dynamic level shifting used in this circuit was seen in Sec. 4.2.4 (Fig. 4-24). The feedforward capacitor and the diode capacitance shown will drive the input to the superbuffer below V_{SS} even if Q_5 were omitted. If the control signals for the multiplexer were know to alternate frequently[2] (at least every 100 ns), then Q_5 would not be necessary.

HOMEWORK PROBLEMS

Several of the problems below require simulation. If one of the MESFET models described in Chap. 2 is available, the typical model parameters provided there can be used for simulation of circuits using only d-mode MESFETs. If an E/D MESFET circuit is to be simulated, the following parameters can be used for the Raytheon MESFET model:

****

.model emes nmf(vto=0.1 beta=1.5e-04 b=0.61 alpha=3.01 lambda=0.101
+ pb=0.8 is=3e-16 rs=1670 rd=1670 cgs=1.6fF cgd=0.32fF)

****

.model dmes nmf(vto=−0.5 beta=1e-4 b=0.47 alpha=3.64 lambda=0.074
+ pb=0.8 is=3e-16 rs=2900 rd=2900 cgs=1.6fF cgd=0.32fF)

****

Of course, any design of an actual GaAs IC should use only those model parameters determined from typical devices from that process.

4.1. Analyze the dc and transient characteristics of an FL inverter using only one level-shift diode, as shown in Fig. P4-1.
   (a) Optimize the width ratios for Q_1, Q_2, and Q_5 so that $V_{NL} = V_{NH}$ for $W_2 = 10$ μm at two representative FET threshold voltages (−0.4 and −0.7 V) (assume that the vto model parameter can be changed as needed without adjusting the other parameters). Simulate the dc characteristics (as in Table 4.1) for both cases.
   (b) Determine the optimum power supply voltages for the inverters in part (a) so that V_{DS} of the pull-up and pull-down FETs has just reached saturation.
   (c) Determine the propagation delay of the one-diode FL inverters by simulation for fan-out = 2 and a load capacitance of 20 fF.

4.2. Analyze the dc and transient characteristics of the CDFL inverter shown in Fig. 4-1. Use the model parameters in Table 2.4 with $V_T = -1.0$ V to describe the 1 μm gate length MESFET.

---

[2] For example, in a data multiplexer for optical fiber communication, several low-speed inputs are repetitively sampled so that a single high-speed serial output is obtained.

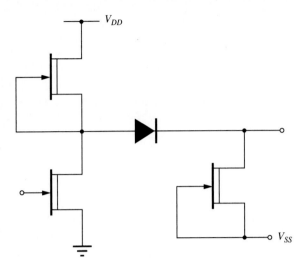

FIGURE P4-1
FL inverter with one level shift diode.

    (a) For $W_2 = 15 \ \mu m$, W_{D1} through $W_{D3} = 10 \ \mu m$, and $W_5 = 2 \ \mu m$, select width W_1 such that $V_{NL} = V_{NH}$. Determine the dc characteristics (as in Table 4.1 for the FL circuit) for the optimized CDFL inverter.
    (b) Determine the optimum power supply voltages for the inverter in part (a) so that V_{DS} of the pull-up and pull-down FETs has just reached saturation.
    (c) Estimate the area of D_4 and C_D required to obtain $V_{out} = 0.9 V_{DS}$ for the pulse edges. Assume that $C_{j0} = 2 \ fF/\mu m^2$, $m = 0.5$, and pb $= 0.8$ V for D_4. Also assume a Si_3N_4 dielectric with $\epsilon_r = 8$ and a thickness of 500 nm. (D_4 and C_D are required to have the same area since they overlay one another.)
    (d) Determine the propagation delay of the CDFL inverter by simulation for fan-out = 2 and a load capacitance of 20 fF.
    (e) Comment on the practicality of this design. How would you compromise to make the circuit more practical?

4.3. Derive an equation for the output impedance of a source follower. Use the midfrequency small-signal model for this analysis.

4.4. Determine the transfer characteristic (V_{out} vs. V_{in}) for the source follower/level-shifter circuit shown in Fig. P4-4. Perform the analysis using the simplified SPICE JFET device equations:

$$I_D = \beta W[2(V_{GS} - V_T)V_{DS} - V_{DS}^2] \qquad V_{DS} \leq V_{GS} - V_T$$
$$I_D = \beta W[V_{GS} - V_T]^2 \qquad V_{DS} \geq V_{GS} - V_T$$

Assume that the diodes can be modeled as voltage sources of 0.7 V each. Let $V_T = -0.86$ V for this problem. Neglect R_S to get equations that can be solved without iteration. Evaluate the transfer function, V_{out} vs. V_{in} over the range 0 to 2.5 V. $V_{DD} = 2.5$ V. $W_{SF} = W_{PD}$.
    (a) Plot V_{out} vs. V_{in}. Determine V_{OL} and V_{OH}. Does the voltage gain depend on V_{in}? Explain.
    (b) Plot the source follower/level-shifter voltage gain as a function of V_{in}. What will change if V_{DD} is increased to 3 V?

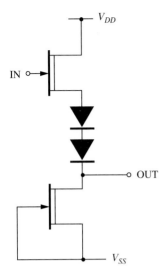

FIGURE P4-4
Source follower with level shifting.

4.5. Check the results of Prob. 4.4 by simulation using SPICE. Use the following parameters for the JFET model: vto $= -0.86$; beta $=$ 1e-4; rs $= 0$; lambda $= 0$; rd $= 0$; is $=$ 5e-16. Explain any differences between the observations. Next, include R_S and lambda in the model parameters on SPICE and note any changes.

4.6. A BFL three-input NOR gate is shown in Fig. 4-5. For $V_T = -1$ V, dc simulation shows that the transfer characteristic will provide the best noise margins when $W_S = 10$ μm and $W_L = 6$ μm. Furthermore, V_{NL} and V_{NH} can be made equal when the width of the level-shift diodes relative to $W_{SF} = W_{PD}$ are properly selected.

  (a) For $W_{SF} = W_{PD} = 10$ μm, use SPICE to find the width W_{LS} such that $V_{NL} = V_{NH}$.
  (b) A load capacitance $C_L = 100$ fF is being driven in addition to an FO $= 2$. Investigate t_{PD} of gate 2 in a chain of three identical gates using SPICE for $W_{SF} = W_{PD} = 10$, 20, and 30 μm. Scale the level-shift diodes in proportion to W_{SF} to avoid shifting the dc characteristics. Which gives the best power-delay product?
  (c) Explain qualitatively the result observed in part (b) through the use of a simplified analytical model for the circuit.

4.7. You can afford a power dissipation of 5 mW for a three-input NOR gate.

  (a) At the 5-mW (± 0.25 mW) power dissipation level, determine by simulation whether a BFL or a CDFL gate will be faster for FO $= 1$, FO $= 2$, and FO $= 3$. Assume that the drain node layout capacitance (node A, Figs. 3-12 and 4-1) and the interconnect capacitance per fan-out is 5 fF. Use typical model parameter values from Chap. 2 for the simulation.
  (b) Repeat for FO $= 1$ with $C_L = 10$, 50, and 100 fF.
  (c) Determine the peak load current–average I_{DD} ratio by simulation for the CDFL gate and compare with BFL.

4.8. Design a BFL gate with a dynamic pull-down. Compare the propagation delay and peak load current–average I_{DD} ratio for static and dynamic BFL for a fixed C_L of 200 fF and fan-out $= 2$.

4.9. Design a depletion mode, source-coupled FET logic gate. Let $I_{CS} = 1$ mA. Use $V_T = -1$ V with the model parameters from Table 2.4.

(a) Determine device widths and appropriate level shifting. Simulate a chain of three gates (connected differentially) and determine the noise margins and logic levels of the middle gate.

(b) Add two more gates to your chain and connect the output of gate 5 to the input of gate 1 to form a five-stage ring oscillator. Determine the average t_{PD} from the oscillation frequency. Show how t_{PD} varies with change in V_{DD}. Justify your observation. Also find P_D vs. V_{DD}.

4.10. Calculate the sensitivity of the maximum width and maximum square noise margins of a DCFL inverter to changes in V_{TE} and V_{TD} using SPICE simulation. Assume $V_{TD} = -0.7$ V. Compare two designs, one with large W_S/W_L and $V_{TE} = 0.25$ V, the other with small width ratio and $V_{TE} = 0.10$ V.

4.11. Use the V_T, R_S, and β in the example with the other parameters in Table 2.4 for this exercise and for Prob. 4.12.

(a) Complete the design of the source follower stage in Example 4.4.
(b) Determine the noise margins of the completed SCFL gate by simulation.
(c) Determine the sensitivity of the noise margins to variation in V_T.

4.12. Find the optimum partitioning of power in the SCFL circuit in Example 4.4 so that the minimum propagation delay is achieved. Total power dissipation should remain constant. Assume that a load capacitance of 50 fF is being driven by each output with a fan-out of 1.

4.13. Design an e/d BFL inverter for maximum noise margins with V_{TE} of both 0.1 and 0.2 V. Let $V_{TD} = -0.7$ V in both cases.

4.14. Ring oscillators (RO) are sometimes used to evaluate the transient performance of logic gates. When lightly loaded, the RO provides some information on the intrinsic speed of the circuit. When more heavily loaded with some fan-out and parasitic capacitance, the RO will provide information on the speed of the circuit in a realistic layout environment. A ring oscillator consists of an odd number of logic gates connected in a closed loop. When power is applied to the gates, they will oscillate at a frequency given by $f = 1/(2n t_{PD})$, where n is the number of stages and t_{PD} is the propagation delay per gate.

(a) Simulate the operation of the seven-stage RO in Fig. P4-14. The inverters can be implemented using superbuffer FET logic (SBFL). Let $V_{DD} = 1.5$ V. You will need to simulate for two complete cycles to make sure that a steady-state condition is reached. Include the device parasitic capacitances and an interconnect capacitance of 5 fF per stage, but do not include fan-out or other capacitive loading. Use the device model parameters given at the beginning of the homework problems.

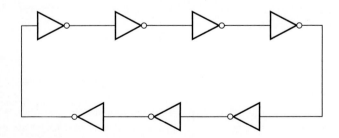

FIGURE P4-14
Seven-stage ring oscillator.

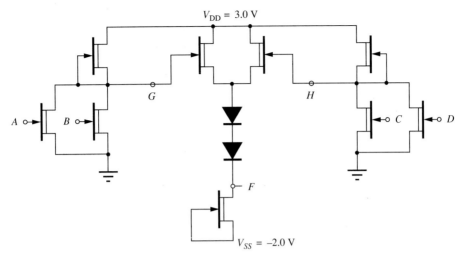

FIGURE P4-15
Two-level logic gate.

(b) Simulate the operation of the seven-stage RO with FO = 2 and fixed-load capacitance (for each stage). Use C_L = 0, 20, and 100 fF. Plot the delay vs. load capacitance.

4.15. The circuit shown in Fig. P4-15 might be occasionally used in a GaAs IC to perform two-level logic.
  (a) Identify the logic function that this circuit would provide.
  (b) Consider the source follower section only. Assume that the device characteristics shown in Fig. 3-11 apply to these devices. Make an estimate (graphical method) of V_{OH} and V_{OL} for the case where input G is driven and input H is low. Sketch the transfer characteristic, V_{out} vs. $V_{node\ G}$. (Assume that node G has a range of voltages typical of the circuit driving it.)
  (c) Determine V_{OL} and V_{OH} by simulation for the following input (logic state) conditions:

  ABCD
  0000
  1000
  1010
  1111

  (d) Simulate a transfer characteristic for input A with all other inputs at V_{OL}. Estimate the noise margins for the worst-case V_{OL} found in part (c).

REFERENCES

1. Barna, A., and Liechti, C.: "Optimization of GaAs MESFET Logic Gates with Subnanosecond Propagation Delays," *IEEE J. Solid State Cir.*, vol. SC-14, pp. 708–715, August 1979.
2. Livingstone, A. W., and Mellor, P. J. T.: "Capacitor Coupling of GaAs Depletion Mode FETs," *1980 GaAs IC Symp.*, Las Vegas, Nev., paper 10, 1980.
3. Namordi, M., and White, W.: "A Novel Low-Power Static GaAs MESFET Logic Gate," *IEEE Elect. Dev. Lett.*, vol. EDL-3, pp. 264–267, September 1982.
4. Eden, R. C.: "Capacitor Diode FET Logic (CDFL)—Circuit Approach for GaAs D-MESFET IC's," *Proc. 1984 IEEE GaAs IC Symp.*, Boston, Mass., pp. 11–14, October 1984.

5. Van Tuyl, R. L., Liechti, C. A., Lee, R. E., and Gowen, E.: "GaAs MESFET Logic with 4-GHz Clock Rate," *IEEE J. Solid State Cir.*, vol. SC-12, pp. 484–496, October 1977.
6. Eden, R. C., Welch, B. M., and Zucca, R.: "Low Power GaAs Digital ICs Using Schottky Diode FET Logic," *Dig. Tech. Papers, 1978 Int. Solid-State Cir. Conf.*, pp. 66–69, February 1978.
7. Helix, M. J., Jamison, S. A., Chao, C., and M. S. Shur: "Fan Out and Speed of GaAs SDFL Logic," *IEEE J. Solid State Cir.*, vol. SC-17, pp. 1226–1231, December 1982.
8. Eden, R. C., Lee, F. S., Long, S. I., Welch, B. M., and Zucca, R.: "Multi-level Logic Gate Implementation in GaAs IC's Using Schottky Diode FET Logic," *Dig. Tech. Papers, 1980 Int. Solid-State Cir. Conf.*, pp. 122–123, February 1980.
9. Vu, T. T., et al.: "Multiple Input and Output OR/AND Circuits for VLSI GaAs IC's," *IEEE Trans. Elect. Dev.*, vol. ED-34, pp. 1630–1641, August 1987.
10. Ekroot, C., and Long, S. I.: "A GaAs 4 bit Adder-Accumulator Circuit for Direct Digital Synthesis," *IEEE J. Solid State Cir.*, vol. SC-23, pp. 573–580, April 1988.
11. Cirillo, N., and Abrokwah, J.: "8.5 ps Ring Oscillator Gate Delay," *43rd IEEE Device Res. Conf.*, paper IIA-7, Boulder, Colo., June 1985. Abstract in: *IEEE Trans. Elect. Dev.*, vol. ED-32, p. 2530, November 1985.
12. Nakamura, H., et al.: "A 390 ps 1000 Gate Array Using GaAs Super-Buffer FET Logic," *Dig. of Tech. Papers, 1985 Int. Solid-State Cir. Conf.*, pp. 204–205, February 1985.
13. Nakamura, H., et al.: "2 GHz Multiplexer and Demultiplexer using DCFL/SBFL Circuit and the Precise Vth Control Process," *Proc. IEEE GaAs IC Symp.*, Grenelefe, Fla., pp. 151–154, October 1986.
14. Katsu, S., Nambu, S., Shimano, A., and Kano, G.: "A Source Coupled FET Logic—A New Current-Mode Approach to GaAs Logics," *IEEE Trans. Elect. Dev.*, vol. ED-32, pp. 1114–1118, June 1985.
15. Saito, S., Takada, T., and Kato, N.: "A 5-mA 1-GHz GaAs Dual-Modulus Prescalar IC," *IEEE J. Solid State Cir.*, vol. SC-21, pp. 538–543, August 1986.
16. Larson, L. E.: "Gallium Arsenide Switched-Capacitor Circuits for High-Speed Signal Processing," Ph. D. Dissertation, UCLA, 1986.
17. Scheinberg, N.: "High Speed GaAs Operational Amplifier," *IEEE J. Solid State Cir.*, vol. SC-22, pp. 522–527, August 1987.
18. Tamura, A., et al.: "High-Speed GaAs SCFL Divider," *Elect. Lett.*, vol. 21, pp. 605–606, July 1985.
19. Davenport, W. H.: "Macro Evaluation of a GaAs 3000 Gate Array," *Proc. IEEE GaAs IC Symp.*, Grenelefe, Fla., pp. 19–22, October 1986.
20. Zuleeg, R., Notthoff, J. K., and Troeger, G. L.: "Double-Implanted GaAs Complementary JFETs," *IEEE Elect. Dev. Lett.*, vol. EDL-5, pp. 21–23, January 1984.
21. Notthoff, J., et al.: "A 4k × 1 Bit Complementary E-JFET Static RAM," *Proc. IEEE GaAs IC Symp.*, Portland, Oreg., pp. 185–188, October 1987.
22. Rasset, T. L., Niederland, R. A., Lane, J. H., and Geideman, W. A.: "A 32-Bit RISC Implemented in Enhancement Mode JFET GaAs," *IEEE Computer*, vol. 19, pp. 60–68, October 1986.
23. Krambeck, R. H., Lee, C. M., and Law, H. F. S.: "High-Speed Compact Circuits with CMOS," *IEEE J. Solid State Cir.*, vol. SC-17, pp. 614–619, June 1982.
24. Rocchi, M., and Gabillard, B.: "GaAs Digital Dynamic IC's for Applications up to 10 GHz," *IEEE J. Solid State Cir.*, vol. SC-18, pp. 369–376, June 1983.
25. Jensen, J., Salmon, L. G., Deakin, D. S., and Delaney, M. J.: "26GHz GaAs Room-Temperature Dynamic Divider Circuit," *Proc. IEEE GaAs IC Symp.*, Portland, Oreg., pp. 201–204, October 1987.
26. Namordi, M., Newman, P. F., Cappon, A. M., Hanes, L. K., and Statz, H.: "A 2.2 GHz Transmission Gate GaAs Shift Register," *Dig. Tech. Papers, 1985 IEEE Int. Solid-State Cir. Conf.*, pp. 218–219, 1985.
27. Yang, L., Chakharapani, R., and Long, S.: "A High Speed Dynamic Domino Circuit Implementation with GaAs MESFETs," *IEEE J. Solid State Cir.*, vol. SC-22, pp. 874–879, October 1987.
28. Yang, L., Yuen, A. T., and Long, S. I.: "A Simple Method to Improve the Noise Margin of III-V DCFL Digital Circuit, Coupling Diode FET Logic," *IEEE Elect. Dev. Lett.*, vol. EDL-7, pp. 75–78, March 1986. Also see "Reply to 'Comment on "A Simple Method . . . " ', " *ibid*, pp. 556–557, September 1986.

CHAPTER 5

INTERCONNECTIONS

5.1 TRANSMISSION LINE THEORY

The interconnections between active circuit elements have a very strong influence on the performance of a very-high-speed IC. The bandwidth requirements of very short pulses force the designer to deal with the transmission line nature of the interconnections on-chip, in the package, and especially on the printed-circuit board. Such concerns as attenuation, reflected waves, crosstalk, and distortion of pulses suddenly limit the performance of a fast circuit well below its true potential. Since the circuit can respond to signals that have durations similar to its response time, the problem becomes more severe as the rise and fall times of the active logic elements decrease. GaAs ICs with 50 to 200 ps t_r and t_f are at least a factor of two faster than silicon ECL and a factor of ten faster than TTL. The bandwidth requirements of a minimum-width gaussian-shaped pulse[1] with full-width at half-maximum of 100 ps is 3 GHz. The spectral components needed to represent square-wave-shaped pulses can be much greater, as the sharp edges will require high-order odd harmonics to avoid rounding.

A second problem which occurs as a consequence of the fast transitions will be the apparent noise on the signal paths and power supply lines which comes about from reflected waves, crosstalk between lines, and inductive spikes due to current transients. The faster the logic family, the more sensitive it will be to transient noise.

---

[1] Assuming that the shape of the minimum width pulse is a gaussian, the bandwidth needed to propagate a pulse of width τ (full-width at half-maximum) is $0.29/\tau$.

These effects can be effectively modeled by the use of the transmission line techniques described in this chapter. Many of these techniques are already well established by those who design and develop wideband analog or microwave circuits (impedance-matched interconnects, terminations, and bypassing of power supplies, for example). Some are rather unique to high-speed digital circuits and are still in a development stage.

The properties of the transmission line play a central role in predicting the influence of interconnections. In this chapter, interconnecting wiring both on-chip and off-chip will be discussed. As will be seen, even connections that are short enough to be considered as lumped capacitors or inductors can be analyzed with transmission line methods. Since the emphasis in this text is on digital IC design, analysis in the time domain will be more useful than the frequency-domain analysis most often found in the microwave literature, assuming that the loss can be neglected. The essential fundamentals of transmission line circuits are reviewed in this section. Specific analytical techniques appropriate for time-domain studies are also presented.

Any interconnection between two nodes may be thought of as a transmission line if there is an identifiable path for current to flow away from the driving node and back through a return path or reference node. If the interconnecting material is metal and its series resistance is small compared with its characteristic impedance (to be defined below), the losses in the circuit can be neglected for first-order analysis. Also, if the interconnecting wire is in a dielectric medium, which presents a permittivity ϵ and permeability μ that does not depend on the amplitude or frequency of the excitation on the transmission line, then the transmission line represents a linear network that is nondispersive and can be analyzed in the time domain. Signals will not be strongly attenuated or distorted by travelling on such an ideal interconnection. Naturally, this idealization does not accurately represent all cases of interest. However, for first-order analysis, the lossless, nondispersive, linear model is quite useful.

The equations describing the propagation of a signal along a conductor are analogous to the equations describing the propagation of a plane electromagnetic wave in an isotropic dielectric [1]. In the latter case Maxwell's equation for the condition in which charge $\rho = 0$ and conductivity $\sigma = 0$ lead to second-order differential equations in $\mathbf{E}$ and $\mathbf{H}$ which can be identified as wave equations:

$$\nabla^2 \mathbf{E} = \mu\epsilon \frac{\partial^2 \mathbf{E}}{\partial t^2} \tag{5.1a}$$

$$\nabla^2 \mathbf{H} = \mu\epsilon \frac{\partial^2 \mathbf{H}}{\partial t^2} \tag{5.1b}$$

The *velocity of propagation* or *phase velocity* of the waves is identified as

$$v = (\mu\epsilon)^{-1/2} \tag{5.2}$$

and solutions to these wave equations are of the form

$$E_y = F_1(x - vt) + F_2(x + vt) \tag{5.3a}$$

$$H_z = (\mu\epsilon)^{1/2}\left[F_1(x - vt) - F_2(x + vt)\right] \tag{5.3b}$$

F_1 and F_2 are functions with arbitrary form which describe transverse waves traveling in the direction of increasing x (for F_1) or decreasing x (for F_2).

For the propagation of a signal on a conductor pair, the direction of propagation is again one dimensional. Here, a circuit such as shown in Fig. 5-1 is considered, where one conductor is designated as a reference or return (line 2) and the other is used for signal transmission. The actual realization of the conductor pair can take many forms, some of which will be described later, but a schematic two-line representation can refer to any of those possibilities.

Also, it is more convenient to refer to voltages and currents on these lines rather than E and H, although the existence of V and I imply E and H which are related to V and I by line integrals. Voltages on line 1 are measured with line 2 as the reference node. In this chapter V and I will be assumed to be functions of x and t (they do not represent phasors).

If we assume that the series resistance per unit length $R_0 \ll \omega L_0$ and that the shunt conductance per unit length $G_0 \ll \omega C_0$, as defined in Fig. 5-1, then the line can be approximated as being without loss, and the voltage and current can be easily described. Over a length dx on the conductors, there will be a change in voltage given by

$$-\frac{\partial V}{\partial x} = L_0 \frac{\partial I}{\partial t} \tag{5.4a}$$

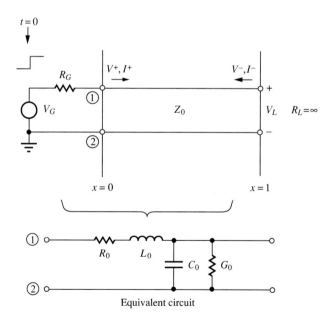

FIGURE 5-1
Open-circuited transmission line of characteristic impedance Z_0. The step function generator has open-circuit voltage V_G and internal resistance R_G.

and a change in current

$$-\frac{\partial I}{\partial x} = C_0 \frac{\partial V}{\partial t} \qquad (5.4b)$$

where L_0 and C_0 are the inductance and capacitance per unit length of transmission line, also defined by the equivalent circuit represented in Fig. 5-1. Solving for V or I by eliminating a variable, wave equations for voltage and current along the line are obtained:

$$\frac{\partial^2 V}{\partial x^2} = L_0 C_0 \frac{\partial^2 V}{\partial t^2} \qquad (5.5a)$$

$$\frac{\partial^2 I}{\partial x^2} = L_0 C_0 \frac{\partial^2 I}{\partial t^2} \qquad (5.5b)$$

Therefore, by analogy with Eqs. (5.1), the velocity of propagation along the line is given by

$$v = (L_0 C_0)^{-1/2} \qquad (5.6)$$

Also, the product $L_0 C_0$ is independent of the geometry of the lines if they are in an isotropic medium and the conductors are lossless. In this case, the velocity v is the same as for a plane electromagnetic wave propagating in the same medium, [Eq. (5.2)]. Thus,

$$v = \frac{c}{\sqrt{\epsilon/\epsilon_0 \cdot \mu/\mu_0}} = \frac{c}{\sqrt{\epsilon_r}} \qquad (5.7)$$

for nonmagnetic media. Here, ϵ_0 and μ_0 are the permittivity and permeability of free space, ϵ_r is the relative dielectric constant, and c is the velocity of light in free space. It will be shown later that for transmission lines fabricated in anisotropic dielectrics (microstrip, for example), an *effective dielectric constant* ϵ_{eff} can be defined and $v = c/\sqrt{\epsilon_{\text{eff}}}$.

Equations (5.5) also imply that waves can propagate in either direction (positive or negative x) on the conductors. Since the transmission line, in its simplest form, is a linear system, then any combination of such waves can exist by superposition at any point on the line. The equation

$$V(x,t) = F_1(x - vt) + F_2(x + vt) \qquad (5.7a)$$

is a solution for Eq. (5.5a). From Eqs. (5.4), it can also be shown that the current on the lines is related to the voltage by a *characteristic impedance*, Z_0:

$$Z_0 I(x,t) = F_1(x - vt) + F_2(x + vt) \qquad (5.7b)$$

$$Z_0 = \sqrt{\frac{L_0}{C_0}} = \sqrt{\frac{\mu}{\epsilon}} \qquad (5.8)$$

Suppose that at time $t = 0$, the voltage generator in Fig. 5-1 produces a unit step function, $V_G = u(t)$, through the internal source resistance R_G to the input of the transmission line, $x = 0$. Then a voltage wave, V^+, is launched on the line and travels in the forward direction. At the same time, current $I^+ = V^+/Z_0$ is also launched. If the line has length l, then at time $T = l/v$, the forward traveling wave will reach the opposite end, $x = l$.

V^+ can be easily related to V_G, the generator open-circuit voltage, at time $t = 0$. The line has a characteristic impedance Z_0, which would appear to be purely resistive if the line were infinitely long or perfectly terminated in a load resistor $R_L = Z_0$. Therefore, for the initial interval $0^+ \le t \le 2T$, during which any reflected waves from $x = l$ cannot have yet returned to $x = 0$, V^+ is related to V_G by a simple voltage divider ratio:

$$V^+ = V_G \frac{Z_0}{Z_0 + R_G} \tag{5.9}$$

When V^+ has traveled to $x = l$ and the line is open-circuited, as in Fig. 5-1, a discontinuity in impedance occurs which will lead to a reflection of V^+ from the interface. That a reflected wave, V^-, must be generated can be seen from the boundary condition $I^+ = 0$ at $x = l$. If no current, I_L, is allowed beyond $x = l$, then a reflected wave of amplitude $V^- = V^+$ must leave the interface at $t = T$ traveling in the $-x$ direction. Then $I^+ = I^-$ and $I_L = 0$. Since the incident and reflected wave voltages add and currents subtract at any point on the line or at the end of the line, the total voltage at the "load" on the open-circuited line V_L is twice the incident voltage V^+.

V^- will arrive at $x = 0$ when $t = 2T$. In the most general case in which $R_G \ne Z_0$, further reflections will occur at this interface that will launch another reflected wave in the forward direction. Reflections will continue to occur at each end of the line until the line charges up to, in this case with $R_L = \infty$, the open-circuit voltage of the source, V_G. If the line is short, T is small, and this process may take place very quickly, as illustrated in Fig. 5-2, curve(a). In this example, the small voltage steps which occur when $R_G \gg Z_0$ continue to accumulate by superposition as more reflections are produced. The steps are closely spaced, and they approach an exponential with time constant $R_G C_0 l$, where $C_0 l$ is the capacitance presented to the source by a line of impedance Z_0 and length l. The short, unterminated line (or stub) can therefore be accurately modeled by a simple RC lumped element approximation. This approximation is acceptable if the round-trip delay time is shorter than the response time of the logic circuits which are interconnected by the line; that is, $2T < t_r$ or t_f, the rise or fall times.[2]

For longer interconnect lines, where $2T > t_r$ or t_f, the voltage on the line should be decomposed into its discrete steps as illustrated in Fig. 5-2, curve (b). Here, it can be seen that the poorly matched ($R_G \gg Z_0$, $R_L \ne Z_0$) long transmission line may introduce significantly more delay in the voltage approaching its maximum level than that which would be expected from its electromagnetic propagation time

---

[2]Although in the quasi-static limit, a voltage change from V_{OL} or V_{OH} to a voltage just beyond the metastable point is sufficient to switch a logic circuit to the other state; for normal, very short signal edges found in real circuits, the dynamic noise margin applies. Thus, the 10 to 90 percent voltages, t_r or t_f, will be more appropriate measures of response time.

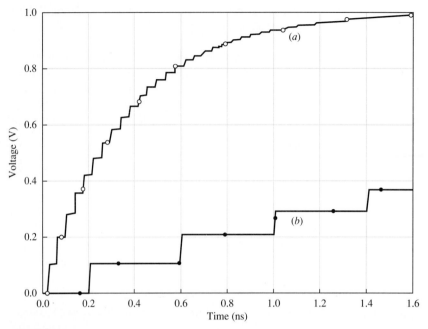

FIGURE 5-2
Charging of (*a*) short and (*b*) long open-circuited transmission lines by a step function voltage source with high internal source impedance. The charging takes place by the superposition of many small steps caused by multiple reflections.

T if $R_G \gg Z_0$. In this case, the voltage at $x = l$ builds up in small steps, and excess delay will be suffered while waiting for V_L to approach the opposite logic state, V_{OH}. Thus, to maintain the high intrinsic speed of GaAs ICs, it is desirable to design buffers with output impedance equal to Z_0 when long interconnect lines must be driven.

It is useful to define a *reflection coefficient*, Γ, which relates the incident voltage and reflected voltage:

$$\Gamma = \frac{V^-}{V^+} \quad (5.10)$$

For resistive terminations or interfaces between transmission line segments, Eq. (5.10) is the ratio of two real functions of time and Γ is a real number varying between -1 and 1. If the termination has a reactive component, then if Γ is utilized it must be used in the frequency domain where Γ will be complex. In the example above, for $R_L = \infty$, $V^- = V^+$, therefore $\Gamma = 1$.

The relationship between Γ and an arbitrary load impedance, Z_L, can be derived from the boundary conditions at the end of the line. At $x = l$, $V_L/I_L = Z_L$. Also, it must be true that

$$V^+ + V^- = V_L \quad (5.11a)$$

and
$$I^+ + I^- = I_L \tag{5.11b}$$

From the above three equations [Eqs. (5.10) and (5.11 a, b)],

$$\Gamma = \frac{Z_L - Z_0}{Z_L + Z_0} \tag{5.12}$$

Therefore, if $Z_L = Z_0$, $\Gamma = 0$, and there is no reflected wave. If $Z_L = 0$, then $\Gamma = -1$ and $V^- = -V^+$ so that $V_L = 0$. In this extreme, $I^+ = +I^-$.

A *transmission coefficient* is also convenient for calculating the voltage transmitted through an interface between lines or between a line and a load. It can be defined as the ratio between the voltage transmitted through the interface to the voltage incident on the interface:

$$\Gamma_T = \frac{V_L}{V^+} = 1 + \Gamma \tag{5.13}$$

This agrees with our intuition, since for a short-circuited line, $\Gamma_T = 0$ and thus $V_L = 0$. It also shows that the voltage at the end of an open-circuited line is twice the incident voltage.

If currents I^+ and I^- are considered instead of voltages, then

$$\Gamma_I = \frac{I^-}{I^+} = -\frac{V^-}{V^+} \tag{5.14}$$

and Eq. (5.13) becomes

$$\frac{I_L}{I^+} = 1 + \Gamma_I \tag{5.15}$$

From Eq. (5.12), the sign of the reflection coefficient at an interface will depend on the direction of propagation through the interface. For example, in Fig. 5-3, two transmission lines of impedances Z_1 and Z_2 are joined together as shown. If a signal approaches the interface from the left, passing from line 1 to line 2, Γ is $(Z_2 - Z_1)/(Z_2 + Z_1)$, whereas with the signal traveling from the right, passing from line 2 to line 1, $\Gamma = (Z_1 - Z_2)/(Z_2 + Z_1)$ is seen. As a result, the polarity of the reflected voltage or current will be opposite depending on the direction of travel.

The reflection coefficients can be used to calculate the time dependence of the voltage or current on a transmission line. The calculation is facilitated through the construction of a position-time diagram, as demonstrated in the following example.

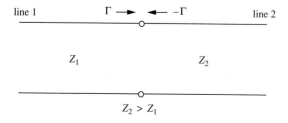

FIGURE 5-3
Junction between two transmission lines with characteristic impedances $Z_2 > Z_1$.

252   GALLIUM ARSENIDE DIGITAL INTEGRATED CIRCUIT DESIGN

Example 5.1. A small logic gate is driving an open-circuited interconnect line on a GaAs IC chip. If the transit time on the line is 50 ps, calculate the voltages at $x = 0$ and $x = l$ of Fig. 5-1 from $t = 0$ to $t = 6T$ if $R_G = 900 \ \Omega$, $Z_0 = 100 \ \Omega$, and $V_G = 1$ V.

Solution. First determine the reflection coefficients as seen by waves traveling on the line. At $x = 0$, $\Gamma_0 = 0.8$. At $x = l$, $\Gamma_l = 1$. At $t = 0$, a wavefront of amplitude

$$V^+ = V_G \left(\frac{Z_0}{R_G + Z_0} \right) = 0.1 \text{ V} \tag{5.16}$$

begins to travel to the right from $x = 0$ with velocity v. In order to simplify the construction, assume that the generator switches with zero rise time (note that a finite rise time can be included in the construction if more care is taken in computing the time evolution of the line voltages).

To follow the progress of the signal $V(x, t)$ on the line, a graph of position (horizontal) against time (vertical) is made as shown in Fig. 5-4. The signal is represented by a diagonal line whose slope is $1/v$. The direction of propagation reverses at each reflection. To the left and right of the graph are plotted voltages $V(0, t)$ and $V(l, t)$ versus time, which are determined by application of superposition and the corresponding reflection and transmission coefficients. The amplitude of the wave traveling on the line is labeled V_i^+ or V_i^-, where the subscript i is an integer representing the number of round-trip cycles completed by the wave and the superscript represents the direction of propagation.

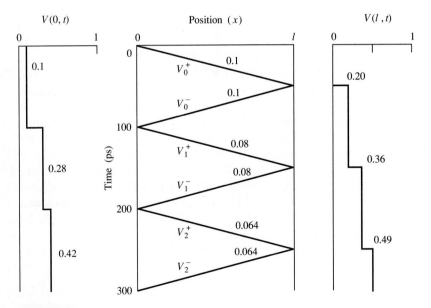

FIGURE 5-4
Position-time diagram for an open circuited transmission line with $Z_0 = 100 \ \Omega$ and $R_G = 900 \ \Omega$. $T = 50$ ps.

Mathematically, the amplitude of $V(l, t)$ progresses according to

$$V(l,t) = \left(\frac{Z_0}{Z_0 + R_G}\right) V_G (1 + \Gamma_l) \sum_{i=0}^{\infty} (\Gamma_0 \Gamma_l)^i \quad (5.17)$$

for $t \geq T$ and $i \geq 0$ and for $t < T$, $V(l, t) = 0$.

Similarly, $V(0, t)$ accumulates according to

$$V(0,t) = \left(\frac{Z_0}{Z_0 + R_G}\right) V_G \left[1 + (1 + \Gamma_0) \Gamma_L \sum_{i=0}^{\infty} (\Gamma_0 \Gamma_l)^{i-1}\right] \quad (5.18)$$

for $t \geq 2T$ and $i \geq 1$. If $t < 2T$,

$$V(0,t) = \left(\frac{Z_0}{Z_0 + R_G}\right) V_G$$

The diagram, however, provides a more intuitive method for determining $V(0, t)$ and $V(l, t)$ by keeping a record of V_i^+, V_i^-, and accumulating $V_i(0, t)$ and $V_i(l, t)$. From Eq. (5.16), it is seen that $V(0, 0) = V_0^+ = 0.1$ V. Since $\Gamma_l = 1$, $V(l, T) = (1 + \Gamma_l) V_0^+ = 0.2$ V. Then $V_0^- = \Gamma_l V_0^+ = 0.1$ V. When V_0^- arrives back at $x = 0$ for the first round-trip, $V(0, 2T) = (1 + \Gamma_0) V_0^- + V(0, 0)$. Continuing for a second cycle ($i = 1$), $V_1^+ = \Gamma_0 V_0^-$ and $V(l, 3T) = V_0(l, T) + V_1^+(1 + \Gamma_l)$. Figure 5-4 illustrates this progression for $0 < t < 6T$.

It can be seen by continuing this process that $V(0, t)$ and $V(l, t)$ both approach V_G asymptotically. In this example, the rate of approach is quite slow and actually requires more than 10 round trips (1 ns) to exceed 90 percent of V_G. This illustrates the limitations of drivers with high internal source resistance (R_G) when driving interconnections with significant transit time.

5.2 ANALYSIS OF INTERCONNECTIONS

Interconnecting wiring on GaAs IC chips generally consists of metal strips deposited on GaAs ($\epsilon_r = 12.9$) or on a dielectric layer such as SiO$_2$($\epsilon_r = 3.9$), silicon nitride or oxynitride ($4 \leq \varepsilon_r \leq 8$), or polyimide ($\epsilon_r = 3$ to 3.5), or even in some cases suspended in air above the GaAs wafer surface (air bridges). These metal strips are often thin compared with their widths. Also, the GaAs semi-insulating substrate is quite thick[3] (typically 600 μm) in comparison with the line-to-line spacing. Thus, any backside ground plane has less influence than neighboring conductors on the surface. These metal interconnect lines also are often short enough that their series resistance is smaller than their characteristic impedance. Therefore, they can sometimes be approximated by ideal transmission lines or segments of transmission line.

Ideally, when working with a well-established and fully characterized IC process, at least the capacitance of interconnect lines of typical widths, spacing, and on

---

[3] GaAs semi-insulating substrates are typically at least 600 μm thick in order to provide mechanical strength during processing. In digital applications, the thickness is generally not reduced following the processing.

all levels will have been measured and will usually be provided by the processing facility. Such measurements require either scale models or interconnect test structures specifically designed for accurate characterization. However, it is also useful to have some approximate models of these interconnections when considering a new process, changing an old process, or interpolating between measured data. Also, line characteristic impedances and inductances are seldom measured or provided, and they can be calculated if capacitances are known.

If the characteristic impedance Z_0 or the capacitance per unit length C_0 and the effective dielectric constant ϵ_{eff} are known for a given transmission line, then other properties of interest can be calculated. The effective dielectric constant allows one to calculate electrostatic quantities of interest for a system of conductors in a nonuniform dielectric using a quasi-TEM approximation. More precisely, if the total capacitance of the conductor imbedded in its dielectric layers is C_2 and the capacitance of the conductor in free space ($\epsilon_r = 1$) is C_1, $\epsilon_{\text{eff}} = C_2/C_1$. Integrated circuit wiring, as mentioned above, can consist of metal strips imbedded in slabs of dielectrics, including air.

The propagation velocity and impedance have been related to L_0, C_0, Z_0, and ϵ_{eff} by Eqs. (5.6), (5.7), and (5.8). From these equations,

$$L_0 = \frac{Z_0 \sqrt{\epsilon_{\text{eff}}}}{c} \tag{5.19}$$

and

$$C_0 = \frac{\sqrt{\epsilon_{\text{eff}}}}{Z_0 c} \tag{5.20}$$

These fundamental relationships apply to any lossless TEM mode structure [or nearly TEM, such as microstrip, coplanar strips (CPS), or coplanar waveguide (CPW)].

Prediction of interconnect line properties on semi-insulating GaAs substrates is more complicated than on silicon ICs because there is no local ground plane under the conductors. Coplanar conductors on or in a dielectric substrate can be strongly coupled if the surface ground return lines are distant from the signal lines or if many signal lines share a ground. The multiple modes which can be supported on a set of coupled lines make calculation of a characteristic impedance difficult because the calculation of the mode velocities and voltage and current eigenvectors is required. Finding the capacitance and inductance matrices and the nodal impedances and propagation velocities in the most general and accurate way will require two-dimensional numerical solution of the electrostatic field equations. If loss and frequency dispersion are considered, the problem is even more difficult. This type of calculation is quite time-consuming, and appropriate software is not widely available; thus it is not very practical as a tool for the circuit designer. A more practical but less accurate approach, which if applied and interpreted conservatively will lead to circuits meeting delay requirements and not adversely affected by noise, is to apply approximate or exact analytical solutions for transmission line geometries most closely resembling the actual IC layout.

For example, multiple interconnect lines running in parallel along a wiring channel occur frequently in many IC layouts (Fig. 5-5a). The transmission line

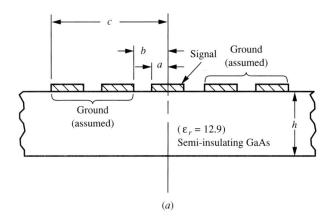

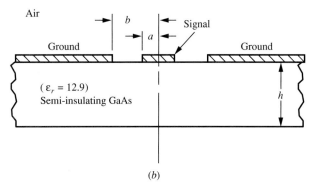

FIGURE 5-5
(a) Section of a typical wiring channel. Interconnect strips are not covering the surface completely and are not semi-infinite in extent. (b) A section of the coplanar waveguide (CPW) transmission line. The usual analytical solution for line impedance or capacitance assumes semi-infinite extention of ground strips and dielectric.

properties of one of these lines (labeled signal) might be approximated by a *coplanar waveguide* (CPW) geometry, as shown in Fig. 5-5b. CPW is a surface-oriented structure which assumes that the dielectric and the reference or ground (earth) strips have semi-infinite extent. Even though on the IC the adjacent conductor strips, assumed to be ground, are not semi-infinite, the error is small (Z_0 is 6 percent low when $b/c = 0.6$), and a correction for this error may be made if desired [2]. Also, the first-order CPW analysis below ignores the influence of the lower ground plane on the substrate back side. This lower ground is significant only if the substrate thickness $h < 5b$ [2].

Thus, the CPW model will provide a good estimate of the line properties if every other line is a ground return or reference line. The CPW model in this case will tend to *overestimate* the total line capacitance and *underestimate* the characteristic impedance; therefore, it leads to a conservative delay estimate although neglecting crosstalk. As will be described in Sec. 5.5, alternating ground and signal lines also greatly reduces the adverse effects of line-to-line coupling (crosstalk). If two or more signal lines share ground returns, coupling will occur which, in the case of a high-impedance driving gate (small FETs for low bias current) and long parallel line lengths, can be very significant.

For the ideal CPW of Fig. 5-5b without a lower ground plane, Ref. 3 provides the following equations:

$$Z_0 = \frac{30\pi}{\sqrt{\epsilon_{\text{eff}}}} \frac{K(k')}{K(k)} \qquad (5.21)$$

$$\epsilon_{\text{eff}} = 1 + \frac{\epsilon_r - 1}{2} \frac{K(k')}{K(k)} \frac{K(k_1)}{K(k'_1)} \qquad (5.22)$$

where

$$k = \frac{a}{b} \qquad (5.23)$$

$$k_1 = \frac{\sinh(\pi a/2h)}{\sinh(\pi b/2h)} \qquad (5.24)$$

For the usual case where $h \gg b > a$, $\sinh x \approx x$ and $k_1 \simeq a/b$. Thus, in most instances, Eq. (5.22) reduces to

$$\epsilon_{\text{eff}} = 1 + \frac{\epsilon_r - 1}{2} \qquad (5.25)$$

where ϵ_r is the relative dielectric constant of GaAs. In these equations, $K(k)$ is a complete elliptic integral of the first kind and $K'(k) = K(k')$ is the complementary function. k' is defined as

$$k' = \sqrt{1 - k^2} \qquad (5.26)$$

An approximation for the elliptic function was provided in Ref. 4 and is reproduced below. For convenience, Fig. 5-6 provides a plot of $K(k)/K'(k)$ for arguments k in a useful range for CPW calculations:

$$\frac{K(k)}{K'(k)} = \left[\frac{1}{\pi} \ln\left(2\frac{1 + \sqrt{k'}}{1 - \sqrt{k'}}\right)\right]^{-1} \qquad (5.27a)$$

for $0.707 \leq k \leq 1$. For $0 \leq k < 0.707$,

$$\frac{K(k)}{K'(k)} = \frac{1}{\pi} \ln\left(2\frac{1 + \sqrt{k}}{1 - \sqrt{k}}\right) \qquad (5.27b)$$

If the CPW is coated with a thin deposited dielectric such as SiO_2, Si_3N_4, or polyimide, which are frequently used to separate first-level from second-level metallizations, an approximate correction factor, B, can be applied to the capacitance in order to account for the increase in effective dielectric constant [5]. In this case,

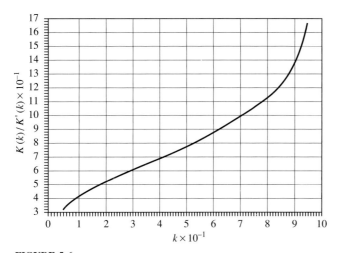

FIGURE 5-6
Elliptic function $K(k)/K'(k)$ where $K(k)$ is the elliptic integral of the first kind and $K'(k)$ is the complementary function.

$C_0' = BC_0$, where C_0' is the capacitance per unit length corrected for the coating:

$$B = 1 + \frac{\epsilon_L - 1}{\epsilon_r + 1}\left[1 - \exp\left(\frac{-4.6t}{a + b}\right)\right] \qquad (5.28)$$

In the equation above, ϵ_L is the relative dielectric constant of the thin dielectric coating and t is the thickness of this coating. ϵ_r is the relative dielectric constant of GaAs.

Another transmission line geometry which may occur occasionally in digital IC layouts is that of the *coplanar striplines* (CPS) shown in Fig. 5-7a. The two strips are assumed to be completely separated from neighboring conductors on the surface. In practice, if the spacing between the coplanar strips and the nearest conductors is at least $10b$, then the CPS formula from Ref. 3 can be applied:

$$Z_0 = \frac{120\pi}{\sqrt{\epsilon_{\text{eff}}}} \frac{K(k)}{K(k')} \qquad (5.29)$$

Equations (5.22) to (5.26) and (5.28) also apply to CPS.

The *microstrip* line configuration shown in Fig. 5-7b would apply to digital IC chips, if the substrate were conductive. Lightly-doped p-type substrates (or semi-insulating substrates with a buried p-layer) might be used as an alternative to semi-insulating substrates for isolation of devices. On semi-insulating substrates, the distance from the line to neighboring conductors is generally less than the substrate thickness so the coplanar models are more appropriate. Microstrip occurs more frequently on circuit boards and on microwave monolithic ICs (MMICs). The impedance and effective dielectric constant for the microstrip have been derived in Ref. 6 and are given below:

258 GALLIUM ARSENIDE DIGITAL INTEGRATED CIRCUIT DESIGN

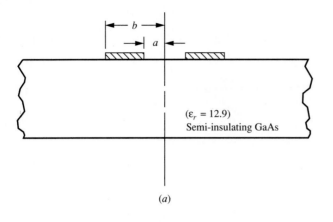

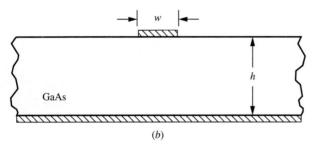

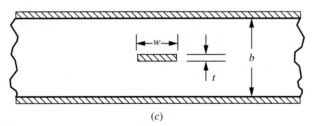

FIGURE 5-7
(a) Coplanar stripline (CPS). (b) Microstrip line. (c) Stripline.

$$Z_0 = \frac{60}{\sqrt{\epsilon_{\text{eff}}}} \ln\left(\frac{8h}{w} + \frac{w}{4h}\right) \quad \text{for} \quad \frac{w}{h} < 10 \quad (5.30)$$

$$\epsilon_{\text{eff}} = \frac{\epsilon_r + 1}{2} + \frac{\epsilon_r - 1}{2}\left(1 + \frac{10h}{w}\right)^{-1/2} \quad (5.31)$$

Finally, the *stripline* geometry of Fig. 5.7c (not to be confused with microstrip line) is a frequent configuration found on or in multilayer circuit boards. Since the interconnection of chips to form a larger digital system is the essence of packaging (Sec. 5.6), the equation for Z_0 for the stripline is also provided below. If the conductor thickness t is negligible compared with the ground plane spacing b, then the following equations apply [7]:

$$Z_0 = \frac{30\pi}{\sqrt{\varepsilon_r}} \frac{K'(k)}{K(k)} \tag{5.32}$$

$$k = \tanh\left(\frac{\pi W}{2b}\right) \tag{5.33}$$

Note that the relative dielectric constant is applicable in this case because the conductor is imbedded in a symmetric, uniform dielectric. These equations are accurate to within 10 percent if $W/b \geq 0.5$ and $t/b \leq 0.05$. Outside these boundaries, the error grows rapidly, and the equations in Refs. 7 or 8 should be used.

Example 5.2. Foundry XYZ provides data on interconnect line capacitance. For second-level metal, as shown in Fig. 5-8, $C = 0.09$ fF/μm for the minimum design rules; width $= 3$ μm; spacing $= 3$ μm. Estimate the characteristic impedance and inductance per micrometer of these interconnections.

Solution. Generally, interconnection capacitance test structures resemble Fig. 5-8 in design (the type of geometry should be described by the foundry). Assuming that neighboring lines are grounded, the CPW model can be applied. Proceed by calculating the capacitance for $\varepsilon_r = 1$; then the effective dielectric constant can be determined:

$$Z_0 = 30\pi \frac{K(k')}{K(k)}$$

In this example, width and spacing are equal (3 μm) so $a = 1.5$ μm and $b = 4.5$ μm; therefore $k = 0.33$. Using Fig. 5-6, $K(k')/K(k) = 1.56$ and $Z_0 = 147$ Ω. From Eq. (5.20),

$$C_0 = \frac{1}{Z_0} c = 0.0226 \text{ fF}/\mu\text{m}$$

and the ratio of the capacitances gives

$$\varepsilon_{\text{eff}} = \frac{0.09}{0.0226} = 4.0$$

This effective dielectric constant is smaller than would be obtained if the interconnect

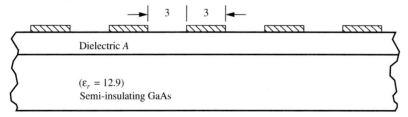

FIGURE 5-8
Typical coplanar transmission lines in wiring channel. Lines are equal widths and spaces (3 μm) in this example and are on top of a thin dielectric layer (A).

lines were directly on the GaAs substrate ($\epsilon_{\text{eff}} \simeq 7$) because of the intermediate dielectric layer. Now, the actual impedance and inductance can be determined from Eqs. (5.20) and (5.8) using $\epsilon_{\text{eff}} = 4.0$:

$$Z_0 = \frac{\sqrt{\varepsilon_{\text{eff}}}}{C_0 c} = \frac{2}{(9 \times 10^{-13}\text{F/cm})(3 \times 10^{10}\text{cm/s})} = 74 \: \Omega$$

$$L_0 = Z_0^2 C_0 = 0.5 \text{ pH}/\mu\text{m}$$

The accuracy of the estimate above depends on how adequately the ideal CPW model matches the actual interconnect geometry. If no other than the nearest neighboring lines contribute to the ground reference and other lines are floating (high impedance drivers), the estimates of impedance and inductance will be low. Applying the correction for finite ground plane extension [3], the error due to this condition is only 4.5 and 8 percent respectively for Z_0 and L_0.

Note that it is a much more difficult problem to predict the transmission line parameters of lines on a two-level dielectric system such as Fig. 5-8 directly from electromagnetic theory. With some measured line characteristics, however, it is possible to determine the other unknown parameters. This technique may not be valid for estimating parameters for other widths and spacings as the dielectric thickness is small and may interfere with the assumption $h >> b$ or a.

5.3 POWER SUPPLY AND GROUND DESIGN

The layout of a chip requires interconnections for power distribution and for signal transmission. High-speed circuits will require more careful consideration of both types of interconnection than is necessary in lower-speed technologies if the best performance is to be obtained. In this section, the problems associated with the distribution of power on-chip will be discussed. The distribution of power to multiple chips on PC boards or chip carriers is covered in Sec. 5.6. It will be assumed that the interconnection is accomplished by two or more levels of metal, rather than by semimetals or semiconductors such as silicides, polysilicon, or diffusion, because the excessive voltage drop on these higher-resistivity materials would produce very serious design difficulties.

The objective of the power distribution is to provide constant and equal power supply and ground potentials to every device on a chip. Since the metal thickness is constrained by the processing technology, line width increases may be needed to reduce voltage variations. In conflict with this goal, one would also like to minimize the chip area consumed by power buses so that the density of circuits and signal interconnections can be as high as possible. These objectives are further complicated by the problems listed below:

1. Resistance: ir drop
2. Electromigration: current density limitations
3. Inductance: $L\,dI/dt$ voltage variations

5.3.1 Resistance

Variations in power supply and ground potentials will occur as a result of the finite resistivity of metal interconnect lines. This happens even though the resistivity of most metals is quite small[4] because the on-chip interconnect lines are very thin (0.3 to 1.0 μm typically). The resistance (R) of a conductor with width W, length L, and thickness t can be calculated from the *resistivity* ρ (ohm-centimeters) or the *sheet resistance* ρ_s (ohms per square):

$$R = \frac{\rho L}{Wt} = \rho_s \frac{L}{W} \qquad (5.34)$$

The sheet resistance is defined as ρ/t. Thus, for a particular thickness of metal line, the resistance of the line can be found quickly by multiplying ρ_s by the number of squares of width W contained in a conductor of length L. Right-angled corners in a conductor count as 0.559 square [11].

The resistivity of a metal is temperature-dependent, depending on the scattering of electrons in the metal by lattice vibrations [12]. As the temperature increases, the resistivity increases in proportion to temperature (kelvin). Therefore, the voltage drop will also increase proportionally. Conversely, if operation at low temperatures is of interest, the resistivity of Al and Au drop to about one-tenth and one-fifth of their room temperature values respectively at 77 K [9], and current levels can be increased accordingly.

The voltage drop on ground lines erodes the noise margin of logic gates with a common-source input stage[5] by shifting the threshold voltage of the input. Voltage drops on V_{DD} or V_{SS} lines are not as serious for these logic gate topologies which connect the supply voltages through active-load current sources or source followers so long as these FETs remain in saturation. Since the FET when in the saturation region tries to maintain a constant current for any supply voltage in spite of variations, the sensitivity of logic voltage levels to power supply voltage variation is reduced. The SCFL circuit should be relatively insensitive to supply variations because of the low common mode gain. Also, it operates with nearly constant current since all FETs should remain in saturation. Other circuits, such as the GaAs domino or the SBFL circuit, have the logic low output tied directly to V_{SS} or ground respectively, so variations in these voltages will also change V_{OL} by a corresponding amount.

If the power wiring were point to point, then the resistance calculation in Eq. (5.34) could be used to estimate the voltage drop on a conductor. However, it is more frequent that a power bus line will distribute current to a number of gates rather than only one gate. If the current provided by the bus to logic gates or to other shorter interconnecting lines were uniformly distributed along the length of the line

---

[4]The bulk resistivities of the most commonly used interconnect metals are: aluminum, $\rho = 2.74 \times 10^{-6}$ Ω·cm, and gold, $\rho = 2.20 \times 10^{-6}$ Ω·cm, both at room temperature [9]. Resistivities of thin films typically used for IC fabrication are often higher by as much as 50 percent due to grain boundaries [10].

[5]FL, BFL, SDFL, DCFL, SBFL and e/d BFL gates, described in Chap. 4, all have common-source input stages.

(of length L), the current will vary with position x according to

$$I(x) = I_T\left(1 - \frac{x}{L}\right) \tag{5.35}$$

and the voltage drop at the far end of the line (ΔV) can be determined from the total current, I_T, by integration:

$$\Delta V = \frac{\rho I_T}{A} \int_0^L \left(1 - \frac{x}{L}\right) dx = \frac{\rho I_T L}{2A} \tag{5.36}$$

Thus, the voltage drop is one-half of what it would be if all of the load were concentrated at the far end of the line.

5.3.2 Electromigration

The second factor which also requires attention to the total current on a line is electromigration [13,14]. This is the tendency for metal atoms to migrate in the presence of electron flow at high dc current densities because of the force of the electric field on metal ions (generally small) and the exchange of momentum caused by the motion of electrons (electron "wind"). The rate of diffusion is increased in thin-film structures by grain boundaries. Under stress at high current densities, the grain boundaries can form vacancies in the crystal structure of the metal which tend to merge to form larger voids. Thus, the current density is further increased in the regions of voids, leading to an increasing rate of migration. Eventually, cracks form which open-circuit the interconnect line. Therefore, electromigration presents a reliability problem over long periods of time.

Up to 125°C, the safe upper limit of current density for aluminum conductors without an overglass coating is about 2×10^5 A/cm² while for gold it is 6×10^5 A/cm² [15]. These numbers apply to the thinnest region of conductor, often at steps (if any) or at the edge of contact cuts with other layers. Also, thicker (second-layer or above) metals require etching (reactive ion etch for Al or ion beam etch for Au), and etching generally reduces the linewidth below the pattern (mask) linewidth by several tenths of a micrometer. Thus, very narrow lines need to have extra allowance for current density. Therefore, for any given process, some safety factor must be allowed in the maximum current density to account for these effects. Very high current density must be avoided, since at current densities above 10^6 A/cm² the mean-time-to-failure (MTTF) of Al interconnections is of the order of tens of hours [14].

Electromigration is essentially a dc phenomenon. Alternating current studies show greatly reduced electromigration. Pulsed dc current will increase the time required for electromigration by the inverse of the duty cycle or, alternatively, the time-averaged current density can be used [15].

The MTTF for electromigration depends exponentially on temperature and superlinearly with current density. This behavior has been modeled by the following equation [14]:

$$\text{MTTF} = AKJ^{-m} \exp\left(\frac{\phi}{kT}\right) \tag{5.37}$$

In this equation, MTTF depends linearly on A, the conductor cross-sectional area, since it requires more time to form a continuous crack through a wider conductor. K is a constant that includes many factors. The rate of electromigration is very sensitive to the mechanism of transport, so the exponent m ranges from 1 to 3 depending on whether the transport is due to a dissimilar junction at the ends of the line or thermal gradients along the line. Finally, the experimentally determined activation energy, ϕ, for electromigration in thin-film conductors is found to be much lower than that required for bulk diffusion in the same material. For aluminum, $\phi = 0.4$ eV [14] and for gold, $\phi = 0.9$ eV [16]. This means that the MTTF changes rather slowly with temperature and thus remains finite even at room temperature for high current density. If low-temperature operation (77 K) is of interest, significantly higher current density is possible, which will help to increase the density of layout.

5.3.3 Inductance

The third factor which is troublesome for power supply and ground interconnections, both on-chip and off-chip, is the self-inductance associated with the transmission line nature of the wiring. If the current supplied by the line changes at the rate dI/dt, the inductance L will lead to a transient change in voltage according to the familiar equation below:

$$V = L\frac{dI}{dt} \qquad (5.38)$$

The inductance of a transmission line can be calculated using the models described in Sec. 5.2. Since L is independent of ϵ_{eff}, it is more efficient and just as accurate if the lines are assumed to be surrounded by air ($\epsilon_r = \epsilon_{\text{eff}} = 1$) rather than on the surface of the dielectric when calculating L.

Example 5.3. Estimate the ground noise generated on-chip by one SBFL clock line driver. The capacitive loading on the output of the driver is 200 fF; t_{PHL} and t_{PLH} should be 100 ps. For illustration, assume that the ground interconnection will use minimum-width (2 μm) first-level gold metalization with thickness of 0.3 μm and length of 2 mm.

Solution. Figure 5-9 illustrates the problem. Since there is only one load at the end of the V_{DD} and ground lines, Eq. (5.34) can be used to find the resistance. The sheet resistance is 0.1 Ω/□ and the ground line is 1000 squares long; therefore the resistance of the line is 100 Ω.

The superbuffer FET logic driver (described in Sec. 4.2.1) is well suited for driving large capacitive loads since the peak-to-average load current ratio is high. The current required to charge or discharge the capacitor with a t_{PLH} or t_{PHL} of 100 ps can be determined from $I = C\,dV/dt$ assuming that the current is constant over half of the logic swing.[6] The logic voltage levels for SBFL are $V_{OH} = 0.7$ V and $V_{OL} = 0$ V; thus half of the logic swing is 0.35 V. Therefore, a load current (I_C) of 0.7 mA is required. Now, the total current on the ground line will change for either transition. For the LH

---

[6]This method was described in Sec. 3.3.

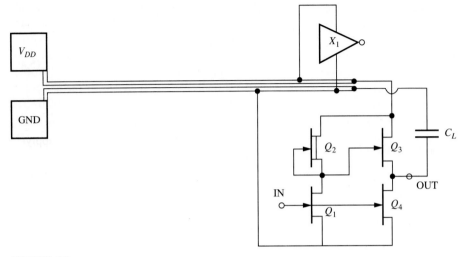

FIGURE 5-9
Superbuffer inverters on 2 × 2000 μm V_{DD} and ground lines. C_L = 200 fF in Example 5.3.

transition, Q_1 and Q_4 are cut off by the input. The static current of Q_4, however, would be quite low once C_L has fully discharged and therefore is neglected when the output is initially low. Therefore, the ground current will decrease only by I_{Q_2}, some fraction of I_C. For this example, assume I_{Q_2} = 0.3 mA. However, in the HL transition, Q_1 and Q_4 both turn on, and the ground current will increase by $I_C + I_{Q_2}$, a total of 1 mA. Therefore, the worst case ΔV due to resistance will be (1 mA) (100 Ω) = 100 mV, which is a significant fraction of the static noise margin (about 40 percent!), and only one gate was assumed to be active on the ground line. In addition, the superbuffer circuit also can generate a current pulse during the HL transition because of the delay of the inverter stage (Q_1 and Q_2) on the input. Both Q_3 and Q_4 will be in conduction for about one propagation delay, generating a momentary increase, I_{spike}, in the ground current. This effect can be seen when simulating the transient response of an SBFL gate.

Other logic gates on the ground line, such as X_1 in Fig. 5-9, will experience the same ΔV even if no change in input is received. If X_1 were part of a flip-flop or connected to a precharged node, then a logic upset might occur. The dynamic noise margin should be used as a predictor of logic failure, since both I_{spike} and the I_C pulse occur for short time periods.

The average current density must also be evaluated to guarantee that electromigration will not cause premature failures. The maximum current density will be 10^3 A/(3 × 10^{-5}) (2 × 10^{-4}) cm^2 = 1.67 × 10^5, and the minimum will be 5 × 10^4 A/cm^2. If a 50 percent duty cycle is used on this clock signal, the average current density is below 10^5 A/cm^2, acceptable for either Al or Au.

Finally, the inductance should be considered. Assume from Fig. 5-9 that the V_{DD} and ground lines are parallel strips as shown. Therefore, the CPS geometry and Eq. (5.29) apply. Substituting Eq. (5.29) into Eq. (5.19) yields

$$L_0 = \frac{120\pi}{c} \frac{K(k)}{K(k')} \tag{5.39}$$

independent of ϵ_{eff}. If the spacing and widths are each 2 μm, then $k = 0.33$ and, from Fig. 5-6, $L_0 = 8.0$ nH/cm. Thus, for a 2-mm-long segment, $L = 1.6$ nH. If the maximum ΔI (1 mA) occurs in 100 ps, then

$$\Delta V = (1.6 \times 10^{-9})\frac{10^{-3}}{10^{-10}} = 1.6 \times 10^{-2} \text{ V} \qquad (5.40)$$

an insignificant perturbation for the short time interval involved.

The example above only deals with a single clock driver circuit, whereas in the design of a complete chip, the power distribution buses must be capable of supplying power to all gates on the chip. Much higher currents will be required, yet voltage variations must still be controlled. The design of the power distribution scheme for a larger chip is illustrated by the next example.

Example 5.4. Determine the number of bond pads and the metal width required to provide power for a 4 × 4 mm DCFL circuit containing 3000 gates. The average power dissipation per gate is 0.5 mW. No more than 50 mV of ir drop can be allowed. Estimate the inductance to the center of the chip. The first- and second-layer metalizations are aluminum with thicknesses of 0.3 and 1.0 μm respectively.

Solution. First, electromigration and resistance will be considered. The overall strategy will be to use the second-layer metal (2M) for the chip-wide power distribution. Since the second metal is over three times thicker than first-layer metal (1M), it will be most suitable for supplying the high current levels required while still maintaining acceptable voltage control with smaller bus widths. If designed properly, most of the voltage drop will occur on the second layer [17]. If a third layer of metal were available it should be used instead of 2M. The first-layer metal will be used to distribute V_{DD} and ground locally, and will be connected to the second-layer distribution system at every opportunity.

The number of bond pads required can be determined by considering the worst-case ir drop and the electromigration current density limitation ($J_{max} = 2$ mA/μm^2 for aluminum). Since the maximum connection width between the power bus and the bond pad is limited by the bond pad width (100 × 100 μm is standard), and a $V_{DD} = 1.5$ V (typical for DCFL) will require $I_{DD} = 1.0$ A for the 3000 gates, at least five bond pads will be needed for V_{DD} and five for ground, just to satisfy electromigration requirements. If connected to bond pads at only one end, a 100-μm bus width is also required. The bus width required to guarantee less than the worst-case ir drop must still be estimated.

Assume that the V_{DD} and ground buses are arranged in 2M stripes running between bond pads at each end, as shown in Fig. 5-10, and that the logic gates are uniformly distributed on the chip. Bond pads will be placed at each end of the stripe to reduce ir drop. While J_{max} would permit reducing the stripe width to 50 μm, we will see that this will not produce an acceptable ir drop in this example.

Five stripes of V_{DD} and five of ground are required to satisfy the current density requirement if the current is assumed to be evenly distributed over the entire chip. With this assumption, each stripe must deliver 200 mA. By symmetry, the largest voltage drop will occur at the center ($l = 2$ mm), and $I = 100$ mA must be provided from each end. Equation (5.36) may be used to evaluate ΔV:

$$\Delta V = \frac{\rho l I}{2Wt} = \frac{(3 \times 10^{-6})(0.2)(0.1)}{2(10^{-2})(10^{-4})} = 30 \text{ mV} \qquad (5.41)$$

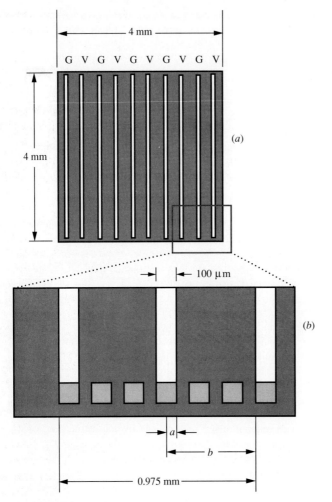

FIGURE 5-10
(a) Plan view of 4 mm square IC chip with V_{DD} and ground wiring distributed evenly in vertical second-level metal strips. (b) Expanded view of lower right corner. $a = 50$ μm; $b = 383$ μm.

Note that a slightly higher value than the bulk resistivity was used for the estimate of ΔV to account for the nonidealities of the deposited fine-grain thin film with thickness t. The voltage drop specification is satisfied with this configuration, but only with twice the minimum stripe width permitted by J_{max}. This illustrates the importance of considering both J_{max} and ir drop when distributing power. Even if the metalization were gold with its higher J_{max}, it would not result in narrower bus widths for this specification ($\Delta V = 50$ mV maximum).

Before discussing the 1M routing, possible variations in the 2M layout should first be considered. There are typically many options available for arranging the power/ground wiring on the chip. If inductance were not of importance, these buses could be arbitrarily located. However, the transmission line models of Sec. 5.2 can provide some guidance

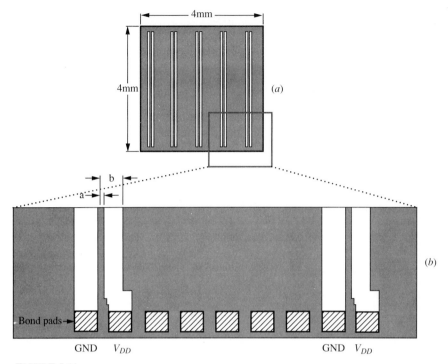

FIGURE 5-11
(a) Plan view of 4 mm square IC chip with V_{DD} and ground wiring distributed evenly in pairs. (b) Expanded view of right corner. $a = 1.5 \ \mu m$; $b = 101.5 \ \mu m$.

on selecting a low-inductance layout. Compare, for example, the layout shown in Fig. 5-11. Here V_{DD} and ground lines run in parallel, grouped as pairs rather than widely spaced as in Fig. 5-10. In Fig. 5-10, the conductor spacing is approaching the substrate thickness, so the ideal CPW model is not strictly valid. However, corrections for finite substrate thickness with a lower ground plane [18] do not strongly affect the impedance in this case. The ideal CPW applied to Fig. 5-10 with $\epsilon_r = 1$ predicts $Z_0 = 205 \ \Omega$ and $L_0 = 6.8$ nH/cm, whereas the more accurate model predicts $Z_0 = 196 \ \Omega$ and $L_0 = 6.5$ nH/cm, a negligible difference considering that the presence of 1M perpendicular to these lines has been neglected. Figure 5-11 is clearly a CPS geometry, and Eqs. (5.29) and (5.19) predict $Z_0 = 106 \ \Omega$ for $\epsilon_r = 1$ and $L_0 = 3.5$ nH/cm, nearly a factor of two lower inductance than the widely spaced CPW. Therefore, it is advantageous to group power and ground conductors together in closely spaced parallel strips to reduce their series inductance.

Multiple 1M lines (gray) perpendicular to these stripes can be used to connect to rows of gates, represented by triangles, as shown in Fig. 5-12. Connections between 1M and 2M are represented by × symbols for vias (or windows). The width required for the 1M buses will depend on the number of gates to be powered on each row between via connections to the second-layer buses. Space is allowed between gates for horizontal first-level metal and vertical second-level metal wiring channels. This number of gates will depend on the space allowed for vertical (2M) interconnect wiring between each gate and on the size of the gate structure itself. Since the maximum span between 2M buses

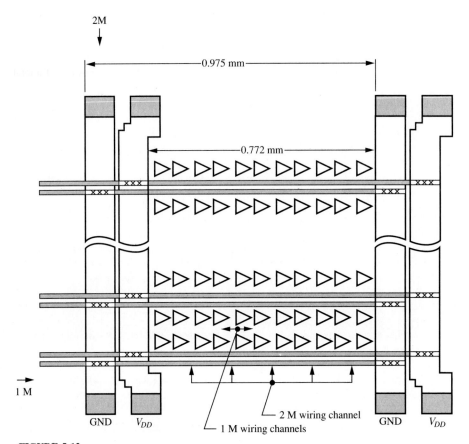

FIGURE 5-12
Partial power bus layout for large IC. First-layer metal runs horizontally; second layer runs vertically. Gates are represented as triangles; vias by ×'s.

is 0.772 mm and the number of gates per row is relatively small, the voltage drop can be kept quite low on the local (1M) buses if the widths are properly selected.

Suppose that 20 gates per row were being powered. This will require a current of 6.6 mA. Then an additional voltage drop to the center of the span can be kept below the required 20 mV if the 1M bus widths, W, are 10 μm or more:

$$\Delta V_{1M} = \frac{\rho l I}{2Wt} = \frac{(3 \times 10^{-6} \; \Omega\cdot\text{cm})(0.0386 \; \text{cm})(6.6 \times 10^{-3} \; \text{A})}{2W(3 \times 10^{-5} \; \text{cm})} = 0.02 \; \text{V} \quad (5.42)$$

Finally, the inductance can also be estimated. From the CPS model, $L_0 = 3.5$ nH/cm. Since the center of the strip (2 mm from pad) is connected to the ground or V_{DD} by two parallel paths, the net inductance to the center is reduced by half:

$$L = \frac{1}{2}(0.2 \; \text{cm})(3.5 \; \text{nH/cm}) = 0.35 \; \text{nH} \quad (5.43)$$

If the chip is connected to its circuit board or chip carrier by bond wires, an additional

0.8 nH is added in series with each bond pad (assuming bond wires of 1 mm length) [19], and $L = \frac{1}{2}[(0.2 \text{ cm}) (3.5 \text{ nH/cm}) + 0.8 \text{ nH }] = 0.75$ nH. Therefore, bond wire interconnections can significantly increase total inductance.

In order to keep the inductive voltage spike below 100 mV, the maximum current slew rate is $dI/dt = 0.1/(0.85 \text{ nH}) = 11.8 \times 10^8$ A/s or $\Delta I = 11.8$ mA out of a total of 100 mA in 100 ps. Since DCFL steers current between the switch FET and the input gate-source diode of the next stage, the I_{DD} and ground current is relatively constant; however, a fractional change in current below 12 percent may be difficult to achieve. A more detailed analysis of the circuit through simulation will be necessary to determine whether this can be achieved.

The above estimate of inductance is pessimistic, as the strapping of 2M strips together with multiple 1M connections will form a mesh which may have less inductance than would be predicted by a simple microstrip model. Also, the presence of 1M strips under the 2M buses will reduce the impedance to some degree and also the inductance.[7] The actual inductance would be very difficult to calculate, however.

Further consideration of inductance models will be helpful in comparing strategies for reduction of on-chip inductance. The example above showed that rearrangement of power/ground bus geometry to reduce the impedance will be an effective tool in reducing inductance. Problem 5.8 compares the inductance vs. conductor width for the CPS and CPW geometries, where it is seen that L_0 does not drop rapidly with width beyond 10 μm. Since very wide power buses consume large amounts of chip area, a better strategy for reducing inductance is to connect the buses to bond pads at more frequent intervals, therefore paralleling inductances. Also, since the bond wires play an important role, extra bond pads will parallel the bond wire inductance as well.

Finally, the inductance of the transmission lines or planes on the circuit board (PCB) or chip carrier (CC) can also be extremely important due to their greater length. The standard approach to decouple the chip from the PCB or CC inductance is to provide a low-impedance shunt (bypass capacitor) very close to the chip boundary. Since these techniques fall into the general category of packaging, further discussion is deferred to Sec. 5.6.

5.4 DELAY ESTIMATION

Useful estimates of the total propagation delay on any given path in a logic circuit can only be obtained if all of the appropriate loading effects influencing the propagation of the desired signal are considered. Since many mechanisms can influence signal propagation, it will be helpful to sort out the most influential mechanisms for a given path so that the most efficient estimate of interconnect delay can be made.

In the section that follows, an approach for the estimation of propagation delays due to on-chip interconnections is presented. The effects considered include transmission line capacitance, electromagnetic wave propagation, and distributed C delay. Neglecting these effects will always lead to an overly optimistic estimate of circuit performance.

---

[7]Gold mesh has been used in lieu of bond wires for years for low-impedance microwave or millimeter wave IMPATT diodes because of its lower inductance.

5.4.1 Crossover Capacitance

The total loading capacitance of all interconnections must include the transmission line capacitance and any excess capacitance that would be caused by lines on one level (physical plane containing metal lines) crossing over or under lines on another level. While this overlap capacitance may be used to advantage at times (for on-chip coupling capacitors, for example), it can result in a major source of additional delay if not minimized through well-informed layout practices. The total capacitance of a small crossover is significantly larger than its parallel plate capacitance due to the significant extension (fringing) of the electrostatic field beyond the edges of the conductors.

An equation suitable for estimating crossover capacitance was reported in Ref. 5 and is reproduced below. Here, ϵ_i is the relative dielectric constant of the insulator. Dimensions h, Z, and W (in micrometers) are defined in Fig. 5-13.

$$C_x = 8.854 \times 10^{-18} \epsilon_i \left\{ \frac{WZ}{h} + 1.393(Z + W) \right.$$

$$\left. + \frac{2}{3}\left[Z \ln\left(\frac{W}{h} + 1.444\right) + W\ln\left(\frac{Z}{h} + 1.444\right)\right] \right\} \quad (5.44)$$

The first term is the familiar parallel plate capacitance; the other terms add the excess fringe capacitance, estimated using a microstrip line model.[8]

An additional source of excess loading capacitance is also contributed by bond pads or probe pads. Bond pads are usually $100 \times 100 \ \mu m^2$ metal pads on the semi-insulating substrate at the edges of the IC chip. Probe pads may be smaller if strictly intended for temporary probe testing and not for wire bonding. Pads as small as $20 \times 20 \ \mu m^2$ can be probed if adequately spaced from surrounding circuitry. Since

---

[8] The equation shows that the parallel plate term dominates only when Z and W are quite large.

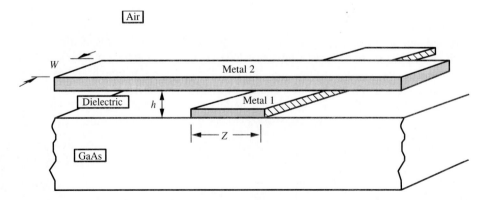

FIGURE 5-13
Cross-sectional view of first-level/second-level metal crossover.

these pads require space from other surface features, their capacitance can be modeled as capacitance to the back side of the substrate (microstrip model). Again, Ref. 5 provides an approximate microstrip-based model for estimating the capacitance of a pad with width W and length Z. The substrate thickness is h. All dimensions are in micrometers. ($2\pi\epsilon_0 = 5.559 \times 10^{-17}$ F/μm.)

$$C_p = 2\pi\,\varepsilon_0 \left[\frac{Z\,\epsilon_{\text{eff}}(W)}{\ln(8h/W)} + \frac{W\,\epsilon_{\text{eff}}(Z)}{\ln(8h/Z)} \right] - \varepsilon_0 \left(\frac{\epsilon_r WZ}{h} \right) \quad (5.45)$$

This formula adds the total microstrip line capacitance of a conductor of width W and length Z to one with width Z and length W. Then one parallel plate capacitance is subtracted, because it had been included twice by the first term in the above equation:

$$\epsilon_{\text{eff}}(W) = \frac{\epsilon_r + 1}{2} + \frac{\epsilon_r - 1}{2} \frac{1}{\left[1 + 12(h/W)\right]^{1/2}} \quad (5.46a)$$

$$\epsilon_{\text{eff}}(Z) = \frac{\epsilon_r + 1}{2} + \frac{\epsilon_r - 1}{2} \frac{1}{\left[1 + 12(h/Z)\right]^{1/2}} \quad (5.46b)$$

where h/W and h/Z must be greater than or equal to unity.

5.4.2 Effective Source Resistance and Attenuation

The effective source resistance and attenuation must first be defined before the proper delay model can be selected. In Sec. 5.1, Fig. 5-1, the internal generator resistance, R_G, was acknowledged but not specifically related to any particular circuit details. However, if the logic gate or inverter (driver) were described by its Thevenin equivalent, V_G would be the open-circuit voltage and R_G the internal source resistance of the generator. This resistance is the same as the output resistance of the driver, which is weakly nonlinear and can be estimated by small-signal analysis. Figure 5-14a shows a circuit diagram and Fig. 5-14b the corresponding small-signal model for a source follower output driver from a BFL gate. If Q_1 and Q_2 are biased in their saturation regions at all times, then the midfrequency output resistance of Q_1 is approximately $1/g_{m,i} + R_S$. Adding in the series resistance of the level-shift diodes and placing r_{ds2} in parallel leads to the Thevenin equivalent shown in Fig. 5-14c. In reality, the dc bias current of Q_1 will vary slightly with the logic level of the output. Therefore, $g_{m,i}$ will depend on whether v_{out} is at high or low logic level and so will R_G.

R_G can also be estimated by circuit simulation methods. With the input v_{in} biased at V_{OH} or V_{OL}, an ac voltage source with amplitude much less than the logic swing (small-signal) can be applied to the output through a dc blocking capacitor as illustrated in Fig. 5-15a. The blocking capacitor is necessary to avoid changing the dc biasing of the driver. The small-signal output resistance will be given by $v_{\text{out}}/i_{\text{out}}$ and should be evaluated at both logic states.

Finally, Fig. 5-15b illustrates another approach toward determining R_G by simulation, which has been described in Ref. 20. Because of the nonlinearity of

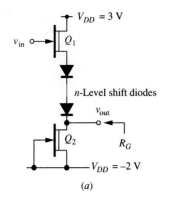

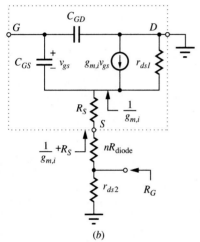

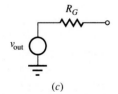

FIGURE 5-14
(a) Source follower with level shift. R_G is the effective source impedance of the Thevenin equivalent circuit [shown in (c)] of the SF driver. (b) Small-signal model of the SF driver.

the driver output, an effective average value of R_G is needed for both logic states. Thus, a large-signal method will be more able to obtain such a compromise value for R_G. In this method, a pulse is launched toward the output of the BFL driver through a transmission line. The line has its characteristic impedance equal to that of the actual interconnections, and the length of the line is long enough so that a full pulse length can be observed without any interference from secondary reflections from the generator. The use of the current source avoids interference with the dc bias point; the clamp diode D_{CL} is included so that the bias point of Q_1 will be correct when the output is at logic high. The test current $i_G(t)$ is determined by $i_G(t) = v_G(t)/Z_0$.

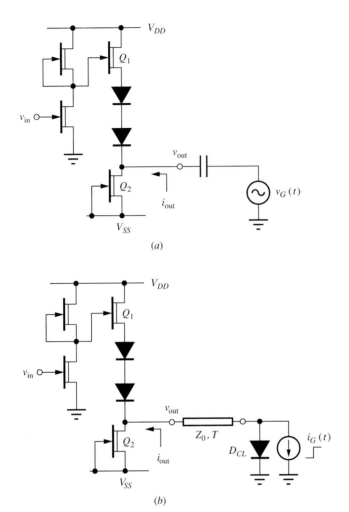

FIGURE 5-15
(a) AC small-signal measurement of $R_G = v_{out}/i_{out}$ (b) Pulse method for finding R_G.

$v_G(t)$ can be modeled as a trapezoid with an appropriate amplitude and with rise and fall times typical of unloaded logic gates. If a primary driven signal line is analyzed, then the amplitude should be $V_{OH} - V_{OL}$. If a coupled line is being considered, then $(V_{OH} - V_{OL})/4$ would be more typical of crosstalk amplitudes. The signal $v_G(t)$ can either pulse down from V_{OH} or up from V_{OL} as required by the v_{in} dc bias condition.

The line with solid square points in Fig. 5-16 shows an incident voltage $v_G(t) = V_3$ as it leaves the source, and the open squares represent V_4, the voltage V_3 after it has travelled to the output terminals of the gate as determined through simulation. In this sketch we have assumed that the gate output impedance is higher than the line impedance, and as a result the reflected wave adds to this incident wave giving a total voltage of amplitude $V_4 > V_3$, as suggested in the figure. The total voltage computed

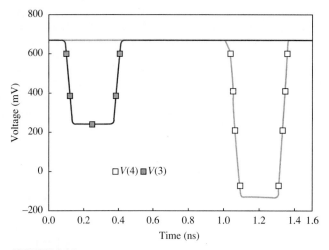

FIGURE 5-16
The line with solid squares shows the voltage $V_G(t) = V_3$ in Fig. 5-14b after is has arrived at the gate terminals, and $V_A(t) = V_4$ is the total voltage at the gate terminals.

may be somewhat distorted in shape due to the nonlinearity of the logic gate and due to the frequency sensitivity and the reactive part of the gate impedance. However, the results of SPICE computations with various buffered-FET logic examples showed surprisingly little distortion, the main distortion observed being, in some cases, some overshoot in the waveform at the leading edge of the waveform crest. Knowing incident voltage V_3 and the computed amplitude V_4, the effective resistance of the gate terminals can be computed from

$$R_G = \frac{Z_0(V_4/V_3)}{[2 - (V_4/V_3)]} \tag{5.47}$$

If it is desired to obtain the effective *input* resistance of a gate, essentially the same procedure can be applied.

The above techniques are best illustrated by an example.

Example 5.5. A small BFL gate is selected from a cell library and is to be used to drive an interconnect line. The source follower Q_1 and pull-down Q_2 are 1×10 μm depletion mode MESFETS with $V_T = -1$ V. The diodes are 2×10 μm level-shift diodes. Estimate the Thevenin equivalent output resistance for this BFL gate.

Solution. Assume that the BFL source follower remains in saturation over the full logic swing. The small-signal model of Fig. 5-14b can be evaluated and R_G found. The intrinsic FET $g_{m,i}$ can be obtained from the model parameters and the model equations or from measured device characteristics. Using the former approach, taking the derivative of I_D [Eq. (2.13b)] with respect to $V_{GS,i}$ yields

$$g_{mi}(V_{GS,i}, V_{DS,i}) = \beta W(1 + \lambda V_{DS,i}) \left\{ \frac{b(V_{GS,i} - V_T)^2 + 2(V_{GS,i} - V_T)}{[1 + b(V_{GS,i} - V_T)]^2} \right\} \tag{5.48}$$

Before we can evaluate g_m, $V_{GS,i} = V_{GS} - I_D R_S$ and $V_{DS,i} = V_{DS} - I_D(R_S + R_D)$ must be determined; thus $I_D(V_{GS}, V_{DS})$ must be known. This could also be obtained from Eq. (2.13b) using an iterative solution (such as Newton's method). Parameters b, β, and λ must be known for the FET to be used in the circuit.

Alternatively, if the FET I-V data is available (Fig. 2-14, for example), then I_D and g_m can be measured directly from the characteristics. Using this latter approach, Fig. 2-14 gives the I_D vs. V_{DS} characteristics for a 1×50 μm MESFET whose model parameters are summarized in Table 2.4. Since Q_1 will be biased with $V_{GS} = 0$ V, $\Delta I_D/\Delta V_{GS}$ is estimated around that characteristic. For this FET, g_m is relatively independent of V_{DS} in the saturation region. $g_m(H) = 3.9$ mS. This is an extrinsic transconductance, and therefore already includes the effect of R_s. If Eq. (5.48) were used instead of the measured characteristics, $g_{m,i}$ would be corrected for R_S by

$$g_m = \frac{g_{m,i}}{1 + g_{m,i} R_S} \tag{5.49}$$

The series resistance of the level-shift diodes can also be determined from the measured data, or, if the parameters in Table 2.1 apply to this diode, $R_{\text{diode}} = R_S/A$ where A is the diode area in square micrometers. In this example, $R_{\text{diode}} = 75$ Ω.

Finally, r_{ds2} should be estimated for Q_2. From Eq. (2.13b), with constant V_{GS},

$$g_{ds,i} = \frac{\partial I_D}{\partial V_{DS}} = \beta W \left(\frac{(V_{GS,i} - V_T)^2}{1 + b(V_{GS,i} - V_T)} \right) \lambda \tag{5.50}$$

Then, correcting for R_S,

$$g_{ds,i} = \frac{g_{ds,i}}{1 + g_{m,i} R_S} \tag{5.51}$$

or, alternatively, if the I_D-V_{DS} characteristics are available, measure the slope of the drain current in saturation for the $V_{GS} = 0$ characteristic. From Fig. 2-14, $\Delta I_D/\Delta V_{DS} = 0.19$ mS for the 50-μm wide device.

Since we are concerned with the small-signal response at frequencies well above 1 MHz, g_{ds} increases by a factor of 2 to 3 for a typical ion-implanted MESFET, as discussed in Chap. 2. Therefore, scaling to $W = 10$ μm and using a factor of 2.5 as an average increase in g_{ds}, $g_{ds2} = 2.5(3.9 \times 10^{-2} \text{ mS}) = 1/r_{ds2}$, or $r_{ds2} = 10$ kΩ.

Therefore, the small-signal effective source resistance is estimated as

$$R_G = \left(\frac{1}{g_m} + 2R_{\text{diode}} \right) \| r_{ds2} = 1100 \, \Omega \tag{5.52}$$

The value of R_G predicted by this method would not change with the logic state of the output since g_m is not strongly dependent on V_{DS}.

A SPICE simulation can also be used to determine the effective source resistance. The approach given in Fig. 5-15a will provide v_{out} and i_{out} through small-signal analysis with the ac mode. Large-signal simulation of R_G can be performed with a pulse technique, illustrated in Fig. 5-15b. The MESFET model parameters in Table 2.4 were used to simulate the BFL gate with SPICE; then the .ac and .tran modes could be used to find R_G in both the high (V_{OH}) and low (V_{OL}) output states of the gate. The results obtained from the three methods are compared in Table 5.1.

It does not come as a great surprise to see that the results differ significantly between analysis methods. The small-signal model includes many assumptions that become less valid as the signal swing increases. The two SPICE methods are in somewhat closer agreement at each logic condition.

TABLE 5.1
Comparison of effective output resistance for output at $V_{OL} = -1.15$ V or $V_{OH} = +0.65$ V

Method	$R_G(V_{OL})$, Ω	$R_G(V_{OH})$, Ω
Small-signal model	1100	1100
Small-signal .ac SPICE	887	585
Large-signal .tran SPICE	1020	677

It is apparent from this example that the series resistance of the diodes and the source resistance of Q_3 present a technological limitation on the minimum Thevenin equivalent output resistance that can be obtained from a MESFET or MODFET driver. This limitation becomes more serious if low impedance drivers are required; much greater device widths are needed to obtain the low R_G because of the parasitic series resistances. Improved device structures and contacts would reduce these resistances. In addition, the shunt capacitance of the logic diodes will provide a feedforward path for high-frequency signal components, improving the signal rise time on the output. Finally, many e/d MESFET or MODFET circuits do not require level-shift diodes at all, an advantage in applications that could benefit from low effective source resistance.

Section 5.1 assumed that the propagation of waves on transmission lines was not affected by losses. While this assumption is good if the line is very short or very wide, there are circumstances in which narrow, thin conductors are considered for interconnection of high-speed signals. The loss, principally in the form of the dc series resistance, will attentuate the signal.[9]

Additional resistive loss due to the *skin effect* [21], the exponential decay of the electric field as it penetrates into a conductor at high frequencies, is relatively unimportant for thin-film conductors. This is because the penetration depth δ is typically greater than the conductor thickness at frequencies of interest:

$$\delta = \left(\frac{\omega\mu_0\sigma}{2}\right)^{-1/2} \tag{5.53}$$

where μ_0 is the permeability and σ is the conductivity. Although it is difficult to associate a single frequency with the high-speed pulses or fast steps found in GaAs digital ICs, it is convenient to consider the propagation of a gaussian pulse. Minimum width pulses in high-speed systems often resemble a gaussian anyway.

---

[9]Not only that, but lossy lines tend to be dispersive (attenuation and velocity of propagation are functions of frequency) because of the series resistance and any asymmetry in the dielectric. While simple attenuation is easily evaluated, dispersion is quite difficult to model. Since it does not strongly influence on-chip wiring delay, it is neglected in the discussion that follows.

$$V(t) = V_0 \exp\left(\frac{-t^2}{2\tau^2}\right) \tag{5.54}$$

In Eq. (5.54), τ is the standard deviation and 2.35τ is the full-width at half-maximum. The Fourier transform

$$V(\omega) = (2\pi)^{1/2}\tau V_0 \exp\left(\frac{-\omega^2 \tau^2}{2}\right) \tag{5.55}$$

is conveniently also a gaussian. The bandwidth required by the pulse can then be characterized by the spectral width of Eq. (5.55). For example, 99 percent of the frequency spectrum required to form the pulse is contained in the bandwidth [22]:

$$f_{99\%} = \frac{0.29}{\tau} \tag{5.56}$$

For a pulse with $\tau = 100$ ps, a bandwidth of about 3 GHz is required. For aluminum or gold, $\delta = 1.5$ μm at 3 GHz. Therefore, for conductors much thinner than 3 μm, the skin resistance is simply the dc resistance. This would include most on-chip interconnections except possibly wafer-scale interconnects.

Analysis of pulse propagation on thin-film microstrip transmission lines has been carried out [23]. These studies show that the skin resistance and inductance are largely negligible, affecting only the leading edge of the pulse for line lengths and pulse widths typically found on IC chips. As τ approaches 1 ps, however, the effects are quite severe [22].

Note that skin effect corrections will be more significant on PCBs because the conductor thickness is usually much greater than δ and the line lengths are longer. This problem is discussed in Sec. 5.6.

Wave propagation on lossy transmission lines has been analyzed and reported [24,25]. The general transmission line equations (5.4a,b) are modified to include series resistance as shown below:

$$-\frac{\partial V}{\partial x} = R_0 I + L_0 \frac{\partial I}{\partial t} \tag{5.57}$$

The corresponding equation (5.4b) remains the same, since shunt conductance is always quite small. In the complex frequency domain, the line voltage at $x = l$ is given by

$$V(l,s) = e^{-\gamma(s)l} V(0,s) \tag{5.58}$$

where the propagation constant, $\gamma(s)$, is now complex due to the loss:

$$\gamma(s) = s(L_0 C_0)^{1/2}\left[1 + \left(\frac{R}{sL}\right)\right]^{1/2} \tag{5.59}$$

When the line is driven by a step function $u(t)$, then $V(0,s) = 1/s$, and the inverse Laplace transform is taken, then the time-domain behavior is shown to consist of an attenuated step function delayed by the transit delay $T = l(L_0C_0)^{1/2}$ plus a function $g(l, t - T)$ which slowly increases from zero beginning at $t = T$ [24, 25], behaving qualitatively like an *RC* lowpass circuit:

$$v(l, t) = e^{-R_0 l/2Z_0} u(t - T) + g(l, t - T)u(t - T) \tag{5.60}$$

This equation shows that the signal propagating down the line is attenuated by a factor $e^{-\alpha l}$ where the *attenuation factor* α is the ratio of line series resistance per unit length, R_0, to twice the characteristic impedance. R_0 and Z_0 are independent parameters since R_0 varies according to Eq. (5.34) with the material (σ) and inversely with the cross-sectional area of the line, while Z_0 depends on the width/spacing ratio. Therefore, at time $t - T$, the voltage $v(l, T)$ will abruptly increase from zero to $e^{-R_0 l/2Z_0}$ corresponding to the first transit of the step function input. If the line is open-circuited at the end, then the attenuated voltage will be doubled. If the voltage source at the input has an effective source resistance R_G, then there is a voltage division at the input ($x = 0$), so that

$$v(l, T) = 2\left(\frac{Z_0}{Z_0 + R_G}\right) e^{-R_0 l/2Z_0} \tag{5.61}$$

Thereafter, the voltage will charge up at a slower rate (*RC*-limited) to the steady-state condition.

The above model can be used to establish a maximum length for an interconnection which will charge its output to an acceptable signal amplitude in a single-signal transit; i.e., for a minimum acceptable $v(l, T)$, a maximum l is determined for the R_0 and Z_0 under consideration. The interconnect delay will be least for a line under this constraint. Since the one-pass delay contributed by the line to the interconnect path is just the transit delay of the electromagnetic wave, lines longer than this limit will suffer additional *RC* delay in order to reach the minimum acceptable output voltage. This extra delay will be approximately the same as required for multiple-transit voltage buildup. If we assume that the minimum $v(l, T) = 0.9$ or 90 percent of the steady-state value, then by Eq. (5.61), $R_0 l/Z_0$ must satisfy the equation below:

$$\alpha = \frac{R_0 l}{2Z_0} = \ln\left(\frac{2.22}{1 + R_G/Z_0}\right) \tag{5.62}$$

Since small α is preferable, small R_G or high Z_0 leads to longer single-transit interconnections.

Attenuation can be beneficial under certain conditions. The returning reflected waves at odd multiples of T are attenuated twice as much as the initial incident wave. Including some absorption in R_G, reflections are well suppressed on lossy transmission lines, eliminating the need to terminate lines if the dynamic noise margin is not exceeded by the largest reflected wave at $3T < t < 5T$. Unterminated lines can save on power dissipation for off-chip GaAs-to-GaAs interconnections if superbuffer drivers can be used.

TABLE 5.2
Metal strip width and spacing

W, μm	S, μm
3	3
6	3
6	6

The criteria presented in Refs. 24 and 25 for lossy interconnects ($2Z_0/3 < R_0 l < 2Z_0$) applies only for the case where $R_G \ll Z_0$ (typical of bipolar). For other effective source resistances, Eq. (5.62) must be used to determine the maximum single-transit interconnect length.

Example 5.6. Consider the maximum single-transit lengths for second-level metal interconnects (Table 5.2) with the widths and spacings shown. This 2M system has a metal thickness of 1 μm, $e_{\text{eff}} = 4$, and $\rho = 3 \times 10^{-6}$ $\Omega\cdot$cm. Assume that the driver source resistance, R_G, is made equal to Z_0 by proper selection of FET widths.

Solution. Equation (5.62) shows that $R_0 l/Z_0 = 0.21$ will be the same for all three cases if $R_G = Z_0$ and the 90% criteria is applied. Therefore, calculate R_0 and Z_0 for each case to determine l. R_0 is given by

$$R_0 = \frac{\rho}{Wt} \ \Omega/\text{cm}$$

and assuming the CPW model applies (wiring channel)

$$Z_0 = \frac{30\pi}{\sqrt{\epsilon_{\text{eff}}}} \frac{K(k)}{K'(k)}$$

The solutions are shown in Table 5.3.

We can see from this example that increasing the width of the conductor will increase $l_{\max}$. However, the characteristic impedance will also be reduced which will increase the power dissipated by the driver. Increasing the spacing as well as the width is even more helpful, but involves further sacrifice in area.

TABLE 5.3
Maximum single-transit interconnect length

W, μm	S, μm	R_0, Ω/cm	Z_0, Ω	$l_{\max}$, cm
3	3	100	74	0.16
6	3	50	60	0.25
6	6	50	74	0.31

5.4.3 Interconnect Delay

There are many mechanisms that affect wave propagation and, hence, propagation delay on IC interconnect wiring. Because of this, the selection of an appropriate model for predicting delay will depend on the relevant conditions. Is the effective source resistance of the driving gate or inverter more than or less than the characteristic impedance of the transmission line? Is the transmission line attenuation too high to charge the output terminals to an acceptable voltage in a single transit? Finally, is the response time of the logic circuit faster or slower than the transit delay on the transmission line? The answers to these questions will be used to guide the reader toward the best model for a particular interconnect situation.

Figure 5-17 illustrates a hierarchy of five of the many possible interconnect models that are suitable for use with circuit simulation tools. The simplest model, the

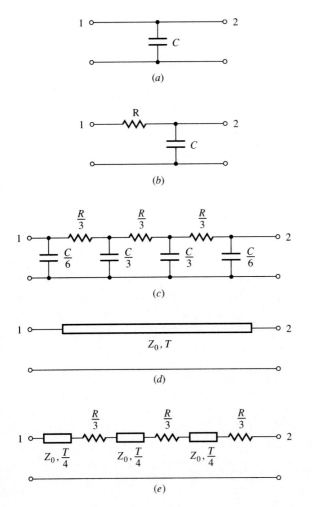

FIGURE 5-17
Hierarchy of equivalent circuit models for calculation of the delay caused by an interconnection. (a) C model. (b) L model. (c) P3 model. (d) TRL model. (e) TRL4 model.

lumped capacitor (this will be referred to as the C model), has been used in Chaps. 3 and 4 to represent the load capacitance contributed by wiring and device parasitics and is discussed first. Section 5.2 has provided tools for analysis of transmission line structures that can provide capacitance per unit length and characteristic impedance estimates. Also, Sec. 5.4.1 described a technique for estimating the load capacitance contributed by crossovers. Therefore, a net value for C can be determined once the layout is fully described. This C will be composed of the sum of line, crossover, and fan-out capacitance.

The C model is accurate only in two limiting conditions:

$$\frac{R_G}{Z_0} \gg 1 \qquad (5.63a)$$

or
$$T \ll t_r \quad \text{or} \quad t_f \qquad (5.63b)$$

where t_r and t_f are the rise time and fall time of the driver. The first case applies when the driver effective source impedance is much larger than the Z_0 of the interconnect. Since Z_0 typically ranges between 50 and 80 Ω, the condition (5.63a) applies whenever a line is driven by small-area MESFETs or MODFETs or an HBT operating at very low current levels. Normally, this will cover most of the on-chip interconnections encountered in an IC design.

This case can be further understood by considering the buildup of voltage on the line by multiple reflections. This was analyzed in Sec. 5.1 and is represented in Fig. 5-2, but neglects attenuation. Since the line initially receives only a small fraction of V_G, many reflections are needed to reach a steady state.

The second condition would apply even for low R_G so long as the line is quite short. At this limit, the signal is reflected from the load, returns to the generator,

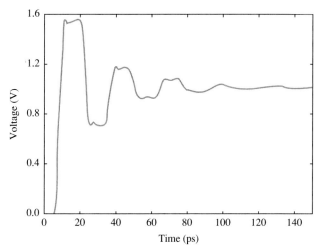

FIGURE 5-18
Line voltage at $x = l$ for the case with $R_G < Z_0$.

and is reflected again back toward the load. A representative waveform calculated for the case $R_G = 0.2Z_0$ is shown in Fig. 5-18. If the time interval of the first reflection ($T \le t \le 3T$) is small compared with the response time of the logic circuit (t_r or t_f), then the dynamic noise margin will not be exceeded. The C model predicts the time required to reach 50 or 90 percent V_G within acceptable error limits, even though it omits the fine details of the reflections. This low R_G case is less common in FET circuits since large FETs are not generally used to drive short lines. However, HBT emitter follower drivers can easily produce $R_G < Z_0$ because of their large $g_m = qI_c/nkT$.

Estimating the propagation delay or rise time of circuits that use the C model is straightforward. For hand analysis, the line, crossover, and effective input capacitance of the load gates are added. The capacitor is then charged through R_G, causing the output voltage $v_2(t)$ to rise according to a single exponential solution

$$v_2(t) = V_G\left(1 - e^{-t/R_G C}\right) \tag{5.64}$$

or in a ramp if a constant current generator is assumed. For the above equation,

$$t_{50\%} = 0.69 R_G C \tag{5.65a}$$

and
$$t_r = t_f = 2.2 R_G C \tag{5.65b}$$

Since logic gates are nonlinear circuits, this linear, single-exponential model is not very accurate. More commonly, a circuit simulator will be used and a fixed C composed of the sum of line and crossover capacitance is connected between the driver and load as shown in Fig. 5-17a.

Example 5.7. A 10-μm BFL gate (Example 5.5) is used to drive a first-level (1M) metal interconnect as shown in Fig. 5-19. This line, 1 mm long, is connected to an identical gate at the load end and runs along a wiring channel. The minimum 1M line width is

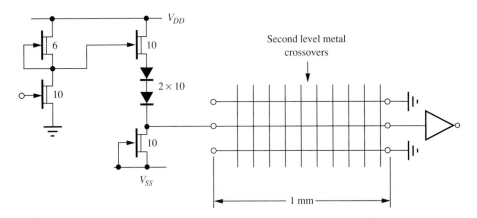

FIGURE 5-19
BFL source follower driving an inverter through a 1 mm length of first-level metal interconnect line with second-level metal overcrossings.

2 μm and line-to-line spacing is also 2 μm. There are ten 2 × 3 μm crossovers. The dielectric is 0.5 μm thick Si_3N_4 ($\epsilon_i = 8$). Using SPICE, find the time required to charge the line to 50 percent of the open-circuit voltage of the driver (propagation delay) and the rise and fall times. Use the simulated results to calculate an effective source resistance, R_G, from Eqs. (5.65a, b). Assume that the effective input capacitance of the load gate is 20 fF.

Solution. The CPW model in Sec. 5.2 can be used to estimate the line capacitance. In this example, $a/b = 0.33$, $K(k)/K'(k) = 0.64$ (Fig. 5-6), $\epsilon_{\text{eff}} = 7$ from Eq. (5.25), so $Z_0 = 56$ Ω. Then the line capacitance is given by

$$C_0 = \frac{\sqrt{\epsilon_{\text{eff}}}}{cZ_0} = 1.57 \text{ pF/cm} \tag{5.66}$$

This should be increased slightly by the presence of the dielectric; according to Eq. (5.28), $C_0' = BC_0 = 1.22 C_0$. Thus, for 1 mm of line, $C = (1.22)(1.57)(0.1) = 0.192$ pF.

The crossover capacitance may be estimated from Eq. (5.44); each 2 × 3 μm crossover will add 1.77 fF. Thus, adding the line, crossover, and load capacitances, the net $C = 0.192 + 0.018 + 0.020 = 0.230$ pF.

When the sum of the line and crossover capacitance (0.21 fF) is used for the C model on SPICE, and the logic gates are simulated using the MESFET model parameters of Table 2.4, the results below are obtained:

$$t_{50\%}(L \rightarrow H) = t_{PLH} = 190 \text{ ps}$$

$$t_{50\%}(H \rightarrow L) = t_{PHL} = 200 \text{ ps}$$

$$t_r = 295 \text{ ps}$$

$$t_f = 440 \text{ ps}$$

From Eq. (5.65b), effective source resistances can be determined from t_r and t_f. For charging and discharging the capacitor, these are 580 and 870 Ω respectively, in reasonable conformance with the predictions of the small-signal ac SPICE model in Example 5.5. However, much higher R_G is calculated from Eq. (5.65a) for the propagation delays (about 1200 Ω). The simulated R_G in Example 5.5 is in the same range, but not in close agreement with this estimate. This lack of agreement between the calculated and simulated Thevenin equivalent source resistances just illustrates that the behavior of the BFL gate cannot be accurately predicted with the simple single-exponential model.

Although the C model (Fig. 5-17a) is useful for many interconnect configurations, it produces a poor estimate of the interconnection delay when the effective source resistance is much less then $10Z_0$. There are two phenomena that invalidate the C model. If $10 > R_G/Z_0 > 1$, the line resembles a distributed RC network in which the resistance and capacitance are spread out over the entire length of the line. In this view, the electromagnetic propagation is neglected, as it was in the C model.

Figure 5-20 illustrates the relationship between RC delay and the electromagnetic propagation time (LC delay). Since both R and C increase linearly with l, the RC delay is proportional to length squared, whereas the LC delay is proportional to length. There will be a crossing over of the dominant delay mechanism at a particular length depending on the width of the line and the ratio R_G/Z_0.

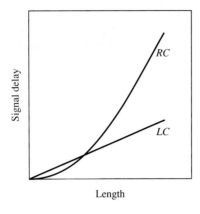

FIGURE 5-20
Qualitative plot of interconnect delay as a function of line length for (a) electromagnetic (*LC*) dominated line and (b) distributed *RC* line.

For extremely long lines in silicon MOS VLSI, repeaters have been used, located at periodic intervals to force the delay to scale more linearly with length. Obviously, for very high-speed GaAs ICs, *RC*-limited lines should be avoided by increasing interconnect width and/or line impedance. Even so, models must be developed to evaluate a particular interconnect delay to determine whether design modifications will be needed.

The theory for distributed *RC* networks shows that the transient response of a distributed *RC* is faster than that of the lumped *RC* equivalent with the same total *R* and *C* (Fig. 5-17b) [26]. In fact, the "L model," where L could symbolize either the shape of the network or could stand for lowpass, is a particularly poor approximation to the step response of a distributed *RC*. The "pi model," shown in Fig. 5-17c, is much more accurate. A three-section pi network equivalent circuit is shown (referred to as P3), although in some circumstances, a two- or even one-section pi will suffice. Since at this point we have left the realm of convenient pen-and-paper solutions anyway, the P3 model with its greater accuracy might just as well be used. The $R = R_0 l$ and $C = C_0 l$ are partitioned as shown in Fig. 5-17c.

Table 5.4 summarizes the ranges of R_G/Z_0 and R/Z_0 (where $R = R_0 l$) over which the P3 model is most accurate. It should be recalled, however, that the P3 model neglects the transmission line nature of the interconnect. This deficiency becomes

TABLE 5.4
Preferred model (see Fig. 5-17) for prediction of interconnect delay within an accuracy of 10 percent.

	R_G/Z_0			
R/Z_0	0.2	1.0	2.0	10
0.2	TRL1	TRL1	C	C
0.5	TRL4	TRL4	P3	C
1.0	TRL4	TRL4	P3	C
2.0	TRL4	P3	P3	C

apparent when $R_G/Z_0 \leq 1$ because the total predicted interconnect delay time becomes shorter as R_G is reduced. This trend can be seen in Example 5.8 where the transmission line is modelled by the C, P3, and TRL4 (a four-section approximation to a lossy transmission line) models. Since $R_G/Z_0 = 0.2$, there will be a negative reflection coefficient at the source end ($R_G < Z_0$) which leads to a damped oscillatory solution which cannot be modeled by any RC passive network.

Therefore, in order to properly predict the interconnect delay for $R_G/Z_0 < 1$, Table 5.4 shows that a transmission line model, such as drawn in Fig. 5-17d and e, will be needed. Figure 5-17d represents a lossless or ideal transmission line (TRL1) suitable only for modeling very short interconnects driven by a low R_G driver (Table 5.4; $R_G/Z_0 = 0.2$ or 1.0; $R/Z_0 = 0.2$). Note that this ideal line will be acceptable as an interconnect only for logic circuits whose response time t_r or $t_f \gg 2T$, the duration of the first negative reflection.

The lossy transmission line is modeled as shown in Fig. 5-17e, breaking up the line into sections (four in this example, hence TRL4) and inserting a proportional fraction of the series resistance R. This is the most accurate model presented, as it will predict the proper interconnect delay under any conditions. The accuracy improves slightly as the number of sections is increased, in which case the duration of the multiple reflections, which show up as discrete steps, is reduced.

The models of Fig. 5-17 can all be easily used with a circuit simulator such as SPICE.[10] Values for R, C, Z_0, l can be estimated from the circuit layout and substituted into the equivalent interconnect circuits. SPICE can be used to solve for the transient response of the network. The execution time for each model increases as we progress from Fig. 5-17a to e; therefore there is a simulation time penalty to be paid for using the most general model (TRL4) when a simpler model would do.

Example 5.8. Compare the predicted 50 and 90 percent delay times for the C, P3, and TRL4 models for the case $R_G/Z_0 = 0.2$ and $R/Z_0 = 1.0$. A first-level metal interconnection is used as described in Example 5.7; $Z_0 = 56$ Ω and $C_0 = 1.92$ pF/cm. Model the generator as a Thevenin equivalent step function.

Solution. The resistance R_0 of the 2 μm wide by 0.3 μm thick line will be $R_0 = (3 \times 10^{-6}$ Ω·cm$) / (2 \times 10^{-4}$ cm$) (3 \times 10^{-5}$ cm$) = 500$ Ω/cm. Thus, if $R = R_0 l = 56$ Ω, $l = 0.11$ cm. Then $C = C_0 l = 0.22$ pF. The electromagnetic propagation delay is given by $T = l/v_p$, where

$$v_p = \frac{1}{Z_0 C_0} = 9.3 \times 10^9 \text{ cm/s}$$

from Eqs. (5.7) and (5.20). Thus, $T = 11.8$ ps.

The SPICE node listings and commands required to simulate these three models are given in Fig. 5-21. The output voltages at $x = l$ are plotted in Fig. 5-22. Note that in this example, considerable error would arise from the use of the wrong interconnect model.

---

[10] Except for the L model which should not be used at all.

```
compare t-line models r/z0=1.0; rg/z0=0.2
** z0 = 56 ohms
*c model
v1 1 0 pulse(0 1 0 0 0 0.3ns 0.3ns)
r1 1 2 15
c1 2 0 0.22pf
*tr1 n=4
rg4 1 41 15
t41 41 0 42 0 z0=56 td=3.0p
r41 42 43 4.9
t42 43 0 44 0 z0=56 td=3.0p
r42 44 45 4.9
t43 45 0 46 0 z0=56 td=3.0p
r43 46 47 4.9
t44 47 0 48 0 z0=56 td=3.0p
*
* p3 model
r3 1 5 15
c3 5 0 .037pf
r4 5 6 18.7
c4 6 0 .074pf
r5 6 7 18.7
c5 7 0 .074pf
r6 7 8 18.7
c6 8 0 .037pf
.tran 5ps 0.3ns
.probe v(2) v(48) v(8)
.options nopage
.end
```

FIGURE 5-21
SPICE input file which compares the C, P3, and TRL4 interconnect models for a line with $Z_0 = 56\ \Omega$ and $T = 11.8$ ps.

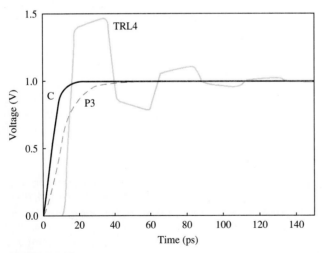

FIGURE 5-22
Voltage at the output of three models of an interconnect line (C, P3, and TRL4) with $Z_0 = 56\ \Omega$ and $T = 11.8$ ps. The lines are driven with a unit step function. The outputs are open-circuited.

These models are suitable for predicting the total path delay when logic gates are involved. The appropriate model is simply inserted into the SPICE file between the driver gate output and the load gate input(s).

5.5 CROSSTALK

(This section is contributed by G. L. Matthaei)

The preceding sections of this chapter have been concerned with the analysis and modeling of wave propagation on single transmission lines used as on-chip or off-chip interconnections. The following section focuses on closely spaced pairs of coplanar, microstrip, or stripline transmission lines which are often found on the surfaces of an integrated circuit chip or circuit board. If the spacing is small, the electric and magnetic fields of the lines overlap sufficiently so that a wave propagating in one line will induce a wave in the adjacent line or lines. If this coupling between lines is strong enough, the voltages and currents induced in the coupled, inactive line have the potential for producing a logic error if the dynamic noise margin of the logic gate is exceeded (see Sec. 3.3.7 for a definition of dynamic noise margin). The coupled voltages and currents are generally referred to as crosstalk. For simplicity, this discussion will be confined to the lossless case.

A simplified, two-line, symmetric analysis which could be used with the coupled microstrip or stripline shown in Fig. 5-23 is presented in order to stress the concept of coupled line effects and provide insight into the important design consequences. The time-domain, two-mode (even and odd mode) superposition analysis described here will be useful in this simplified situation where symmetric geometry prevails. The analysis also requires that the logic gate terminal impedances be approximated by linear "effective resistances" (described in Sec. 5.4) and that all such effective source resistors are the same and all load resistors are the same. Examples of this condition can be found in packages, on circuit boards, and occasionally on IC chips. The line impedances and propagation velocities must be known for the coupled lines in order to apply this method, but these quantities can be found from straightforward analytical solutions for microstrip and stripline[27].

This linearized, mode-superposition approach can be extended if needed to treat the more general asymmetric case commonly found with on-chip interconnections as described in Ref. 28. Asymmetry prevails when the conductors are coplanar and the ground return strip is asymmetrically located with respect to the signal lines as illustrated in Fig. 5-24. Whereas the analysis of a two-coupled line case with coplanar ground is a straightforward extension of the symmetric mode-superposition analysis to be described, if the velocities of propagation and self and mutual capacitances or even/odd mode impedances can be found, these cannot be found by a simple analytical solution. A two-dimensional numerical analysis is generally required, making the modeling not feasible unless such a tool is available.

A very general equivalent circuit approach has been developed by Tripathi and Rettig [29] which makes it possible to analyze coupled transmission lines using the SPICE circuit simulation-program and an equivalent circuit for the transmission lines involving various dependent sources and lines without mutual coupling along their

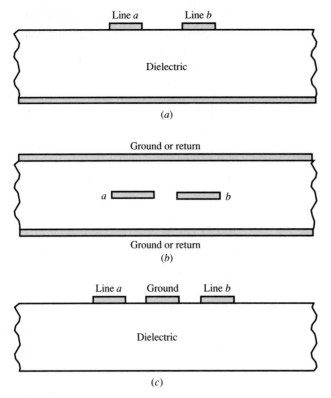

FIGURE 5-23
Cross-sectional view of symmetric coupled transmission lines: (*a*) coupled microstrip lines, (*b*) coupled striplines, and (*c*) coupled coplanar lines.

lengths. This has the advantage that nonlinear device models attached to the lines can be included but has the disadvantages that the computation of the equivalent circuit parameters for a given case is relatively abstract and lengthy and the computer time consumed in obtaining responses can be large. In fact, while the method is not limited *a priori* in the number of coupled lines it can model, the computer time required to obtain convergence of the interactive dependent sources is such that even two coupled lines may take considerable computer time. Nevertheless, arbitrary coupled line cases can be modeled accurately if the capacitance and inductance matrices can be determined and sufficient computer time is available.

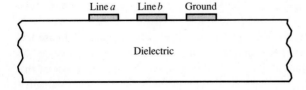

FIGURE 5-24
Cross-sectional view of asymmetric coupled coplanar transmission lines which are typical of on-chip interconnections.

A full analysis of all of the effects that can be found in multiple-line systems with loss and dispersion can become arbitrarily complicated and is beyond the scope of this text. In this regard, Ref. 30 gives a very comprehensive coverage of various general methods for analysis of coupled transmission lines.

5.5.1 Symmetrical Even and Odd Modes and Their Parameters

Recall that the characteristic impedance of a single TEM mode transmission line is given by

$$Z_0 = \sqrt{\frac{L_0}{C_0}} \quad (5.67)$$

where L_0 and C_0 are, respectively, the inductance and capacitance per unit length of the line. Meanwhile, the velocity of propagation on the line is given by

$$\nu = \frac{1}{\sqrt{L_0 C_0}} \quad (5.68)$$

which can also be computed from

$$\nu = \frac{1}{\sqrt{\mu_0 \epsilon_{\text{eff}}}} \quad (5.69)$$

where μ_0 is the magnetic permeability and ϵ_{eff} is the "effective dielectric constant" of the medium surrounding the conductors. An effective dielectric constant is often needed because in many situations, such as for microstrip, the fields are partly in air and partly in dielectric. Thus, the propagation is only quasi-TEM because of the distorted field lines, and the wave velocity is somewhere between that for air and that for the dielectric. By use of Eqs. (5.67) to (5.69) it is seen that as long as the mode is TEM or quasi-TEM, Z_0 can be computed using

$$Z_0 = \frac{1}{\nu C_0} \quad (5.70)$$

which is often convenient.

For simplicity, we will confine our attention to the use of pairs of *symmetrical* coupled lines such as those in Fig. 5-25a and b. In such situations we can analyze the operation of the lines in terms of two TEM modes commonly known as the "even" and "odd" modes. In the case of the even mode, the two lines have the same potential with respect to the ground plane, as suggested in Fig. 5-25a, while for the odd mode the potential for one line with respect to ground is the negative of that for the other line, as suggested in Fig. 5-25b. In this case there are three capacitances per unit length to be considered, as shown in Fig. 5-26a, though for this symmetrical case the capacitances C_a and C_b to ground are equal. The even- and odd-mode impedances of a pair of coupled lines are defined as the impedance of *one line* in the presence of the other, when the lines are operated in the even or odd mode, respectively. It is readily

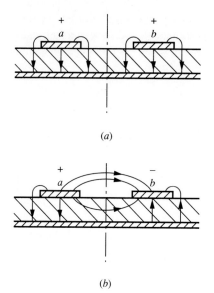

FIGURE 5-25
Symmetrical microstrips excited in (a) the even mode and (b) the odd mode.

seen from Fig. 5-26a that when the lines are excited in the even mode that C_{ab} has no effect and the even-mode impedance for line a must be

$$Z_0^e = \frac{1}{v^e C_a} \tag{5.71}$$

where v^e is the even-mode velocity, and by symmetry the even-mode impedance for line b must be the same. The circuit in Fig. 5-26b will be seen to be equivalent to

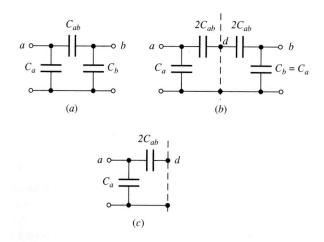

FIGURE 5-26
(a) and (b) Two equivalent configurations for the distributed capacitances per unit length for two symmetrical, coupled transmission lines. (c) The configuration (b) is bisected.

that in Fig. 5-26a. For symmetrical lines the odd-mode impedance for line a can be computed by splitting capacitor C_{ab} into two capacitors in series as shown in Fig. 5-26b and Fig. 5-26c, where node d is at zero potential for the odd mode. Thus the odd-mode impedance for line a must be

$$Z_0^o = \frac{1}{v^o(C_a + 2C_{ab})} \tag{5.72}$$

where v^o is the odd-mode velocity, and likewise for line b. The wave velocities for the even and odd modes are given by

$$v^e = \frac{1}{\sqrt{\mu \epsilon_{\text{eff}}^e}} \quad \text{and} \quad v^o = \frac{1}{\sqrt{\mu \epsilon_{\text{eff}}^o}} \tag{5.73}$$

where ϵ_{eff}^e and ϵ_{eff}^o are the effective dielectric constants for the even and odd modes. For lines such as microstrip these will be somewhat different because of the presence of two different dielectrics and the differences in the even- and odd-mode field configurations as seen in Fig. 5-25a and b. For this symmetrical case the even-mode voltage on line a at a given point x is related to that on line b for the same x by

$$v_{eb}(t,x) = v_{ea}(t,x) \tag{5.74a}$$

while the odd-mode voltages on the lines are related by

$$v_{ob}(t,x) = -v_{oa}(t,x) \tag{5.74b}$$

and similarly for the line currents. Let us now consider the nature of the terminal voltages at the ends of two coupled lines as viewed in terms of these modes.

5.5.2 Analysis by Superposition of Modes on Symmetric Lines

Consider the pair of symmetrical coupled lines in Fig. 5-27a which are driven at port 1 and where R_1 and R_3 are arbitrary. The excitation in Fig. 5-27a can be viewed as a superposition of the even-mode excitation in Fig. 5-27b and the odd-mode excitation in Fig. 5-27c. (Note that if we superimpose the generators in Fig. 5-27b and c, the generator voltages at port 1 add up to v_g and those at port 2 add to zero, giving total voltages as in Fig. 5-27a.) Let T_e be the time for the even mode to travel the length l of the lines and T_o to be the corresponding time for the odd mode. Then if v_g starts at $t = 0$, for $t < 2T_e$ and $2T_o$, no reflected waves will yet have arrived back at the input. Then in Fig. 5-27b the even-mode voltage vs. time at node 1 will be

$$v_{e1}^t = \frac{v_g}{2}\left[\frac{Z_0^e}{Z_0^e + R_1}\right] \tag{5.75a}$$

and in Fig. 5-27c the odd-mode voltage will be

$$v_{o1}^t = \frac{v_g}{2}\left[\frac{Z_0^o}{Z_0^o + R_1}\right] \tag{5.75b}$$

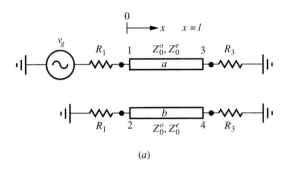

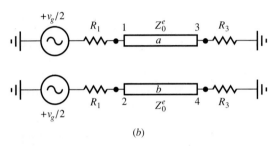

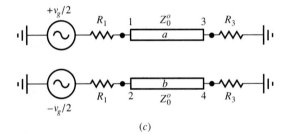

FIGURE 5-27
(a) Two coupled lines are excited at port 1. This excitation can be regarded as a superposition of the even-mode and odd-mode excitations shown at (b) and (c) respectively.

(Here the superscript t's imply wave components transmitted from the generator onto line a.) The total voltage at node 1 in Fig. 5-27a is then

$$v_1 = v_{e1}^t + v_{o1}^t \tag{5.76a}$$

and using Eqs. (5.74a,b) with $x = 0$, at node 2 the voltage must be

$$v_2 = v_{e1}^t - v_{o1}^t \tag{5.76b}$$

Using Eqs. (5.75a,b) in Eqs. (5.76a,b) gives

$$\begin{Bmatrix} v_1 \\ v_2 \end{Bmatrix} = \frac{v_g}{2}\left[\frac{Z_0^e}{Z_0^e + R_1}\right] \pm \frac{v_g}{2}\left[\frac{Z_0^o}{Z_0^o + R_1}\right] \tag{5.77}$$

where the upper sign refers to v_1 and the lower sign refers to v_2.

Note from Eq. (5.77) that for $t < 2T_e$ and $2T_o$, so that no reflected waves from the far ends can have arrived, we obtain

$$\frac{v_2}{v_1} = \frac{Z_0^e - Z_0^o}{(2Z_0^o Z_0^e / R_1) + Z_0^e + Z_0^o} \tag{5.78}$$

which shows two interesting points. One is that there is an output at node 2 the same instant there is an input at node 1, and the other is that this initial output is proportional to $(Z_0^e - Z_0^o)$.

The even and odd modes excited at the left end of the lines in Fig. 5-27a at time $t = 0$ will propagate to the right on these lines at the velocities v^e and v^o and arrive at the far end at times T_e and T_o respectively. Thus, the even- and odd-mode voltages *incident* on node 3 are simply delayed versions of the even- and odd-mode voltages transmitted from node 1 and so

$$v_{e3}^i(t) = v_{e1}^t(t - T_e) \tag{5.79a}$$

$$v_{o3}^i(t) = v_{o1}^t(t - T_o) \tag{5.79b}$$

For simplicity, we focus on the effects of the modes at nodes 1 and 3 at the ends of line a; then the corresponding mode voltages at nodes 2 and 4 at the ends of line b can be easily deduced by use of Eqs. (5.74a,b).

Depending on the size of the terminating resistors, the even and odd modes incident on the right ends of the lines in Fig. 5-27a may be partially reflected by the R_3 terminations, and these mode components will travel back toward the left end in Fig. 5-27a and be partially reflected by the R_1 terminations. These waves will then travel to the right and again be partially reflected by the R_3 terminations, and so on. The reflection coefficients for the even and odd modes as referred to terminals 1 and 3 are

$$\Gamma_{em}\big|_{m=1 \text{ or } 3} = \frac{v_{em}^r}{v_{em}^i} = \frac{R_m - Z_0^e}{R_m + Z_0^e} \tag{5.80a}$$

$$\Gamma_{om}\big|_{m=1 \text{ or } 3} = \frac{v_{om}^r}{v_{om}^i} = \frac{R_m - Z_0^o}{R_m + Z_0^o} \tag{5.80b}$$

After the first reflection at the right ends of the lines, the total voltage at nodes 3 and 4 is the sum of the incident and reflected components

$$\left.\begin{array}{c} v_3 \\ v_4 \end{array}\right\} \quad v_{e3}^i + v_{e3}^r \pm \left[v_{o3}^i + v_{o3}^r\right] \tag{5.81a}$$

or

$$\left.\begin{array}{c} v_3 \\ v_4 \end{array}\right\} \quad v_{e3}^i(1 + \Gamma_{e3}) \pm v_{o3}^i(1 + \Gamma_{o3}) \tag{5.81b}$$

where the upper sign refers to v_3 and the lower sign refers to v_4.

After enough time has elapsed for the modes to travel from $x = 0$ to $x = l$ and back N times, the total voltage at nodes 1 and 2 will be a superposition of the incident waves in Eqs. (5.75a,b) plus the sum of all of the waves reflected from the $x = l$ end of the line and re-reflected at the $x = 0$ end. The result is

$$\begin{Bmatrix} v_1 \\ v_2 \end{Bmatrix} = \frac{Z_0^e}{2(Z_0^e + R_1)} \left[v_g(t) + \sum_{n=1}^{N} v_g(t - 2nT_e)\Gamma_{e3}^n \Gamma_{e1}^{n-1}(1 + \Gamma_{e1}) \right]$$

$$\pm \frac{Z_0^o}{2(Z_0^o + R_1)} \left[v_g(t) + \sum_{n=1}^{N} v_g(t - 2nT_o)\Gamma_{o3}^n \Gamma_{o1}^{n-1}(1 + \Gamma_{o1}) \right] \quad (5.82)$$

which is exact for N equal to the largest integer in $t_{\max}/(2T_{sm})$, where $t_{\max}$ is the maximum value of t and T_{sm} is the smaller of T_e and T_o. However, Eq. (5.82) may be accurate even if fewer terms are used in the summations because the wave components may diminish in amplitude rapidly due to absorption at the loads. The corresponding result for the $x = l$ ends of the lines is

$$\begin{Bmatrix} v_3 \\ v_4 \end{Bmatrix} = \frac{Z_0^e}{2(Z_0^e + R_1)} \left\{ \sum_{n=0}^{N} v_g[t - (2n+1)T_e](\Gamma_{e3}\Gamma_{e1})^n (1 + \Gamma_{e3}) \right\}$$

$$\pm \frac{Z_0^o}{2(Z_0^o + R_1)} \left\{ \sum_{n=0}^{N} v_g[t - (2n+1)T_o](\Gamma_{o3}\Gamma_{o1})^n (1 + \Gamma_{o3}) \right\} \quad (5.83)$$

In the above equations the plus sign goes with v_1 or v_3 while the minus sign goes with v_2 or v_4.

5.5.3 Case of Purely Backward Coupling

TEM-mode microwave directional couplers are designed to exhibit purely "backward coupling," and the conditions that result in this phenomenon [31,32] can provide insight into coupling phenomena. It can be shown that if $T_e = T_o = T$ (pure TEM transmission line such as stripline) and we set

$$R_1 = R_3 = \sqrt{Z_0^o Z_0^e} \quad (5.84)$$

that $\Gamma_{o1} = \Gamma_{o3} = -\Gamma_{e1} = -\Gamma_{e3} > 0$ and $v_{e1}^t/v_{o1}^t = \sqrt{Z_0^e/Z_0^o}$. It turns out that these conditions will cause all of the wave components in Eq. (5.83) to cancel at port 4 in Fig. 5.27a and all of the reflected components in Eq. (5.82) to cancel port 1. Thus, the input impedance seen looking into port 1 is matched to $R_1 = \sqrt{Z_0^e Z_0^o}$, and no reflections will be observed there in either the time or frequency domains. Also, port 4 always has zero output, while there is output at port 3 and "backward-coupled" output on line b at port 2. If all of the ports are terminated as in Eq. (5.84) we will always see a matched impedance R_1 looking into any of the ports of the coupler. Nevertheless, as can be seen from Eq. (5.84), the even- and odd-mode impedances within the coupler are not matched to the terminations. In agreement with Eq. (5.78), the amount of signal backward coupled to port 2 is proportional to the difference $(Z_0^e - Z_0^o)$.

As indicated above, for *purely* backward coupling the coupled lines must be terminated with resistors having the value in Eq. (5.84) and the velocities of the odd and even modes must be the same so that $T_e = T_o$. However, even if these conditions are violated by use of other terminations and by unequal mode velocities, backward coupling will still be present as long as Z_0^e is different from Z_0^o. However, then there will be output at all three ports 2, 3, and 4 for an input at port 1, and the input impedances will not, in general, be R_1.

5.5.4 Case of Purely Forward Coupling

In the backward-coupler example discussed above we assumed that the even- and odd-mode velocities of the coupled lines were the same, but there were reflections between the terminating resistances and the odd- and even-mode impedances. Conversely, if we can eliminate the reflections at the ends of the lines, while the even- and odd-mode velocities on the lines are different, we will achieve purely "forward coupling." In order to better understand this coupling mechanism we briefly outline the operation of forward couplers.

Forward coupling is most commonly utilized in waveguide couplers but it can also be realized in microstrip as discussed in Ref. 33. In the couplers discussed in that paper the two microstrip lines are brought together gradually and then after some distance are separated gradually. When the two lines are in close proximity to each other they have different even- and odd-mode impedances and also somewhat different even- and odd-mode velocities. However, since the lines are brought together and separated gradually, the impedance variations are slow and no significant reflections occur. As a result, the effect is much the same as though the even- and odd-mode impedances were equal and matched to the terminations. Thus, here any coupling effects are due purely to the difference in the velocities of the two modes.

From the frequency-domain point of view it can be shown that due to the differences in phase shift for the odd and even modes, the voltages for the two modes as seen on the coupled line will tend periodically to add at some points and cancel at other points. For this reason a forward coupler can give complete power transfer between the input end of the driven line and the far end of the coupled line at certain frequencies [33]. However, for digital applications we are more interested in the effects of forward coupling as seen in the time domain.

Let us suppose that a voltage step with a rise of duration t_r is applied at port 1 of a forward coupler at time $t = 0$. (The configuration can be thought of as being like Fig. 5-27a except that lines with tapered spacings are added at the left and right in order to make possible the simultaneous matching of both the odd and even modes.) Then since only the left end of line a is driven, the odd and even modes will be excited equally. [See Eqs. (5.75a, b) with $Z_0^e = Z_0^o$.] At the input both mode voltages will have a positive sign on the upper line, but will have opposite signs on the lower line. Figure 5-28a shows the resulting even- and odd-mode voltage waveforms vs. distance after the two modes have propagated a distance along line b. Note that in this case since the terminations effectively match the mode impedances, the amplitudes of the even mode on line b in Fig. 5-27a will be $v_g/4$. Since the odd mode travels somewhat faster than does the even mode, there is a region near the leading edge of the odd mode on line b where its voltage is not cancelled by that of the even-mode voltage. This is further illustrated in Fig. 5-28b where the even- and odd-mode

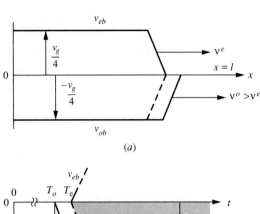

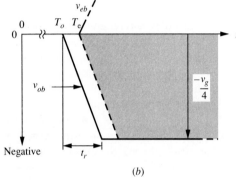

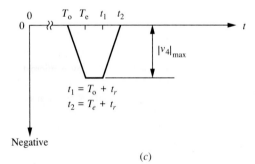

FIGURE 5-28
This figure illustrates how unequal even- and odd-mode velocities lead to forward coupling. The shaded region in (b) indicates odd-mode voltage on line b which is cancelled by even-mode voltage.

voltages seen at port 4 are shown vs. time. The shaded region indicates the negative odd-mode voltage which is cancelled by the positive even-mode voltage on line b. Thus the net output at port 4 is the negative pulse indicated in Fig. 5-28c.

From the geometry in Fig. 5-28b and c it is easily seen that if $(T_e - T_o) < t_r$,

$$|v_4|_{max} = \left(\frac{v_g/4}{t_r}\right)(T_e - T_o) \qquad (5.85a)$$

$$= \left(\frac{dv_1}{dt}\right)_r \left(\frac{l/v^e - l/v^o}{2}\right) \qquad (5.85b)$$

where $(dv_1/dt)_r$ is the time derivative of the rising portion of the step seen at port 1, v^o and v^e are, of course, the odd- and even-mode velocities, and l is the line length.

Equation (5.85a) can be found as a term in equations in other papers dealing with coupled lines [34], but use of the mode superposition point of view greatly clarifies its meaning and derivation. It can be further seen that if $(T_e - T_o) > t_r$ then $|v_4|_{max}$ will be $v_g/4$. Since at port 2 in Fig. 5-27a the even and odd modes have not yet traveled any distance they completely cancel each other. Thus the only output on line b is the "forward-coupled" output at port 4.

5.5.5 Forward and Backward Coupling

In typical cases of interconnects in circuits, both forward and backward coupling will occur. However, except for some cases where the coupled lines are quite long, the forward coupling due to the differences in wave velocities will be small. Nevertheless, there can still be sizable output at the forward-coupled port due to mismatches at the terminations causing reflected waves on the coupled lines. Let us examine an example that illustrates the nature of the outputs due to the differences in wave velocities and differences in mode impedances.

Example 5.9. A pair of microstrip interconnect lines runs in parallel between two GaAs IC chips for a distance of 3 cm on a Teflon-fiberglass circuit board with ($\epsilon_r = 2.8$). The conductors are 0.125 mm wide and are spaced by the same distance. The dielectric separating the conductors from the ground plane is 0.5 mm thick. Referring to Fig. 5-27, plot the voltages at terminals 1, 2, 3, and 4 as a function of time for the first two reflected returns if one line is driven by a unit step function. Assume that $R_1 = 50 \, \Omega$ and $R_3 = 500 \, \Omega$ in this example.

Solution. The interconnect consists of two conductors over a ground plane; therefore the even- and odd-mode impedances and velocities can be determined by a microstrip coupled line model [26]. From the equations described in this reference, we determine that $Z_0^o = 88 \, \Omega, Z_0^e = 215 \, \Omega, v^o = 2.24 \times 10^{10}$ cm/s, and $v^e = 2.08 \times 10^{10}$ cm/s.
The reflection coefficients at each end can be determined using Eqs. (5.80a,b):

$$\Gamma_{e1} = \frac{R_1 - Z_0^e}{R_1 + Z_0^e} = -0.622$$

$$\Gamma_{o1} = \frac{R_1 - Z_0^o}{R_1 + Z_0^o} = -0.275$$

$$\Gamma_{e3} = \frac{R_3 - Z_0^e}{R_e + Z_0^e} = -0.399$$

$$\Gamma_{o3} = \frac{R_3 - Z_0^o}{R_3 + Z_0^o} = -0.701$$

The propagation delay time for each transit of the lines is given by

$$T_o = \frac{l}{v^o} = 134 \text{ ps}$$

$$T_e = \frac{l}{v^e} = 144 \text{ ps}$$

TABLE 5.5
Development of voltages at nodes 1 and 2 [$v_1(t)$ and $v_2(t)$] with time. Δv_e and Δv_o are the incremental changes in even- and odd-mode voltages which occur at the arrival of the next reflected wave. Index n is the number of reflections. t is the time of arrival. $v_2(t)$ is plotted in Fig. 5-29

n	t, ps	Δv_e, V	Δv_o, V	$v_1(t)$	$v_2(t)$
0	0^+	0.41	0.32	0.73	0.09
1	268		0.16	0.89	−0.07
1	288	0.06		0.95	−0.01
2	536		−0.03	0.92	0.02
2	576	−0.02		0.90	0.00

Equations (5.82) and (5.83) can be used to calculate the voltage at terminals 1, 2, 3, and 4. For the first two reflected returns at $x = 0$,

$$\left.\begin{matrix}v_1(t)\\v_2(t)\end{matrix}\right\} = 0.41\left[u(t) + 0.151\,u(t - 2T_e) - 0.037\,u(t - 4T_e)\right]$$
$$\pm\, 0.32\left[u(t) + 0.508\,u(t - 2T_o) - 0.098\,u(t - 4T_o)\right]$$

At $x = l$, the equations of (5.83) yield

$$\left.\begin{matrix}v_3(t)\\v_4(t)\end{matrix}\right\} = 0.41\left[1.4\,u(t - T_e) - 0.348\,u(t - 3T_e) + 0.086\,u(t - 5T_e)\right]$$
$$\pm\, 0.32\left[1.7\,u(t - T_o) + 0.327\,u(t - 3T_o) - 0.063\,u(t - 5T_o)\right]$$

The line voltages vs. time are summarized in Tables 5.5 and 5.6. Here, Δv_e and Δv_o are the incremental changes in even- and odd-mode voltages which occur at the time of arrival of the next reflected wave. The index n indicates the number of reflections. Note for, say, $n = 1$ that the odd-mode voltage arrives first. This is then closely followed by the even-mode contribution.

TABLE 5.6
Development of voltages at nodes 3 and 4 [$v_3(t)$ and $v_4(t)$] with time. These voltages are plotted in Fig. 5-29

n	t, ps	Δv_e, V	Δv_o, V	$v_3(t)$	$v_4(t)$
0	134		0.54	0.54	−0.54
0	144	0.57		1.11	0.03
1	402		−0.11	1.00	0.14
1	432	−0.14		0.86	0.00
2	670		0.02	0.88	−0.02
2	720	0.04		0.92	0.02

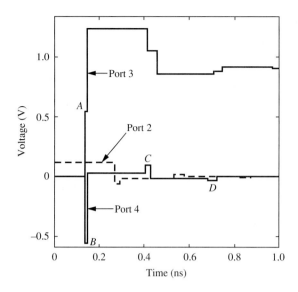

FIGURE 5-29
Outputs at ports 2, 3, and 4 due to a step input at port 1 for coupled lines having both forward and backward coupling. (Example 5.9).

Voltages $v_2(t)$ through $v_4(t)$ are plotted in Fig. 5-29. If v_g at port 1 in Fig. 5-27a is a unit step function applied at $t = 0$, the outputs at ports 2, 3, and 4 obtained using Eqs. (5.82) and (5.83) are as indicated in Fig. 5-29. Observe that there is immediately an output at the coupled port 2. The small step indicated by A in the response for port 3 in the figure results from the odd mode arriving at the $x = l$ ends of the line before the even mode. Meanwhile, the sharp negative pulse at port 4 indicated by B in the figure occurs for the same reason. The port 4 pulse at C results after the reflected waves have traveled to the $x = 0$ ends of the lines and returned again to the $x = l$ end. Note that this pulse is three times as wide as that at B because the two modes have now traveled three times as far. The next round-trip of reflected waves results in the small pulse D, which is five times as wide as that at B. Note that even though these lines are moderately long (3 cm), the pulse at B is quite narrow, and if the step had a finite rise time as in Fig. 5-28, its total amplitude would be considerably reduced. For these reasons, except for very long runs of coupled lines where the delay difference between the even and odd nodes can get to be quite sizable, coupling due to the difference in mode velocities will usually not be of much importance because the resulting output pulses may be too narrow and possibly too small in amplitude to trigger a gate. Note that, besides the forward-coupled pulse output at port 4 shown in Fig. 5-29, there is also some other output at port 4 due to uncanceled reflections at the terminations, along with the difference between the mode velocities. If the mode velocities were equal, and if the terminations were all 137.5 Ω, as is called for by Eq. (5.84), there would be no output at all at port 4, and the input impedance seen looking into all of the ports would be 137.5 Ω at all frequencies [31, 32].

5.5.6 RC Approximation

In many cases where the signal rise time is sizable compared to the transit time on the interconnect line and where the lines are terminated with impedances significantly greater than the mean characteristic impedance $\sqrt{Z_0^o Z_0^e}$, it is possible to predict coupled line voltages with a simple RC model. In this limit, the interconnect lines are replaced by their total, lumped self and mutual capacitances. The source and load impedances are again approximated by their linearized effective resistances. A circuit diagram describing this model is shown in Fig. 5-30. This situation is analogous to the conditions under which the C model of Sec. 5.4 was applied. Thus, if the driving logic gates are physically small and the interconnect line round-trip propagation delay, $2T$, is less than t_r or t_f, the RC model is simpler to use than the mode-superposition model and is reasonably accurate. However, for more advanced technologies (submicrometer lithography or heterostructure transistors) which have very short rise times, or for driver gates with low effective source resistance (such as HBT emitter followers), the accuracy of the RC approximation will be poor.

Example 5.10. To illustrate the limitations of the RC model, the conditions described in the previous example (Example 5.9) will be repeated using the RC approximation to calculate the voltages $v(3)$ and $v(4)$. Referring to the schematic in Fig. 5-30, $R_1 = 50\ \Omega$ and $R_3 = 500\ \Omega$ as before, whereas the mean characteristic line impedance [Eq. (5.84)] is 137.5 Ω. Thus, the effective source resistance is significantly lower than the line impedance. The self and mutual capacitances can be calculated from Eqs. (5.71) through (5.73) from the even- and odd-mode impedances and velocities of propagation. These equations give $C_a = 0.672$ pF and $C_{ab} = 0.426$ pF for the 3-cm lengths required. Using SPICE to predict the step response (worst case, since the rise and fall times are zero), the response can be calculated, and the result is plotted in Fig. 5-31 (dotted lines) along with the response for the mode-superposition method [dashed line for $v(3)$, solid line for $v(4)$].

Note that there are significant differences in the responses predicted by the two methods. The RC model cannot predict the overshoot on $v(3)$, the negative voltage on $v(4)$, or the line transit delay on either line, which is predicted correctly by the mode method.

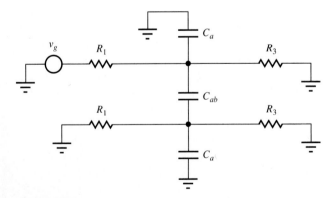

FIGURE 5-30
RC approximation for calculation of coupled line voltage.

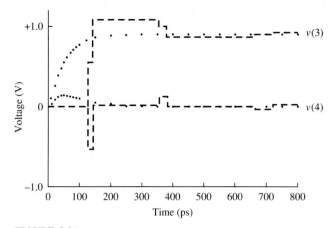

FIGURE 5-31
Comparison of node voltages $v(3)$ and $v(4)$ as predicted by the RC model (dotted lines) and the mode-superposition method (dashed lines).

5.6 PACKAGING AND CIRCUIT BOARD INTERCONNECTIONS

Packages and circuit boards are the interfaces between the high-speed GaAs digital IC chip and the system that uses the chip. An ever-increasing sophistication in packaging is needed in order to avoid losing a significant percentage of the increased speed of the GaAs technology just by packaging it. There are many reasons why extra care in packaging is required with high-speed circuit technologies. The interfaces can easily contaminate the signal through impedance mismatches which lead to reflected waves (Sec. 5.1), power supply inductance which leads to voltage spikes on the supply and ground lines (Sec. 5.3), attenuation due to resistive losses (Sec. 5.4), and crosstalk between closely spaced conductors (Sec. 5.5); therefore, the design of packages and boards is a critical and integral part of the overall system design and cannot be ignored by the IC designer. The difficulty in designing interfaces suitable for GaAs ICs is a consequence of the fast rise and fall times and the high current slew rates found in many GaAs logic circuits and output drivers. It is those applications requiring that the inter-chip interconnect delay be minimized that create the greatest challenge. Some applications that can benefit strictly from reduced on-chip logic delays (e.g., a 16 or larger bit multiplier) might be able to function with lower speed interfaces and still provide an overall improvement in system performance. Also, the heat generated by the IC because of its static and dynamic power dissipation must be conducted away safely to a heat sink so that the reliability and performance of the chip is not sacrificed by high temperature operation. Thus, the packaging scheme must provide low thermal resistance. Finally, complex digital systems in GaAs can at present only be realized by interconnecting large numbers of chips, typically with LSI circuit integration density. VLSI levels of integration will reduce the total number of circuit board-to-package connections, but the number of pins per chip will increase. In most applications, the number of I/O pins increases with the chip size at a sublinear rate according to Rent's rule [35] (which applies to arbitrarily partitioned logic blocks), and the package and board must be capable of providing 50 to 200 pins per chip in many cases.

As on-chip delay continues to decrease with advances in devices and circuit designs, the propagation delay time in the wiring of the package and circuit board becomes a larger fraction of the total cycle time of the system. Thus, in order to capture the performance improvement provided by the higher speed logic, the entire system must be made physically smaller in order to reduce the length of the interconnections and therefore the interconnect delay time. This size reduction and consequent density increase can be implemented by reduction in line widths, spacings, and package lead pitch. In addition, low dielectric constant materials are preferred because the velocity of propagation, given by Eq. (5.7), varies inversely with the square root of the relative or effective dielectric constants.

There are several factors that limit the minimum physical dimensions in any given dielectric material. As discussed in Sec. 5.5, the crosstalk amplitudes will be sensitive to the spacing between lines and also the arrangement of the lines relative to the ground return. If a return can be provided between each signal line, then the crosstalk can be negligible in most cases. However, the wiring density will be reduced by 50 percent in this case. Therefore, in designing the wiring width and spacings for a circuit board or package, the coupling between lines must be calculated as a function of spacing to determine the minimum acceptable dimensions and minimum occurrence of ground or power supply lines in the signal plane to improve isolation where needed.

With some PCB materials and fabrication technologies (such as glass-epoxy and polyimide), 50 Ω microstrip and stripline transmission lines are wider than the minimum width allowed for an etched metal conductor by the process. Thus, the interconnect density will be further reduced in the cases where low impedance lines are required. Higher impedance levels may be used when the process allows linewidths to be reduced subject to the constraints imposed by attenuation (higher in narrow conductors since R_0 increases faster than Z_0) and crosstalk (line-to-line capacitance is a larger fraction of C_0).

The physical properties of the dielectric material itself will occasionally constrain the minimum via size. For example, Teflon ($\epsilon_r = 2.2$) is desirable because of its low dielectric constant, but its tendency to deform under sustained stress limits the minimum via size, and thus the wiring density as well. In the case of multilayer ceramics, where conductors are deposited as a thick film paste, the resistivity and the attenuation that it produces (Sec. 5.4) will limit the minimum linewidth. A useful technique for the comparison of the wiring density provided by various board materials has been described [36].

A complete presentation and discussion of packaging and circuit boards and a survey of present applications would take an entire book in itself. Therefore, the material presented in this section will stress only the application of the basic principles discussed in Secs. 5.1 to 5.5. From an understanding of these principles, packages and boards can be analyzed or designed without the need to study all of the details of the many application examples.

5.6.1 Impedance Discontinuities in Point-to-Point Interconnections

Conventional IC packages consist of a die cavity which is bounded by a dielectric (usually ceramic or plastic) containing thin-film metal lines or a somewhat thicker metal lead frame (traces). The IC chip (die) is generally wire-bonded to the metal traces inside the package. Outside the package, the leads are formed into pins, as in the familiar dual-in-line package (DIP) or leaded flatpack, or into solder bumps or "gull wings" on the leadless chip carriers (LCC). The length of these traces varies with the package type and the position of the lead. A particularly bad example might be the 64-pin DIP in which the longest trace has nearly 50 nH of series inductance because of its excessive length [37] (greater than 3 cm).

The point to be made on all of these conventional packages is that the package trace resembles a transmission line, whether or not that was intended. There is a series inductance due to the pin, trace, and wire bond and a distributed capacitance (trace-to-trace or trace-to-ground plane) which will give the trace some equivalent characteristic impedance and velocity of propagation. Figure 5-32 shows a cross section of a small LCC package along with the equivalent circuit (T model). For the model shown, which was calculated for a 24-pin LCC, the effective trace impedance is high, approximately 100 Ω. Alternatively, to estimate the trace impedance, transmission line models could be applied. For instance, an LCC or flatpack package without a metal floor, as above, might fit the coplanar waveguide model (CPW, see Sec. 5.2), or if the package were constructed with a metallized ceramic floor and lid, the stripline model might be more suitable.

The fact that these package traces can be represented as short transmission line sections with an average characteristic impedance, say Z_p, will allow us to study

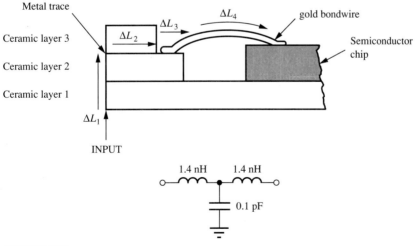

FIGURE 5-32
Cross section of a small LCC package and its equivalent circuit (T model).

304   GALLIUM ARSENIDE DIGITAL INTEGRATED CIRCUIT DESIGN

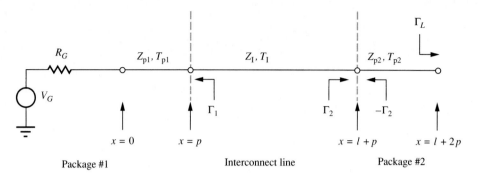

FIGURE 5-33
Schematic diagram of two packaged ICs interconnected by a transmission line of impedance Z_I with length l and propagation delay T_I. The internal package traces are of length, p, impedance Z_{p1}, Z_{p2} and delay T_{p1}, T_{p2}.

the effect which the package will have on the voltages at the IC chip bond pads. A transmission line model, such as shown in Fig. 5-33, can be used for this purpose. Here, we have a model for two packaged ICs with trace impedances Z_{p1} and Z_{p2} connected across a PC board by a line with impedance Z_I. The time of propagation on the lines is represented by T_{p1}, T_{p2}, and T_I respectively. Reflection coefficients can be calculated at each interface as shown. Terminal and interface voltages can then be determined as a function of time by the use of a position-time diagram.

The technique is illustrated by a somewhat simpler example where the driver IC is assumed to be an ideal voltage source whose internal effective source resistance $R_G = Z_I$. Then, any reflected waves from the package or chip back to the source will be absorbed and can be forgotten about. Of course the actual case with both driver and load ICs in packages will not be so ideal. However, the same method can be extended to treat this case as well at the expense of some additional accounting of wavefronts. This activity is best handled by a computer, and, fortunately, is not difficult to accomplish when linearized effective source and load resistances are utilized [20].

Example 5.11. Determine the input voltage waveform at the bond pad of a packaged IC chip (V_L) when the PC board interconnect line is driven by an unpackaged ideal voltage source (unit step function), as illustrated in Fig. 5-34a. Here, $Z_p = 100$ Ω, $T_p = 50$ ps, $Z_I = 50$ Ω, and $T_I = 100$ ps. Assume that the effective source resistance $R_G = Z_I$ and that the effective input resistance of the load IC is high, $R_L = 10$ kΩ. Therefore, the reflection coefficients can be found from Eq. (5.12) to be $\Gamma_1 = 0$, $\Gamma_2 = 0.33$, and $\Gamma_L = 1.0$. Neglect any loss in the transmission lines.

Solution. The amplitudes of the forward and reflected waves have been calculated by the position-time diagram technique described in Sec. 5.1. The source voltage is divided according to Eq. (5.9), and voltages transmitted through each interface ($x = l$ and $x = l + p$) are calculated by Eq. (5.13). Symbols representing these wave components

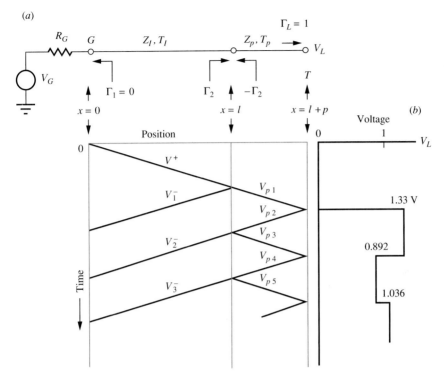

FIGURE 5-34
(a) Schematic diagram of packaged load IC (trace length p) driven point to point from an ideal (unpackaged) voltage source with internal source resistance $R_G = Z_I$, the impedance of the circuit board interconnection trace, showing (bottom) the position-time diagram of a step propagating from the source to the load on the circuit. (b) Plot of load voltage vs. time (right).

are shown adjacent to the rays tracing the time evolution of the signals propagating on the transmission lines, and the voltages are summarized in Table 5.7. The calculated terminal voltage waveform V_L is plotted in Fig. 5-34b to the right of the position-time diagram.

Note that, in this example, the main reflections occur at the PC board package pin interface and the IC input terminal. The wave undergoes multiple reflections within the package itself with a period of $2T_p$. Since the reflection coefficient from package to board is negative (the wave travels from the high-impedance to the low-impedance line), the voltage will oscillate about the steady-state value (1 V). The minimum voltage is seen to be 0.892 V, which lasts for 100 ps ($2T_p$). Transient simulations would need to be carried out with this waveform and the logic gate in use in order to determine if the dynamic noise margin has been exceeded in this interval.

From the time T_p, the transmission line model can be used to determine the length of the trace being considered. For example, if CPW were used to model the package trace, Eq. (5.25) can be used to find the effective dielectric constant. If the package material were alumina ($\epsilon_r = 9.6$), then $\epsilon_{\text{eff}} = 5.3$ and the propagation velocity is $v = c/\sqrt{\epsilon_{\text{eff}}} = 1.27 \times 10^{10}$ cm/s. Thus, 50 ps corresponds to a trace length $p = 0.64$ cm, a rather long trace for use in a high-speed package.

**TABLE 5.7
Calculated line voltages corresponding to the symbols shown in Fig. 5-34b**

Line symbols	Voltages, V
V^+	0.500
V_1^-	0.165
V_{p1}	0.665
V_{L1}	1.33
V_{p2}	0.665
V_2^-	0.439
V_{p3}	−0.219
V_{L3}	0.892
V_{p4}	−0.219
V_3^-	0.144
V_{p5}	0.073
V_{L5}	1.036

Some additional insights can be obtained from further consideration of the example. It is common practice to terminate high-speed interconnections with *termination resistors* whose resistance $R_T = Z_l$ [38]. These resistors are to be located as close to the package pins as possible and should be selected for their low series inductance; hence chip resistors are sometimes used. If in Fig. 5-34 a resistor is connected to the line at $x = l$, the package pin, it can be seen that the reflection coefficient at that interface, Γ_2, will be increased, not decreased as was intended by this practice. If $R_T = 50$ Ω, then Γ_2 increases from 0.33 to 0.60. Hence, the reflections V_p will increase in amplitude, making the package ringing problem almost twice as bad as it was in the unterminated case.

A better solution, if terminating resistors were to be used, would be to locate the resistor at $x = l + p$ inside the package as close to the boundary of the chip as possible. The termination resistance should equal the package trace impedance, $R_T = Z_p$ to produce $\Gamma_L = 0$. This would solve the internal ringing problem, but might produce other unrelated problems. Locating the terminating resistors inside the package will increase the heat generated within the package. Therefore a lower package thermal resistance will be required with this solution. The signal amplitude at the bond pad will be reduced by a factor of two. Also, the termination resistors increase the dI/dt, which must return through the package pins to the power supply. If the inductance cannot be reduced to very low levels, locating the terminating resistors inside the package will increase the power supply noise significantly. If terminations are located inside packages, distributed interconnections (more than one input connected to a transmission line—a bus-type interconnect) will not be possible unless both terminated (for line ends) and unterminated (for taps) versions of each circuit are provided. Finally, note that if the ideal source were placed inside another identical package (see Fig. 5-33), the signals reflected from the pin of package 2 will be reflected again from the impedance discontinuity at the pin of package 1 ($\Gamma_1 = 0.33$) and will return to the load IC to further distort the voltage waveform.

Matching the package trace impedance(s) Z_{p1} and Z_{p2} to the PC board line impedance Z_l provides a much better solution. Referring to Fig. 5-33, the impedance-matched package traces would eliminate interface reflections by making $\Gamma_1 = \Gamma_2 = 0$. Then, if the driver impedance were reasonably well matched to the line impedance, no termination at the chip interface is necessary. Signal wavefronts would reflect off the chip bond pad since $\Gamma_L = 1$, but would be absorbed in the source, eliminating multiple reflections and ringing. Since there are no terminating resistors in the package, the excessive dI/dt problem is also eliminated. Note that a different driver circuit (with pull down) will be necessary in this case since most source follower drivers or open drain drivers depend on the terminating resistor to complete the circuit to the power supply. If a static pull down were included as part of the driver circuit, an increase in power dissipation at the source would be expected. However, if non-ECL signal levels are acceptable, as is the case when connecting GaAs to GaAs, a superbuffer might be used that would reduce the static power dissipation significantly.

Finally, we might imagine one additional possibility. Based on the discussion of attenuation in lossy transmission lines in Sec. 5.4, it would be possible to use an unterminated interconnect even if the source resistance does not perfectly match the transmission line provided that the round-trip loss on the interconnection is high enough to attentuate the reflected wave. This approach would require a driver with a pull down device, as in the other unterminated possibilities discussed above. There would be a range of interconnect lengths that would be suitable for this condition. At one extreme, the line would be too short to effectively attenuate the reflection. This may still be acceptable, however, if the duration of the reflected signal is short enough that the dynamic noise margin will not be exceeded. This possibility could be evaluated by simulation. At the other extreme, if the line is too long, the primary signal (V^+) would be attenuated to the extent that the full logic swing could no longer be obtained, even with the factor $(1 + \Gamma_L)$ being considered at the load end of the line. The range will depend on the ratios R_G/Z_l and R_0l/Z_l where R_0l is the series resistance of the interconnect line of total length l.

While considering the effects of attenuation, the skin effect phenomenon (Sec. 5.4) should be reconsidered. Although it was not an important correction for on-chip interconnections, circuit board traces are frequently thicker than the penetration depth, δ, given by Eq. (5.53). For copper, $\delta = 1.2$ μm at 3 GHz (representing a gaussian pulse with 100 ps full-width at half-maximum) whereas conductor thicknesses of 25 μm are not uncommon. Therefore, a calculation of attenuation based on the dc resistance of the conductor will not be accurate.

The surface impedance Z_S of a plane conductor has been calculated for a sinusoidal electric field as a function of frequency for the case in which thickness is much larger than skin depth [21,22]. The real (resistive) and imaginary (inductive) parts are found to be equal in magnitude. They are given by the equation below for a unit width and length of conductor (ohms per square):

$$Z_s = (1 + j)\left[\frac{\omega\mu_0}{2\sigma}\right] \tag{5.86}$$

In the above equation, σ is the conductivity as defined previously. As the frequency is reduced, the excess resistance and inductance will approach the dc values because, as δ approaches half of the conductor thickness, the boundary conditions change from a semi-infinite conductor to a conductor with finite thickness d for which Z_s becomes [22]

$$Z_s = \frac{1}{\sigma d} + \frac{j\omega\mu_0 d}{3} \tag{5.87}$$

This equation applies whenever the electric field is uniform across the conductor thickness.

The attenuation of the signal on the transmission line can be calculated if the characteristic impedance is known. From Eq. (5.60), the signal is attenuated according to

$$V_L = V_0 e^{-\alpha l} \tag{5.88}$$

where the attenuation factor α consists of the sum of the contributions due to dc series resistance, the skin effect resistance, and dielectric losses. Thus, $\alpha = \alpha_{dc} + \alpha_s + \alpha_D$. The attenuation factor per unit length for the dc resistance was shown in Sec. 5.4.2 to be

$$\alpha_{dc} = \frac{R_0}{2Z_0} \tag{5.89a}$$

The attenuation factor per unit length for the skin effect can be found from the real part of the propagation constant γ. For a microstrip transmission line in the limit where $\delta \ll d$ or s (s is the spacing between the strip and ground plane) and strip width $w \gg s$, Kautz [22] has approximated the attenuation factor with the following equation:

$$\alpha_s = \frac{1}{s}\sqrt{\frac{\omega\epsilon_{\text{eff}}\epsilon_0}{2\sigma}}\left(1 - \frac{\delta}{2s}\right) \tag{5.89b}$$

This equation shows that the skin effect attenuation increases as $\sqrt{\omega}$ and as $1/s$ at high frequencies where the above restrictions apply. Thus, thin circuit board or package dielectrics will result in greater attenuation. Since the skin depth is usually much less than the conductor thickness, this term will usually dominate over the dc contribution to α at high frequencies.

Finally, the dielectric losses also contribute to attenuation when $\epsilon = \epsilon' + j\epsilon''$ has a nonzero imaginary term. Usually, $\epsilon' \gg \epsilon''$. From Ref. 22, the attenuation factor for dielectric loss is approximately

$$\alpha_D = \frac{1}{2}\sqrt{\mu_0\epsilon'\epsilon_0}\frac{\epsilon''}{\epsilon'}\omega \tag{5.90}$$

This contribution to the attenuation is smaller than the other factors for all good dielectrics, except if the frequency is very high (α_D is proportional to ω whereas α_s scales with $\sqrt{\omega}$).

A comparison of the sizes of attenuation factors for several typical circuit board materials can be found in Ref. 36.

5.6.2 Distributed Interconnections

While the point-to-point interconnections discussed above can yield controlled interfaces and minimum reflections if properly designed, the distributed or bus type of interconnect is much more difficult to manage. It is preferable to avoid distributed buses altogether since they can often generate multiple reflections, but, since they are firmly ingrained in the style of computer architectures evolving from lower-speed prototypes (see Chap. 7), some discussion of the problems they present is appropriate.

First consider a simple 50 Ω distributed network, shown in Fig. 5-35a, where there are only two packages on the transmission line, one in the center of the line at A and the other at the end of the line at B. Assume that the driver source resistance is matched to the line and that impedance-matched packages with internal terminating resistors are used. The wavefront leaving D will encounter an impedance discontinuity at A where the line and package are joined in parallel. As illustrated in Fig. 5-35b, the signal will be split three ways at node A: (1) a reflection back toward the source with reflection coefficient $\Gamma_1 = -0.33$, which will be absorbed by the effective source

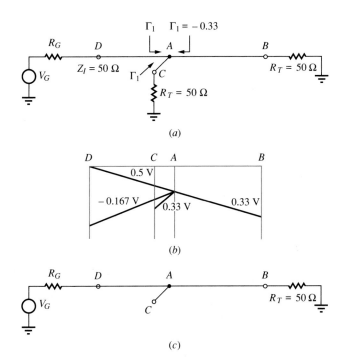

FIGURE 5-35
(a) Schematic diagram of distributed interconnection in which two impedance-matched, in-package terminated ICs are driven from an impedance-matched voltage source. (b) Position-time diagram for the above circuit. (c) Same circuit modified by removing the terminating resistor from the package at node C.

resistance, and (2 and 3) transmission through the node toward both C and B, each with the transmission coefficient $1 - 0.33 = 0.66$. If we assume, for convenience, that the open-circuit source voltage is 1 V, then V^+ will initially be 0.5 V. The circuit will reach a steady state after the reflection reaches the source, when the voltage at all terminals (B, C, and D) will be 0.33 V.

Note that the source voltage has been reduced to one-third of its open-circuit value by placing two terminating resistors in parallel; this lower voltage might not be adequate for driving the load ICs. While the initial line voltage could be increased by reducing the source resistance of the driver, this would also introduce a re-reflection from the source that would distort the signal at B and C and might cause a dynamic noise margin failure. Also note that the time of arrival of the signals at B and C is not the same; there is a difference in path length which will skew the timing of the signal and could also lead to timing failures or force a reduction in the clock frequency to compensate for the skew.

If the terminating resistor at C were removed as shown in Fig. 5-35c, most of the signal in this branch would be reflected from C and would be split at A with the 0.66 transmission coefficient in each direction on the main transmission line. The signal would also reflect a second time from A toward C with negative polarity. Thus, the voltages at C and at B will now oscillate about the steady-state value with period equal to the round-trip propagation time in the package. This may be acceptable if the trace is very short, but if there were many packages on the transmission line, the noise amplitude could become substantial due to the superposition of many multiply-reflected signals. In addition, the time skew between packaged ICs still remains. If the connecting taps (open-circuited stubs) were made long enough to equalize delays, the duration of each oscillation would become significantly long to cause severe logic stability problems. Thus, a point-to-point, equal-length interconnection network is vastly superior to buses in several respects, as discussed further in Chap. 7.

Since the finite package trace length has been identified as one of the troublemakers in this system, let us consider what would happen with a zero-length package, that is, no package. This situation could be realized if the IC chips were bonded upside-down on the PC board (flip-chip bonding). Even in this improved approach, the input capacitance of the pad receiver, bond pad, PC board via (interlayer contact), and solder bump will load the transmission line at the point of connection. Thus, since the L/C ratio of the line is decreased at these taps by the extra capacitance, the local line impedance is reduced at these points, and there will be an impedance discontinuity which causes reflections wherever an IC is bonded to the transmission line. If there is only one tap on the line, the reflection produced will be absorbed in the source resistance if this is properly matched to the transmission line. If two or more taps are on the line, signals will multiply reflect between the taps, causing an oscillating noise voltage on the signal at the IC bond pads. Therefore, to effectively use unpackaged devices on a bus, the input receivers must be designed to present a very low input capacitance compared with the capacitance of the transmission line in the segment where the connection is made.

One further idea should be mentioned which may accommodate a fan-out of two distributed interconnections if the range of characteristic impedances available for the PCB technology can approach 100 Ω. This scheme is shown in Fig. 5-36

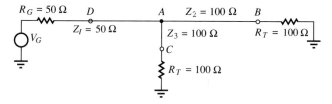

FIGURE 5-36
Schematic diagram of distributed interconnection in which a transition is made from one 50-Ω to two 100-Ω lines in parallel. In-package 100-Ω terminations are shown.

where a branch is made on the transmission line from a 50-Ω feeder to two 100-Ω terminated stubs. Since the 50-Ω line will not see a significant discontinuity at point A, no reflections will be induced by this impedance matched divider.

5.6.3 Power Supply and Ground Connections

The proper management of power supply and ground inductance on circuit boards and in packages is an essential part of the design of any high-speed digital system. As was discussed in Sec. 5.3, current transients will produce voltage spikes (noise) because of the series inductance of the power interconnections. The current transients can be especially large in the power supply connections to output drivers because of the low line impedances being driven. Even a small-signal voltage swing will require a large step in the current, often within 100 to 200 ps for a very-high-speed interconnection. In the previous discussion, techniques for reducing on-chip inductance were discussed; in this section, the focus is off-chip design techniques.

The traditional method used to reduce voltage spikes on circuit boards is to provide a low impedance path to ground through a capacitor. The *bypass* capacitor acts as a local energy source which can deliver charge into the IC(s) at a faster rate than can the power supply, usually separated from the board by cables which exhibit large series inductance. If the capacitor were ideal, and there were no inductance between the capacitor and the chip, the equivalent circuit in Fig. 5-37 would apply,

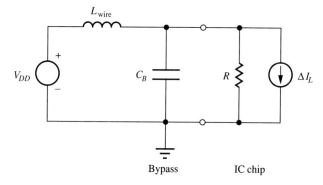

FIGURE 5-37
Equivalent circuit of power supply connection to IC chip with bypass capacitor.

where the constant current portion of the IC chip is modeled as a resistor, R, and the time-varying current as a current source ΔI_L. The transient response of this circuit to a step function in current will be an exponentially damped sine wave with decay time constant

$$\tau = 2RC_B \tag{5.91}$$

and the maximum peak-to-peak amplitude of the damped sine wave is given by

$$\Delta V = 2\Delta I_L \sqrt{\frac{L_{\text{wire}}}{C_B}} \tag{5.92}$$

Here, we see that a larger bypass capacitance leads to reduced amplitude of the ringing, but a longer decay time [39].

Typically, real capacitors also have parasitic series inductance L_S and an equivalent series resistance R_S of their own, and therefore exhibit a series (*RLC*) resonant behavior. Below the resonant frequency, $f_C = 1/(2\pi\sqrt{L_S C})$, the capacitor exhibits capacitive reactance and above, inductive reactance. The magnitude of the impedance reaches a minimum value of R_C at the resonant frequency, as illustrated in Fig. 5-38. The behavior of this resonance therefore depends on the type of capacitor, with microwave chip ceramic capacitors generally having the highest useful frequency of operation and lowest R_S. In the time domain, for frequency components above f_C, the capacitor is ineffective in reducing the predominantly high-frequency noise spikes. Of course, f_C can be extended to higher frequencies by using parallel combinations of smaller-valued fixed capacitors at some cost in area and materials.

The loss tangent (tan δ), also known as the dissipation factor (DF), is often specified by the manufacturer and is directly related to the equivalent series resistance by

$$R_S = \frac{\tan \delta}{2\pi f C} \tag{5.93}$$

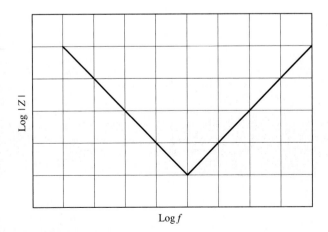

FIGURE 5-38
Bode plot (relative scale) of the impedance magnitude of a real capacitor with series *RLC* equivalent circuit.

TABLE 5.8
Terminology and constants used in Chapter 5

Symbol	Name	Meaning or value	Unit
c	Velocity of light	2.99×10^{10}	cm/s
MTTF	Median-time-to-failure	50% of devices "dead"	h
q	Electron charge	1.6×10^{-19}	C
$u(t)$	Unit step function	$u(t) = 1$ if $t > 0$; otherwise $= 0$	
Z_0	Characteristic impedance of a transmission line	Ratio of V/I at any point on line	Ω
Z_0^e	Even-mode impedance (two coupled lines)	Characteristic impedance determined for common mode line voltages	Ω
Z_0^o	Odd-mode impedance (two coupled lines)	Characteristic impedance determined for differential line voltages	Ω
α	Attenuation	Distance at which the signal is attenuated to $1/e$ of its initial amplitude	cm^{-1}
δ	Skin depth	Penetration depth of electric field into a conductor at high frequencies	cm
ϵ	Permittivity of a dielectric material	$D = \epsilon E$	F/m
ϵ_0	Permittivity of free space	8.85×10^{-12}	F/m
ϵ_{eff}	Effective dielectric constant	Ratio of the capacitance of a conductor in dielectric to its capacitance in free space	
ϵ_r	Relative dielectric constant	ϵ/ϵ_0	
Γ	Reflection coefficient	Ratio of reflected voltage to incident voltage	
Γ_T	Transmission coefficient	Ratio of transmitted voltage to incident voltage	
μ	Permeability of a magnetic material	$B = \mu H$	H/m
μ_0	Permeability of free space	$4\pi \times 10^{-7}$	H/m
ρ	Resistivity	Aluminum: $\rho = 2.74 \times 10^{-6}$ Gold: $\rho = 2.20 \times 10^{-6}$	$\Omega \cdot \text{cm}$
ρ_s	Sheet resistance	$\rho/\text{thickness}$	$\Omega/\square$
σ	Conductivity	$1/\rho$	$(\Omega \cdot \text{cm})^{-1}$

The loss tangent increases with frequency for most dielectrics; for certain dielectrics intended for use at lower frequencies, it can increase by several orders of magnitude. While the equation indicates that R_S can be reduced through increasing C, it is also true that f_C decreases as C increases. Hence, there is a tradeoff necessary which can only be resolved by microwave s-parameter measurements [40].

Note that the bypass capacitor will not be effective if it is separated from the IC chip by even a small inductance, which might arise from the package trace or circuit board trace. This inductance will lower the resonant frequency of the bypass so as to make it increasingly less useful as the distance from the chip to bypass becomes larger. Therefore, it is essential that the bypass capacitors are incorporated into the package itself (when used) and the circuit board as well. One of the most effective methods for providing a low-impedance bypass capacitor is to use the power planes of the board to form a large parallel plate capacitor [39]. Then, a disturbance caused by ΔI_L will propagate as a wave radially into a region of larger and larger area and hence larger capacitance. This approach can be used in multilayer packages or chip carriers as well. Another approach toward in-package bypassing has been to incorporate the bypass capacitor as a reverse-biased pn junction into a chip carrier made of silicon [41]. The chip carrier is then placed into the die well of an ordinary package.

HOMEWORK PROBLEMS

Sections 5.1 to 5.3

5.1. Prove that $v = (\epsilon\mu)^{-1/2} = (L_0 C_0)^{-1/2}$ for a coaxial conductor with inner radius a and outer radius b. Repeat for parallel cylinders with radius a and spacing $2d$.

5.2. Derive $Z_0 = \sqrt{L_0/C_0}$ from Eqs. (5.4).

5.3. Derive Eq. (5.12).

5.4. Repeat Example 5.1 for an input signal with a rise time of 100 ps.

5.5. (a) Determine $V(0, t)$ and $V(l, t)$ for the transmission line system shown in Fig. 5-1 if $R_G = 10 \; \Omega$ and $Z_0 = 50 \; \Omega$.
(b) In what way might this interconnect system be less than ideal for high-speed logic circuits?

5.6. A long, unterminated interconnect line with $Z_0 = 50 \; \Omega$ is to be driven such that the rise time and fall time at the load are to be as short as possible, but with not more than 10 percent overshoot.
(a) What is the proper source resistance if the driver is modeled by its Thevenin equivalent and provides an open-circuit rise or fall time of $T/4$? Use the position-time diagram to estimate the load voltage for $t \leq 8T$.
(b) Assume that a source follower is used to drive the line. Determine the correct channel widths using 1-μm GaAs MESFETs with $g_m = 150$ mS/mm and $I_{DSS} = 100$ mA/mm. What will be the power dissipated by the driver?
(c) Repeat part (b) for an HBT emitter follower. Assume that the ideality factor for collector current vs. V_{BE} is 1.00.

5.7. (a) Construct a graph of L_0 and C_0 vs. spacing for CPW and CPS interconnections on semi-insulating GaAs. Assume that the conductor width is fixed at 2 μm while the spacing is varied from 2 to 20 μm.

(b) Now assume that the lines are coated with 1 μm of Si_3N_4 with $\epsilon_r = 8$. Plot the new L_0 and C_0 on the graph of part (a).

5.8. (a) Construct a graph of L_0 and C_0 vs. conductor width for CPW and CPS interconnections of semi-insulating GaAs. Assume that the spacing is fixed at 3 μm while the width is varied from 3 to 100 μm.

(b) Now assume that the lines are coated with 1 μm of Si_3N_4 with $\epsilon_r = 8$. Plot the new L_0 and C_0 on the graph of part (a).

5.9. Prove that the inductance of a transmission line is independent of ϵ_r.

5.10. A multilayer printed circuit board with $\epsilon_r = 2.7$ is designed on a 0.05 inch grid.

(a) If the minimum feature size is 0.004 in. and spacing between the planes is 0.010 in., what is the maximum impedance possible for a signal line on the surface and for one in an interior plane?

(b) What is the minimum impedance and inductance for power supply lines on the surface? How could the internal planes be used to reduce inductance?

Section 5.4

5.11. Make a plot of crossover capacitance as a function of conductor width for square crossovers ($W = Z$). Compare this capacitance with the parallel plate model on the plot. Assume an Si_3N_4 insulator with $\epsilon_r = 8.0$ and thickness 0.5 μm.

5.12. Compare the predicted crossover capacitances for Si_3N_4 ($\epsilon_r = 8; h = 0.5$ μm), SiO_2 ($\epsilon_r = 4.0; h = 0.5$ μm), and air bridges ($\epsilon_r = 1; h = 2$ μm).

5.13. Estimate the crossover capacitance for coincident 1M and 2M lines as shown in Fig. P5-13. Assume that the insulator is Si_3N_4 with $\epsilon_r = 8.0$ and $h = 0.5$ μm). Compare with the capacitance of minimum-geometry (assume 2 μm width and spacing) first-level metal coplanar strips on the same media with the same length.

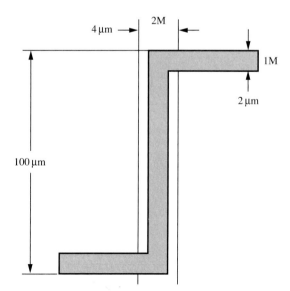

FIGURE P5-13
Coincident 1M and 2M lines.

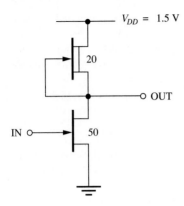

FIGURE P5-14
DCFL driver.

5.14. Determine the effective source resistance, R_G, for the DCFL driver shown in Fig. P5-14 by:
(a) Small-signal model
(b) AC SPICE simulation
(c) Transmission line method (assume $Z_0 = 60\ \Omega$)
Use the typical e/d FET model parameters of Chap. 4 for this exercise.

5.15. Repeat Prob. 5.14 for an SBFL driver, as shown in Fig. P5-15.

5.16. A driver with $R_G = 100\ \Omega$ and $V_G(t) = u(t)$ is used to drive a CPW first-level metal interconnect with line width 4 μm and spacing 2 μm. The metal is aluminum with 0.3 μm thickness. The dielectric coating is Si_3N_4 with $\varepsilon_r = 8.0$ and $h = 0.5\ \mu$m.
  (a) Calculate the amplitude $V_0(t - T)$ and time of arrival, T, at the open-circuited end of the line for the first-signal transit if the line length is 1 mm.
  (b) Find the maximum single-transit interconnect length for an output voltage of $0.9\ u(t - T)$.
  (c) What would be the maximum single-transit interconnect length if the width and spacing were scaled to 1 μm?
  (d) If the chip temperature were reduced to 77 K, $\rho_{Al} = 2.5 \times 10^{-7}\ \Omega\cdot$cm. Repeat parts (b) and (c) for this temperature.

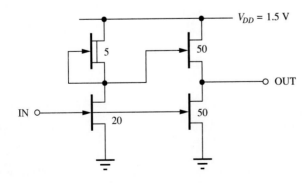

FIGURE P5-15
SBFL driver.

5.17. (a) A large BFL driver with $R_g = 70\ \Omega$, $V_{OL} = -1.1$ V, $V_{OH} = +0.7$ V, and $t_r = t_f = 50$ ps is used to drive an interconnect line with $Z_0 = 70\ \Omega$. A 10-μm BFL gate with $C_{\text{eff}} = 20$ fF is connected to the output. If the line is gold with width = 3 μm, thickness = 1 μm, and length = 3 mm, determine the propagation delay for the signal on this path.
(b) Repeat for $R_G = 140\ \Omega$.

Section 5.6

5.18. Calculate the voltage as a function of time at the output of the source (node $G; x = 0$) and the input to the IC chip (node $T; x = l + p$) in Fig. 5-34a. Assume that the parameters of Example 5.11 apply and that $Z_{p1} = Z_{p2}$ and $T_{p1} = T_{p2}$. Follow the reflections at least until the first reflection from the source returns to node T.

5.19. An HBT driver circuit is designed which has an effective source resistance of 20 Ω, a logic voltage swing of 1 V, and rise/fall times of 30 ps. Impedance-matched packages (50 Ω and PC board interconnect line) are used with an unterminated interconnection. The effective input resistance is 500 Ω.
(a) Calculate the maximum interconnect length allowable such that a full logic swing is just sustained for a copper interconnect line with a 100 × 25 μm cross section.
(b) Using the typical HBT model parameters from Chap. 2, find the dynamic noise margin of the HBT ECL gate shown in Fig. P5-19. Use this to determine the minimum interconnect length (if any) if the input to the ECL gate were located at the load end of the circuit.

5.20. The PCB shown in Fig. P5-20 has two buried layers for power and ground and two external layers (component side and solder side) for microstrip interconnect. Total PCB thickness is 60 mils. All metallization is 0.0014 in. thick, and all conducting layers are spaced apart equally. The PCB is constructed from G-10 epoxy which has a dielectric constant (ϵ_r) of 5.0. Fabrication tolerances limit microstrip width and spacing to a minimum of 8 mils.

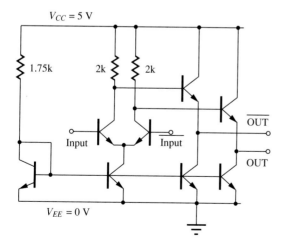

FIGURE P5-19
HBT ECL gate.

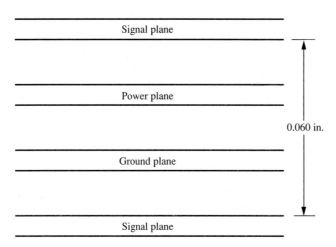

FIGURE P5-20
PC board cross section.

(a) Find the characteristic impedance (Z_0) of a minimum-width microstrip transmission line.
(b) Find the width of a microstrip transmission line that will provide a good match to a 50 Ω source and load.
(c) How many minimum-width microstrip transmission lines will fit in an area 1 in. wide, assuming all lines are parallel to each other?
(d) How many 50 Ω microstrip transmission lines will fit in an area 1 in. wide, assuming all lines are parallel to each other?
(e) Find the propagation velocity for a 50 Ω line.
(f) Find the inductance per unit length (L_0) of a 50 Ω line.
(g) Find the capacitance per unit length (C_0) of a 50 Ω line.

5.21. Repeat Prob. 5.20 for a PCB made of polyimide. ϵ_r is 3.5, PCB thickness is 10 mils, conductor thickness is 1 mil, and the minimum width and spacing for microstrip is 2 mils.

REFERENCES

1. There are many textbooks on electricity and magnetism which give useful, detailed explanations of electromagnetic wave propagation. For example, see B. Bleaney and B. Bleaney, *Electricity and Magnetism*, Oxford University Press, 1965, or S. Ramo, J. Whinnery, and T. Van Duzer, *Fields and Waves in Communication Electronics*, Wiley, New York, 1965.
2. Ghione, G., and Naldi, C. U.: "Coplanar Waveguides for MMIC Applications: Effect of Upper Shielding, Conductor Backing, Finite-Extent Ground Planes, and Line-to-Line Coupling," *IEEE Trans. Microwave Theory and Techniques*, vol MTT-35, pp. 260–267, March 1987.
3. Ghione, G., and Naldi, C.: "Analytical Formulas for Coplanar Lines in Hybrid and Monolithic MICs," *Elect.Lett.*, vol. 20, pp. 179–181, 16 Feb. 1984.
4. Hilberg, W.: "From Approximation to Exact Relations for Characteristic Impedances," *IEEE Trans. Microwave Theory and Tech.*, vol. MTT-17, pp. 259–265, May 1969.

5. Estreich, D. E.: "Some Simple Equations for the Computation of GaAs IC Capacitances," course notes from EE392, Stanford University, 1983. Personal communication.
6. Schneider, M. V.: "Microstrip Lines for Microwave Integrated Circuits," *Bell Syst. Tech. J.*, vol. 48, pp. 1421–1444, May/June, 1969.
7. Gupta, K. C., Garg, R., and Chadha, R.: *Computer-Aided Design of Microwave Circuits*, chap. 3, Artech House, 1981.
8. Howe, H., Jr.: *Stripline Circuit Design*, chap. 2, Artech House, 1974.
9. *American Institute of Physics Handbook*, 3d ed., pp. 9-39–9-43, McGraw-Hill, 1972.
10. Maissel, L., and Glang, R.(Eds.): *Handbook of Thin Film Technology*, McGraw-Hill, 1970.
11. Glaser, A. B., and Subak-Sharpe, G. E.: *Integrated Circuit Engineering*, chap. 4, Addison-Wesley, 1977.
12. Kittel, C.: *Introduction to Solid State Physics*, 6th ed., chap. 6, Wiley, New York, 1986.
13. D'Huerle, F. M.: "Electromigration and Failure in Electronics: An Introduction," *Proc. IEEE*, vol. 59, pp. 1409–1418, October, 1971.
14. Black, J. R.: "Electromigration Failure Modes in Aluminum Metallization for Semiconductor Devices," *Proc. IEEE*, vol. 57, pp. 1587–1594, September, 1969.
15. Mil.Spec. MIL-M-385105.
16. Mukherjee, S.: "Interactions of Metal Films on Semiconductors," in *Reliability and Degradation*, (Eds. M. Howes and D. V. Morgan) Wiley, 1981.
17. Song, W. S., and Glaser, L. A.: "Power Distribution Techniques for VLSI Circuits," *IEEE J. Solid State Cir.*, vol. SC-21, pp. 150–156, February, 1986.
18. Ghione, G., and Naldi, C.: "Parameters of Coplanar Waveguides with Lower Ground Plane," *Elect.Lett.*, vol. 19, pp. 734–735, 1 Sept. 1983.
19. Schlosser, W. O., and Sokolov, V.: "Circuit Aspects of Power GaAs FETs," *GaAs FET Principles and Technology*, (Eds. J. V. DiLorenzo and D. D. Khandelwal), p. 545, Artech House, 1982.
20. Matthaei, G. L., Shu, C.-H., and Long, S. I.: "Simplified, Linear Representation of Logic Gate Terminal Impedances for Use in Interconnect Crosstalk Calculations," *IEEE J. Solid State Cir.*, vol. SC-24, pp. 1468–1470, Oct. 1989.
21. Ramo, S., Whinnery, J., and Van Duzer, T.: *Fields and Waves in Communication Electronics*, Secs. 4.12 and 5.14, Wiley, New York, 1965.
22. Kautz, R. L.: "Miniaturization of Normal-State and Superconducting Striplines," *J. Res. Nat. Bur. Stand.*, vol. 84, pp. 247–259, May/June 1979.
23. Ghoshal, U., and Smith, L. N.: "Finite Element Analysis of Skin Effect in Copper Interconnects at 77K and 300K," *1988 IEEE Microwave Theory and Techniques Symposium Proceedings*, paper OF-2-13, pp. 773–776.
24. Ho, C. W.: "High-Performance Computer Packaging and the Thin-Film Multichip Module," in *VLSI Electronics: Microstructure Science*, vol. 5, chap. 3, Academic Press, 1982.
25. Ho, C. W., Chance, D. A., Bajcrek, C. H., and Acosta, R. E.: "The Thin Film Module as a High Performance Package," *IBM J. Res. Dev.*, vol. 26, pp. 289–291, May 1982.
26. Sakurai, T.: "Approximation of Wiring Delay in MOSFET LSI," *IEEE J. Solid State Cir.*, vol. SC-18, pp. 418–426, August, 1983.
27. Garg, R., and Bahl, I. J.: "Characteristics of Coupled Microstriplines," *IEEE Trans. Microwave Theory and Techniques*, vol. MTT-27, pp. 700–705, July 1979.
28. Matthaei, G. L., Shu, C.-H., and Long, S. I.: "Some Fundamental Concepts for Wave-Coupling Between Lines in High-Speed Integrated Circuits," to be published.
29. Tripathi, V. K., and Rettig, J. B.: "A SPICE Model for Multiple Coupled Microstrips and Other Transmission Lines," *IEEE Trans. Microwave Theory and Tech.*, vol. MTT-33, pp. 1513–1518, December 1985.
30. Djordjevic, A. R., Sarkar, T. K., and Harrington, R. F.: "Time-Domain Response of Multiconductor Transmission Lines," *Proc. IEEE*, vol. 75, pp. 743–764, June 1987.
31. Oliver, B. M.: "Directional Electromagnetic Couplers," *Proc. IRE*, vol. 42, pp. 1686–1692, November 1954.

32. Levy, R.: "Directional Couplers," in *Advances in Microwaves* (Ed. L. Young), vol. 1, Academic Press, New York, 1966.
33. Ikalainen, P. K., and Matthaei, G. L.: "Wideband, Forward-Coupling Microstrip Hybrids with High Directivity," *IEEE Trans. Microwave Theory and Tech.*, vol. MTT-35, pp. 719–725, August 1987.
34. Feller, A., Kaupp, H. R., and Digiacomo, J. J.: "Crosstalk and Reflections in High-Speed Digital Systems," *Proceedings Fall Joint Computer Conference*, 1965, pp. 511–525.
35. Price, J. E.: "VLSI Chip Architecture for Large Computers," in *Hardware and Software Concepts in VLSI* (Ed. G. Rabbat), Van Nostrand-Reinhold, 1983.
36. Gheewala, T., and MacMillan, D.: "High-Speed GaAs Logic Systems Require Special Packaging," *EDN*, Cahners Publishing Co., 17 May 1984.
37. *Electronics*, vol. 55, p. 135, 14 July 1982.
38. *ECL Data Book*, Fairchild Semiconductor Corp., 1977.
39. Doubrava, L.: "Proper Handling of Voltage Spikes Safeguards Circuit Designs," *EDN*, Cahners Publishing Co., pp. 83–87, 5 Mar. 1979.
40. Perna, V. F.: *The RF Capacitor Handbook*, American Technical Ceramics, 1979.
41. Gheewala, T. R.: "Packages for Ultra-High Speed GaAs ICs," *1984 IEEE GaAs IC Symposium Proceedings*, paper 3.3, pp. 67–70, October 1984.

CHAPTER 6

TEST METHODS FOR VERY-HIGH-SPEED DIGITAL ICS

Very-high-speed digital ICs require considerably more care and planning during the design stage to facilitate testing than do similar circuits implemented using lower-speed technologies. As was discussed in Chap. 5, the IC/outside world interfaces must be engineered properly to avoid noise and spurious logic responses. Testability, in addition, will demand that circuit functionality can be assessed by exercising a minimum of high-speed input and output lines, since the number of high-quality signal interfaces available for testing is greatly limited during wafer probing. Maintaining signal quality also requires attention to the number and placement of power supply and ground connections so that noise and crosstalk can be minimized during the test. Direct electrical delay measurement across a chip should also be avoided due to the difficulties involved in calibration. Therefore, the use of on-chip (built-in) test paths is recommended. Accurate measurement of propagation delay per gate can be performed on test structures designed to minimize measurement error. The use of non-contacting electrooptical probing has provided a more direct measurement of on-chip timing. It has proven to be very useful for evaluating the condition of otherwise inaccessible high-speed circuit nodes, and its excellent time resolution makes it capable of performing direct delay measurement across a chip.

6.1 WAFER PROBING TECHNIQUES

The conditions necessary to properly operate and therefore to test a very-high-speed digital IC in a well-designed package were described in Chap. 5. In this section, we are concerned with the need to test these chips before they are separated from their wafer and bonded into an IC package. The dicing and bonding process is time-consuming; since the packages themselves are expensive as well, it is essential that some dependable technique for high-speed testing at the wafer level be implemented before packaging so that the cost of manufacture is kept to a minimum. In addition, the package and wire bonds distort the performance of the circuit so that intrinsic circuit operation or device characteristics can only be indirectly inferred. Finally, correlating the performance of a chip with its position on the wafer may provide useful feedback to the processing staff.

The important considerations for wafer-level (wafer probe) testing are closely related to those described in Chap. 5. The goal is to provide electrical interfaces to the chip which yield signals that are free of glitches from uncontrolled reflections and power/grounds which are quiet electrically and stable in potential. Therefore, the principles already developed will be adapted to this special requirement.

Accurate high-speed wafer-level measurement of timing, circuit delay, or maximum clock frequency is limited by the test interface. Electrical contact to IC chips on their wafers often utilizes a probe card, a conventional printed circuit board which is used to rigidly support and electrically contact short wire probes, usually tungsten wires with length on the order of 1 to 2 cm. These standard probe cards are widely used for parametric testing of devices at dc and low frequencies but are unsuitable for testing all but the most insensitive circuits at high frequencies. These probe cards are relatively inexpensive due to a well-established manufacturing technology and low-cost materials.

The problem with these probe cards for high-frequency testing is the lack of a controlled impedance from the boundary of the card to the end of the wire probe. Some cards make no attempt in this direction, using arbitrary printed circuit trace widths without ground planes, edge connectors, and long, unterminated probe wires. The probes may be thought of as short segments of high-impedance transmission line (air dielectric) which will behave similarly to the non-impedance-matched package traces discussed in Sec. 5.6. Even if the probe card traces are cut and coaxial lines are soldered as close to the probe wires as possible, the uncontrolled reflections on the probes themselves will lead to ringing whose period of oscillation corresponds to the round-trip delay on the probes. An equivalent circuit of this type of probe card is shown in Fig. 6-1a [1]. The input to the chip consists of a generator matched to a 50-Ω transmission line which is terminated with a 50-Ω resistor. The probe wires are represented by transmission line segments with an impedance on the order of 100 Ω. Because the parallel terminating resistance on a clock or data input line is separated from the IC pad by the high Z_0 probe, signal edges reflected from the chip input are re-reflected from the terminating resistor with negative amplitude, as discussed in Sec. 5.6, and can lead to logic upset if the delay time on the probe exceeds the response time of the logic gates. Delay times of approximately 50 to 100 ps would be expected on 1 to 2 cm probes. The output circuit will experience a reflection at the coax/probe interface. Therefore, output signals will not appear as clean pulses on an oscilloscope. In general, the output amplitudes dependence on frequency and noise is not harmful to the operation of the chip and is therefore less critical.

A better approach for control of reflections on high-speed inputs is shown in Fig. 6-1b [1]. Here a source termination is used to minimize the reflection coefficient

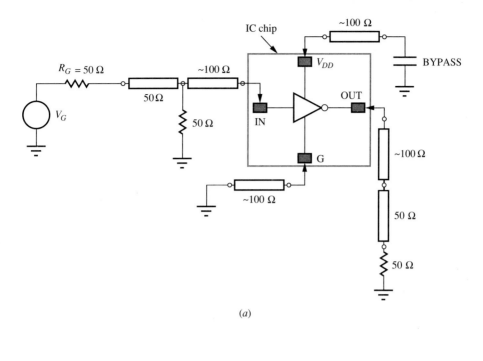

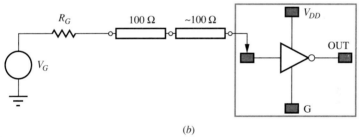

FIGURE 6-1
(a) Schematic representation of IC under wafer-level test. Four probes (input, output, ground, and supply) are shown. Input is parallel terminated at the probe card/probe wire interface. (b) Series termination at the source and high-impedance cable will reduce the amplitude of reflections at the IC interface. (*From Long* [1], © *IEEE. Used by permission.*)

at the source end. Higher impedance coaxial line is used to minimize reflection at the coax/probe interface. Therefore, the reflected wave from the probe/IC interface will be largely absorbed at the source and will not re-reflect the signal. The shunt capacitance at the point where the probes are attached to the board will also lead to an impedance discontinuity at this point (reduction) which is more sensitive now due to the higher impedance line, but this will be much less severe than the reflection from a 50-Ω resistor in parallel with a 50-Ω coaxial line.

The high Z_0 probes also appear as inductors in series with the low-impedance power supply (to the bypass capacitor) and ground circuit if the length of the probes is less than one-quarter wavelength. This series inductance, approximately 10 to 15 nH for a 2-cm probe, will increase these impedances, thereby promoting transient switching noise on the chip. Bypass capacitors can be located on the probe card near the probe wires, but the 1 to 2 cm of wire probe reduces their series resonant frequency, possibly making them ineffective. Multiple power and ground probes will help reduce this effect, but the performance of the card is ultimately limited by the probe length.

Finally, the closely spaced, high-impedance probe wires are subject to crosstalk. Electromagnetic coupling between adjacent probes will be high unless power supply or ground lines are located between all of the critical signal lines. Even in this event, the common mode ground inductance provides a mechanism for coupling between signal lines. The equivalent circuit for common mode crosstalk is shown in Fig. 6-2. In this two-probe example, Z_{h1} and Z_{h2} are the probe inductance, Z_{g12} the common mode ground inductance, and Z_{L1}, Z_{L2} the input of the device under test. If in the worst case all of the probes were shorted together, then the voltage at the probe tips would be the same as the voltage drop across the ground return inductance.

In spite of the above deficiencies, some circuits, such as ring oscillators and to some degree frequency dividers, are relatively insensitive to the signal interfaces and can be tested with conventional probe cards. The operation of these circuits is discussed in the next section. On frequency dividers, it is common to experience "dropouts" which are frequencies at which the power supply or ground noise becomes excessive or the signal amplitude diminishes and the proper divide ratio is no longer obtained. Large-amplitude clock feedthrough can be observed on the output when this condition occurs. Often, an increase in frequency beyond this condition will restore correct operation and testing can be continued since these circuits do not require any initialization.

Some manufacturers of wafer probe equipment have attempted to shorten the probe wires and connect them to microstrip lines on ceramic supports. Coaxial cables are connected directly to the ceramic supports, which can also accommodate chip resistors for line termination or chip capacitors for bypass when needed. This approach has helped increase the maximum useful frequency of the card.

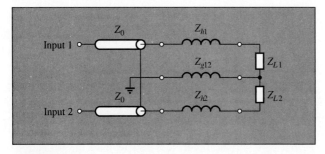

FIGURE 6-2
Schematic diagram illustrating the common ground inductance Z_{g12} that can produce crosstalk. (*From E. Strid, Cascade Microtech. Personal communications.*)

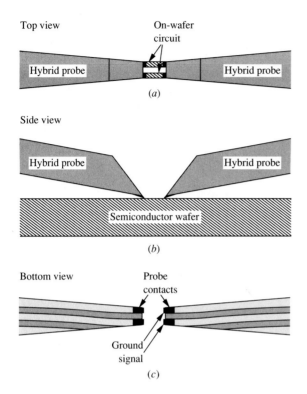

FIGURE 6-3
Coplanar waveguide probe interface for microwave wafer probing: (*a*) top view, (*b*) side view, and (*c*) bottom view. (*From Strid, Gleason, and Reeder* [3]. © *Academic Press. Used by permission.*)

A better solution to the wafer probing interface problem has been provided by an improved mechanical and electrical design of the probe structure [2,3]. Coplanar waveguide transmission lines on thin alumina substrates have been used to transition between the coaxial connectors (SMA or K connectors) and the extremely short, plated-up bumps used for the probe tips. A schematic drawing of this approach is shown in Fig. 6-3 [3]. Probes of this type are routinely used for direct *S*-parameter measurements on microwave transistors up to 26 GHz. The probes can also be configured with terminating resistors or bypass capacitors close to the probe tips as needed. The wide ground return connections help to reduce the common mode ground inductance to very low levels (50 pH or so) and also help shield the signal lines electromagnetically. Thus, crosstalk between signal inputs is kept very low.

6.2 ON-CHIP (BUILT-IN) TEST APPROACHES

There are two types of on-chip test approaches which will be discussed in this section. Firstly, structures suitable for accurate measurement of logic gate propagation delay will be described. These test structures would be useful for process control measurement or for assessment of circuit design considerations on gate speed. Secondly, built-in test paths can be included in many larger logic circuits to convert an otherwise difficult test problem (direct delay measurement of many outputs, for example) into a relatively easy measurement (maximum operating frequency, for example).

6.2.1 Propagation Delay Test Structures

Propagation delay test structures must be designed to be both easy to measure and to provide good correlation between delay of the test structure and delay measured in real logic circuits. Several useful test structures have been reviewed in the literature [1] and are described here.

The "classic" asynchronous test structure has been the ring oscillator (RO) [4]. The ring oscillator, shown in Fig. 6-4, consists of an odd number (n) of inverting logic gates connected in a closed loop. When power is applied, the circuit will oscillate at frequency $f = 1/(2nt_{pd})$, so the propagation delay per gate can be found from a single measurement. The number of gates in the ring should be large enough to guarantee that each output reaches the proper steady-state logic levels before beginning to switch to the other state. Also, a large n reduces the output frequency. The influence of the off-chip interface can be minimized by connecting to the ring through one of the fan-out gates or by weakly coupling to the ring capacitively, driving the output pad with a Darlington buffer, and using a sensitive measurement tool such as a spectrum analyzer to determine the frequency. A larger power and ground inductance can be tolerated than would be acceptable for other circuits because the RO is inherently asynchronous, with at most m gates switching simultaneously, where m is the number of the fan-out. The ring oscillator can be designed with on-

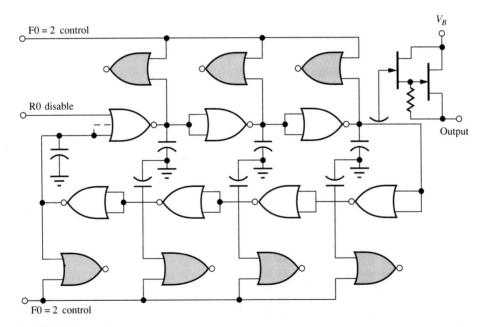

FIGURE 6-4
Schematic diagram of a seven-stage NOR-gate ring oscillator. Several circuit options are shown which allow the basic circuit to be configured during layout as fan-out = 1 or = 2, fan-in = 1 or = 2. The load gates are shaded. Optional metal insulator-metal load capacitors can also be attached to the ring (*From Long [1]*. © IEEE. Used by permission.)

chip loading as shown in Fig. 6-4 with realistic fan-in, fan-out, and load capacitance. The ring oscillator is formed by the inner loop of seven NOR gates. Fan-out gates are shaded. Good correlation between the RO t_{pd} and that of real combinational logic circuits is expected. The loading can consist of additional identical logic gates and lumped metal-insulator-metal fixed capacitors attached to each intermediate node to simulate the loading of fan-out and interconnects. However, if the RO is designed with minimum layout area and no loading, correlation with complex logic circuit delay is quite poor. Nevertheless, the circuit is widely used to determine intrinsic performance of a device/circuit combination and to evaluate the influence of circuit or device design changes on intrinsic circuit speed.

Synchronous circuits such as frequency dividers and shift register-counters are also suitable for use as propagation delay test structures. The maximum clock frequency of correct operation (output data pattern or signature is properly related to the input, usually the clock) can be related to a multiple N of the average logic gate propagation delay through transient simulation or timing analysis. Proper divide ratios can be verified and measured accurately with a frequency counter; however, the interpretation of the measurement is somewhat unclear. Since the loading and perhaps the sizing of gates in the basic flip-flop (FF) which comprises these dividers and counters is not necessarily the same for each gate, the gate delay extracted from this measurement cannot be associated with any particular size or load condition for the logic family in use. In addition, some types of flip-flops achieve their maximum frequency of operation with an asymmetric clock waveform. For example, the timing analysis of the single-phase clocked, NOR implemented, D-type FF in Fig. 6-5 shows that $f_{max} = 1/5t_{pd}$, where $3t_{pd}$ is the logic 1 state and $2t_{pd}$ is in logic 0 [1]. If the clock is symmetric with respect to the logic threshold, the maximum clock frequency of the D-FF decreases to $1/6t_{pd}$. The preferred clock asymmetry arises from the inherent asymmetry in delay produced by a NOR or NAND implemented latch circuit. Slight errors in the clock symmetry or offset lead to large errors in t_{pd}. This sensitivity to clock waveform makes it more difficult to perform the measurement accurately.

A better synchronous propagation delay test circuit can be realized through the

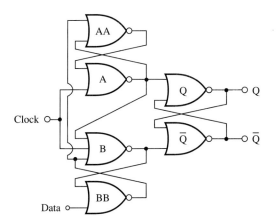

FIGURE 6-5
Schematic diagram of a single-clocked, NOR implemented, D flip-flop.

use of a Johnson counter rather than a ripple divider. A two-stage version of this counter, shown in Fig. 6-6a, is a shift register with an inverting feedback connection. It will produce an output code of alternating pairs of bits 00110011, etc., and therefore divides the clock waveform by four. As it stands, however, it still has the delay uncertainty associated with the unequal loading inside the FF cells and will still require an asymmetric clock waveform. However, if two versions of the counter are provided on the same chip with different amounts of delay added to their data paths, the situation changes markedly.

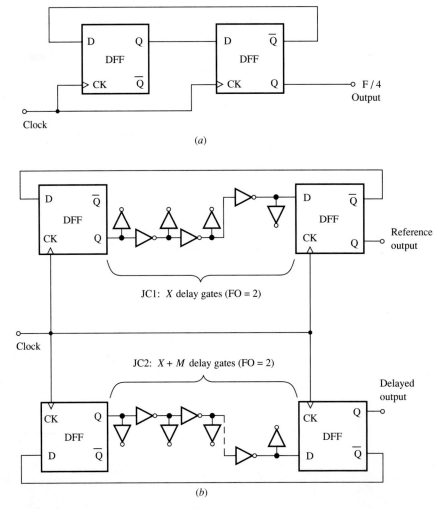

FIGURE 6-6
(a) The two-stage Johnson counter provides a divide-by-four output. (b) Improved accuracy in determining t_{pd} can be obtained from the dual delayed Johnson counter. Reference version (JC1) has X gates in the data path; the delayed version (JC2) has $X + M$ gates in the data path.

This situation is illustrated in Fig. 6-6b. Call the intrinsic FF delay Nt_{pd}; for the D-FF mentioned above, $N = 5$. JC1, the reference counter, has X gates added in the delay path, while JC2, the delayed counter, has $X + M$. These counters would operate at the highest clock frequency for an extremely asymmetric clock: $3t_{pd}$ at logic 1 and $(2 + X)t_{pd}$ or $(2 + X + M)t_{pd}$ at logic 0 respectively, as shown in Fig. 6-7a. It is more convenient for testing, however, if the clock waveform driving the counters is symmetric. Therefore, filling out the logic 1 side of the clock, the maximum frequency of operation of JC1 becomes

$$f_1 = \frac{1}{N + 2X - 1} \qquad (6.1)$$

illustrated by the waveform in Fig. 6-7b. The minus 1 in the denominator comes from the $\frac{3}{2}$ asymmetry of the D-FF in Fig. 6-5. If a symmetric (two-phase clocked) FF were used, it would not be necessary. Similarly, JC2 operates with a maximum

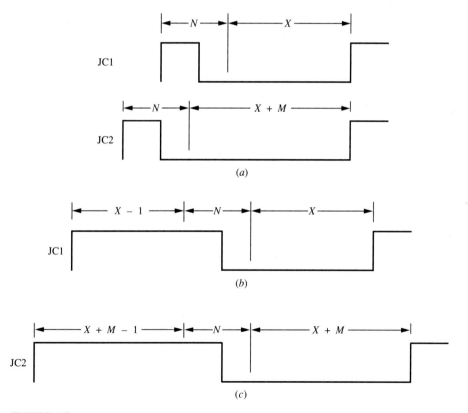

FIGURE 6-7
(a) Highly assymetric clock waveforms are needed to operate the delayed Johnson counters at their highest possible clock frequencies. Waveforms are made symmetric by increasing the time that the clock waveform is at logic 1. (b) Symmetric clock for JC1. (c) Symmetric clock for JC2.

clock frequency of

$$f_2 = \frac{1}{N + 2(X + M) - 1} \quad (6.2)$$

as shown in Fig. 6-7c. Using Eq. (6.1) to eliminate $(N + 2X - 1)$ in Eq. (6.2) we obtain the following result:

$$t_{pd} = \frac{f_1 - f_2}{2M f_1 f_2} \quad (6.3)$$

Now we see that the delay extracted from the measurement depends only on the number of excess delay gates on JC2 and the measured maximum clock frequencies. The measurement can have very good accuracy if M is a large number. Then, any uncertainties in N or in the clock symmetry or, equivalently, offset become unimportant. Also, the delay is dominated by a specific type of gate, the gate used in the delay path. Thus, fan-in, fan-out, and loading can be selected at layout time to represent the case of interest. Inverters with FO = 2 were used for the example in Fig. 6-6b.

Finally, common to both synchronous propagation delay test cells is the requirement for low noise on the power supply and ground lines. Most of the logic gates change state following one or both clock edges. Therefore, the rate of change of supply or ground current dI/dt is larger than in asynchronous circuits, and the control of inductances becomes more crucial to a successful measurement.

6.2.2 Built-in Test Approaches

The preceding section has discussed the design of SSI propagation delay test structures which are useful for process monitoring and design optimization and verification. The determination of functionality of a complex IC at its rated speed of operation is a much more difficult task to accomplish at the wafer probe level. Often the number of inputs and outputs that should be tested will exceed the number of controlled impedance probe interfaces available with the probe card technology described in Refs. 2 and 3. Therefore, whenever possible, additional circuitry (extra gates, latches, feedback paths, etc.) should be provided on-chip to allow for a built-in delay test mode or pattern generation mode which will make the high-speed functionality measurement feasible [1]. Of course, the nature of this additional logic will depend on the type of circuit to be tested.

This approach has been employed in the testing of GaAs parallel multiplier ICs[5–7]. The fault coverage implications have been discussed in Ref. 8. The combinational network of half and full adders composing the multiplier is converted into an asynchronous oscillator through optional on-chip feedback connections which can be selected from off-chip. When selected, the feedback produces a distinct signature on each of the output pins which can be monitored even through low-speed interfaces. Low-speed interfaces are acceptable because the output pins are insensitive to reflections generated during wafer probe testing. Reflected output signals will not create a logic upset due to noise margin failure unless they are driving another GaAs IC chip input. The repetition frequency of the outputs can be related to the propagation delay through the selected circuit path.

Example 6.1. Design a built-in selectable feedback path that will cause oscillation in an 8× 8 bit parallel multiplier. Specify the inputs (*A* and *B*) that are required to produce the oscillation and relate the frequency of oscillation to the number of propagation delays in the selected path. Assume each full adder produces a two-t_{pd} carry delay and a three-t_{pd} sum delay.

Solution. One solution to this requirement is illustrated in Fig. 6-8 [6]. The gates associated with the built-in test path are cross-hatched, and the feedback lines are wide. If we set $A = 1111\ 1111$ and $B = 1000\ 0000$, and the p_{15} output is inverted and fed back to the b_0 (LSB) input, the circuit will oscillate as shown in Fig. 6-9 because the product is unstable.

The oscillation period is determined by the propagation delay of the signal around the loop. The complete 8 × 8 bit multiplication is formed in a parallel or array multiplier by summing the partial products $(a_i b_j)$ as shown in Fig. 6-10. The partial products (logical AND) are formed in the central block of Fig. 6-8 by "NORing" the inverted

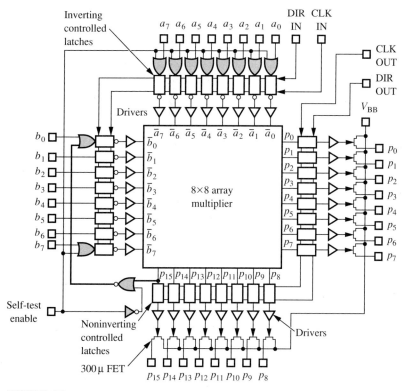

FIGURE 6-8
Block diagram of an 8 × 8 bit parallel multipler with on-chip feedback for built-in testing. (*From Lee et al.* [6]. © *IEEE. Used by permission.*)

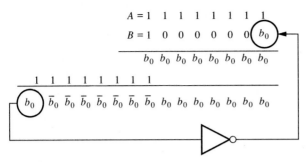

FIGURE 6-9
Diagram illustrating the 8 × 8 bit product (11111111)(1000000b_0) as performed by the multiplier in Fig. 6-8. The inverter represents the built-in, on-chip feedback for self-testing.

input bits. This operation will require one t_{pd}. Figure 6-10 illustrates the multiplication process that occurs inside the central block by binary addition of the partial product terms (as would be done in the more familiar decimal multiplication). The binary addition is performed by half adders (which add two bits with no carry input) and full adders (add two bits plus the carry input); each produces a sum bit and a carry bit. In the feedback path shown, b_0 is the input bit that is toggled by the output p_{15}. The signal path from b_0 to p_{15} will proceed down the eighth column (with $a_7 b_0$ on the top). Each addition requires three t_{pd} for a sum bit and two t_{pd} for the carry bit in the NOR implementation described in Ref. 6. In the diagram in Fig. 6-10, sum outputs are directly under or to the right of the partial product term; carry outputs are shown with left-facing diagonal arrows. Thus, sums are formed vertically, and the carry bits, if any, are added into the column to the left.

From Fig. 6-10, the maximum delay from b_0 to p_{15} can be seen to be six sum delays and eight carry delays for a total of 34 t_{pd} proceeding along the path enclosed by the shaded box. Another one t_{pd} is required for the partial product formation, one t_{pd} for the inversion in the feedback path, and four t_{pd} for the controlled latches (set to pass data through in the test mode); a total delay of 40 t_{pd} for one half-cycle of oscillation is required. The frequency of oscillation will therefore be $1/(80 t_{pd})$. When tested, the multiplier [6] oscillated at 83.1 MHz for a delay per gate of 150 ps.

Unfortunately, the fault coverage of this approach is usually incomplete. Additional low-speed functional testing must also be performed to guarantee error-free operation of the entire circuit. Alternate feedback modes might be provided which would increase the coverage somewhat. When the feedback method was compared with direct delay measurement on a packaged device, agreement was within 10 percent [7].

The second example of modifying multiple-input high-speed circuits to simplify wafer-level testing is illustrated by Fig. 6-11. This technique uses input and output registers to interface from the high-speed on-chip environment to a low-speed off-chip environment for generation of test vectors (inputs) and analysis of test results (outputs). The registers can be activated independently from off-chip by a single strobe pulse. The input register will at that time transfer its input to its output. At some time later determined by the delay line (or perhaps an inverter chain on-chip), the outputs

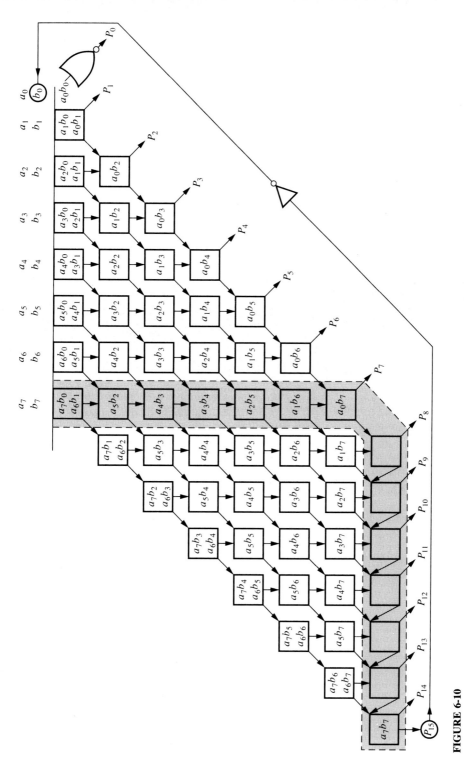

FIGURE 6-10
Parallel multiplication algorithm for an 8 × 8 bit multiplier.

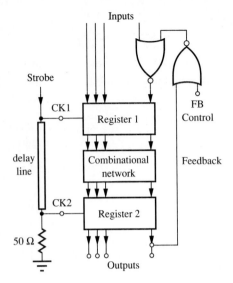

FIGURE 6-11
A representative multiple-input, high-speed digital IC as modified for wafer-level testing. The off-chip strobe pulse and delay line can facilitate measurement of propagation delay time from input to output. Data inputs and outputs can be accessed at low speeds. (*From Long [1]*. © *IEEE. Used by permission.*)

will be latched. This approach allows a low-speed test computer to conduct high-speed functionality testing as well as verifying delay from input to output.

Another variation is possible if the circuit under test can be configured as a finite state machine through a selectable on-chip feedback path. This possibility is also shown in Fig. 6-11. By clocking the registers at high speed for a predetermined number of counts, the correct operation of the chip can be verified from the contents of the output register. This checking of the output can be done with a low-speed test computer. The extent of fault coverage would need to be determined from analysis of the circuit and its response to the test vectors driving it.

6.2.3 Scan-Path Techniques

It is often possible to organize the state-holding elements of a chip so that assertion of a special *test mode* signal reconfigures them into the form of a shift register. Such reconfigurability requires significant modification to the basic flip-flop circuits used throughout a design. Once this has been done, however, the testability of the entire chip is greatly enhanced.

A shift register with access to both input and output ends can provide complete controllability and observability of all state bits. With the ability to control and observe the state of each flip-flop, the sequential testing problem reduces in complexity to that of testing a *combinational* circuit. Combinational circuits offer a much greater opportunity for high coverage[1] yet require a relatively small number of test vectors as

---

[1] *Coverage* is a measure of testability. It is defined as the ratio of detected failures from a given class to the total possible failures in that class.

compared with the number needed to obtain similar coverage from a fully sequential system.

The so-called "scan-path" approach to testability was first proposed by Michael Williams in 1973 [9]. The approach is largely technology-independent and can be applied to any system that is organized as a finite state machine. Figure 6-12 depicts the general scan-path testing schema at a high level.

The circuit modifications needed to reconfigure the state-holding elements into a shift chain can be designed without inserting significant delay into the normal signal path. There are many tradeoffs involved in these flip-flop designs. Figure 6-13 shows the internal logic diagram of one commonly used scan element. For GaAs implementations a flip-flop of similar functionality should be made from NORs (See Prob. 6.5) as NORs are strongly preferred. The flip-flop of Fig. 6-13 is a simple master-slave structure with an additional data input (D1). Each data input has its own enable signal—EN1 for D1, EN2 for D2. The two data inputs essentially provide a two-to-one multiplexer on the flip-flop input, one which has no added delay. The cost of using this type of flip-flop over a simple master-slave lies in area and in fan-out loading. Normal operation uses two alternating clocks on the lines labeled EN1 and EN. Scan-path shifting (assuming an interconnection pattern as shown in Fig. 6-12) is accomplished by alternating clocks on lines EN2 and EN. Alternatively, the slave enable signal (EN) can be derived as the NOR of EN1 and EN2. If this is done, the system is edge-triggered using a periodic signal on EN1 for normal operation and on EN2 during scan-path shifting. When the slave latch enable signal is derived, special

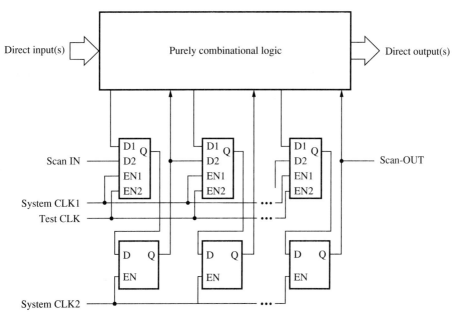

FIGURE 6-12
Overview of scan-path technique using dual-latch design.

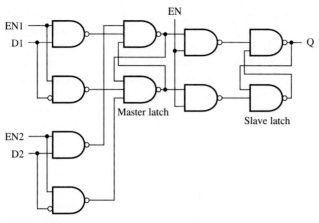

FIGURE 6-13
Scan-path flip-flop internal logic. Data selection is done by asserting the appropriate enable (EN1 or EN2). This action latches the selected data value into the master latch. Asserting EN latches data into the slave portion. For normal use, D1 is selected via EN1. For scan-path shifting, D2 is selcted via EN2. Slave enable (EN) can be derived as the NOR of EN1 and EN2.

care must be taken to ensure correct operation; i.e., the various enabling signals must never be high (or near the logic threshold) simultaneously.

The sequence for test application with such a system is:

1. Verify that all flip-flops along the scan path are functional and that the scan path is correctly established. This test involves shifting a fixed pattern from "Scan-IN" through the on-chip scan path register and observing the delayed pattern at the output "Scan-OUT" (refer to Fig. 6-12).
2. Scan in the next test vector using alternating clocks on the lines labeled "System-CLK2" and "Test-CLK."
3. Apply an appropriate portion of the test vector to the direct circuit inputs (these are signals from off-chip that feed directly into the on-chip combinational logic).
4. Observe the direct circuit outputs. This may require off-chip latches. If off-chip componentry is undesirable, such outputs can instead be latched on-chip and accessed via the scan path.
5. Apply one normal system clock cycle ("System-CLK1" followed by "System-CLK2") to step the state machine into the next state.
6. Scan out the resulting state. Note that this step can be overlapped with step 2. Check the observed state and direct outputs (from step 4) against expected values. Mismatches indicate detected failures.
7. Repeat steps 2 through 6 for all test vectors.

As seen in Fig. 6-12 the overhead for testability involves at least three dedicated I/O pins ("Scan-IN," "Scan-OUT," and "Test-CLK") as well as the additional area required for flip-flop elements due to implementation of the scan path. There are several well-known scan-based methodologies in current commercial use [10–13]. Though the techniques can be used for any state machine-based system implemented in any circuit technology, application of scan testing to GaAs integrated circuits is still a new frontier. As GaAs densities improve, the incorporation of this sort of on-chip testability will become increasingly attractive. For further reading, an excellent presentation of scan-path methods can be found in Ref. 14.

6.2.4 On-Chip Pseudo-Random Testing

If, instead of a shift register, the state-holding elements of the chip are reconfigurable as a linear feedback shift register (LFSR), then a pseudo-random test pattern sequence can be generated and applied on-chip at full circuit speed. The combinational (or sequential) "core" circuits under test respond at normal operational speed, producing an output sequence that is presented to another LFSR. The job of the output LFSR is to compress the circuit response sequence; this compressed result is called the *signature*. At the end of a high-speed test sequence, the signature can either be compared with a fixed on-chip pattern or it can be read out for off-chip comparison. There are many possible variations of LFSR-based on-chip testing. The general setup is depicted in Fig. 6-14.

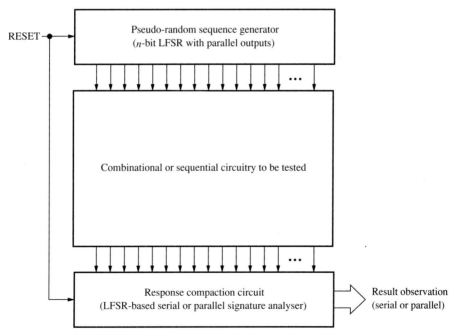

FIGURE 6-14
General structure for pseudo-random on-chip testing.

The two greatest advantages to this form of on-chip testing are:

1. The approach is quite general; no fault model is needed. Both combinational and sequential circuits can be exercised and reasonably high (though unknown) coverage can be obtained.
2. The circuitry under test is operated at full speed. This allows timing problems caused by delay faults and undesired transient conditions to be discovered as well as the easier-to-detect permanent (or dc) faults.

Disadvantages include the fact that coverage is unknown. Because the output response sequence is often compacted by a signature analysis circuit, the effect of the information loss inherent in compression and the circuit and pattern-dependent mapping between faults and changes in output sequencing is that error-detection coverage cannot be precisely computed.

In pseudo-random testing, a long (pseudo-random) input sequence is generated and applied to the circuits under test. Parallel outputs from a maximal-length linear-feedback shift register are suitable for this purpose. The term "pseudo-random" is used because input patterns of the sequence are not truly random. Purely random patterns are not required for testing and would be inconvenient for simulation purposes, making checking of the output sequence extremely difficult. If the pseudo-random pattern is sufficiently long, it has been shown to have excellent randomness properties, as seen by the circuit under test [14].

There are in general many choices of maximal-length LFSRs for a given shift register length n; e.g., for $n = 16$ there are 1024 such LFSRs; for $n = 24$, there are over 100,000. Tables of maximal-length LFSRs and their primitive polynomials (tap positions) can be found in Refs. 15 and 16 and elsewhere.

Figure 6-15 shows two different forms of LFSRs. Both forms are basic shift registers with XOR feedback taps. In the first style (Fig. 6-15a) the feedback taps are XORed to compute the FFO data input. The XOR can be implemented either as a series of cascaded two-input XOR gates or as a single multiple-input XOR, if such a gate can be made in the chosen technology. The second form (Fig. 6-15b) feeds back the output of FFO, tapping it into selected positions within the shift register.

In order to achieve both stimulus generation and output checking on-chip without excessive circuit overhead, an output compression technique such as signature analysis is often used. In this approach, the output sequence is applied (either serially or in parallel) to an LFSR. The serial form is discussed here though parallel signature analysis is also commonly done [17].

An LFSR such as the one shown in Fig. 6-16 effectively performs polynomial division modulo 2. During this division, bits of the quotient are shifted out and lost; the remainder is the only result kept. The remainder (or *residue* in LFSRs of the form shown in Fig. 6-15a) is the signature. The goal of output compression is to be able to discriminate between an erroneous output sequence and a normal output sequence, even though a large number of response patterns are presented to the compression circuit. Clearly, since information is lost in the compression process, the detection can never be perfect. Because the quotient from the division process is discarded,

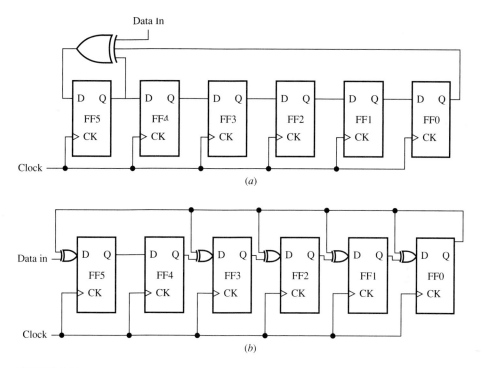

FIGURE 6-15
Two styles of linear feedback shift register: (*a*) LFSR with one, multiple-input XOR; (*b*) LFSR with in-line taps.

sequences that have identical remainders are indistinguishable from one another. This *aliasing* effect makes coverage computations extremely difficult. In fact, no general coverage properties relating circuit faults to error detection have yet been derived. Nevertheless, signature analysis has been found to be effective on many systems since it was first proposed in 1975 [17]. Further discussion of the coverage of LFSR compaction can be found in Ref. 18.

Example 6.2 As an example of output response compression as implemented with an LFSR, consider the circuit of Fig. 6-16. The divisor polynomial for this shift register is

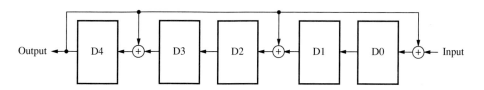

FIGURE 6-16
LFSR to divide by $x^5 + x^4 + x^2 + 1$.

$x^5 + x^4 + x^2 + 1$. The table below illustrates the input, current content, and discarded (quotient) output after each shift step for the division of $x^7 + x^6 + x^5 + x^4 + x^2 + 1$ by $x^5 + x^4 + x^2 + 1$:

				LFSR contents				
Output	←	D_4	D_3	D_2	D_1	D_0	←	Input
–		0	0	0	0	0		11110101
–		0	0	0	0	1		1110101
–		0	0	0	1	1		110101
–		0	0	1	1	1		10101
–		0	1	1	1	1		0101
–		1	1	1	1	0		101
1		0	1	0	0	0		01
10		1	0	0	0	0		1
101		1	0	1	0	0		–
Quotient				Remainder				

The quotient ($x^2 + 1$) can be seen at the output and the remainder (x^2) is found in the LFSR at the end of the sequencing. To illustrate that binary polynomial division has indeed been done, check by performing

$$(\text{divisor} \times \text{quotient}) + \text{remainder}$$

using long-hand multiplication and addition. Note that all additions are XOR (i.e., they are modulo 2 sums, without carry).

```
          110101    (divisor)
    ×        101    (quotient)
          ──────
          110101
    +    1101010    (addition mod 2)
         ───────
         1100001
    +      10100    (remainder)
         ───────
        11110101    (dividend)
```

The following sequence of actions describes an on-chip test scenario using pseudo-random testing:

1. Assert global RESET line to clear and initialize all internal state.
2. Apply appropriate external controls to establish test mode, i.e., to reconfigure the system so that inputs come from the pseudo-random generator LFSR and outputs go to the signature analysis LFSR.
3. Apply a fixed (predetermined) number of system clocks at full operational speed. During this step the circuitry under test is exercised and its outputs are compressed.

4. Check the signature for agreement with predetermined expected signature. This step may be done on-chip or off-chip, depending on pin-out limitations and other considerations.

6.3 ELECTROOPTICAL METHODS

There are many instances where the direct electrical probing of internal nodes in an IC cannot be performed. Either the access to the internal node may be impossible due to layout constraints or the probe itself will load the circuit sufficiently to invalidate the measurement. Recently, a new technique based on the electrooptic properties of the GaAs crystal has been reported [19–21] which allows accurate, noncontacting probing of an IC using a finely focused infrared optical probe beam. This technique will provide access to otherwise inaccessible nodes while at the same time providing bandwidth that is unattainable in direct electrical measurements. The benefits of the technique include:

1. Circuit is not electrically contacted
2. High sensitivity of measurement to sample voltage
3. Picosecond time resolution; wide measurement bandwidth

In the following section, we briefly describe the use of electrooptic sampling for characterization of GaAs IC chips.

Gallium arsenide forms crystals with the cubic zincblende structure, and as a consequence is optically birefringent. An applied electric field changes the index of refraction of the crystal and includes this birefringence which can be described by the intersection of the index ellipsoid and the plane normal to the direction of propagation of the light [22]. The ellipsoid is described by the following equation:

$$\frac{x^2 + y^2 + z^2}{n_0^2} + 2r_{41}(E_x yz + E_y xz + E_z xy) = 1 \qquad (6.4)$$

The variables x, y, and z are defined along the [100], [010], and [001] axes respectively. r_{41} is the only nonzero electrooptic tensor component for this structure. n_0 is the index of refraction for GaAs. Since most GaAs IC wafers are fabricated on wafers with surface orientation (100), the light will be incident in the [100] (x) direction if the surface is illuminated, and the equation is simplified. In the $x = 0$ plane,

$$\frac{y^2 + z^2}{n_0^2} + 2r_{41}E_x yz = 1 \qquad (6.5)$$

The allowed polarizations are along the axes of the ellipse described by Eq. (6.5) and yield indices of refraction that are linearly proportional to the component of the electric field in the x direction, the direction of (longitudinal to) the optical beam. Thus, for the typical coplanar geometry used in GaAs digital ICs shown in Fig.

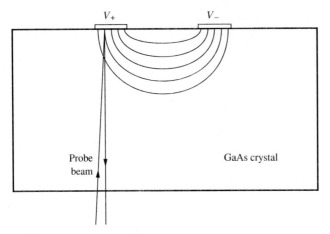

FIGURE 6-17
Sampling geometry for electrooptically probing planar digital GaAs IC. (*From Kolner and Bloom [19].* © *IEEE. Used by permission.*)

6-17, the change in phase induced between the y and z axes depends only on the potential under the line that is being probed, the distant back side being regarded as zero potential. The phase change is detected by polarizing the incident beam and analyzing the transmitted beam such that the light amplitude will respond in a nearly linear manner with the change in electric field. This is the principle used for Pockel's cell light modulators.

The time resolution of the technique comes from the use of very short optical pulses for sampling the electric field. In the papers that describe the method [19–21], these light pulses have been generated by mode-locked YAG and mode-locked semiconductor lasers. Pulse compression techniques are applied to reduce the optical pulsewidth to only 2 ps. Since the GaAs is highly transparent at the wavelengths used and the wafer is back-side illuminated, it is easy to focus the light to a small spot and position it under the line to be probed by viewing the spot through a microscope with an infrared camera at the front surface of the substrate. The optical pulses are usually much faster than the electrical signal being measured. By synchronizing the clock or input signal to the IC chip with the mode-locking generator, and then changing the frequency of the generator to some multiple of the mode-locking frequency plus a small increment, Δf, the electrical signal can be sampled and the waveform viewed on a low-bandwidth oscilloscope. A block diagram of an electrooptic sampling system is shown in Fig. 6-18 [21]. Note that the signal to the circuit under test can also be modulated at some moderate frequency (10 MHz in this system). The modulation appears on the sampling beam and allows the use of synchronous detection in a narrow bandwidth in order to avoid the low-frequency noise generated by the laser.

The voltage sensitivity of the technique has been shown to be very high, 22 $\mu V/\sqrt{Hz}$. The measurement bandwidth exceeds 100 GHz, and its application has been demonstrated on several GaAs digital and analog ICs.

TEST METHODS FOR VERY-HIGH-SPEED DIGITAL ICS 343

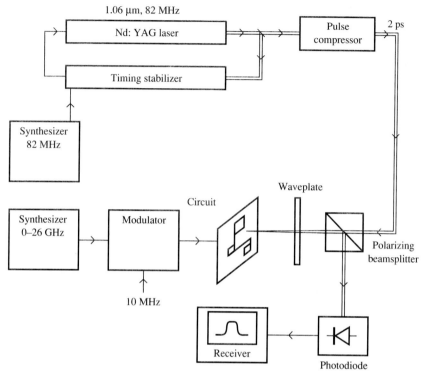

FIGURE 6-18
Block diagram of electrooptic sampling system. (*From Rodwell* et al. [*21*]. *Used by permission.*)

A related technique which could be used on a similar system utilizes a frequency-doubled laser pulse to excite an electrical signal in a photoconductive detector. A split-beam approach can then be used to excite a signal on a circuit and measure its response at some adjustable delay time later. This avoids any problems with timing jitter in the laser pulse generation since both drive and sample pulses come from the same source.

HOMEWORK PROBLEMS

6.1. A standard probe card is modified by cutting traces and adding coaxial cables, a terminating resistor, and a bypass capacitor as shown in Fig. 6-1a. Assume that the probe impedances are 100 Ω, $\epsilon_{\text{eff}} = 1.0$, and the length of each probe is 2 cm.
   (a) If V_G is a unit step generator, plot the voltage at the IC input bond pad as a function of time. Compare this response with that of the circuit in Fig. 6-1b.
   (b) If the dI/dt on the ground resulting from the input step is 10 mA/ns, calculate the amplitude of the voltage spike on ground that will result from the probe.
   (c) If the bypass capacitor is 1 nF, find the resonant frequency of the capacitor and probe.

6.2. Refer to Fig. 6-2. Assume that the two signal lines are shorted to the common ground return. Neglecting any coupled line effects, find the maximum common lead inductance that would result in less than 20 mV (rms) of crosstalk in line 2 if line 1 were driven by a generator at 2 GHz with 50 Ω internal source resistance and with 1 V (rms) open-circuit voltage amplitude.

6.3. A dual Johnson counter is constructed with four delay gates on the reference counter and 14 gates on the delayed counter. If $t_{pd} = 100$ ps, what is the maximum frequency of operation possible for each counter? If the clock waveform were made symmetric (50 percent duty cycle), find the maximum clock frequencies for each counter.

6.4. Perform a timing analysis on the D-type flip-flop of Fig. 6-5. Connect the Q bar output to the D input so that it is "T-connected" and divides by 2, and assume equal delay (t_{pd}) for each gate. Show that the maximum clock frequency will occur with a 3/2 clock asymmetry as discussed in the text.

6.5. Design a scan-path flip-flop similar to the one of Fig. 6-13 using NORs rather than NANDs.

6.6. Using the LFSR-based signature analyzer shown in Fig. 6-16:
   (a) Find two input sequences that produce identical signatures (remainders).
   (b) Propose a single-output combinational circuit, a test sequence, and a fault (or set of faults) that produces an undetectable error.

REFERENCES

1. Long, S. I: "Test Structures for Propagation Delay Measurements on High-Speed Integrated Circuits," *IEEE Trans. Elect. Dev.*, vol. ED-31, pp. 1072–1076, August 1984.
2. Gleason, K. R., Reeder, T. M., and Strid, E. W.: "Precise MMIC Parameters Yielded by 18-GHz Wafer Probe," *Microwave Systems News*, vol. 13, pp. 55–65, May 1983.
3. Strid, E. W., Gleason, K. R., and Reeder, T. M.: "On-Wafer Measurement of Gigahertz Integrated Circuits," in *VLSI Electronics: Microstructure Science*, vol. 11, chap. 7, pp. 266–287, Academic Press, New York, 1985.
4. Buehler, M. G.: "Microelectronic Test Chips for VLSI Electronics," *VLSI Electronics: Microstructure Science*, vol. 6, pp. 529–576, Academic Press, New York, 1983.
5. Long, S. I., Lee, F. S., Zucca, R., Welch, B. M., and Eden, R. C.: "MSI High-Speed Low-Power GaAs IC's using Schottky Diode FET Logic," *IEEE Trans. Microwave Theory and Tech.*, vol. MTT-28, pp. 466–471, May 1980.
6. Lee, F. S., et al.: "A High-Speed LSI GaAs 8×8 Parallel Multiplier," *IEEE J. Solid State Cir.*, vol. SC-17, pp. 1110–1115, August 1982.
7. Nakayama, Y., Suyama, K., Shimizu, H., Yokoyama, N., and Shibatomi, A.: "A GaAs 16 ×16b Parallel Multiplier Using Self-Alignment Technology," *1983 ISSCC Tech. Dig.*, pp. 48–49, February 1983.
8. Buehler, M. G., and Sievers, M. W.: "Off-line, Built-in Test Techniques for VLSI Circuits," *Computer*, vol. 15, pp. 69–82, June 1982.
9. Williams, M. J. Y., and Angel, J. B.: "Enhancing Testability of Large Scale Integrated Circuits via Test Points and Additional Logic," *IEEE Trans. Computers*, vol. C-22(1), pp. 46–60, January 1973.
10. Eichelburger, E. B., and Williams, T. W.: "A Logic Design Structure for LSI Testability," *Proceedings of the 14th IEEE Design Automation Conference*, pp. 462–468, June 1977.
11. DasGupta, S., Walther, R. G., and Williams, T. W.: "An Enhancement to LSSD and Some Applications of LSSD in Reliability, Availability, and Serviceability," *Proceedings of the 11th IEEE Symposium on Fault-Tolerant Computing*, pp. 32–34, June 1981.
12. Stewart, J. H.: "Future Testing of Large LSI Circuit Cards," *Proceedings of IEEE Semiconductor Test Conference*, pp. 6–15, October 1977.
13. Stewart, J. H.: "Application of Scan/Set for Error Detection and Diagnostics," *Proceedings of IEEE Semiconductor Test Conference*, pp. 152–158, October 1978.
14. McCluskey, E. J.: *Logic Design Principles*, Prentice Hall, Englewood Cliffs, N.J., 1986.
15. Golumb, S. W.: *Shift Register Sequences*, Aegean Park Press, Laguna Hills, Calif., 1982.

16. Peterson, W. W., and Weldon, E. J.: *Error-Correcting Codes*, 2d ed., Colonial Press, 1972.
17. Benowitz, N., Calhoun, D. F., Alderson, G. E., Bauer, J. E., and Joeckel, C. T.: "An Advanced Fault Isolation System for Digital Logic," *IEEE Trans. Computers*, vol. C-24(5), pp. 489–497, 1975.
18. Smith, J. E.: "Measure of the Effectiveness of Fault Signature Analysis," *IEEE Trans. Computers*, vol. C-29(6), pp. 510–514, 1980.
19. Kolner, B. H., and Bloom, D. M.: "Electrooptic Sampling in GaAs Integrated Circuits," *IEEE J. Quantum Electronics*, vol. QE-22, pp. 79–93, January 1986.
20. Freeman, J. L., Diamond, S. K., Fong, H., and Bloom, D. M.: "Electro-optic Sampling of Planar Digital GaAs Integrated Circuits," *Appl. Phys. Lett.*, vol. 47(10), pp. 1083–1084, 15 Nov. 1985.
21. Rodwell, M. J. W., Weingarten, K. J., Freeman, J. L., and Bloom, D. M.: "Gate Propagation Delay and Logic Timing of GaAs Integrated Circuits Measured by Electro-optic Sampling," *Elect. Lett*, vol. 22(9), pp. 499–501, 24 Apr. 1986.
22. Kaminow, I. P.: *An Introduction to Electrooptic Devices*, Academic Press, New York, 1974.

CHAPTER 7

USING GaAs IN SYSTEMS

It is apparent from the material presented in previous chapters that gallium arsenide devices and the circuits made from them can be extremely fast. The challenge is to harness that speed and couple it productively to an application area. Since it is unlikely that entire systems will be built from GaAs componentry, one must also address the myriad issues of embedding GaAs circuits into systems with other logic. This chapter discusses the architectural concepts pertaining to very-high-speed systems covering both on-chip and off-chip issues such as subsystem interconnection methods, transaction flows, clocking, and synchronization. Many architectural techniques found in MOS VLSI are also successful in GaAs designs. The circuits and on-chip design building blocks are the subject of Chap. 8. Here the unique architectural aspects of very-high-speed GaAs systems as well as the use of GaAs circuits with other logic families is discussed. Interface circuits and chip-to-chip I/O issues are detailed.

It is important to understand that there are no absolutes, no hard and fast rules, and few existing systems available for case study in GaAs computer architecture. The material presented herein is based on modern high-performance computer architecture as adapted to fit the special requirements of gallium arsenide circuitry. Attention is given to high-speed *computer* architecture as a vehicle to describe how to use GaAs in systems. Many of the initial applications of GaAs were not in computer systems, but today's modern computers, with a need for very high levels of performance, must consider GaAs componentry.

7.1 ARCHITECTURAL CONSIDERATIONS

This section focuses on *computer* systems, though much of the philosophy and most of the resulting guidelines pertain to a much broader class of digital systems.

There are three principal methods for increasing the speed of computation in a digital computer. These methods all attempt to maximize performance *(P)* by attacking different factors of the very-high-level model given by

$$P = \frac{\text{computations}}{\text{second}} = \frac{\text{computations}}{\text{instruction}} \times \frac{\text{instructions}}{\text{cycle}} \times \frac{\text{cycles}}{\text{second}} \quad (7.1)$$

The three speed-up methods are:

1. Improved circuit technology (higher clock rate, increased *cycles/second*)
2. Simplified instruction set (trades decreased *computations/instruction* for lower *cycles/instruction*)
3. Parallelism (use additional hardware to increase *instructions/cycle*)

IMPROVED CIRCUIT TECHNOLOGY. The first method utilizes faster circuit technology in order to enable the use of higher system clock rates. Since GaAs circuits comprise the fastest digital technology commercially available, the use of GaAs clearly falls within this performance improvement category. Strictly speaking, this approach provides a direct, linear speed-up with no changes in architecture—an n-fold clock speed-up produces an n-fold performance improvement. In practice, however, changes in architecture are typically necessary in order to enable use of the highest possible clock rates. For example, in order to scale up the speed of an entire system the effects of a dramatically higher clock rate are felt in nearly every part of the system. Clearly one cannot speed up the CPU alone without also matching the memory timing. Without proportional speed-ups, one portion of the system would become faster than other parts yet would still be required to exchange data and synchronize with the slower parts. When the faster part is a CPU and the slower part is a memory, for example, wait states (idle CPU cycles) are necessary and much of the potential performance improvement due to the faster CPU is lost. This important speed disparity is discussed in detail and suggestions for dealing with it are offered; refer to Sec. 7.1.9.

SIMPLIFIED INSTRUCTION SET. The second method for increasing computer performance is to streamline the instruction set, changing the number of cycles per instruction and/or the instruction granularity so as to accomplish a greater number of useful computations per unit time. Instruction set streamlining is largely technology-independent and can often be used to complement other speed-up approaches. The popularity of reduced instruction set computers (RISCs) is testimony to the success of this technique. RISCs are based on a design style that attempts to enhance useful throughput by decreasing the number of cycles per instruction—ideally all the way to a single cycle per instruction. This is often done in concert with pipelining and with an increase in the system clock rate—not due to use of faster circuit technology but due primarily to strict adherence to a design philosophy that limits the amount of processing per clock. This style moves optimization and sometimes even correctness-preserving interlock checking out of the hardware and into the compiler. Programs tend to be larger in code size but execution speed can be significantly better since clock rates are high and most RISC systems issue one instruction per clock. Several successful RISC implementations exist. (cf. [1–3]).

PARALLELISM. The third performance improvement method—parallelism—attempts to use additional hardware operating in parallel in order to increase the number of instructions executed per cycle. Architectures for parallel and distributed computing are the subject of intense research. Since improvements incorporating parallelism are largely *architectural*, rather than technological, they can be implemented in any of several circuit technologies, including GaAs. It is typical to find this method used together with the first speed-up method—improved circuit technology.

In this chapter the focus is on the use of GaAs in systems. It is evident that the first speed-up method will be the primary entreé for GaAs but appropriate adaptation and awareness of *all* available high-performance techniques should occur.

The early use of GaAs in digital systems has been for implementation of critical sections of systems where a significant speed-up had a direct impact on the entire system, e.g.:

1. GHz sample/hold
2. GHz D/A converter
3. Very-high-speed multipliers
4. Telecommunications multiplexer/demultiplexers
5. Bit serial processing elements

The next step in the introduction of digital GaAs was the implementation of ECL- or TTL-compatible chip sets for processors or controllers (e.g., bit-sliced processors such as the AMD 2900 [4]). Also, static RAMs with sufficient bandwidth to keep up with high clock rate processors were developed. These RAMs have also found use as high-speed caches and register files in very high clock rate systems.

In the future not only will the level of integration continue to increase but the effects of *optical* communications will be seen. Because of its direct bandgap, its inherent circuit speed, and the level of integration possible, GaAs is the material of choice for integrating optics and electronics.

7.1.1 Architectural Fundamentals

The classical Von Neumann computer consists of the five functional unit types as shown in Fig. 7-1. In the figure, solid lines represent data channels while dashed lines show controls. Though this model has been known for over 40 years and many improvements and nuances have been added, it still describes in essence most modern-day computers. The units shown in Fig. 7-1 exist in all computer systems and their

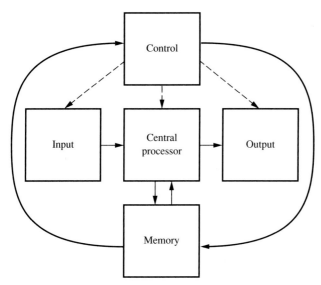

FIGURE 7-1
Components of the classical Von Neumann computer.

implementation and interconnection strongly influences the computing capability of the implemented machine.

There are many ways to interconnect the functional units of a digital system. For the most part data interconnections are *parallel*, transferring a full word of data at one time. System throughput is normally maximized when parallel interconnect is used though there are a large number of interconnecting lines in such systems. Throughput may usually be traded in exchange for a decreased number of interconnections. The design space of this tradeoff is quite large, involving virtually all integral numbers of lines from the purely serial single-line case up through a fully parallel organization.

7.1.2 Architectural Problems Unique to GaAs Digital Systems

The architectures of most contemporary computers and digital systems are influenced more by function and application than by implementation. If this approach is taken with a gallium arsenide system, the result will quite probably be a less than optimal design because GaAs digital systems can suffer from several technology-related problems. These problems are only partially solvable during the implementation phase of a project. Therefore, if a design is to be efficiently implemented in GaAs, these technology-related issues must be taken into consideration during the *architectural* design phase of the system.

Virtually all of the problems and complexities associated with using GaAs in a system stem from the extremely fast signal transitions that propagate between chips. Rise times on the order of 100 to 150 ps are not unusual. When signal path propagation times are of the same order as the rise time of the signals that are traveling on the paths, transmission line theory applies and it is necessary to consider the use of impedance-controlled structures for interconnections. The implications of this are many and the solutions require significant effort in the architecture and planning phases of system design. A few of the most important problems due to the use of transmission lines and matched-impedance structures are listed below:

1. *Relatively low density packaging.* Impedance-controlled structures require more space than equivalent noncontrolled structures. As an example, consider the space occupied by two unrelated stripline connections and compare that to the space occupied by two unrelated ordinary printed circuit traces. Coaxial cables and connectors are also large compared with ribbon cable, back-plane connectors, and conventional printed circuit board connectors.
2. *Signal attenuation.* The loss factors of common printed circuit or hybrid packaging materials (e.g., glass-epoxy, polyimide) are rather high as compared with free air. At MOS or TTL speeds the attenuation over short distances is negligible. At GaAs signaling speeds or over increased distances the story is different. Signal attenuation at high frequencies causes pulse dispersion and begins to degrade noise margins. Design and analysis techniques for dealing with signal attenuation (both skin effect and dielectric losses) are covered in Chap. 5.
3. *Crosstalk.* Whenever signal lines carrying gigahertz frequencies must be routed in parallel for even a few centimeters, crosstalk (undesired coupling between lines) must be considered. This effect is proportional to the length of interconnection lines and to signal speeds as well as other parameters. Crosstalk is inversely proportional to the distance between two traces. A detailed treatment of crosstalk and its analysis can be found in Sec. 5.5.
4. *Signal propagation delay.* Signals traveling in transmission line structures typically have a propagation time somewhat greater than 1 ns/ft. For a system running a 1-GHz clock, such a signal would take a substantial percentage of the entire clock period merely traversing even a 4 in. printed circuit card. Similar problems occur for signals traveling in coaxial cabling.

The message here is that the extremely fast rise/fall times found in GaAs circuits necessitate the use of impedance-controlled structures off-chip. Such structures are large, however, so their number and length must be minimized in order to obtain maximum performance from a GaAs system. The best approach to the problem of the relatively low density of high-speed interconnect is to design an architecture that *minimizes the amount of high-speed interconnect.* It will not be possible to eliminate all of the high-speed clocks and control signals, but structures such as off-chip high-speed busses should definitely be avoided. Section 7.1.4 discusses high-speed busses in more detail and includes some alternatives to this design style.

Another technology-related problem affecting implementation of a GaAs computer is the relatively low density of GaAs ICs as compared to MOS. Large-scale integration density is now becoming practical for GaAs logic but VLSI has some serious obstacles yet to overcome (e.g., fabrication yield, accurate control of threshold variation, power dissipation). The requirement that all GaAs FET gates be aligned in the same direction is a particularly severe one affecting the density ultimately achievable in circuit layouts. All MESFET gates must be oriented in the same direction because GaAs is a piezoelectric material. The strain caused by the differential expansion of the gate/GaAs interface and the GaAs/dielectric interface can give rise to a threshold shift [5]. The amount of shift depends on the crystal orientation since GaAs is anisotropic. Problems like these work against a GaAs digital system because more chips are required to implement a specified function, and more chips means more interconnect, which is also undesirable. Advanced GaAs processes are offering two orientations for MESFETs, one preferred and a second perpendicular orientation allowed. As such processes mature it is natural to expect a noticeable improvement in GaAs layout density.

A layout density comparison between CMOS and depletion mode GaAs is shown in Fig. 7-2. The circuit implemented in each case is an edge-triggered D flip-flop made from AND/OR/Invert (AOI) and OR/AND/Invert (OAI) gates. The comparison favors 2-μm CMOS by 2.5 to 1 ($\approx$ 6000 μm^2 vs. 14,600 μm^2), and 1-μm CMOS by nearly 10 to 1 ($\approx$ 1500 μm^2 vs. 14,600 μm^2). The lower density of GaAs is primarily due to the presence of level-shift networks, an extra supply rail, less dense contacts, and gate sizing asymmetries. In this particular comparison (because of the CMOS layout selected), gate *orientation* restrictions do not affect the results. There are many

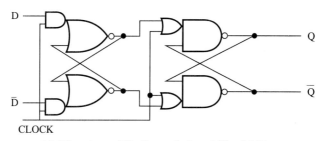

(*a*) Edge-triggered flip-flop made from AOI and OAI gates

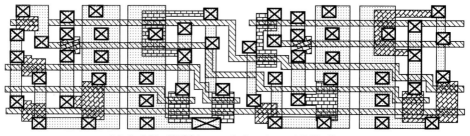

(*a*) Scalable CMOS layout of edge-triggered D flip-flop

FIGURE 7-2
Layout density comparison between scalable CMOS and GaAs: (*a*) Edge-triggered flip-flop made from AOI and OAI gates; (*b*) Scalable CMOS layout of edge-triggered D flip-flop.

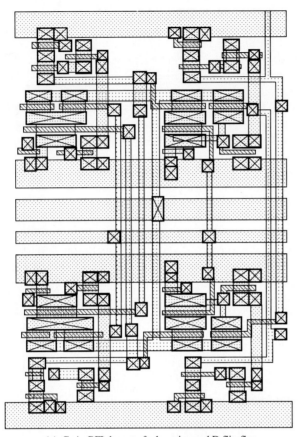

(c) GaAs BFL layout of edge-triggered D flip-flop

FIGURE 7-2 (*continued*)
(c) GaAs BFL layout of edge-triggered D flip-flop.

layouts, however, where optimal area can only be achieved by taking advantage of the ability to orient gates both horizontally and vertically. Since such nonuniform gate orientation is illegal in GaAs, layout densities will ultimately be suboptimal when compared with MOS. Additional area is required for ohmic contacts in GaAs. The need to maintain low source resistance on high g_m devices does not permit nonlocal contacts—ones that are tens of micrometers away from devices—as are typical in CMOS.

One comprehensive approach to solving all of the above-mentioned problems is to design an architecture that is *modular*. Ideally, modules should be designed such that all of the high-speed components that need to communicate with each other are in the same module. The use of this technique will minimize the high-speed interconnect between modules. The implementation goal will be to get all parts of a module on a single chip. Each module that cannot be realized within a single chip can be packaged as a hybrid semiconductor module, e.g., like the one shown in Fig. 7-3. Using hybrids can minimize the length and maximize the density of the high speed interconnect between components. There is also an area benefit that comes from elimination

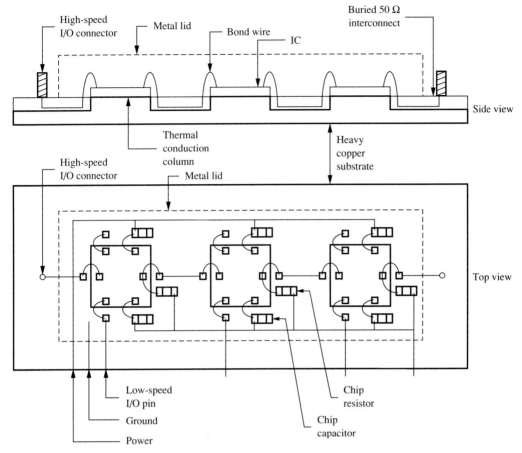

FIGURE 7-3
Hybrid semiconductor module. (*From* Fouts et al. [16]. *Used with permission.*)

of extra packages since smaller driver circuits can be used. The hybrid semiconductor modules contain all of the chips, the high-speed interconnect, and termination for the interconnect. The hybrid packaging also contains provisions for distributing power and ground, bypassing the power supplies, and dissipating heat.

A form of hybrid packaging is used in the CRAY-3 implementation [7] (refer to Fig. 7-4). This packaging concentrates up to 1024 GaAs die per module while allowing short interconnection paths within the module. Cooling is accomplished by immersion in a liquid coolant—fluorinert. The physical principles affecting packaging structures and techniques are discussed in Chap. 5.

Even with a carefully designed architecture and hybrid semiconductor modules, it may still be necessary to have some high-speed signals travel significant distances, perhaps as much as 6 in. or more. Such is often the case for clock and control signals. This problem can be overcome by using asynchronous communication techniques or by using carefully engineered synchronous communications.

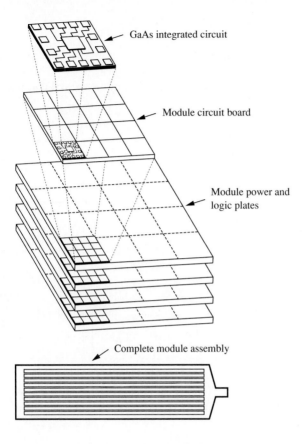

FIGURE 7-4
CRAY-3 computer plug-in module assembly containing module power/logic plates and multilayer circuit boards housing 1024 integrated circuit die. (*From Kiefer and Heightley* [7]. *Used with permission.*)

7.1.3 Synchronous Design

A system is said to be *synchronous* if all signals change at very nearly the same time—all edges being derived from a single widely distributed clock. Such systems are very common because they are easy to understand, to design, to analyze, and to debug. Since all events happen only at clock edges, the main clock period directly determines system speed.

For GaAs, a synchronous design style poses some unique problems.

1. *Clock distribution*. Since the synchronizing clock signal affects nearly all parts of the system, it has an unusually large fan-out. Distributing high-frequency, high-fan-out clock signals to many different packages and subsystems is a nontrivial design and implementation problem.

2. *Clock skew*. The difference in time of arrival of fanned-out clock signals within a synchronized system is known as *clock skew*. In a TTL or CMOS system the interconnect, buffering, and stray packaging delays make up only a small portion of the clock cycle. Five nanoseconds of packaging and stray wiring delay, for example, would add only a 10 percent skew in a 20 MHz TTL or MOS system. In a 1-GHz GaAs system, on the other hand, the entire clock period is 1 ns. Stray

wiring capacitance and other packaging delays must be understood and carefully controlled at GHz speeds.
3. *Power distribution.* Since current transients on power and ground busses are also synchronized with clock edges, inductive voltage spikes can be severe. This concern introduces additional constraints on layout and density in an attempt to reduce inductance.

The approach to solution of these problems must be undertaken early in the design cycle. Two approaches are offered here.

BALANCED DISTRIBUTION. If synchronous communication techniques are to be used, then special steps are required to maintain synchronization. A set of a clock distribution guidelines follows. These "strawman" rules are based on an attempt to carefully balance the loading of clock signals.

1. All synchronized signals should originate from a single board, which is centrally located.
2. The cables from the sending board to the receiving boards should be made as nearly as possible the same length—the length of the longest required cable.
3. Loading of the high-speed signals should be kept balanced. If possible, the loading should be exactly the same for every signal.

In addition to these guidelines, a method of adjustment may still be required. A fan-out and phasing unit may be a necessary addition to the clocking scheme. This approach inserts a programmable delay with high-fan-out drivers in each clock or high-speed control signal path. Figure 7-5 shows a schematic for such a design.

When a synchronous signal is received on a board or hybrid module, it is routed through the fan-out and phasing unit. The phasing unit can be programmed to select an experimentally determined amount of delay such that the received signal is kept in phase with the rest of the signals in the system. Since all fan-out drivers within a fan-out and phasing unit are fabbed on the same die, the skew problems due to differing driver characteristics are minimized. With well-balanced and matched loads, this technique can be used successfully with clock rates above 1 GHz.

DIVIDE AND CONQUER. This approach to synchronous design uses multiple synchronous subsystems—each having its own private clock—globally connected via first-in-first-out (FIFO) queues. This technique allows the best features of synchronous design to be applied where they are most effective—in relatively small subsystems. In this approach, the small subsystems interact in a loosely coupled manner. First-come-first-served queues maintain the sequence of transactions between subsystems without requiring lock step or so-called "hard synchronization." No global clock signals are required. The system is more complex than in the case of the balanced distribution setup because it requires queues, careful initialization, and flow control (*start/stop* control signals in response to *empty/full* indications from the FIFOs).

Tighter coupling can be used by removing the FIFO queues and incorporating an

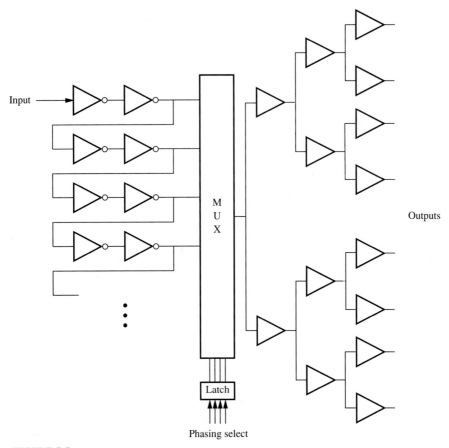

FIGURE 7-5
Phasing and fan-out unit. (*From Fouts* et al. [16]. *Used with permission.*)

asynchronous handshake between communicating subsystems. Asynchronous signaling is discussed in a comprehensive framework in Sec. 7.3. The advantage of using a divide and conquer style, whether implemented in a loosely coupled fashion with FIFOs or in a more rigid manner using an asynchronous handshake, is that one does not need to worry about issues of skew of clock, control, or data signals except *within* a given subsystem. By this technique, global timing problems can be converted into more manageable subsystem-local issues.

7.1.4 Bus-Type Connections

Because of its efficient use of package pins as well as its generality and modularity, bus-style interconnect has long been the preferred organization for computer systems. The busses within a system are the backbone of the computer, playing a central role in many architectural decisions during the design of the machine. Figure 7-6 shows a typical single-bus system. Consider the events occurring as block "A" sends

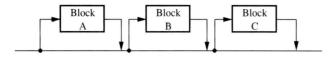

FIGURE 7-6
A single-bus system.

a value to block "C" which subsequently responds by an asynchronous "handshake." Futhermore, consider the time involved in such a transaction and remember that the bus is a *shared* resource—no other bus-connected units can communicate while the bus is in use. The typical sequence of events involved in an asynchronous handshake is given in Table 7.1.

TABLE 7.1
Return to zero signaling—four-cycle bus protocol

Block A (master and originator)	Block C (slave)
Request bus mastership, waiting if necessary for a previous transaction to complete or for a higher priority module to finish bus usage	
Prepare data value, enable bus-drivers	
Drivers charge/discharge the bus lines according to the data values	
After waiting an appropriate length of time, block A asserts the "data request" line to signal data availability	Receive and latch the data on leading edge of "data request"
	Acknowledge receipt of the data by asserting a bussed "data acknowledge" control line
Deactivate the data bus drivers upon leading edge of "data acknowledge"	
Deassert the "data request" line	
	Deassert the "data acknowledge" line
Relinquish bus mastership	

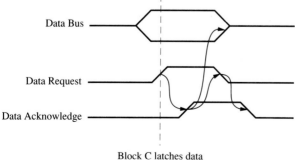

Block C latches data

The protocol presented in Table 7.1 is known as *four-cycle signaling*. In this technique the handshake between two communicating systems is explicit. Formally, the full handshake involves two control lines: *request* and *acknowledge*. In Table 7.1 these signals are referred to as *data request* and *data acknowledge*. The protocol is termed *return to zero* because each of the two handshake lines is asserted, seen by the other party, then de-asserted, i.e., returned to zero. The term *four-cycle signaling* arises because there are four distinct subcycles to the protocol, each involving an edge generated by one party and observed by the opposite party. In addition to the time used for the data-exchange handshake, time is also required to request bus mastership and to arbitrate such requests.

As an alternative, a faster (but more complex to implement) protocol known as *two-cycle signaling* can be used. In this protocol both sides of the conversation must keep track of the previous logical state of each control line and only *edges* on that line are used to make a handshake. In one transaction a leading edge on *request* will be acknowledged by a leading edge on the *acknowledge* line. The next transaction will start with a trailing edge on the *request* line which will be acknowledged by a trailing edge on the *acknowledge* line. Figure 7-7 illustrates this technique. Two-cycle signaling is clearly a more efficient protocol than four-cycle signaling since it involves half as many signal transitions — in fact it has been shown to be optimal from an energy point of view.

Both four-cycle and two-cycle protocols are *asynchronous* techniques for implementing intermodule communication via busses. These protocols are often used in systems where flexibility and modularity are most important, e.g., in general-purpose computers where interfaces and subsystems from a variety of vendors must work together effectively. In specialized systems or very-high-speed systems it is more common to find *synchronous* bus structures or no bus structure (point-to-point communications).

Figure 7-8 depicts a dual-bus system. Such a system has more complexity than a single-bus system but nearly twice the bandwidth. Multiple-bus systems involving two, three, or more busses are possible. Such organizations are more typical on-chip than off and generally use synchronous rather than asynchronous signaling.

In a synchronous bus system, all signaling on the shared bus lines is coordinated by a global clock which is distributed as one of the bus lines. All bus transactions are preplanned on a cycle-by-cycle basis by assigning each transaction a coded *type* and

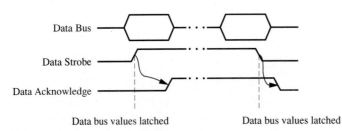

FIGURE 7-7
Non-return to zero signaling (NRTZ) — two-cycle bus protocol (two complete handshake transactions shown).

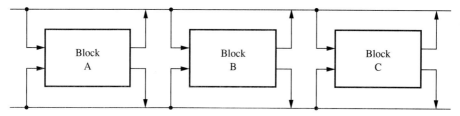

FIGURE 7-8
A dual-bus system.

by subdividing each into a small number of subphases. Any given bus transaction is identified by placing its coded type and its subphase signals on special bus control lines. Each participant in a transaction decodes the transaction type signals to ensure maintenance of the preplanned protocol throughout every phase of the transaction.

Because of the tight synchronization requirements and the extra controls and complexity, synchronous busses are most often found *within* a single chip or module rather than spread across chip and/or module boundaries. If signaling speed is to be in the gigahertz range then this type of bus will *only* be found on-chip.

There are several reasons why high-performance systems are moving away from bus-based organizations. Because GaAs components will most often be found in the portions of a system operating at the highest speed, this trend away from bus-type interconnect is particularly prevalent in GaAs subsystems. Among the most compelling reasons for this trend are the following:

1. *High clock rates.* Busses are inherently *shared* channels. As system clock rates climb, the ability of a bus to serve all connected modules without itself becoming a bottleneck diminishes.
2. *Parallelism.* Modern digital systems are performing an increasing number of operations in parallel. More simultaneously active hardware means more data movement per clock. Moving operands to and from data transformation circuitry is fastest with point-to-point connections rather than via bus-type shared interconnect.
3. *Transmission line effects.* Reflections and crosstalk problems get worse as clock frequencies climb. Controlled impedance structures are most often required in order to remedy such problems. Bidirectional taps, such as are used with bus-type interconnect, cause reflections because they present an impedance discontinuity on the transmission line. The only structure that avoids impedance discontinuities completely is point-to-point—one source driving one receiver with a matched termination at the receiver (refer to Chap. 5 for a discussion of transmission line theory and practice).
4. *Pipelining.* High clock rate pipelines are efficiently implemented in GaAs. Not only are bus structures incapable of GHz transfer rates but the full generality offered by bus structures is often not needed anyway. Pipelines have a few well-defined entry/exit ports and extremely predictable behavior. If the flexibilities of modularity and random access provided by bus structures are not required, then interconnection networks are often a more appropriate choice.

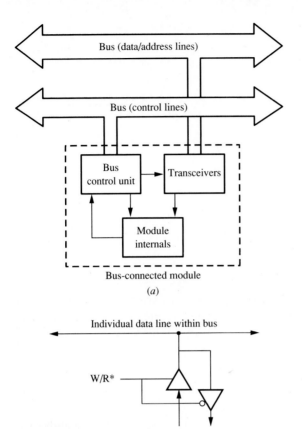

FIGURE 7-9
A typical bus-connected module.
(*a*) Block diagram. (*b*) Transceiver detail.

The packaging and cabling/interconnect methods have a strong influence on the organization of GaAs systems. The requirement of controlled-impedance circuits for chip-to-chip interconnections along with the rather high sensitivity to fan-out makes bus-type organizations ill suited for very-high-speed technologies. Figure 7-9 illustrates bus interconnections and their transformation to multiplexer-style structures. The bus control unit logic remains the same across this transformation since only the tristate driver enable signals (**W/R***) are of interest. In this transformation the bus lines each form a distributed OR (or NOR) gate. If several on-chip modules are positioned sufficiently close to one another, the bus "line" can be implemented on the input side of one gate. The advantage in such a case is that the gate is never tristated—it never switches the output drive on/off—and the bus is driven all of the time by a single set of devices. This brings down the amount of loading on such lines and enables higher overall rates for synchronizing clocks. This bus-to-gate mapping technique attempts to keep the flexibility of busses but without the use of tristate drivers and slow bidirectional lines. If the capabilities of a bus are required in a system, this mapping can be used to get a gate-logic equivalence for GaAs implementation.

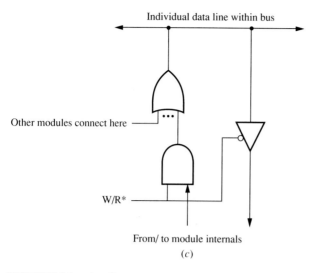

FIGURE 7-9 (*continued*)
(*c*) Equivalent transceiver (bus never tristated, single distributed-OR driver).

7.1.5 Interconnection Networks

When a bus structure is used to interconnect the subsystems of a computer, transfers are possible between all module pairs; i.e., each unit can (at least in principle) initiate as well as receive a transfer to/from any other unit on the bus. In practice, many of the available transfers are never required. In fact, units are often simplified by specializing them to support only the type(s) of transfer that are required. As an example, consider the single-bus system shown in Fig. 7-10a. In the figure there are four processors (P_0, P_1, P_2, P_3) and four memory modules (M_0, M_1, M_2, M_3). The dominant traffic in this system flows between processors and memories. It may be the case in some systems that processors *never* communicate directly with other processors and memories never with other memories. The price for the excessive generality of a bus-type interconnection is that each bussed line must have high-performance *bidirectional* driver and receiver circuits. Bidirectional circuits can never perform as well as specialized unidirectional circuits because when part of the circuit is inactive (e.g., consider the driver transistors which are not active when the line is in "input" mode) it adds considerable unwanted stray capacitance and also presents an impedance discontinuity. The capacitance loads the devices driving the line and the discontinuity in impedance causes reflections.

There are significant differences in maximum throughput between bus-organized systems and ones that use interconnection networks. To illustrate these differences, consider first the bus network of Fig. 7-10a where the basic "clock" rate c is determined not by a synchronizing clock signal but by the intrinsic round-trip delay t of the bus system. Conventional wisdom says to operate the bus no faster than one "event" per two times the round-trip delay from end to end of the bus, that is, $c \leq 1/(2t)$.

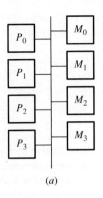

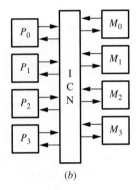

FIGURE 7-10
Two interconnection systems. (a) Bus-style interconnection. (b) 16-port interconnection network.

With this model, the maximum throughput (T_{BUS}) using four-cycle signaling is given by

$$T_{\text{BUS}} = \frac{1}{2t} \frac{\text{events}}{\text{second}} \times \frac{1}{4} \frac{\text{transfer}}{\text{events}} = \frac{1}{8t} \frac{\text{transfers}}{\text{second}} \quad (7.2)$$

Contrasting the best throughput of an n-module bus with that of a $2n$-port interconnection network (ICN) illustrates the reason why high-performance systems are moving increasingly toward ICNs. The system of four processors and four memories shown in Fig. 7-10b requires a 16-port ICN so that each processor and each memory port becomes unidirectional. In order to present a fair comparison with the bus system of Fig. 7-10a the assumption of an equal clock rate is made, though in typical situations much higher ICN clock rates are possible. This is so because the considerable capacitances associated with bus lines are not present in ICN designs. In the best case with no blocking, all processors could be simultaneously transferring data to or from all memories. Thus, the maximum throughput (T_{ICN}) is given by

$$T_{\text{ICN}} = \frac{1}{2t} \frac{\text{clocks}}{\text{second}} \times 1 \frac{\text{transfer}}{\text{clock}} \times 8 \text{ port pairs} = \frac{4}{t} \frac{\text{transfers}}{\text{second}} \quad (7.3)$$

The comparison of best-case throughputs between a bus and an interconnection network for the given example shows 32 times better performance in favor of the ICN. More careful analysis requires details of ICN substructure as well as typical address distributions. In virtually all but specially chosen circumstances, the ICN will outperform the bus system, especially when higher ICN clock rates are used. Achievable bus clock rates are limited by global properties such as bus size and capacitive loading.

Equations (7.2) and (7.3) describe at a high level the maximum throughput attainable with bus and ICN interconnections respectively. In order to fairly evaluate the two interconnection styles it is important to consider both the worst case and the average case. The worst case for a bus system occurs when a module wishes to send data but cannot gain bus mastership due to the use of the bus by other higher-priority modules. In a properly designed system the *arbitration delay* is a transient condition and thus should not influence throughput computations. Worst-case throughput ($T_{\text{BUS-worst}}$) occurs for a CPU module of Fig. 7-10a when the other three CPUs compete simultaneously for bus mastership. Assuming a fair arbitration policy (such as round-robin) and ignoring arbitration delay, the $T_{\text{BUS-worst}}$ is

$$T_{\text{BUS-worst}} = \frac{1}{2t}\frac{\text{events}}{\text{second}} \times \frac{1}{4}\frac{\text{transfer}}{\text{clock}} \times \frac{1}{4}\text{bus masterships} = \frac{1}{32t}\frac{\text{transfers}}{\text{second}} \quad (7.4)$$

Once bus mastership has been obtained all transfers require the same handshake time regardless of destination. Since a bus system can only support one transfer at a time, the worst-case throughput is the same as the best-care throughput for the highest-priority module with degraded levels of throughput falling off according to the arbitration algorithm for lower-priority modules. There are many approaches to avoiding the worst case for the lowest-priority module, e.g., using a rotating or a first-come-first-served priority scheme.

For an ICN-type system such as that shown in Fig. 7-10b the worst case occurs when all sending modules have the same receiver. In this case the ICN can deliver only one transfer per cycle and the throughput becomes

$$T_{\text{ICN-worst}} = \frac{1}{2t}\frac{\text{clocks}}{\text{second}} \times 1\frac{\text{transfer}}{\text{clock}} = \frac{1}{2t}\frac{\text{transfers}}{\text{second}} \quad (7.5)$$

There are many workarounds that can be imagined for this "hot-spot" problem. One popular technique incorporates first-come-first-served queues within the ICN switching modules. This approach has the dual benefits of buffering out local hot-spots (provided they occur with a reasonably uniform distribution and with sufficient delay between each burst occurrence) as well as offering the possibility of asynchronous rather than lock-step synchronous signaling between the modules of the system. The reader is referred to Ref. 8 or Ref. 9 for details and for other approaches.

7.1.6 Point-to-Point Connections

If the generality of a bus is not required and if high performance is the primary design goal, point-to-point interconnections may be appropriate. There are several advantages in using point-to-point interconnect:

1. *Minimal fan-out loading.* Since only the required connections are implemented in a point-to-point philosophy, the fan-out is minimum.

2. *Contention-free.* Since each transaction has its own private channel, there will never be any contention for interconnect.
3. *Low control complexity.* A point-to-point system is essentially a pure register-transfer model for control purposes; each register-transfer is independently enabled by a single control signal.
4. *Less centralized control.* Since there are no busses, there is no need for centralized arbitration logic.

The primary disadvantage of point-to-point interconnect is *cost* in space—either chip, hybrid package, or printed circuit board area. For most systems, a substantial fraction of all possible interconnections are needed. For a system with n modules, there are $n(n-1)$ possible point-to-point interconnections. Following is a simple recipe for evaluating the needs using a bus-type model and, if appropriate, for converting it into one with point-to-point interconnect:

1. List each of the required bus transactions by considering the various transfers used during operation of the system. Use a register transfer style, e.g., block B ← block C.
2. Identify each port of each block explicitly and use only a single direction per port.
3. The set of unique register transfers of part 1 is a list of required point-to-point interconnections. Trade off the size and implementation cost of this point-to-point list against the size and cost (and potentially lower speed) of bus-type interconnect.

If a large number of connections are required then direct point-to-point interconnection may not be appropriate for the system at hand. Instead, a single- or multistage interconnection network can be considered [8,9].

7.1.7 Transaction Flows

An analysis of the time sequence of transactions in a bus-organized system model can help to identify a set of single-function modules that could be connected in a producer-consumer assembly-line style *without* a bus. If such a processing sequence can be found and generalized to cover most of the necessary instructions, then the benefits of very high clock rates can often be realized. Figure 7-11 illustrates several basic producer-consumer relationships that can be found in the Von Neumann fetch/execute cycle.

Systolic systems [11] are extremely regular arrays of very simple processing units that have been designed to take advantage of producer-consumer transactions with neighbors. This philosophy works well for digital filtering and for other similarly regular problems.

7.1.8 Pipelined Systems

The concept of *pipelining*—organizing functional circuitry in an assembly-line style—is becoming increasingly important to high-performance computing. In a pipeline, a

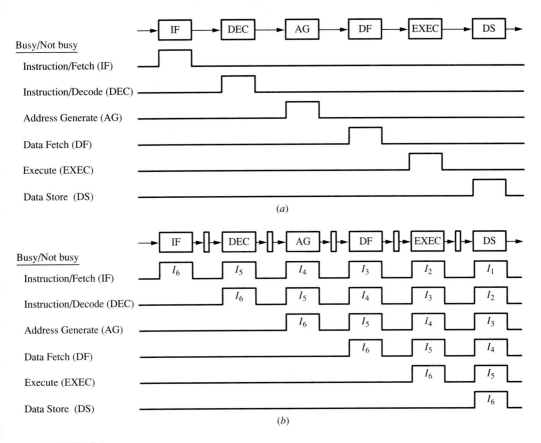

FIGURE 7-11
Unit utilization with and without pipelining. (a) Low unit utilization in a nonpipelined system executing a single Von Neumann machine instruction. (b) Latches inserted to form a pipelined system—six instructions in-progress simultaneously.

new operation can be started before in-progress operations are complete. The technique allows up to an n-fold increase in throughput for an n-stage pipeline.

Figure 7-11a shows a nonpipelined processor performing the traditional steps of the Von Neumann cycle: instruction fetch (IF), instruction decode (DEC), address generation (AG), data operand fetch (DF), instruction execution (EXEC), data operand store (DS). The busy/nonbusy state of each processing stage is plotted as instructions and operands move through the sequence of processing stations. In six clocks, one instruction is accomplished; thus, the throughput is

$$\frac{1}{6} \frac{\text{instruction}}{\text{clock}}$$

From the busy/nonbusy plot, it is clear that the hardware is not fully utilized. The processing stations are actually idle most of the time (in this example, each is idle exactly $\frac{5}{6}$ of the time). If the stages are organized as a pipeline, as in Fig. 7-11b, by placing latches between stages and carefully synchronizing the flow of information, then the units can be more nearly fully utilized. In the best case (once the pipeline has been filled) an instruction will be completed every time the pipeline advances, i.e., every clock cycle, giving a throughput of

$$1 \frac{\text{instruction}}{\text{clock}}$$

As long as the pipeline is full, the unit utilization will be 100 percent.

The goal of a pipeline is to achieve results at the rate of one per clock regardless of the length of the pipeline. If this goal of performance proportional to clock rate can be achieved then pipelines are very favorable for GaAs, since GaAs circuits can be operated with clocks running well into the GHz range. Pipelining is nearly ideal for very-high-speed technologies for several reasons:

1. The interconnectivity of a pipeline makes efficient use of area. Only neighboring stages interconnect. Since the stages are neighbors they can be made extremely close together. In the event that a pipeline cannot be contained within a single chip, the off-chip interconnections (because they are point-to-point) are short and efficient as well.
2. Processing can be finely divided so as to keep the number of levels of logic per stage small. As pipelines grow in length, however, the problems of keeping the pipeline full become more difficult. A complex tradeoff is necessary between pipeline length (i.e., how finely the processing stages are divided) and interdependency issues of the instruction set architecture.
3. Dynamic logic circuits fit well with pipelining. Their use allows substantial savings in space and in power as well as increased speed. The self-latching, edge-triggered property of some of the dynamic circuits of Sec. 4.3 makes them attractive for pipeline implementations. Because of the regularity of pipelining, many of the complexities and overheads of using dynamic logic are hidden or masked by the strict synchronous clocking and the overlapped execution of the pipeline.

Pipelines only deliver their promise of performance in direct proportion to clock rate as long as they are kept filled. Hardware and instruction set interrelationships, data dependencies, memory performance, cache effectiveness, and other factors strongly interact to affect the ability of a system to keep its processing pipelines filled. Pipelined system design is a reasonably well understood subject; the design and analysis techniques for pipelines are not the focus of this book, however. The reader is instead referred to the excellent coverage of pipelines in Stone [8] and Kogge [10].

7.1.9 RISCs and Pipelining—
the CPU/Memory Speed Disparity

Because of the dramatic payoff of pipelining in high clock rate systems, a great deal of attention has been paid to the interrelated architectural issues affecting pipeline performance. There are many implications of using a high clock rate pipeline in a system. Memory system bandwidth is perhaps the single most important consideration.

Pipelining works well with certain types of processing but does not universally apply to all types. Processors are normally quite amenable to pipelining because many of the data transformations performed in processors are already broken down into a sequence of microsteps or single-clock operations. These transactions can often be described in a producer-consumer style as is required for pipelining. Examples that fit well are multiplication, certain techniques for division, many floating-point operations, trigonometric function evaluations, and polynomial expansions as well as the basic Von Neumann instruction fetch/execute cycle.

Memory systems, on the other hand, do not fit well within the pipeline model. Some pipelining of the address decoding is possible, but because of the highly capacitive bit busses that are inherently present in any memory, their cycle times will always be limited. Hence, memories are likely to run slower than the processors they serve. Unfortunately, this speed difference is likely to continue to grow in the future. This imbalance must be directly addressed in the machine architecture or wait states will be necessary and the potential advantages of high clock rates will be partially lost. Unfortunately processor pipelining works against this problem by placing even higher bandwidth demands on memory systems. In the example given by Fig. 7-11*b* up to a sixfold increase in throughput is possible due to pipelining, *provided that the memory system can cycle at a rate six times faster than in Fig. 7-11a.* Furthermore, with a decreased amount of processing per clock there is a strong desire to increase the clock rate as high as possible. All such efforts have a direct impact on required memory bandwidth.

Reduced instruction set computers (RISCs) have been developed independently during the early 1980s by several universities and corporations [1–3, 11–12]. The RISC architectural style is the result of the observation that a high percentage of the complex instructions normally implemented in hardware on a microprocessor are rarely used by compilers. The RISC approach trades a penalty in the time required for software-implemented execution of such instructions for a large savings in hardware complexity and a significant overall speed-up in CPU clock rate. Since the penalized instructions are rarely needed anyway, the resulting machine runs faster and utilizes its simpler hardware resources more fully.

The RISC machines utilize pipelines to closely approach instruction-per-clock throughput. A great deal of attention is paid to memory bandwidth requirements. Most RISC machines use a LOAD/STORE architecture—an approach wherein only LOAD and STORE instructions directly access memory, with all others operating solely on registers. The LOAD and STORE instructions may take longer to execute— typically twice as long—since off-chip memories tend to have slower cycle times. By separating data memory accesses from execution operand accesses and incorporating a large register file on-chip, the bulk of the compute-bound load can execute at high clock rates without making many off-chip accesses.

Another typical RISC feature is that pipeline dependencies relating to control flow are "optimized out" at compile time by code rearrangement—moving useful instructions immediately *after* branches. This is done so that the hardware can execute through pipelined conditional branches without checking for dependencies and interlocks. This approach requires that some small constant number of instructions *after* a branch instruction are always executed before the branch target is executed (independent of the outcome of the conditional branch decision) since they are loaded into the pipeline before the decision about the branch is actually made. If, for a given branch instruction, no useful branch-trailer instructions can be found by the compiler, NO-OPs must be placed following the branch. Since many branches are encountered in iteration or loop structures, however, it is frequently possible to fill in the branch trailers with useful loop overhead instructions.

7.2 I/O AND CHIP-TO-CHIP ISSUES

One of the costs of using GaAs in systems composed of multiple technologies involves the conversion between logic levels. Not only does the conversion cost significantly in circuit area but also in power consumption and in performance. The number of foreign signal levels that must be interfaced in a given system should naturally be minimized. Consider the number of different on-chip circuits that are potentially required in an ECL/GaAs system:

- GaAs-to-GaAs driver (short haul—between chips within a hybrid package)
- GaAs-to-GaAs driver (long haul 50 Ω—for driving packages, cards, or cables)
- GaAs-to-GaAs receiver (from short-haul hybrid driver)
- GaAs-to-GaAs receiver (from long-haul 50 Ω driver)
- GaAs-to-ECL driver
- ECL-to-GaAs receiver
- GaAs-to-GaAs open-source distributed NOR bus driver
- GaAs-to-GaAs bus receiver

The distinction between *short-* and *long-haul* drivers is made to emphasize that 50 Ω interconnect need not necessarily be used between chips within the same hybrid. In that situation a relatively high impedance interconnection is possible allowing significantly smaller output devices to be used.

In addition to each of the driver/receiver circuits above, the line terminations must be carefully considered. The tradeoffs of on-chip vs. in-package terminations are complex. The material on packaging found in Chap. 5 provides a starting point for making such tradeoffs.

7.2.1 GaAs-to-GaAs Driver

The minimum circuit for driving an off-chip GaAs receiver load is given in Fig. 7-12. This simple output driver is designed to operate with power rail voltages of +2.5, 0.0, and –2.0 V. On the input of this driver a voltage swing of approximately 1.5 V is best, with logic '0' being –1.0 V and logic '1' being +0.5 V. The output has a dc return to ground through the terminating resistor.

The circuit of Fig. 7-12 is very basic, but it has been fabricated and tested at the University of California, Santa Barbara, and found to work well. The input signal directly gates a FET logic inverter and a medium-sized (width 31.5 μm) pull-

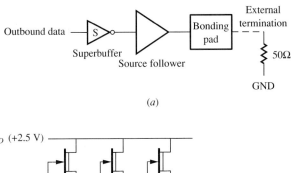

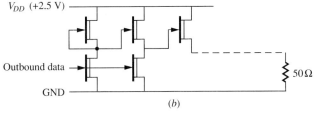

FIGURE 7-12
Simple GaAs-to-ECL pad driver. (*a*) Logic diagram. (*b*) Schematic.

down transistor. The output of this inverter is not level-shifted. It directly drives a medium-sized (width 31.5 μm) pull-up transistor. The intermediate stage of medium-sized devices directly drives the large-output transistor which has a width of 160 μm. The use of switching transistors rather than load-type transistors for both the pull-up and pull-down provides better drive and allows the output transistor to be switched at a faster rate. The first-stage inverter and second-stage switching transistors form the GaAs equivalent of an NMOS superbuffer. The large output transistor forms a source follower circuit when connected to an external terminating resistor pull-down. The output logic '0' voltage is +0.6 V; the logic '1' voltage is +1.4 V.

The driver's output impedance is 50 Ω. This provides a match for the 50-Ω transmission line and prevents reflected signals from being reflected again. The 50-Ω transmission line should be terminated to ground with a 50-Ω resistive load. If GaAs input receiver circuits are very high impedance (as are the ones of Sec. 7.2.3) then the fan-out capability of this driver will be limited primarily by interconnection considerations.

There are several problems with this simple driver. One is the fact that when the input switches from low to high the pull-down transistor turns on before the pull-up device turns off. This event causes approximately 3 mA to flow from V_{DD} to GND. A current spike this size can easily cause noise problems if the power and ground system is not properly designed. Clearly, the simultaneous switching of several output drivers due to a common event could cause a significant current spike. Another problem with this simple circuit is the lack of temperature compensation. As temperatures rise, the output levels change. There is often insufficient noise margin in the system to absorb changes that can be caused by temperature. Thus, some type of temperature compensation network is also desirable.

7.2.2 GaAs-to-Silicon ECL Driver

A GaAs-to-silicon ECL driver circuit is shown in Fig. 7-13. This circuit is a slightly modified version of the GaAs-to-GaAs driver discussed in the previous section and shown in Fig. 7-12. The main challenge in designing a GaAs-to-silicon ECL driver is in translating voltage levels and logic swings between the two different technologies without using additional power supplies and without using an excessive amount of power and area.

The GaAs-to-silicon ECL output driver shown in Fig. 7-13 is designed to connect between a GaAs BFL gate and a bonding pad. This driver is designed to operate with power rails of 0.0, −2.5, and −4.5 V. These voltages were chosen so that the GaAs and ECL chips can share a common −4.5 V power supply. Another reason for this choice is that conversion of GaAs BFL logic levels to ECL levels is easier and faster if the GaAs chips are operated at potentials that are below ground.

The circuit shown in Fig. 7-13 uses a superbuffer input stage driving a voltage-clamped source follower output stage. Input signal swing is intended to be 1.5 V with GaAs logic '0' at −3.5 V and logic '1' at −2.0 V. Output voltage swing and levels are compatible with Fairchild F100K ECL (−1.81 V logic '0', −0.88 V logic '1') [14]. The superbuffer input stage devices have a width of 6 μm for the pull-up transistor and 10 μm for the pull-down device. The second-stage devices have widths of 36.5 μm each. The large-output transistor is 225 μm wide with 14 μm^2 voltage-clamping diodes. These sizes are chosen to correctly set the output voltage levels when using a typical D mode, −1 V threshold technology.

The fan-out capability when driving ECL gates is quite limited. The ECL-to-GaAs receiver presented in Sec. 7.2.4 is compatible with this driver. Because of the

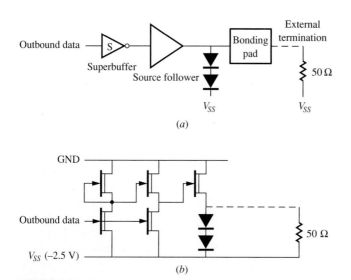

FIGURE 7-13
Simple GaAs-to-ECL pad driver. (*a*) Logic diagram. (*b*) Schematic.

receiver's high input impedance, fan-out is not a problem when it is used. As with the GaAs-to-GaAs driver, fan-out will be limited primarily by the interconnect when driving ECL-to-GaAs receivers.

7.2.3 GaAs-to-GaAs Receiver

A simple non-inverting GaAs-to-GaAs receiver circuit is shown in Fig. 7-14. This circuit is designed to connect an off-chip bonding pad with the input of a GaAs FL or BFL gate. The receiver is designed to operate with power supply rails of +2.5, 0.0, and −2.0 V. An input voltage swing of 0.8 V is best with logic '0' being +0.6 V and logic '1' being +1.4 V. The output voltage swing is from −1.0 V (logic '0') to +0.5 V (logic '1'). The output levels were chosen to be compatible with GaAs FL or BFL logic.

The input impedance of the receiver is very high. As such, it should not limit the fan-out of other chip's output drivers. This receiver circuit is quite basic. The input signal goes initially to a source follower stage to provide high input impedance and a shift in the voltage levels of approximately −1.4 V. The signal then goes through two stages of amplification to increase the voltage swing from 0.8 to 1.5 V. The final stage is another source follower voltage shifter that adjusts the output voltages down to correct BFL levels.

The device width for all source follower MESFETs is 10 μm. Input stage diodes have an area of 7.5 μm^2, level-shifter diodes in the intermediate BFL stage are 24 μm^2, output stage diodes are 18 μm^2. The pull-up and pull-down MESFETs in the intermediate BFL stage have widths of 12 and 24 μm respectively. The remaining pull-up and pull-down MESFETs are 6 and 12 μm wide.

This circuit is the "bare-bones" GaAs-to-GaAs receiver. Incorporation of temperature compensation or other threshold shift adjustments requires significant additions. Refer to Sec. 7.2.7 for a discussion of an adjustable receiver circuit.

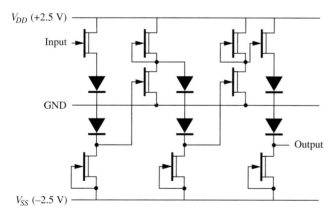

FIGURE 7-14
GaAs-to-GaAs receiver schematic.

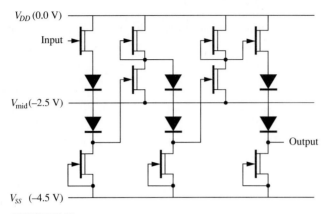

FIGURE 7-15
Silicon ECL-to-GaAs receiver schematic.

7.2.4 Silicon ECL-to-GaAs Receiver

Figure 7-15 shows the schematic for a minimal GaAs circuit to receive silicon ECL signals. The circuit is similar in character to that of the previous section though device sizes and power supply rails are different. This receiver is compatible with F100K ECL logic levels ('0' = -1.475 V, '1' = -1.165 V). The output voltage swing is designed for GaAs FL or BFL using V_{DD} of 0.0 V, $V_{mid} = -2.5$ V, and $V_{SS} = -4.5$ V. The output levels are -3.5 V for logic '0' and -2.0 V for logic '1'.

The input source follower has 10-μm MESFETs and 13-μm^2 diodes. The first BFL stage has a 6-μm wide pull-up and a 10-μm wide pull-down. The source follower on the output of this stage has 10-μm wide MESFETs and 9.5-μm^2 diodes. The second BFL stage also has a 6-μm wide pull-up and a 10-μm pull-down, but the source follower has 24-μm MESFETs and 22-μm^2 diodes. The larger devices on the output source follower allow the receiver to drive several on-chip loads or long interconnects without losing a significant amount of speed.

7.2.5 GaAs-to-GaAs 50-Ω Distributed-NOR Bus Driver

Figure 7-16 presents the schematic for a GaAs-to-GaAs 50-Ω bus-type pad driver circuit. This is not for driving large, distributed-bus systems. It is intended to support distributed-NOR connections between adjacent chips within the same hybrid package. The main application for this driver is wherever the flexibility of a bus is desired and speed is not critical. The speed of the driver and its matching receiver is not slower than the other drivers and receivers presented in this chapter. Transmission line effects (as discussed in this chapter, as well as in Chap. 5) make systems using busses slow. The described bus driver and companion receiver are fast when the entire bus is in a hybrid package so that the bus can be kept at most a few millimeters in length.

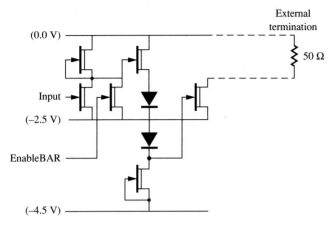

FIGURE 7-16
GaAs-to-GaAs 50-Ω distributed-NOR bus driver schematic.

Several such drivers can be bussed together. Only one load resistor is used. When the output is high the output transistor is completely cut off and thus has no dc effect on the output node. There is, however, still a parasitic capacitance (and therefore an impedance discontinuity) associated with the disabled driver.

The driver is non-inverting with inputs: '0' = −3.5 V, '1' = −1.0 V. The output logic swing is compatible with the receiver shown in Sec. 7.2.6. The output swing is '0' = −1.8 V, '1' = 0.0 V. In the particular circuit given, the enable signal is active low.

The input stage is a BFL NOR gate composed of 10 μm pull-down devices and a 6-μm pull-up MESFET. The source follower level-shift network has 42.5 μm MESFETs and 65 μm^2 diodes. The BFL NOR drives a large output MESFET switch which is 300 μm wide.

7.2.6 GaAs-to-GaAs Bus-Type Receiver

A receiver compatible with the distributed-NOR bus-type driver of Sec. 7.2.5 is shown in Fig. 7-17. The circuit has a source follower input stage with a BFL inverter output circuit. Because of the source follower's high input impedance, the bus driver can have high fan-out i.e., dc receiver loading is very low. There is, of course, a small amount of parasitic capacitance associated with the input. This can cause an impedance discontinuity and a reflection on the bus.

The input logic levels are the same as the output levels of the companion driver ('0' = −1.8 V, '1' = 0.0 V). The output levels are as follows: '0' = −3.5 V, '1' = 1.0 V. The device sizes in the first-stage source follower are: pull-up 10 μm wide, current source 15 μm wide, diodes 11 μm^2 in area. The BFL inverter has a pull-up MESFET size of 6 μm and a pull-down size of 10 μm. The final-stage level-shifting source follower has 24-μm wide MESFETs and 25-μm^2 diodes.

374 GALLIUM ARSENIDE DIGITAL INTEGRATED CIRCUIT DESIGN

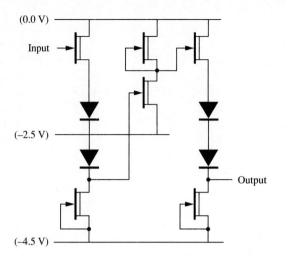

FIGURE 7-17
Schematic of receiver for GaAs-to-GaAs 50-Ω distributed-NOR bus driver.

7.2.7 Threshold-Adjustable GaAs Receiver

Interfacing with silicon ECL logic levels with a total voltage swing of only 800 mV is a challenge. In fact, the voltage swing can be as low as 300 mV and still meet ECL-100K specifications. This wide variation is particularly difficult since GaAs fabrication capabilities cannot guarantee tight control over MESFET threshold

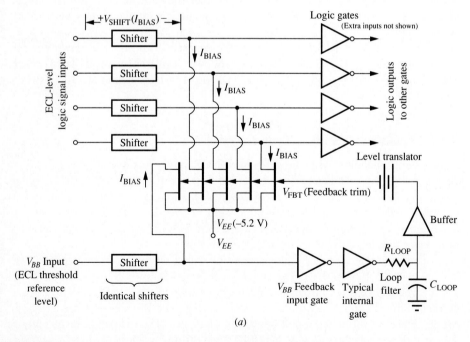

FIGURE 7-18
ECL receiver with adjustable threshold. (*a*) Block diagram. (*From Eden* et al. [15]. *Used by permission.*)

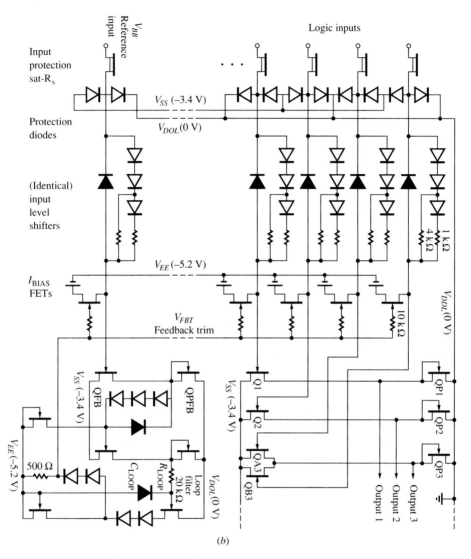

FIGURE 7-18 (*continued*)
(*b*) Schematic.

voltage. The capability to adjust the receiver threshold is extremely desirable. In Ref. 15, Eden *et al.* give a practical ECL-to-GaAs receiver circuit with adjustable threshold voltage. The block diagram and device-level schematic are shown in Fig. 7-18.

The approach uses an off-chip ECL threshold reference level (nominally -1.3 V) and one "V_{BB} feedback circuit" which serves as the threshold shift control for several on-chip receivers. All of the logic inputs to the chip, plus the extra "V_{BB} threshold control input" have identical level shifters biased from identical pull-down FETs to V_{EE} which are all controlled from the same gate potential—the feedback trim voltage—so that they all generate identical and controllable shifter bias currents (I_{BIAS}). The goal of the V_{BB} feedback approach is to ensure that the input threshold voltage of all input gates is at the V_{BB} potential.

7.3 ASYNCHRONOUS TECHNIQUES

Perhaps the greatest barrier to overcome in the face of increased operating speed of GaAs circuits is global clock distribution and synchronization. As feature sizes grow smaller and systems grow larger and more complex, the difficulty is compounded. A solution to this problem is the use of self-timed systems—systems that are "self-synchronizing" rather than under lock-step control of a global clock. The self-timed concept is not new. Indeed, a number of working self-timed systems have been reported in recent years.

The application of the self-timed discipline presents some unique problems to GaAs implementations. First of all, the low propagation delays of GaAs circuits invalidate the MOS assumption of negligible interconnection delays. This suggests that delay-insensitive signaling be used in GaAs systems, which requires validity to be encoded within the data. The MOS approach is to use double-rail coded data, with each information bit encoded on two "wires." Of the four states encodable with a 2-bit representation, only three are needed: logic '0', logic '1', and "no data." The "no data" condition is necessary as a spacer between successive defined data values. In GaAs this double-rail code may not be an acceptable solution, suggesting some other type of encoding.

An alternative to double-rail encoding is *ternary* encoding—using three distinct logic levels on each single "wire" rather than two levels on each of two wires. In this section the development of a family of GaAs ternary logic elements is sketched. These elements, when connected properly, will result in working ternary self-timed systems. The approach is similar to the way in which mature families of logic circuits (CMOS, TTL, ECL, . . .) allow engineers to construct complex systems without great concern for such issues as level restoration, delay hazards on circuit inputs, etc. The central issues in this development are the creation of a suitable ternary algebra which allows the use of *binary* description and minimization techniques (i.e., the ternary aspects of a combinational network are transparent to the designer), circuits for implementing the algebra, extension of the combinational signaling conventions to sequential circuits, and finally self-timed control and system specification.

7.3.1 Self-Timed Signaling

The correct operation of a self-timed system requires the elements of the system to communicate with each other (self-synchronization) according to a strict protocol. Signaling is divided into two general categories: equipotential self-timed signaling and delay-insensitive self-timed signaling. Generally speaking, an *equipotential region* is defined as any physical area of a system in which Kirchoff's laws remain valid. More colloquially, an equipotential region is an area where a signal can be treated as identical at all points along a wire. A quick study of signaling options yields but two possibilities: two-cycle (a.k.a. non-return-to-zero, NRZ) and four-cycle (return-to-zero, Muller, RTZ). Both approaches are discussed in their non-distributed, parallel form in Sec. 7.1.4 in conjunction with bus-connected systems.

Equipotential signaling in either flavor requires the temporal relationship of the input data/request and output data/acknowledge signal pairs, generated at a source to remain valid at the destination. In the modern MOS world, it can be (and usually is) safely assumed that the entire chip resides in an equipotential region. When this relationship between data and control signals cannot be guaranteed, delay-insensitive signaling must be used. This becomes a significant factor in GaAs systems where propagation delays are in the 80-ps range and interconnection delays can approach this order of magnitude. The general solution to delay-insensitive signaling is to embed validity information into the data, essentially distributing the *request* and *acknowledge* signals across any number of data lines. For instance, consider operands applied to a combinational multiplier from a source in a different equipotential region. Rather than triggering on a single-line *request* signal (which would have to be correctly synchronized with the arrival of parallel data), the multiplier performs the required operation *as data become valid*. The arrival of defined data on all inputs implies *request*. The operation is complete only when all output data have become defined. When all output data become defined, the distributed *acknowledge* signal is implied.

Figure 7-19 illustrates four-cycle delay insensitive signaling; the distributed *request* and *acknowledge* signals are asserted when all input and output data, respectively, are defined. The maxima and minima on the diagrams indicate all data defined and no data defined respectively. As mentioned earlier, the encoding of data for delay-insensitive signaling has traditionally been accomplished using a double-rail binary code indicating the datum as zero, one, or undefined. The most common encoding is 01 = '0', 00 = 'undefined', and 10 = '1'.

In MOS technologies the cost of two wires per datum is not prohibitively high.

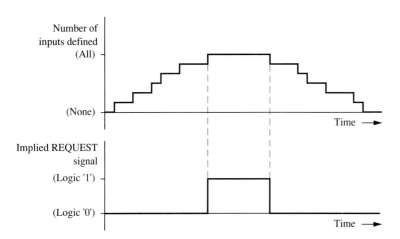

FIGURE 7-19
REQUEST signal in distributed form (shown at top) and implied form (shown at bottom).

Such is not the case in GaAs, however, where die acreage is at a premium and the power budget is already strained. A technique identified in the literature but, to the authors' knowledge, never investigated in any detail, except in Ref. 16, is the use of ternary data to realize the distributed control signals. The obvious cost of this solution is in the complexity of the circuitry. As will become clearer in the following sections, the elegance of binary logic circuits is difficult to achieve in a ternary environment. However, when considering a double-rail system, one can reasonably assume that the complexity of the circuitry exceeds twice that of a standard, synchronous system performing a functionally similar operation. With this in mind, any ternary circuit of roughly twice the complexity of a binary gate constitutes a distinct advantage over the double-rail case because of lower interconnection costs.

7.3.2 Ternary Algebra and Combinational Network Synthesis

This section discusses the development and adaptation of a ternary algebra for self-timed purposes (the reader is referred to Ref. 16 for a detailed treatment). The attributes most appropriate for a ternary algebra are: first, the algebra (and consequently the basic circuits) should obey the laws and theorems of boolean algebra with valid data applied; second, the algebra should allow for the generation of the third, intermediate value ('u' = undefined) when less than the minimum number of inputs are defined to determine the output. A simple example of this second heuristic is the output of a two-input AND gate: 0 when any input is 0, and u when one input is 1 and the second is undefined.

A great deal of work has been done in the area of multiple-valued logic, as witnessed by the annual IEEE symposium on the subject, leaving no shortage of both theoretical and practical algebras. In the context of ternary self-timed systems, three functions are most appropriate. These are: diametrical inversion, $\overline{\text{Max}(X,Y)}$ (the ternary NOR operation) and $\overline{\text{Min}(X,Y)}$ (the ternary NAND operation). The truth tables for these three functions are provided in Table 7.2. Note that the attributes

TABLE 7.2
Truth tables for ternary elements

Inversion		NOR – $\overline{\text{Max}(A,B)}$			NAND – $\overline{\text{Min}(A,B)}$		
In	Out	A	B	Out	A	B	Out
0	1	0	0	1	0	0	1
u	u	0	u	u	0	u	1
1	0	0	1	0	0	1	1
		u	0	u	u	0	1
		u	u	u	u	u	u
		u	1	0	u	1	u
		1	0	0	1	0	1
		1	u	0	1	u	u
		1	1	0	1	1	0

identified above are satisfied in all cases. Additionally, it can be shown that both the Max and Min operations are logically complete in the boolean sense, allowing any combinational network to be constructed from a single type of gate.

Application of these functions to combinational network synthesis is straightforward, assuming the gates can be built. The binary truth table for the function is defined in the usual manner and any binary minimization technique is applied. In adapting this particular subset of multiple-valued logic functions, the generation of undefined outputs is accomplished in hardware and is of no specific concern to the designer. Because NOR gates (binary or ternary) are much more easily realized in GaAs than NAND gates, a product-of-sums expression is desired. Notationally, the standard NOR gate symbol is used for the ternary $Max(X,Y)$ function. Likewise, a binary product-of-sums expression implies a two-level, NOR-NOR implementation.

7.3.3 Ternary Logic Circuits

In developing a family of ternary logic circuits, a number of goals and constraints have been applied. First, the circuits have been created with an eye towards implementation in GaAs. Early prototype versions were designed, fabricated, and tested in NMOS, which at sufficiently low speed is similar to GaAs and for which fabrication was readily available. Second, the complexity of the circuits should be roughly the same as that of "double-rail" gates performing a similar function. Finally, they should be realizable in available GaAs processes.

The basic circuit configuration chosen makes use of device width ratios for correct level restoration and switching points. Figure 7-20 illustrates the ideal voltage transfer curve for a diametrical inverter. The three distinct levels represent logic true ('1'), undefined ('u'), and logic false ('0'), and their ordering is such that '0' < 'u' < '1'.

A depletion mode GaAs circuit for a ternary inverter is shown in Fig. 7-21. This circuit has less than twice the device count of a BFL inverter, is composed exclusively of depletion mode MESFETs and diodes, and, thus, meets all the first-order criteria. The circuit uses two inverters (inputs at T_2 and T_4) with a current-summing source follower output stage. A level-shift diode shifts the inverter threshold of T_4 to be

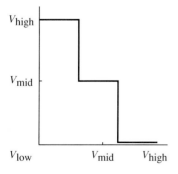

FIGURE 7-20
Ideal V_{out} vs. V_{in} transfer function for a ternary inverter.

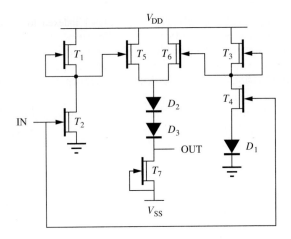

FIGURE 7-21
GaAs (D mode) ternary inverter.

sensitive to the '1' level only whereas the T_2 inverter turns on when its input is either 'u' or '1'. Both switching points and level restoration are critically dependent upon device sizing. Additional inputs can easily be incorporated to form a ternary NOR gate.

7.4 SUMMARY

As the speed and density of GaAs integrated circuits continue to improve, system architects and designers must develop effective techniques for dealing with the unique problems of extremely high-speed digital systems. No longer can the commonplace lock-step synchronous design philosophy be universally applied across entire multi-board systems. Simple, flexible, modular structures such as busses may no longer provide the most appropriate interconnection methodology. The gap in raw speed between GaAs processing elements and high-density non-GaAs memory subsystems is also growing. These and other trends raise significant issues and problems for engineers to study and solve.

Asynchronous techniques, though more complex, are one approach to solving system timing problems in GaAs systems or in systems with embedded GaAs components. Computers will take on a more decoupled character, possibly making use of heavily interleaved memories and/or interleaved instruction streams to avoid wasted wait-state cycles on high clock-rate processors. There is much room for innovation. Harnessing the very-high-speed processing capabilities available through the use of digital GaAs ICs is a formidable challenge for digital designers of the 1990s.

HOMEWORK PROBLEMS

7.1. Use SPICE to validate the behavior of the GaAs-to-GaAs driver circuit given in Fig. 7-12. Find the dc properties using the device sizes discussed in Sec. 7.2.1.

7.2. A short instruction sequence is given below to implement a high-level indexed loop construct. The instruction set is intended to be in the RISC style (i.e., register-to-register, two-address instructions). The "#" introduces an immediate data item (specified in decimal notation); register names are given as R1, R2, and so on. The MOVE and LOAD instructions each have two operands; the first is the source operand and the second is the destination.

```
       LOAD  #1,R1              ;initialize iteration counter
       LOAD  #1000,R2           ;initialize loop termination value
LOOP:  MOVE  R1,R5              ;first instruction of loop body
       ADD   R5,R5              ;second instruction of loop body
       LOAD  ARRAY (R5),R6      ;third instruction of loop body
         .     .
         .     .
         .     .
       BLE   R1,R2,LOOP         ;go to LOOP if R1 ≤ R2
       LOAD  #1234,R1           ;first instruction after loop end
       LOAD  #5678,R2           ;second instruction after loop end
       MOVE  R6,R7              ;third instruction after loop end
```

(a) Without performing any code rearrangement, manually insert NO-OPs to guarantee correct execution with a four-stage pipelined execution unit. Assume that there are no shortcut paths for operands; i.e., an output operand *must complete* execution before it can validly be involved as a source operand in succeeding instructions. The machine is assumed to use pipelined branches without hardware checking for control flow dependencies.

(b) Using a 500 MHz clock, find the total running time for the 1000-iteration loop, assuming that the loop contains 20 instructions (counting the one labeled "LOOP" through and including the "BLE" conditional branch instruction). Your time computation should be exact, including the loop initialization, the NO-OPs necessary from part (a), and the three instructions shown after the loop.

(c) Find the execution penalty ratio due to the insertion of NO-OP instructions. This metric is a ratio—instructions actually executed divided by functional instructions (not including NO-OPs).

(d) Rearrange the code to take advantage of branch trailers (the three instructions following a branch that are executed regardless of the outcome of the conditional branch).

(e) Compute the execution penalty ratio using the modified program segment resulting from code rearrangement [part (d)].

7.3. Pick a microprocessor-based computer and study the details of its processor/memory interface; specifically:

- What is the processor clock rate?
- What is the cycle time of the main memory?
- Does the processor require wait states and, if so, how are they implemented?

  (a) Make an estimate of the effective speed-up of the central processor could be run at a clock rate ten times its current value. Assume for this section that the speed of all other components stays the same.

(b) What change(s) would you propose to the organization of the memory or of other parts of the machine in order to approach a tenfold system-wide performance improvement?

7.4. A short-haul driver/receiver pair for use between closely spaced chips within a hybrid package can be made as shown in Fig. P7-4. The driver is a simple BFL inverter that is broken apart at the source follower to provide a low impedance output. The receiver is the rest of the source follower (the level-shifter diodes and current source) followed by another inverter to clean up and restore the signal. Voltage rails are $V_{DD} = +2.5$ V and $V_{ss} = -2.0$ V. Nominal driver input and receiver output logic levels are approximately -1 V for logic '0' and $+0.5$V for logic '1'. Using SPICE, determine all device sizes such that T_{plh} and T_{phl}, from driver input to receiver output, are 250 ps or less. The driver input should not excessively load the circuit driving it. Use a T-network model for the interconnect between the driver and the receiver. Assume a bond wire inductance of 0.35 nH and interconnect shunt capacitance of 2 pf.

7.5. Two GaAs BFL ICs are located next to each other in a hybrid module. They are connected together using the driver in Fig. 7-16 and the receiver of Fig. 7-17. A 1-GHz signal is to be sent from one chip to the other. Rise and fall times at the driver's input are 100 ps. The rise and fall times of the edges at the receiver input must be kept under 250 ps. The source chip is connected via a bonding wire to a hybrid module trace that is, in turn, connected to another bonding wire that connects to the receiving chip. The bonding wires have a

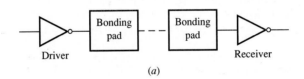

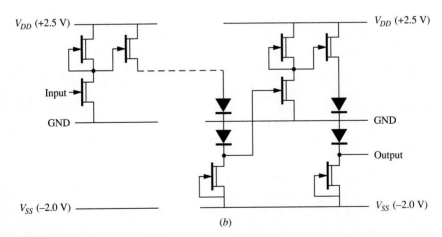

FIGURE P7-4
GaAs-to-GaAs in-hybrid driver/receiver: (a) In-hybrid driver/receiver logic diagram. (b) In-hybrid driver/receiver schematic.

series inductance of 0.35 nH each, and the module trace has a parasitic shunt capacitance (to ground) of 2 pf. Using SPICE and a T-network model for the interconnect, determine the proper sizes for the devices in the driver and receiver.

REFERENCES

1. Patterson, D. A., and Sequin, C. H. : "A VLSI RISC," *IEEE Computer*, vol. 15, no. 9, pp 8–21, September 1982.
2. Patterson, D. A.: "Reduced Instruction Set Computers," *Communications of the ACM*, pp. 8–21, January 1985.
3. Hennessy, J. L.: "VLSI Processor Architecture," *IEEE Trans. Computers*, vol. C-33, no. 12, pp. 1221–1246, December 1984.
4. Bartolotti, R., and Deyhimy, I.: "A GaAs Bit-Slice Microprocessor Chip Set," *Proceedings of the IEEE GaAs IC Symposium*, pp. 197–200, October 1987.
5. Asbeck, P. M., *et al.*: "Piezoelectric Effects in GaAs FETs and Their Role in Orientation-Dependent Device Characteristics," *IEEE Trans. Elect. Dev.*, vol. ED-31, pp. 1277–1380, October 1984.
6. Fouts, D., Johnson, J., Butner, S., and Long, S.: "System Architecture of a GaAs 1 GHz Digital IC Tester," *IEEE Computer*, May 1987.
7. Kiefer, D., and Heightley, J.: "CRAY-3: A GaAs Implemented Supercomputer System," *Proceedings of the IEEE GaAs IC Symposium*, pp. 3–6, October 1987.
8. Stone, H.: *High-Performance Computer Architecture*, Addison-Wesley, Reading, Mass., 1987.
9. Hwang, K. and Briggs, F.: *Computer Architecture and Parallel Processing*, McGraw-Hill, New York, 1984.
10. Kogge, P.: *The Architecture of Pipelined Computers*, McGraw-Hill, New York, 1981.
11. Katevinis, M.: "Reduced Instruction Set Computer Architectures for VLSI," Ph.D. dissertation, University of California, Berkeley, report UCB/CSD 83/141, October 1983.
12. Radin, G.: "The IBM 801 Minicomputer," *IBM J. Res. Devel.*, p. 237, May 1983.
13. Kung, H. and Leiserson, C.: "Systolic Arrays (for VLSI)," *Proceedings of the Symposium on Sparse Matrix Computations and Their Applications* (ed. by I. Duff and G. Stewart), pp. 48–53, 1978.
14. *F100K ECL Data Book*, Fairchild Camera and Instrument Co., Mountain View, Calif., 1982.
15. Eden, R. C., *et al.*: "V_{bb} Feedback Approach for Achieving ECL Compatibility in GaAs ICs," *Proceedings of the IEEE GaAs IC Symposium*, pp. 123–127, October 1986.
16. Johnson, J. M.: "Theory and Application of Self-Timed Integrated Systems Using Ternary Logic Elements," Ph. D. dissertation, University of California, ECE Department, November 1988.

CHAPTER 8

PHYSICAL DESIGN

This chapter presents geometric and electrical design rules for two different GaAs fabrication processes, the selective implant depletion mode process and the self-aligned enhancement/depletion process. Use of each set of rules is illustrated by presentation of design examples. The physical design material found here is an application of the theory presented in Chaps. 2 through 4. In addition to coverage on two leading GaAs fabrication processes, the issues and details concerning the interface between circuit layout and wafer fabrication are discussed.

8.1 FOUNDRY INTERFACE

There are many levels of representation used during the design and fabrication of a digital integrated circuit. Some examples are behavioral specifications, architectural block diagrams, register-transfer diagrams, and logic gate networks as well as circuit schematics, switch-level networks and mask-level geometric representations. It is critical that a clean, unambiguous, and well-defined interface exists between the circuit/layout design effort and wafer fabrication. A set of design rules serves to define this interface. Because it narrows the breadth of requisite expertise, existence of such an interface allows a larger set of designers to be involved in IC design, thus vastly increasing IC design and development throughput.

There are two varieties of design rules, *geometric* and *electrical*. Geometric design rules are constraints on the layout of an integrated circuit chip. These rules take the form of a set of allowed or forbidden geometries for mask features. The rules are stated in terms of minimum linewidths, minimum separations, and required, allowed, or disallowed extensions and overlaps between various mask layers. Electrical design rules are constraints on device widths, voltage and current ranges, and other circuit properties. These rules are necessary in order to keep the designer within the "four corner" limits of the typical wafer processing line.

Design rules essentially define the capabilities of a fabrication line; they do not represent a hard boundary between working and non-working circuits. The more aggressive the rules, the greater the potential for high-performance circuits. On the other hand, if the rules tend toward the conservative, then the resulting circuits have a higher probability of working, but at reduced performance levels. Thus, performance usually comes at the expense of yield. Design rules represent a formally verifiable compromise between circuit performance and process yield. The objective is to obtain circuits with high yield while using the smallest possible geometries in order to maximize performance.

8.1.1 Scalable Design Style

The concept of λ-based design—geometric layouts based on an elementary unit of length, λ—was popularized by the NMOS VLSI textbook authored by Mead and Conway [1]. Since that time, many computer-aided design tools and systems have become available. In the university community the λ-based *scalable design style* is dominant. This style is termed *scalable* because, by changing the assigned value of λ, full chip designs can be scaled down quite simply and mechanically as fabrication technology advances. Because of the presence of vastly more complex design rules as well as an increased need for high levels of efficiency and performance, this simplified and conservative design style is less prevalent in industry. Industrial rule sets tend to be specified with absolute (non-scalable) feature lengths, widths, separations, and overlaps given in micrometers.

There are some problems, however, with λ-based scaling. The main difficulty is that because *all* lengths and widths are specified in terms of the same λ value, when it changes all features are scaled together. The scalability model is too simplistic. It is more typical that as a fabrication process evolves and matures certain features can be reduced in size while others must stay at their previous size. Thus, computer-aided design programs must deal with λ-based scalability in a careful and nontrivial way. For some foundry processes it is necessary to bloat or shrink features *asymmetrically* depending on the type and orientation of other neighboring features. For this reason CAD programs that support scalable designs typically have facilities to support bloating and shrinking (as well as other manipulations such as union and intersection) on a layer-by-layer basis. A layer's neighbor on each edge may, in general, influence the generated masks.

GaAs fabrication processes are still evolving. It is thus vitally important to be moderately conservative in design so that circuits are likely to work across multiple foundries and across variations from run to run within the same foundry, but not overly so such that performance is poor. In this section the design rules are stated in terms of λ, but the value of λ is set to a constant value (0.5 μm for depletion mode and 0.4 μm for the E/D process). The design rules are not claimed to be scalable, though they may appear so due to the use of λ. The λ-based representation is employed primarily to facilitate use of both university and industrial CAD packages.

The design rules and design examples in this book utilize purely *Manhattan layout*. This design style is named for its rigid grid-like structure where all features align with the X and Y axes. Manhattan design is used here because it is simple, clean, and compatible with a wide variety of CAD systems. Since it is a typical requirement that all MESFETs are horizontally aligned in most GaAs technologies, insisting that all *other* features be Manhattan is not a severe limitation. Except for occasional minor gains in layout compaction, non-Manhattan layout is not advantageous and, hence, is not needed. The payoff for using this slightly restricted layout style is compatibility with many CAD systems as well as efficient manipulation and checking of layer-to-layer relationships.

Intermediate in complexity between the rectilinear Manhattan layouts and fully general polygon-based layout is so-called *Boston geometry*. This style has most of the properties of Manhattan layout with the added flexibility that features can contain 45° angles. Often this style can be supported efficiently, giving many of the benefits of Manhattan with most of the flexibilities of fully general layout.

8.1.2 Caltech Intermediate Format—CIF

An intermediate level of representation—occupying a position midway between the CAD data base of user's layouts and the physical masks—is desirable. One such representation is Caltech Intermediate Format or, more conveniently, CIF. Because of its presence early in the VLSI age and its adoption by most of the university VLSI community, CIF has become an important widely accepted standard representation for NMOS, CMOS, and even for printed circuit boards. Most university-generated computer-aided design tools as well as many industrial CAD packages "understand" CIF and can read it and/or write it.

Since CIF has the basic primitives to deal with the needs of physical integrated circuit design and the generality to readily adapt to new layers and unit scalings, it is natural to select it for the specification and interchange of physical GaAs designs. There are, of course, other formats that are suitable for representing mask-level information. It is important at this stage of maturity of GaAs that a widely accepted format be used. With this approach, designs can be exchanged and moved from one CAD system to another with a minimal number of problems. There are translation programs available to convert a CIF-based design to any of several industry-standard CAD data base formats, e.g., CALMA [2], EDIF [3], Mann [4].

The official description of the CIF standard is found in Hon and Sequin [5]. An essentially equivalent description is given in Ref. 1. Though it is not strictly line-oriented, CIF resembles assembly language. It is low-level, with each primitive describing one feature or attribute. Statements end with a semi-colon (;). Comments, which are considered to be *statements*, are contained in parentheses. All CIF descriptions end with the **E** (end) primitive.

For the purpose of representing a flat (i.e., nonhierarchical) Manhattan design, only two CIF primitives are required. The syntax of these primitives is

L *layer-name* ;   (set current layer) ;
B *dx dy* x_c y_c ;   (box on current layer) ;

The **L** (layer) primitive sets the current layer to the one specified as its argument. All features following a layer command are placed on the named layer. Layers are *modal*; i.e., they remain in effect until changed explicitly by a subsequent **L** command. Table 8.1 lists the CIF layer names for GaAs D-mode designs. The **B** (box) primitive specifies a rectangle centered at (x_c, y_c) of size $dx \times dy$. All boxes define rectangular regions on the current layer—the one most recently named in an **L** primitive. The default unit utilized for all sizes and coordinates is 0.01 μm.

In order to support hierarchical designs, several other CIF primitives exist. These allow the definition and instantiation of arbitrarily complex subcells, each with their own coordinate system, unit rescaling, and optional translation, rotation, and mirroring. The three CIF primitives supporting hierarchy are

DS *symbol#* [A] [B] ;   (definition start) ;
DF ;   (definition finish) ;
C *symbol#* [*transformation*] . . .; (symbol "call") ;

The **DS** primitive marks the beginning of the definition of a subcell—called a *symbol* in CIF. Symbols are referred to by integer identification numbers called symbol numbers. All geometric primitives between the **DS** primitive and the next **DF** (definition finish) primitive are associated with the designated symbol. The optional **A** and **B** fields of the **DS** primitive specify a multiplicative scale factor (**A**) and a divisor (**B**). Within the symbol definition, all coordinates and sizes are first multiplied by **A**, then divided by **B**, with the result having the standard 0.01 μm units. The default values for **A** and **B**, when unspecified, are 1 and 1.

A symbol can be instantiated—sometimes this is termed "calling" the symbol—via the **C** (call) primitive. The **C** primitive refers to the given symbol by its integer symbol number, which must have been defined previously via the **DS** and **DF**

TABLE 8.1
CIF layers and colors for the GaAs D-mode process

Layer name	Type	Color	CIF code
schottky-metal	Primitive	Red	GS
n-minus	Primitive	Green	GN
n-plus	Primitive	Green, stippled	GI
ohmic-metal	Derived	—	GC
schottky-contact	Contact log	Red, with X	GS, GC
via	Contact log	Violet, with X	GV, GS, GM
second-metal	Primitive	Blue	GM
glass	Primitive	Black, dot stipple	GO

primitives. The optional *transformation* argument(s) allow mirroring about either the x or y axes (specified as **MX** or **MY** respectively) as well as translation and/or rotation. These transformations apply to all features within the instantiated symbol. Translations are specified as

T x y

The above translates the called symbol's coordinate origin to location (x, y) within the coordinate reference frame of the caller, which may itself be within another symbol definition.

Rotations are specified by giving the x and y coordinates of a single point which represents the tip of a vector originating at $(0,0)$ in the coordinate frame of the caller. The direction of the vector defines the x axis of the called symbol. The length of the vector is not significant, as long as it is nonzero. As an example, the following is a call of symbol 16, translating it to (14500, 16780), mirroring about the x axis, and finally rotating 90° counterclockwise. Note that the order of application of transformations is significant; they are always applied in left-to-right order.

C 16 T 14500 16780 **MX** R 0 1 ;

Several other CIF primitives exist but are not described here since they are not required in Manhattan layouts. The list includes path/polygon ("**P**"), wire ("**W**"), and round ("**R**") primitives plus a set of user extensions that allow local enhancements to be hooked into programs that process CIF.

8.2 DESIGN RULES FOR THE SELECTIVE IMPLANT D-MODE PROCESS

The selective implant depletion mode process, though still maturing, is one of several major GaAs industrial fabrication technologies currently available. In this process, the designer has as active devices depletion mode MESFETs, level-shift diodes, and logic diodes. MESFETs are the principal functional devices. Level-shift diodes serve primarily as voltage-dropping elements in level-shift networks. Logic diodes are available for building Schottky diode FET logic circuits (SDFL), as discussed in Sec. 4.1.4. In the selective implant D-mode process there are two independent layers of metalization. One of the layers, *schottky-metal*, is a required participant in all active devices; the other metal layer is general-purpose.

8.2.1 Mask Layers

At the physical design level there are only five primitive mask layers: *n-minus, n-plus, schottky-metal, second-metal,* and *glass*. These are primary layers, corresponding one-to-one with the masks used for fabrication. The two implant layers, *n-minus* and *n-plus*, form the active device regions—channels, sources, drains, cathodes, anodes. The *schottky-metal* layer—formed from a thin film sandwich of titanium, platinum, and gold—is used for the gates of MESFETs, for limited-distance signal routing, and for source and drain connections. The *second-metal* layer is a thicker, general-purpose metalization layer, usually made of gold or aluminum. The final primary layer is *glass,* a negative mask specifying the location of cuts in the protective overglass. Protective overglass passivation is the topmost layer of the final fabricated wafer. Images appearing on the *glass* design layer specify regions that are to have the overglass taken away. Such structures are called *overglass cuts* or *passivation contacts* and are necessary to allow probing or wire bonding to the GaAs die.

In addition to the primary design layers there are several composite layers:

Implant (the combination of both *n-minus* and *n-plus*)
Mesfet (*schottky-metal* over *n-minus*)
Logic-diode (the combination of *schottky-metal* over *n-plus*)
Level-shift-diode (*schottky-metal* over *implant*)

The primitives and composites are augmented by several higher-level abstraction layers which specify the location and size of logical structures—so-called "logs"— which are replaced before mask-making and fabrication by regular, automatically derived arrangements of lower-level physical mask layers. The set of "logs" includes the following layers:

1. *Source-drain* (an ohmic connection between one of the various active layers, *n-minus, n-plus,* or *implant*, and a special metal—so-called "ohmic" metal—which is normally an alloy of Au–Ge/Ni/Au).
2. *Via* (a contact between *schottky-metal* and *second-metal*).
3. *Schottky-contact* (the name given to the portion of a *source-drain* with a *schottky-metal* overlay).[1] The effect of this composition is a contact between *schottky-metal* and any one of the various active implants.

This design specification style—where "log" structures are treated simply as ordinary layers—is used in the computer-aided design tool "MAGIC" [6,7] which is the basis upon which much university and industrial research in GaAs has been built. Appendix A contains a so-called "technology file" for MAGIC, specifying the details of a representative GaAs D-mode process in terms that this technology-independent CAD program can use. This rule set (which is described in the following sections) has been designed to be relatively conservative and uncomplicated so that it will be suitable for an instructional MPC environment. The technology file is not unique, however. The implemented rules can be made more aggressive for higher performance or can be relaxed somewhat to improve yield.

---

[1] *Schottky-contact* is used here as the name of a design "log." This special usage should not be confused with the traditional solid-state physics terminology referring to a physically rectifying metal/semiconductor diode.

Figure 8-1 (color insert) summarizes the mask layers in the GaAs D-mode technology and shows the stipple patterns and colors used in this book to represent each.

8.2.2 Width and Separation Rules for $\lambda = 0.5$ μm

The minimum widths and separations are summarized pictorially in Figure 8-2 (color insert). Enforcement of these (and other) design rules is supported interactively by the MAGIC computer-aided design package, using the GaAs depletion mode technology file found in App. A. Further discussion of computer-aided design tools and techniques for GaAs can be found in Chap. 9. Readers who may have occasion to modify the given technology file or who may be creating a new one will find that material essential.

Most of the width and separation rules are straightforward and easy to learn. Active regions (*n-plus, n-minus,* and *implant*) must always be at least 2.5 μm (5λ) wide and must be separated from other active areas by a minimum of 3.5 μm (7λ). The separation rule is a compromise between density and isolation.

Because *schottky-metal* serves multiple purposes, its width rules are the most complex. There are different minimum width and separation rules for each type of device formed with *schottky-metal*. When used as gate metal, *schottky-metal* can be as narrow as 1 μm (2λ). This rule effectively sets the minimum channel length for MESFETs. The minimum width of *schottky-metal* in logic diodes is 1.5 μm (3λ). In level-shift diodes, the minimum width is 2 μm (4λ). When used for interconnect, *schottky-metal* wiring must be at least 2 μm (4λ) in width, subject to the current density and resistance limitations described in Sec. 8.2.6 and in Chap. 5. The rules for transition between use of *schottky-metal* as gate and as wiring are discussed in the next section on active devices.

No matter how it is used, *schottky-metal* must always be separated from other *schottky-metal* by at least 2.5 μm (5λ). In addition, *schottky-metal* must be separated from active areas (*n-plus, n-minus,* and *implant*) by at least 1 μm (2λ). *Schottky-metal* and *second-metal* are independent layers but coincident edges must be avoided. This rule is not enforced by the present CAD implementation (i.e., the technology file given in App. A). This is left unenforced not because it is impossible to check but rather because coincident edges for very short distances are *allowed*. As a strawman rule, edges of the two metals may be coincident for up to a maximum of 10 μm (20λ). This practice may bring a small benefit to layout density but almost certainly with grave results on speed.

The *second-metal* minimum width rule is 3 μm (6λ). Minimum separation of unrelated *second-metal* features must be $\geq 3\mu$m (6λ). The width and separation design rules are summarized in Table 8.2.

8.2.3 Active Devices

Figure 8-3 (color insert) illustrates the geometrical layer relationships for forming active devices. These relationships are very simple; every device is a composite made from an active layer—*n-plus, n-minus,* or both in combination (which is called

TABLE 8.2
Summary of minimum width and separation rules

Minimum widths by layer—GaAs D-mode process

Layer	Minimum width (μm)	(λ)
n-plus	2.5	5
n-minus	2.5	5
implant	2.5	5
schottky-metal (gate)	1.0	2
schottky-metal (wiring)	2.0	4
second-metal	3.0	6

Minimum layer-to-layer separations—GaAs D-mode process
(active = n-plus, n-minus, or implant)

Layer	active	schottky-metal	second-metal
active	3.5 μm (7λ)	1.0 μm (2λ)	Independent
schottky-metal	1.0 μm (2λ)	2.5 μm (5λ)	Independent
second-metal	Independent	Independent	3 μm (6λ)

simply *implant*)—with an overlay of *schottky-metal*. The *schottky-metal* forms the gate of MESFETs or the anode terminal of diodes. A MESFET is formed at the intersection of *n-minus* and *schottky-metal*. Where the *n-plus* layer intersects *schottky-metal* a signal diode (also known as a logic diode) is formed, with the *schottky-metal* acting as the anode terminal and the *n-plus* implant implementing the cathode. Where both *n-plus* and *n-minus* implants intersect *schottky-metal* a level-shift diode is formed. In this structure, as in the logic diode, the *schottky-metal* forms the anode while the *implant* is the cathode.

Figure 8-4 (color insert) illustrates two layout styles for diodes. The top style works well for in-line diodes such as are required in level-shift networks. The bottom style is somewhat larger in application but it can be used to turn a corner since the gate connection and cathode connections are in a 90° relationship. Note that it has three terminals, a fact that proves problematic for circuit extractors and electrical rules check or analysis tools since one terminal apparently "floats" (it is not connected to any other circuitry). Though not essential, it is good practice to short the two cathode connections together.

Both the in-line and corner diode layout styles are useful and typical in practice. Though only the *level-shift* diode is illustrated, the same layouts can be used for signal diodes as well. The only change necessary in order to make signal diodes is to use *n-plus* rather than *implant* for the underlying active layer.

The design rules allow many variations on the shapes and sizes of the basic diode. A few general guidelines relative to diode layout may be helpful.

1. Because of the high resistance of implanted regions, it is critical for performance reasons that a Schottky contact be placed as close as possible to the diode on the implant (cathode) side. Such contacts should be sized to have the same width as the diode.
2. The always-horizontal gate orientation rule (see Sec. 8.2.5) does not apply to diodes. This extra freedom can help in many layout situations.
3. Even though diode sizes refer to *area* (i.e., the intersections of *schottky-metal* over *implant* or *schottky-metal* over *n-plus*) the shape of the intersection region is significant. The active region of the diode begins at the *schottky-metal* edge nearest the anode terminal and extends approximately 1 µm into the device (toward the cathode terminal). Any extra length past this distance is inconsequential to performance. Thus, diode area specifications should always be interpreted as *width* specifications with diode length set to the design rule minimum.[2]

The two diode layout styles of Fig. 8-4 appear to differ with respect to their ability to tolerate mask misalignment. The 90° style retains the same area even with mask misalignments up to 1λ while the straight-line style does not. Because the diode's electrical properties depend primarily on the active edge, the possible changes in overall area in the straight line style do not have any adverse effect. Both diode layout styles are essentially equal in layout robustness.

Extensions are required in all active device layouts to ensure that the two interacting layers completely cross one another. These rules are totally analogous to the extension rules between polysilicon gate material and underlying diffusions in the NMOS and CMOS technologies. The extension rules for active devices in the GaAs selective implant depletion mode process are illustrated in Fig. 8-5a (color insert). *Schottky-metal* must always extend 1 µm (2λ) past any edge of an active region. The active region must, in turn, extend at least 1.5 µm (3λ) past the gate. In practice this extension is not a limiting factor in layouts since it is always necessary to attach metal to the source, drain, or cathode terminals. Figure 8-5b (color insert) illustrates some typical source/drain connections to a MESFET. It is important for performance reasons to make these ohmic source/drain/cathode connections as close as possible to the active devices. This is so because the source/drain and cathode regions are quite resistive (on the order of 800 to 1000 Ω/□).

There are two minimum-width rules for *schottky-metal*. One rule applies when *schottky-metal* is used for gate material and a wider rule applies when it is used for wiring. The transition between these two distinct roles is also governed by a design rule. The transition from gate to wiring width should occur no closer than 1 µm (2λ) to the active device. The transition should be gradual—ideally 45° angles would be used. With Manhattan layout there are only two permissible forms; the transition should take one of the exact forms depicted in Fig. 8-6 (color insert). When the wiring

---

[2] The reader is cautioned that diode models in SPICE use AREA specifications which are internally converted to a device having WIDTH × *minimum* LENGTH chosen so as to equal the specified AREA.

metal turns a corner (i.e., has a 90° relationship to gate metal) the transition requires an intermediate square $1.5 \times 1.5\,\mu\mathrm{m}$ ($3 \times 3\lambda$), juxtaposed to the full width $2\,\mu\mathrm{m}$ (4λ) wire. Past this transition the wiring metal can become arbitrarily wider; only the design-rule minimum wiring width is depicted in Fig. 8-6 (color insert). When the wiring metal does not turn a corner, the width must change symmetrically by $1\,\mu\mathrm{m}$ (2λ) on each side of the gate metal, as depicted in the right portion of Fig. 8-6.

8.2.4 Contact Structures

Contact structures are specified in a design through the use of "logs"—high-level abstractions appearing to the designer as simple mask layers. The use of "logs" can greatly simplify a layout because the designer has fewer details to be concerned with. Because contacts are regular, their lowest-level details can be generated automatically from the "logs." Figure 8-7 (color insert) shows two logical layouts for contacts. The contact on the top connects *schottky-metal* to *second-metal* and is called a *via*. The design rules for this structure are simply the minimum width and separation rules governing each of its component layers. The *via* layer must be at least $5\,\mu\mathrm{m}$ (10λ) wide in both dimensions. *Second-metal* and *schottky-metal* can approach and intersect in any manner at the *via* as long as their individual minimum widths are maintained.

The *via* structure is implemented as a traditional contact between two metals. Essentially, a cut is made in the insulating intermediate layer so that the second-level metalization can fill in the cut and short the two metal layers together. The location and size of the cut are automatically derived from the location and size of the *via* by shrinking the *via* outline by $1.0\,\mu\mathrm{m}$. The shrink operation automatically centers the cut within the *via* area and guarantees the correct mask overlap amounts. The *via*'s minimum width rule has been chosen so as to provide sufficient internal space for *via* cut generation.

The contact shown in the bottom portion of Fig. 8-7—a so-called *schottky-contact*—forms an ohmic connection between *n-minus* and *schottky-metal*. Schottky-contacts can be used with any type of implant, not just with *n-minus*. The *schottky-contact* structure is implemented differently than the *via*. Correct construction of the contact involves the interplay of three distinct layers: *n-minus*, *schottky-metal*, and *source-drain*. The contact location and size are defined by the use of two design "logs"—*schottky-contact* and *source-drain*. These "logs" define regions which are logically ORed together, then shrunk by $1.0\,\mu\mathrm{m}$ to produce a mask for the patterning of *ohmic-metal* (CIF layer "GC"). The *ohmic-metal* is put down directly on the surface of the GaAs wafer, making contact with the implanted region below (one of the layers *n-minus*, *n-plus*, or *implant*). The *schottky-contact* part of the structure contains *schottky-metal*, thus providing a contact between any of the types of implanted regions and *schottky-metal*. Photograph 8-3 shows some *schottky-contacts* after fabrication. The rough surface of the *ohmic-metal* makes the contact area quite easy to recognize in the photograph.

The rules for forming Schottky contacts are simply composites of the minimum width and separation rules of their individual component layers. The *source-drain*

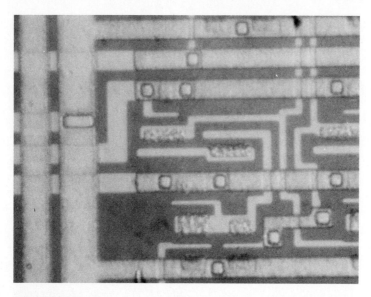

PHOTOGRAPH 8-3
Microscope photograph of GaAs D-mode circuits.

portion of the Schottky contact has a design rule minimum width of 5.5 μm (11λ). The *schottky-metal* covering it (known within the contact structure as the *schottky-contact* layer) has the standard minimum width for wiring of 2 μm (4λ). The only extra rule concerning this contact is that *schottky-metal* and *ohmic-metal* must either have coincident edges or edges that are *exactly* 1 μm (2λ) apart. There must be at least a 2 μm (4λ) wide overlap area at the intersection of these two layers. These somewhat unusual design rules are simply statements of the layer-to-layer relationships that are necessary to ensure successful automatic derivation of the entire contact structure with correct surrounds, overlaps, and extensions.

8.2.5 Miscellaneous Rules

There are several important rules that do not fall cleanly into any one category. Because of their global nature and implementation-related restrictions of CAD systems, the checking of these rules is often difficult to implement. Such rules are not supported in the MAGIC technology file given in App. A.

HORIZONTAL GATES. All MESFET gates must run horizontally due to the anisotropic nature of GaAs. Logic and level-shift diodes are not affected by anisotropic threshold voltage as the forward characteristic is controlled by barrier height and series resistance and not by doping and thickness. Though the requirement for horizontal MESFET gate alignment may appear to be a major layout restriction, in fact its effect is nowhere near as severe as the effect of the required inclusion of level-shifters on the output of each gate.

PLATE 8–1
A GaAs D-mode multiproject chip wafer containing 1987 UCSB class projects. (*Processing courtesy of Rockwell International, Thousand Oaks, Calif.*)

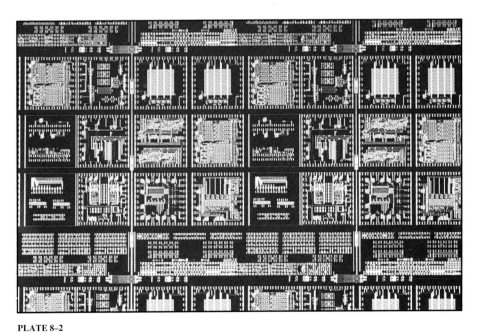

PLATE 8–2
Close-up of GaAs multiproject reticle.

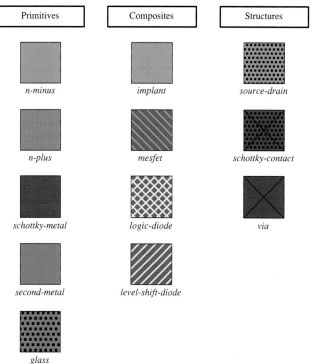

FIGURE 8–1
Design layers of the GaAs D-mode process.

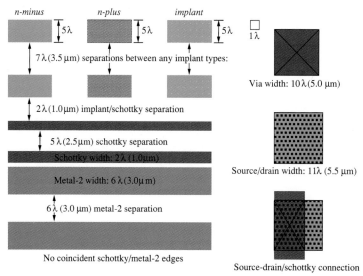

FIGURE 8–2
Minimum width and separation design rules of the GaAs D-mode process.

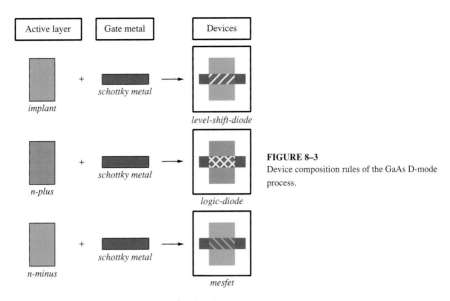

FIGURE 8–3
Device composition rules of the GaAs D-mode process.

Schottky-metal extensions: 2λ (1.0 µm)
Active layer extensions: 3λ (1.5 µm)

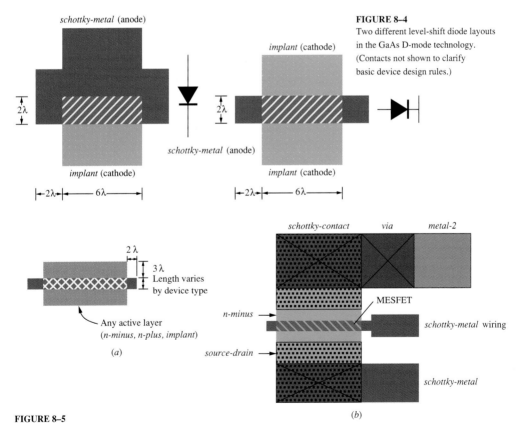

FIGURE 8–4
Two different level-shift diode layouts in the GaAs D-mode technology. (Contacts not shown to clarify basic device design rules.)

FIGURE 8–5
GaAs D-mode extension rules and contact techniques. (a) Extension rules for active devices. (b) Typical MESFET with source/drain connections.

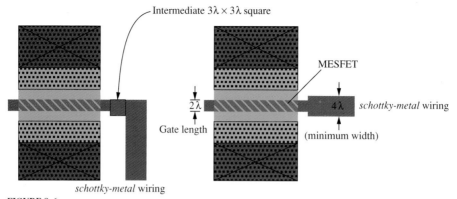

FIGURE 8–6
Gate metal-to-wiring metal transition rules.

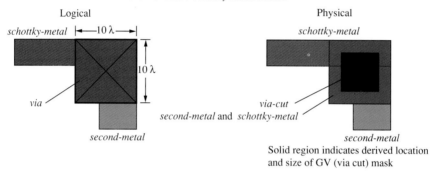

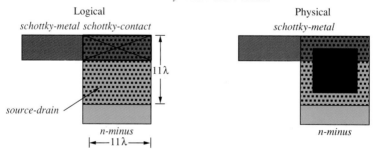

FIGURE 8–7
Logical and physical contact structures in the GaAs D-mode technology.

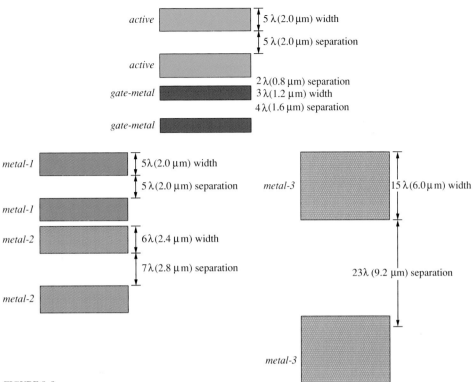

FIGURE 8–8
Minimum width and separation design rules for GaAs E/D process.

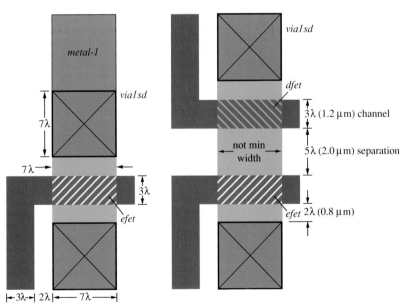

Minimum device channel length is 3λ (1.2 μm), minimum width 5λ (2 μm).
Devices are not shown with minimum width.

FIGURE 8–9
Design rules for active devices in the GaAs E/D process.

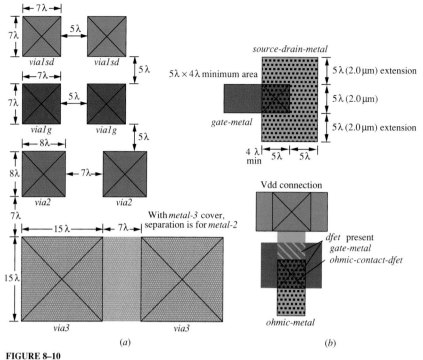

FIGURE 8–10

Minimum width and seperation rules for contacts in the GaAs E/D process. (*a*) Rules for *via*-type contacts. (*b*) *Contact-0* formation rules.

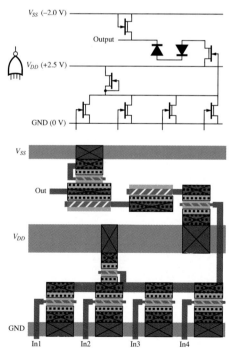

FIGURE 8–11

GaAs D-mode four-input NOR (logic symbol and layout-oriented schematic).

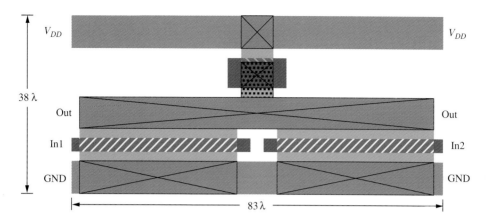

FIGURE 8–12
Two-input NOR—GaAs E/D process.

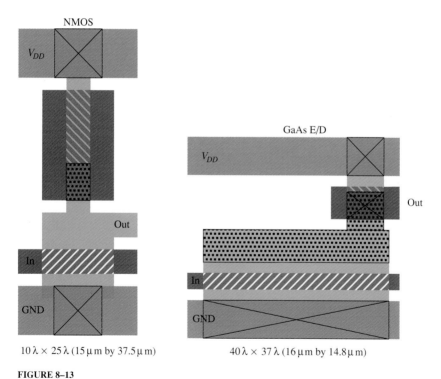

$10\lambda \times 25\lambda$ (15 μm by 37.5 μm)　　　$40\lambda \times 37\lambda$ (16 μm by 14.8 μm)

FIGURE 8–13
Side-by-side comparison of NMOS and GaAs E/D inverter layouts.

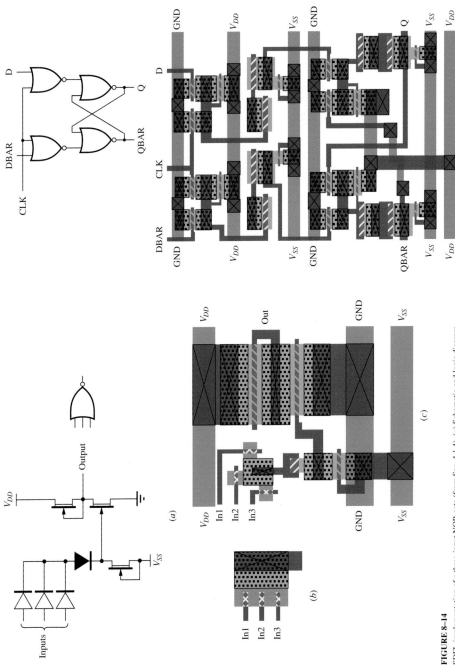

FIGURE 8-14
SDFL implementation of a three-input NOR gate (from Sec. 4.1.4). (*a*) Schematic and logic diagram for SDFL three-input NOR gate. (*b*) Alternate layout for logic diode cluster. (*c*) Layout of SDFL three-input NOR gate.

FIGURE 8-15
Clocked D flip-flop made from GaAs FL NORs.

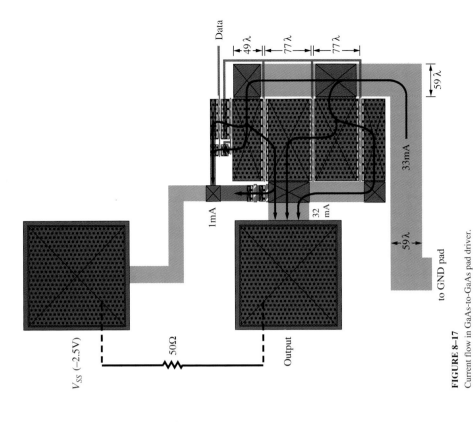

FIGURE 8-17
Current flow in GaAs-to-GaAs pad driver.

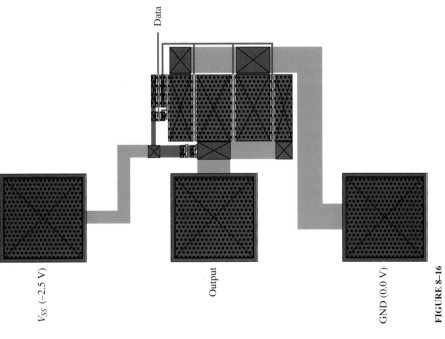

FIGURE 8-16
Simple GaAs-to-GaAs pad driver (from Sec. 7.2.1).

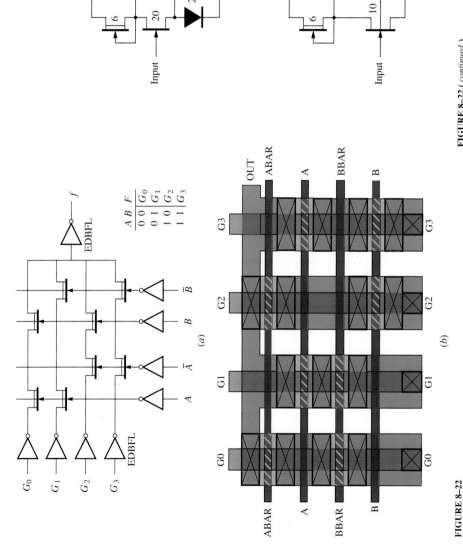

FIGURE 8-22

GaAs general function block using D-mode pass transistors. (*a*) Logic diagram of two-variable general function block. (*b*) Layout of pass transistor network (rotated 90° and sans drivers).

FIGURE 8-22 (*continued*)

(*c*) Schematic diagram of EDBFL driver for general function block data lines (all device widths are given in micrometers, lengths are minimum). (*d*) Schematic diagram of SBFL driver for general function block select lines (all device widths are given in micrometers, lengths are minimum).

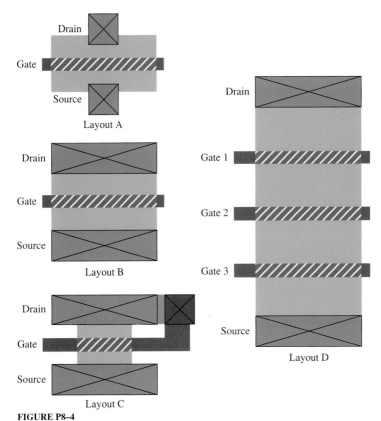

FIGURE P8–4
Four design-rule correct MESFET layouts.

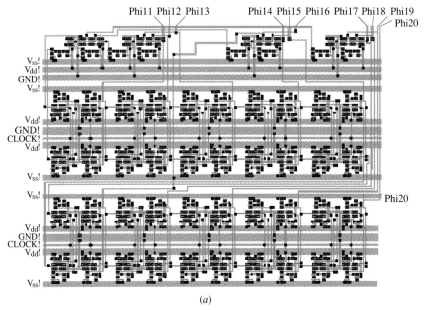

FIGURE 9-16a
A session with startG ad esim. 10-stage shift register to be simulated.

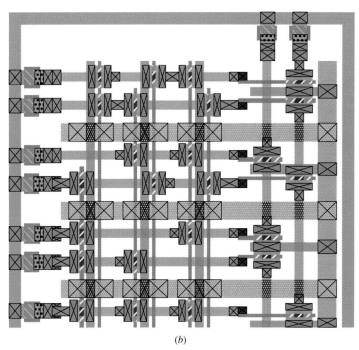

FIGURE 9-17b
PLA Layout as generated by "mpla" (sans drivers).

CLOSED PATHS IN SCHOTTKY-METAL. Because the fabrication process involves deposition of layers of titanium and platinum/gold and then lift-off using photoresist, closed paths on this layer are strongly discouraged. It is not good practice to "box in" regions of a circuit by surrounding them with densely packed *schottky-metal*. During fabrication it is important to be able to lift off the unwanted metal without breaks occurring.

METAL OVER ACTIVE DEVICE REGIONS. Use of *second-metal* directly above MESFETs is not desirable. Capacitive coupling problems are likely if two high-frequency signals occur in such close proximity. The silicon nitride isolation layer between *schottky-metal* gates and *second-metal* is nominally only 0.6 μm thick. Electric fields caused by voltages on such metal runs can shift the effective thresholds of active devices that are covered. Because excessive capacitance can substantially impact the performance of any high-frequency circuit, the practice of running *second-metal* across any dynamic node—regardless of the layer(s) involved—is not encouraged. This is not a design rule, but rather a layout guideline for achieving optimum performance.

COINCIDENT *SECOND-METAL* AND *SCHOTTKY-METAL* EDGES. Specification of *second-metal* and *schottky-metal* features with coincident edges should be avoided. The maximum-distance guideline for coincident edges is 10 μm (20λ). This guideline arises from the fact that a lift-off process is used for patterning the *schottky-metal* features. Lift-off tends to leave "rabbit-ears"—sharp upward-pointing edges—that can cause shorts when a coincident *second-metal* edge occurs above. In addition, there are several other less significant but contributing effects that necessitate this noncoincident edge rule. This guideline is not enforced by the MAGIC rule set implemented by the technology file given in App. A.

8.2.6 Electrical Design Rules

In addition to general circuit design and GaAs-specific circuit theory details, there is a small set of electrical design rules that must be followed. Current density limits must be strictly observed or electromigration may occur. Electromigration is a runaway metal migration process that, once started, will result in an open circuit in the conductor carrying the excessive current. Unless the circuits in the affected area are redundant, this process destroys the capabilities of the chip within which the subject circuit resides. The maximum current density limit (J_{max}) for the metal systems of interest is given by

$$J_{max} = 2 \times 10^5 \text{A/cm}^2$$

Since the computation of current density requires a known cross-sectional area, knowledge of nominal metal thicknesses is required. For the selective-implant D-mode process, representative metal thicknesses are

Schottky-metal   0.3 μm
Second-metal   0.6 μm

Using the conservative metal migration rule for J_{max} given above, the guideline for each of the two metals is

Schottky-metal $\leq 300\ \mu A$ per λ of width
Second-metal $\leq 600\ \mu A$ per λ of width

Another electrical rule relates to the maximum size of any single MESFET transistor. This rule arises primarily from fabrication difficulties in making very wide, but minimum length MESFETS. A single MESFET cannot have a gate wider than 75 μm (150λ). If a larger device is needed, such as might be the case in a pad driver or receiver circuit, multiple MESFETs can be connected in parallel. This rule is not enforced by the MAGIC technology file given in App. A.

Whenever large devices are used, particularly when several are connected in parallel, one must know how much current flows and where it flows. This information is critical in sizing the source/drain and other metalization regions so that metal migration rules are followed. Where large currents flow, the metal supply lines (including the source/drain metalization) must be adequately sized. Section 8.4.5 discusses a simple output pad driver, including the considerations for sizing the features in order to satisfy all of the electrical design rules.

8.2.7 CIF Layers and Colors

The CIF layer names used for the depletion mode GaAs process are given in Table 8.1. Associated with each layer is a color that is used when that layer appears on the screen of a CAD system or in a color plot. The colors in Table 8.1 have been selected in order to be similar to the colors and associated semantics used for NMOS and CMOS. Specifically, the metal layer is always shown in blue and the gate material (in this case *schottky-metal*) is shown in red. Active areas are green with distinct stippling to lend details concerning the specific type of active area. Contact structures are shown with black crosses. Because *source-drain* and *schottky-contacts* resemble the NMOS buried contact, their appearance on CAD systems has been composed of distinct stipple patterns. Such rules help to remind the designer—by consistent on-screen appearance—about the details of which layers are involved in the various structures. This sort of assistance can greatly aid hand-done layout and compaction.

8.3 DESIGN RULES FOR THE SELF-ALIGNED E/D PROCESS

The self-aligned enhancement/depletion process uses tightly controlled implant depth and dose to provide both enhancement and depletion mode MESFETs. As in the selective implant depletion-mode technology, level-shift diodes can also be made. E/D designs bear a superficial resemblance to similar functions implemented in NMOS. In this process, it is not necessary that all gates on the wafer be aligned to match the GaAs crystal orientation. Gates may be oriented either horizontally (preferred) or vertically with the horizontal orientation defined as that one which is aligned with the major flat of the wafer. The effect of this relaxation in layout rules is further enhanced by the availability of a second and third layer of general-purpose metal. Overall, layout density is generally superior to D-mode due to these enhancements as well as the lack of a requirement for level-shift networks on the output of each gate.

The design rules discussed in this section are derived from the Vitesse E/D self-aligned GaAs process rules [8] (adapted and printed with permission).

8.3.1 Mask Layers

The primary design layers in the GaAs E/D process are *active, gate-metal, metal-1, metal-2,* and *metal-3*. These layers correspond one-to-one with the masks used for fabrication. The *active* layer is used to form the active device regions—sources, drains, and channels. The *gate-metal* layer is used for the gates of MESFETs and diodes, for limited-distance signal routing, and for source and drain connections. There are three general-purpose layers of metalization: *metal-1, metal-2,* and *metal-3*.

There are several other masks needed for fabrication. In all, eleven are used for the self-aligned E/D process. Three masks are required for identifying the location of via holes used to interconnect between the three metalizations. These masks are automatically derived from the various "via" design layers which are discussed in Sec. 8.3.4. Another derived mask is the depletion-implant mask, which determines which MESFETs become depletion type. This mask is derived automatically from relationships among the various device design layers. In order to make contact with device terminals, ohmic metal is used. Hence another mask, also automatically generated, is needed. Finally, a passivation-cut mask is used to identify the positions of cuts in the protective passivation oxide to allow bonding or probing. This mask is derived from the *pad* design layer.

8.3.2 Width and Separation Rules for $\lambda = 0.4$ μm

Figure 8-8 (color insert) illustrates the minimum width and separation rules for the GaAs self-aligned enhancement/depletion process. The MAGIC technology given in App. B implements interactive checking of the stated design rules.

The minimum width rules for GaAs E/D are similar to those for the GaAs depletion mode process. *Active* regions must be at least 2 μm (5λ) wide and must be separated from other *active* regions by a least 2 μm (5λ). When *active* is used as a resistor, it may be 1.6 μm (4λ) wide. *Gate-metal* has a minimum width of 1.2 μm (3λ). This rule sets the effective minimum channel length for this technology. *Gate-metal* must be separated from *active* by at least 0.8 μm (2λ) and from other *gate-metal* by at least 1.6 μm (4λ).

The three general-purpose metalization layers are completely independent. *Metal-1* has minimum width and separation rules both set at 2 μm (5λ). *Metal-2* has a minimum width rule of 2.4 μm (6λ). The *metal-2* to *metal-2* minimum separation is 2.8 μm (7λ). A third metalization layer—*metal-3*—is available, but because it runs over much more rugged topography, its minimum width must be at least 6 μm (15λ) and its minimum separation from other *metal-3* must be 9.2 μm (23λ). Because of these large design rules, *metal-3* is used primarily for power and ground supply connections. The coincident edge guidelines and layer-to-layer contact structures for the three general-purpose metals are discussed in Secs. 8.3.4 and 8.3.5.

8.3.3 Active Devices

The layout design rules for active devices in the GaAs E/D process are depicted graphically in Fig. 8-9 (color insert). The layout rules are quite simple; there are two varieties of MESFET, depletion mode *(dfet)* and enhancement mode *(efet)*. The underlying layer in these composite structures is *active*. An enhancement mode MESFET is formed when *gate-metal* crosses *active*. This composition effect is implemented in the MAGIC layout editor—whenever *gate-metal* is placed over *active*, or vice versa, the intersection region automatically becomes *efet*. Depletion mode devices, however, must be explicitly created as such by the user; i.e., there is no automatic composition as implemented for *efets*.

Gates may either be horizontal—aligning with the crystal structure of the wafer—or vertical. There may be no bends or angles in the gates. Minimum gate lengths are 1.2 μm (3λ) for both types of MESFETs. Two MESFETs may not be closer than 2 μm (5λ). In order to ensure that devices are well formed, even with up to 1 λ of mask misalignment,[3] extension rules are required. *Gate-metal* must extend 0.8 μm (2λ) past the edge of *active* and *active* must extend 1.2 μm (3λ) past the edge of the *active/gate-metal* intersection. In practice these extensions are not a limiting factor in layouts since it is always necessary to attach metal to the source and drain terminals. It is important for performance reasons to make these source/drain connections as close as possible to the active devices. This is necessary because the source/drain regions are quite resistive.

8.3.4 Contact and Bonding Pad Structures

Figure 8-10 (color insert) illustrates the minimum width and separation rules for contact structures in the GaAs E/D process. As in the CAD technology description for the selective implant D-mode process, contacts involving via cuts are "logs" which are replaced during CIF generation with lower-level derived structures. Because it does not involve a via cut and also because it is considerably less regular than the other contacts, the so-called *contact-0* is not implemented as a design "log." It is simply made by the user through careful combination of its component parts.

A *contact-0* is a relative of the *buried-contact* of NMOS. It connects *gate-metal* to *active*. The via-type contacts are named according to the layers they interconnect. *Via1sd* connects first-level metal to source/drain areas. *Via1g* makes a connection

---

[3] In a scalable rule set, lambda is chosen to represent the resolution of the fabrication process. Generally speaking, the design rules allow errors in mask alignment, etching, and other processing steps to the extent of 1λ. Though the rules described here are not scalable, the same principle holds.

between *metal-1* and *gate-metal*. *Via2* connects second-level metal *(metal-2)* to first-level metal. Finally, *via3* connects the top-level of metal, *metal-3*, to second-level metal.

The *contact-0* structure can take a variety of forms. Two design layers are involved in its specification: *gate-metal* and *source-drain-metal* (a design layer that represents the coexistence of *active* and ohmic metal[4]). Whenever *gate-metal* and *source-drain-metal* intersect, a *contact-0* is formed. The intersection area (which itself is a design layer, *ohmic-contact*) must always be at least $1.6 \times 1.6 \,\mu\text{m}$ ($4 \times 4\lambda$). *Source-drain-metal* must extend on at least one side of the intersection area by 2 μm (5λ). *Gate-metal* must extend at least 1.2 μm (3λ) from the *ohmic-metal* intersection region.

Because *source-drain-metal* implicitly includes *active* and *contact-0* also includes *gate-metal*, a MESFET (with gate shorted to source) exists under this structure. If the *contact-0* is used only for connecting *gate-metal* to *active*—acting simply as a layer change similar to the *buried-contacts* of NMOS—then the MESFET remains unused since no connections are made to its third terminal. In some NMOS-like circuits it is desirable to build a depletion load, a depletion-mode MESFET with $V_{gs} = 0$ and drain usually connected to V_{dd}. In such cases a *contact-0* is placed directly on top of the *dfet* just at the source end (refer to Fig. 8-10b in color insert). When *ohmic-contact* and *dfet* coexist like this, the design layer in this region becomes *ohmic-contact-dfet*. The channel length of the depletion load in such structures includes the *dfet* as well as the *ohmic-contact-dfet* regions, since the gate-to-channel short occurs where *gate-metal* ends at the *ohmic-contact-dfet/gate-metal* edge.

The minimum width and separation design rules for via-type contacts can be read from the annotations given in Fig. 8-10a. *Vialsd* has a minimum width of 2.8 μm (7λ). This contact must be separated from other *vialsd* and *vialg* contacts by 2 μm (5λ). These minimum separations are dominated by the minimum separation rules for the metals from which they are composed. There are no rules governing how large a *vialsd* can be. Extremely large ones should be subdivided into multiple smaller contacts, however. The *vialg* contacts must be at least 2.4 μm (6λ) in width and length. Like the *vialsd* contact, there is no rule for maximum size. *Vialg* contacts must be separated from other *vialg* contacts by at least 2 μm (5λ).

Via2 contacts have a slightly larger minimum size, $3.2 \times 3.2 \,\mu\text{m}$ ($8 \times 8\lambda$). Second-metal features and associated contacts are independent of first-metal features, but stacked contact structures are not allowed. *Via2* can touch but cannot overlap other first-level contacts, *vialsd* and *vialg*. Because of the rougher terrain on the topmost surface of the fabricated wafer, *via3* contacts are required to be much larger in size. Their minimum size is $6 \times 6 \,\mu\text{m}$ ($15 \times 15\lambda$). *Via3-to-via3* separation is dominated

---

[4] In early versions of the E/D GaAs technology, the *source-drain-metal* design layer was known as *ohmic-metal*, which is perhaps a better description of what it physically represents. For compatibility purposes the two design layer names are allowed to be used interchangeably, referring to the same composite of *active* and *ohmic-metal*.

by the separation rule for *metal-3*—9.2 μm (23λ). When two *via3s* are covered by common *metal-3*, the *metal-2* separation rule (7λ) limits their proximity. *Via3* cannot overlap or stack with any of the other contact types.

One final design layer, *pad*, exists. This layer is used to specify the location and size of bonding or probe pads. Like the various contacts, *pad* is a design "log." When mask-level details are generated, *pad* produces images on the top two metal layers, on their interconnecting via layer, and on the passivation-cut layer. The design rules for *pad* are simple; its minimum size is 110 μm (275λ) and the minimum *pad-to-pad* separation is 40 μm (100λ). This allows a pad center-to-center spacing of 150 μm. Of course, since *pad* includes the top two metal layers implicitly, the minimum separation rules of these layers are also enforced for all *pad*-to-metal relationships.

Bonding pads should always be located around the periphery of the die. Circuitry may be placed between pads, though the designer is cautioned against this practice. The die bonding procedure may introduce local crystal damage which can adversely affect nearby circuit performance. For the same reason, bonding pads must never by placed directly over active circuitry.

8.3.5 Miscellaneous Rules

COINCIDENT EDGE RULES. Because etching is used (rather than a lift-off technique as in the selective-implant D-mode process), there are relatively few coincident edge rules. There are still significant effects, however, that discourage the use of coincident edges. The largest single effect is fringe capacitance, which has a direct impact on performance as well as on crosstalk between signal conductors.

Most of the coincident edge restrictions for the self-aligned E/D process are not seen at the user design level because the design "logs" and other rules allow MAGIC to generate masks that are correct by construction. Two user-enforced rules remain. These are not checked by the MAGIC technology file given in App. B.

When discussing any of the general-purpose metal layers, edges are considered coincident if they occur within 1 μm of one another. There are two coincident edge guidelines. The first guideline pertains to edges of *metal-1* and *metal-2*. Such edges can be coincident, but may only remain so up to a distance of 6 μm. The second user-enforced edge rule concerns parallel *metal-1* and *metal-2* edges bridging a space between adjacent *metal-3* lines. This situation would presumably arise only in grid-like array applications, the kind of thing that would likely be generated by a CAD tool rather than from hand layout. In any case, when such a situation arises, there must be at least 3 μm between one of the *metal-3* lines and the beginning of the coincident edge or a break of at least 2 μm in the coincident edge.

MAXIMUM WIDTH MESFET. The maximum width of any single MESFET is 75 μm (about 188λ). Larger device equivalents must be made by a parallel connection of multiple MESFETs. As in the case of the selective-implant D-mode process, one must pay particular attention to current density, sizing source/drain regions appropriately (as in Sec. 8.3.6).

SECOND-METAL OVER ACTIVE NODES. There is no rule against running *second-metal* (*metal-2*) over active nodes, but there are performance-related reasons for avoiding it. In particular, crosstalk will be worsened due to increased coupling.

8.3.6 Electrical Design Rules

The basis for the electromigration rule is identical to that given for the selective-implant depletion-mode process (refer to Sec. 8.2.6). However, because of differences in metal composition as well as varying thicknesses and lambda settings, the maximum current density values are different. Specifically, the rule for avoiding metal migration in each of the four metal systems is

$$
\begin{aligned}
\textit{Gate-metal} &\leq 2 \text{ mA per } \lambda \text{ of width} \\
\textit{Metal-1} &\leq 400 \text{ }\mu\text{A per } \lambda \text{ of width} \\
\textit{Metal-2} &\leq 560 \text{ }\mu\text{A per } \lambda \text{ of width} \\
\textit{Metal-3} &\leq 1.1 \text{ mA per } \lambda \text{ of width}
\end{aligned}
$$

Because of its rather high resistivity it is unwise to approach the electromigration limit above for *gate-metal*. The IR drop would most likely be unacceptable in such situations, i.e., when *gate-metal* is being used for high-current interconnections.

8.4 DESIGN EXAMPLES

This section contains layouts of several different types of circuits. The circuits are kept small so that all layout details can be clearly seen. More highly integrated systems can be built up from stacking and interconnecting instances of such layouts.

8.4.1 D-Mode Four-Input NOR

Figure 8-11 (color insert) shows a layout of a four-input NOR using the selective-implant D-mode process. Photograph 8-3 shows a photomicrograph of this layout after fabrication. This cell is designed to allow efficient horizontal stacking. There are three second-metal supply rails running across the cell horizontally. The inputs enter from the bottom and the output is available along the top edge. The bottom half of the layout implements the NOR function. The top half is a two-diode source-follower level-shifter. It is abundantly clear from the layout that the level-shift network occupies a significant area. In fact, the area occupied by the level-shifter is, in this layout, approximately the same as the area required for the NOR logic.

This gate has been used extensively in several MPC runs fabricated both by Rockwell International and by Gigabit Logic. Speed measurements indicate approximately 80 ps of gate delay for this layout when loaded with 2 fanouts to similar gates.

8.4.2 E/D Two-Input NOR

The physical design of a two-input NOR gate using the GaAs enhancement/depletion technology is shown in Fig. 8-12 (color insert). Comparison with the D-mode four-input NOR shown in Fig. 8-11 reveals the layout density improvement that is possible due to the elimination of the level-shift network.

The layout of an E/D gate looks remarkably close to the layout for a similar gate in NMOS. Figure 8-13 (color insert) shows an NMOS inverter side by side with a GaAs E/D inverter. Layout details appear to be very similar to NMOS, but there are some differences. The practice of using longer-than-minimum gate lengths on the D-mode pull-up in NMOS is normally not done in GaAs. It is more typical to use a minimum length pull-up device with the pull-down MESFET widened appropriately. The bottom power rail (labeled "GND" in Fig. 8-13) may, in fact, be connected to some voltage other than ground potential in a GaAs E/D circuit.

8.4.3 SDFL Three-Input NOR

Figure 8-14 (color insert) illustrates a three-input SDFL NOR gate circuit and its layout. The input diode cluster is the most unique aspect of the layout. It is critical to performance that each diode be positioned and oriented so that its widest edge is spaced as closely as design rules allow to the *source-drain* metalization. The layout is constrained by the rule for *schottky-metal*-to-*schottky-metal* minimum separation. Use of *second-metal* over active circuit nodes has been avoided. In fact, the only place where *second-metal* overlaps any other circuit node is where the ground line (labeled "GND") runs over the input stage current source. The point where the two features cross is constantly at ground potential so additional stray capacitance does not adversely impact circuit performance.

An alternate layout for the three-diode cluster is shown in Fig. 8-14b (color insert). This alternate layout style is more regular and could be expanded to make a four- or five-input NOR gate. Though the layout is design-rule legal it does not follow the preferred diode layout guideline with the widest edge minimally spaced from the *source-drain* contact. In this orientation the diode voltage drops will not scale with area as they do normally. The common source-drain node is also much larger than the irregular one of Fig. 8-14c (color insert). Thus it has more stray capacitance and the input stage of the circuit will suffer a small performance degradation. Nevertheless, from a layout point of view, the alternate structure is cleaner, more regular, and more easily generalized to add more inputs.

Device sizes are 3×2 μm for all diodes as well as for the pull-down (current source) MESFET. The output stage MESFETs are 17 and 10 μm for the pull-up and switch MESFETs respectively.

8.4.4 FET Logic Latch

Figure 8-15 (color insert) illustrates the physical design of a clocked D-type latch. The design is constructed from simple FET-logic NOR gates. Whenever CLK = 0, the latch is transparent with Q (QBAR) following D (DBAR). With CLK = 1 the cross-coupled gates in the output section are logically isolated, becoming essentially a pair of cross-coupled inverters.

The layout is strongly dominated by the influence of three supply rails. In the layout shown, the two-level logic is implemented as two rows of gates stacked vertically. Each row has its own supply rails and corresponds directly to the logic diagram with two gates per row. The supply rails are organized so that the latch can be arrayed horizontally with other identical latches or with other FL or BFL gate layouts sharing the same supply rails. The layout designer has attempted to avoid running supply rails over active nodes, though in cross-coupled circuits a few such interactions are inevitable.

Device sizes are chosen for low fan-out applications. The NOR-logic pull-down FETs are $2 \times 16\lambda$ with a pull-up that is 20λ wide. The level-shift diodes are $4 \times 26\lambda$ and all current sources are $2 \times 10\lambda$. The latch has been used in several designs, each with a measured performance well above 1 GHz.

8.4.5 A Simple GaAs-to-GaAs Pad Driver

Figure 8-16 (color insert) illustrates the layout of a simple GaAs output pad driver using the selective-implant depletion-mode GaAs technology. The bonding pads shown in the figure are 100 μm square and are located on 200 μm centers. The three adjacent pads are used to attach to V_{ss} (-2.5 V: top pad), the output signal (middle pad), and ground (0.0 V: bottom pad). Close proximity of bonding pads for signal and termination voltages facilitates the use of a high-speed probe card for wafer performance measurements as well as impedance-matched packaging for wire-bonded packaged parts.

Device sizes are as given in Sec. 7.2.1 with metal runs and contacts sized according to the metal migration guidelines. Most of the dc current flows from the GND pad through three large output MESFETs into the output pad, and finally through the 50-Ω external termination to the V_{ss} supply. The dc current for this design is 33 mA. Each section of the current-carrying metalization has been sized using the metal thicknesses and metal migration limits given in Sec. 8.2.6. Figure 8-17 (color insert) illustrates the major current flows and the feature sizes. Note that output driver device widths determine the width of the main circuit block. The height of each section (for example, 77λ) is determined primarily by the amount of current flowing through the thin *schottky-metal* interconnection that feeds from *metal-2* down to the large *schottky-contact* areas.

8.5 SYNTHESIS USING GENERIC STRUCTURES

GaAs circuits are incorporated in a system primarily for reasons of increased performance. Naturally, the attention of the designers is drawn to circuits and structures that are capable of top-end speed with only secondary consideration given to layout issues.

It is often the case within a high-performance integrated circuit that only a relatively small set of paths are actually required to operate at optimum speed. For supporting circuits, other criteria such as area, shape, fan-out loading, or ease of design and layout may be dominant factors in their design. Whenever speed is lesser in importance than design and layout ease, the designer may wish to consider the use of generic structures that can be automatically generated—with correctness guaranteed by construction—to realize a given specification.

8.5.1 Programmable Logic Arrays

Programmable logic arrays—more simply, PLAs—are collections of device and interconnect tiles which are arrangeable in a regular grid-like pattern for detailing ("programming") by a CAD program. By choosing from a small set of building block tiles at each grid point, the PLA can be made to implement any boolean function.[5] Modern CAD packages support a wide variety of types of PLAs, some allowing user function(s) to be specified in symbolic equation form [9]. The user interface and high-level description of one widely known PLA CAD program suite is discussed in Chap. 9.

Figure 8-18 shows the design schema for the logic planes of an E/D GaAs NOR/NOR PLA. In the NOR/NOR style, both the AND (product) plane and the OR (disjunctive sum) plane are realized from NOR gates. A built-in feature of this organization is that the AND-plane inputs must be complemented (since a NOR gate can be equivalently viewed as an AND gate with complemented inputs) and the outputs arrive inverted. Since double-rail inputs are used in the AND plane (for generality) and considerable fan-out drive is needed, particularly for a function with large numbers of products, the need for inverted inputs is of little consequence. It is advisable to buffer the PLA output as well—thus implementing the final, required inversion—so that circuits external to the PLA can use its output(s).

Figure 8-19 gives more detail of the PLA implementation. The AND-plane inputs are driven by E/D superbuffers which must be sized for driving high fan-out— 10 gates is a sufficient bound for most cases of interest. If the PLA CAD software can deal with parameterized drivers (or if it can select from a library based on actual fan-out in the implemented function) then the size and power requirements of a given PLA can be made to closely fit the needs of the application. If this is not the case then worst-case design must be used, resulting in PLAs that potentially use more power and occupy more area than the application requires.

Similar sizing considerations occur for the buffers located between the AND and OR planes, though experience shows that the fan-out situation in the OR plane is often less than that found in the AND plane. The AND and OR planes are "programmed" by

---

[5] There are, of course, limits on the size of function that can be realized in a given PLA technology. In GaAs, the critical limiting factors are fan-out loading (which ultimately affects performance), overall PLA size, and power.

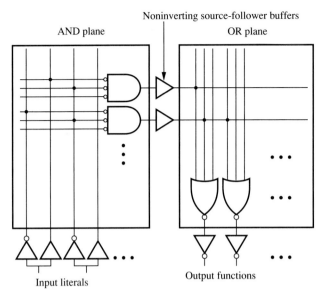

FIGURE 8-18
NOR/NOR PLA strategy for GaAs (E/D process).

placing MESFETs as required into the pull-down part of the distributed NOR gates. In the AND plane a device is placed near the intersection of an inverted input line and the product term (the distributed NOR gate's output node). The dashed boxes in Fig. 8-18 indicate candidate positions—the actual decision is based on whether or not a given literal participates in formation of the given implicant. As an example, if the product $A\overline{B}C$ is to be realized, NOR pull-down devices will be attached to $\overline{A}$, B, and $\overline{C}$ inputs (where they intersect the product line) in order to implement $NOR(\overline{A}, B, \overline{C}) = \overline{\overline{A} + B + \overline{C}} = A\overline{B}C$. Unpopulated sites in the AND plane occur where inputs provided for use in the computation of other products cross product lines that do not depend on them. For some functions, one side of the dual-rail inputs representing a given literal may not be required. Such functions are said to be *unate* with respect to the given input—*positive unate* denotes functions in which only active-high inputs appear; *negative unate* denotes functions referencing only the complemented form of the literal. Some CAD systems are capable of optimizing the PLA by omitting the drivers and unused input lines altogether, thereby reducing power and making the AND plane narrower.

In the OR plane, a MESFET is placed near the site at which an output line and an implicant (member product) line intersect. No device is placed where product lines cross output functions in which they do not participate (i.e., for which they are not implicants). Since a single PLA is typically used to implement several functions in parallel, any given product computed in the AND plane is used only with a subset of the output functions implemented by the PLA. It should never be the case, however, that a product line is unused across *all* output functions—in that case the product need not have been created at all!

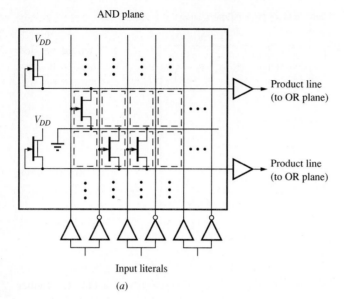

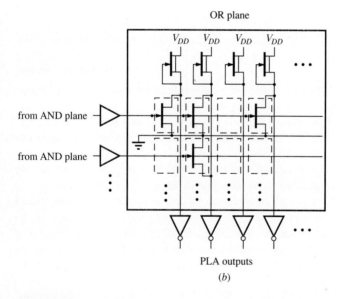

FIGURE 8-19
Detailed view of PLA planes.

Because of the potential of high internal fan-out, it is unlikely that GaAs can support PLA implementation of functions which approach the size and complexity achievable in MOS PLAs. Even if large PLAs could be built, they would be slow and power-consumptive due to the high parasitics and large buffers needed to drive fan-out. Nevertheless, PLAs have a place in GaAs IC design because they offer design closure—an approach that applies to *any* function—and correct-by-construction implementation.

8.5.2 Finite State Machines

The capability to build a PLA realizing any set of combinational logic functions makes the automated implementation of finite state machines possible. Both Mealy and Moore-type state machines can be realized by using a PLA to implement the combinational logic associated with the next state and output functions and by attaching state-holding memory elements to the structure. The two types of state machines are represented schematically in Fig. 8-20. The shape of the PLA portion, with inputs and outputs located on the same side, facilitates direct state input and next state output connections. Most of the now standard MOS PLA and state machine techniques apply to GaAs as well. The limitations and differences over MOS implementations are primarily in scale and, of course, in speed.

The set of flip-flops holding the current state of the finite state machine—the state register—can be implemented in several ways. The choices in this regard depend upon the type of clocking used. Single-phase systems require master/slave flip-flops so that a simple single-sided timing relationship can be used. Since combinational circuits can be generated automatically in PLA form, virtually any type of flip-flop storage

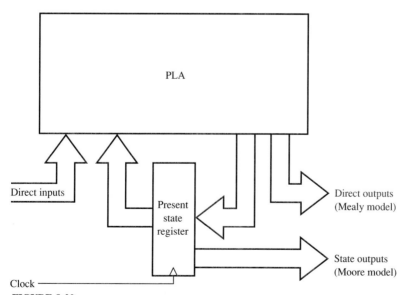

FIGURE 8-20
Finite state machine based on GaAs PLA.

element can be used. The single-input D-type flip-flop is perhaps the most common, though for some state machines other types of storage elements (for example, JK, SR, toggle) may be better suited. With master/slave devices the only timing relationship that must be satisfied is that the clock period t_{CLK} must be greater than the sum of the combinational delay of the PLA's next-state functions t_{NS} and the setup and hold times of the state flip-flops, t_{SH}; that is,

$$t_{CLK} \geq t_{NS} + t_{SH}$$

Master/slave flip-flops require more than twice the area and power of simple latches. It is tempting to consider the use of latches in a state machine implementation, *but beware*. Using latches with a single-phase clock is extremely risky. In order to work correctly and reliably, a two-sided timing relationship must hold (the clock period is bounded from both directions). If the clock runs too slowly, then multistepping can occur. If it is too fast, then the next state functions may not arrive in time to be correctly latched. In either case the state machine will not run correctly. While such two-sided clocking relationships may be attainable in MOS systems, it is unlikely that they would be controllable with sufficient accuracy at GaAs circuit speeds. For this reason, single-phase latch-based state machine implementations are not recommended.

Multiple-phase clock systems can be used quite successfully in the realization of PLA-based state machines. The most common multiphase system is undoubtedly the two-phase non-overlapping scheme. Such a clocking system is somewhat difficult to generate at GHz speeds but, if available, the implementation of safe, robust state machines is facilitated.

As illustrated in Fig. 8-21, the AND-plane input drivers and OR-plane output inverters can be modified slightly to incorporate latching, each activated by different

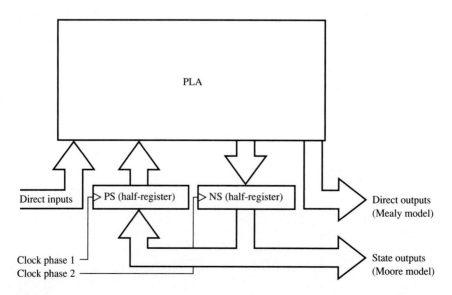

FIGURE 8-21
Finite state machine made with half-registers and two-phase clocks.

nonoverlapping clock phases. The two half-registers (as they are called) together form essentially master and slave state latches. At sufficiently high clock frequencies, dynamic storage can be used. This approach makes the conversion to half-register form both area- and power-efficient with only a single pass transistor added to each existing state line to isolate the gate of the buffer or inverter. The charge residing on the gate becomes trapped there whenever the pass transistor cuts off and thereafter it gradually leaks away. Rewriting each dynamic node is necessary on each clock cycle in order to restore the charged (or discharged) states. The time constants for dynamic logic are very different from those found in MOS systems. Clock rates above 1 MHz are generally required to insure safe dynamic circuit operation.

8.6 SOME ARCHITECTURAL BUILDING BLOCKS

With the basic gates and generic structures as background, the construction of higher-level building blocks can now be addressed. Though integrated circuit implementations in MOS technologies typically involve a wide variety of circuit styles, including many dynamic circuits as well as switch- or steering-logic components, GaAs implementations are primarily composed from gate designs. This is naturally true since GaAs technology is selected primarily for performance reasons. Gate-oriented designs offer clean, fast implementation with few circuit tricks and an excellent possibility for high yield.

8.6.1 Steering Logic

There are frequently portions of an integrated circuit that are not critical to the overall speed performance of the part. It is typical to use an alternative implementation strategy in such regions so that optimization of space or power (or both) is possible without adversely affecting the high-speed paths of the part.

In the GaAs E/D process, steering logic—directly similar to the switching logic (and also known as *pass transistor logic*) used in NMOS—is possible. The MESFET gate has different properties than the gate of a MOSFET, however. Because of the very high impedance of the MOSFET gate circuit, NMOS steering-logic circuits are simple to design and use. No power supply connections are required in the center of a steering-logic block. This makes function blocks and other pass transistor logic structures dense and layout efficient. The MOSFET gate circuits are largely independent of the source/drain steering logic they control.

MESFET-based steering logic must be designed more carefully than its MOS counterpart. Gate current flows in the control circuits whenever gate-to-source voltages in those control signals rise above 0.7 V. This possibility of gate current strongly influences the design of the drivers for control signals. The turn-off properties of MESFET-based steering logic are also more complex to design than their NMOS counterparts.

Figure 8-22 illustrates the parts of a two-variable general function block implementation. In Fig. 8-22a (color insert) the logic block is shown in diagram form. The layout of the pass transistor cell using depletion mode MESFETs is given in Fig.

8-22b (color insert). Several points are worthy of note in this layout. Firstly, the source/drain contacts dominate the vertical pitch of the layout, strongly influencing the ultimate size. Multiple-gate MESFETs could be used, but if the regularity of the layout style were to be preserved, relatively long runs of *active* would be required; these must be avoided for performance reasons. In the horizontal dimension, the device widths dominate block size. As in NMOS steering logic, dual-rail control inputs are required. They are easily accessible on both sides of the layout. This organization allows multiple general-function blocks to be abutted and driven by a single set of control driver circuits.

Using a steering-logic approach one can design with a multiplexer-oriented, table look-up style. Any boolean function, $f(A, B)$, can be realized with this circuit merely by placing the truth table of the desired function on the G lines: $G_0 = f(0,0), G_1 = f(1,0), G_2 = f(0,1)$, and $G_3 = f(1,1)$, and using the two function inputs, A and B, as selectors for the multiplexer. Since the multiplexer does not build-in any given function, one can use this block to implement *programmable* logic, with the truth table values (the G's) coming from a microprogram.

The data paths of the multiplexer are driven by EDBFL circuits as shown in Fig. 8-22c (color insert). The control drivers work best if they are superbuffer (SBFL) circuits, shown in Fig. 8-22d (color insert). Using 10-μm wide pass transistors, this block has a select input to data output delay of 570 ps. The delay from data input (G line) to output is slightly less. These delay figures come from SPICE runs using $V_{te} = 0.1$ V and $V_{td} = -0.5$ V and the device sizes given in the Fig. 8-22c and d schematic diagrams.

Blocks may be stacked together horizontally to implement several (separate and programmable) functions of the same bits, sharing the A and B drivers across all function blocks. Vertical stacking can be used to share the G drivers; in this case the same programmable function can be applied to multiple independent bits at the same time. If vertical stacking is not required, the vertical *metal-2* lines should be removed from the layout. Their only purpose is for vertical stacking—to provide access to the G programming lines at both the top and bottom of the function block cell. When multiple blocks are stacked, no matter whether horizontally or vertically, the output capability of the control and data drivers must be carefully studied. The EDBFL and SBFL drivers in the Fig. 8-22 example have been chosen to drive only a single four-to-one function block.

The *number* of variables used in a steering block is severely limited in GaAs by the size of the logic swing. Blocks larger than four-by-four, such as might be attempted in implementing a three-variable function via an eight-to-one multiplexer, are unlikely to be reliably realizable. This is due to the degradation in logic swing resulting from the use of multiple passive devices in series.

8.6.2 Registers

Digital design is often performed at the register-transfer level. The design paradigm is popular because (1) the synchronous model with state separated from combinational logic is clean and robust, and (2) direct implementation is possible by "plugging in" circuit primitives for each of the small set of design-level primitives. The term *register* is used in a couple of different contexts. Because both refer to digital systems, there is occasionally confusion. In the register-transfer model, a single bit or several bits in parallel (a word) are held in one or several latches and then synchronously transferred through combinational logic to be latched into another such register. This is the lowest-level usage of the term "register." At a higher level—often as part of a "programmer's model" of a computer system—a "register" means one word selected by address from a set of processor-resident storage cells, a "register file." This structure is a random access memory, very different in structure from its apparent namesake because of the execution-time binding (selection by address) property.

Conventional low-level registers are implemented with latches, with flip-flops, or sometimes as dynamic half-registers as described in Sec. 8.5.2 on finite state machines. The physical design of registers is strongly influenced by several factors:

1. *Read/write philosophy*. Can a given register be simultaneously read and written? If so, it must be implemented as a set of master/slave flip-flops.
2. *Clocking philosophy*. The use of single-phase clocking generally forces the use of master/slave flip-flops or other edge-triggered storage circuits. With multiphase systems, simpler latch-based registers are typical and appropriate.
3. *Number of data sources*. One, two, or three sources of data can often be multiplexed without the need for an explicit multiplexer by using a latch or flip-flop design with an enable signal for each potential input. This integral multiplexer flip-flop style is illustrated in the scan-testable circuits depicted in Chap. 6. Such latches trade a small amount of fan-out loading for reduced setup time (compared with approaches using a separate multiplexer). This type of approach only applies when there are a small number of possible data sources. Applications with larger numbers will require a full multiplexer-based implementation.
4. *Wired-OR bus versus point-to-point transfers*. The number of circuits that a given register feeds has a significant influence on the design. In a point-to-point system, the fan-out loading is small; device sizes can be kept small and simple control can be used. If a wired-OR bus-oriented organization is desired, then more complex control is required and device sizes must be carefully chosen to achieve the required level of performance.

The basic circuits for constructing gates, latches, and flip-flops are presented in Chap. 4. Choice of one style over another is dictated by register environment and usage.

Register *file* implementations require a random access memory structure, with address decoder, memory array, read sense, and write driver circuits, as well as output controls. GaAs RAM design involves sense amplifiers, bit-line biasing and balancing, and numerous nondigital circuits. Fortunately, most on-chip register files will be small or modest in size (though there is often a desire to increase capacity and hence, push density to its limits). The size, access, and cycle time requirements are the dominant factors in selecting an implementation strategy for register files. The detailed considerations necessary for designing GaAs random access memories (an important specialty area) are not treated in this text.

8.6.3 Control and Decode

The implementation of circuitry to decode microinstructions (or other encoded representations) and control the operation of a data path is often *ad hoc* and unstructured. PLAs, as presented in Sec. 8.5, provide structure and generality but their performance is not optimal and their shape is inflexible. Given a suitably minimized function to be implemented, its PLA is fixed in both size and shape since these are directly set by the number of inputs, products, and outputs of the function. Furthermore, small *changes* in function (such as might occur as improvements are made in later versions of a given chip) may translate into radically different PLAs.

Certainly one can always implement decoding circuits by designing them at gate level. This will be necessary in areas that require the highest possible performance. Without a higher-level plan or schema, however, such implementations can be difficult to organize for efficient layout.

Intermediate in area cost and performance between large, slow PLAs and unstructured gate-based implementations is the "half-PLA." By organizing the required decoding so that it can be accomplished by a single level gate, the structure of the PLA's AND plane can be used. By choosing between inverting or noninverting buffer stages, the distributed NOR implementation can implement multi-input OR or NOR functions using positive-sense inputs and multi-input AND or NAND functions using active-low inputs.

The half-PLA has several advantages over full-PLA implementations. Firstly, it has less than half of the delay. In addition, its shape is much less constrained than the full PLA, being fixed only in a single dimension. Since the AND and OR plane of the full PLA are each made from NORs and the layouts used are available in both horizontal and vertical orientations, a half-PLA can be subdivided whenever needed into two or more disjoint planes oriented as required to fit with the data path being controlled.

HOMEWORK PROBLEMS

8.1. Lay out a four-to-one multiplexer made purely from NOR gates. Use depletion-mode devices only.

8.2. Use a steering-logic approach to implement a four-to-one multiplexer. Use SPICE to validate its correct operation.

8.3. Trade off the layout density of the design of a programmable truth table block implemented via a four-to-one multiplexer made from gates as in Prob. 8.1 versus a four-to-one multiplexer made from steering logic.

8.4. All of the MESFETs shown in Fig. P8.4 (color insert) are design-rule correct, but not all of them are desirable high-performance layouts. Describe the merits and/or drawbacks of each device layout.

8.5. Draw a schematic diagram for the combinational part of a PLA implementing the functions required to realize a four-bit Gray code counter using D-type master/slave flip-flops. The Gray code counter sequence is given below:

Present state	Next state
0000	0001
0001	0011
0011	0010
0010	0110
0110	0111
0111	0101
0101	0100
0100	1100
1100	1101
1101	1111
1111	1110
1110	1010
1010	1011
1011	1001
1001	1000
1000	0000

REFERENCES

1. Mead, C., and Conway, L.: *Introduction to VLSI Systems,* Addison-Wesley, 1980.
2. CALMA Corporation, *GDS-II Stream Format,* July 1984.
3. Electronic Design Interface Format Steering Committee, *EDIF—Electronic Design Interchange Format Version 1 0 0,* Texas Instruments, Dallas, Tex., 1985.
4. GCA / D. W. Mann, "3600 Software Manual, Appendix B: 3600 Pattern Generator Mag Tape Formats," GCA Corporation, IC Systems Group, Santa Clara, Calif.
5. Hon, R., and Sequin, C.: *A Guide to LSI Implementation,* 2d ed., XEROX Palo Alto Research Center, Technical memo SSL-79-7, January 1980.
6. Ousterhout, J. K., Hamachi, G. T., and Mayo, R. N.: "The Magic VLSI Layout System," *IEEE Design and Test,* pp. 19–30, February 1985.
7. "Magic Tutorials," in *Berkeley VLSI Tools: More Works by the Original Artists,* Technical report UCB/CSD/85-225, University of California, Berkeley, Calif., February 1985.
8. Vitesse Semiconductor Corp., *Foundry Service Design Manual,* Version 4.0, chap. 5, March 1989.
9. "Eqntott Manual," in *Berkeley VLSI Tools: More Works by the Original Artists,* Technical report UCB/CSD/85-225, University of California, Berkeley, Calif., February 1985.

CHAPTER 9

COMPUTER-AIDED DESIGN TOOLS AND TECHNIQUES

The design of a GaAs integrated circuit is a complex task involving literally thousands of details. Because of the enormity of the task of tracking the interrelationships of the many details and design parameters, computer-aided design tools are routinely used. The tools and techniques that are used for GaAs integrated circuit design are adaptations of those used to support integrated circuit design for other technologies: namely, silicon MOS, silicon bipolar, and general analog circuits.

This chapter begins with a look at the usual integrated circuit design sequence, setting the stage for a more detailed presentation of several major computer-supported design and analysis activities of particular importance to GaAs IC design. In order to present the CAD techniques that pertain to GaAs, detailed coverage of at least one CAD system is required. Throughout the chapter one particular CAD package is explained. It is realized that not every reader has access to the described system. However, most commercially available CAD systems have components or features that bear resemblance to the particular CAD system covered here.

For the most part, computer-aided design support for GaAs follows the styles that exist for MOS circuits. Where necessary, modifications are made to MOS tools and techniques.

9.1 NEEDS OF THE DESIGNER

Because computer-aided design tools facilitate and support the efforts of human designers, any study of CAD must begin with a look at the needs of designers. Figure 9-1 depicts a typical integrated circuit design cycle. The cycle for GaAs chips is not fundamentally different from that used for developing any type of integrated circuit. There is perhaps a stronger emphasis on analog circuit simulation and electrical rules

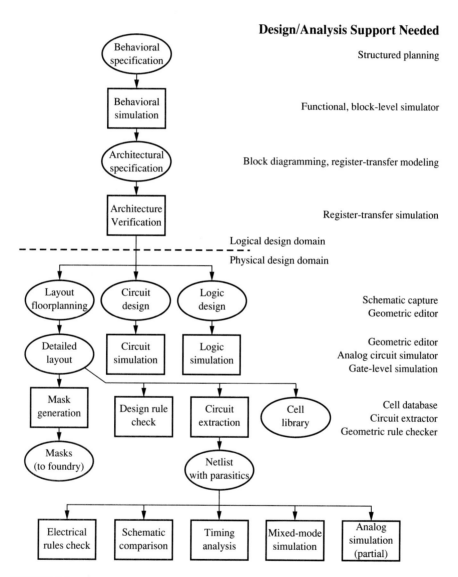

FIGURE 9-1
Design flowchart for GaAs integrated circuits.

check, however. This is due in part to the fact that GaAs circuits do not have the isolation seen in MOS circuits—the non-insulated gates of MESFETs appear as diode circuit elements rather than as simpler capacitive loads. This fact requires more careful device sizing, particularly as fanout details vary. For the same reasons, building blocks and programmable logic circuits become more difficult than their MOS counterparts to design and use. The relatively small voltage swing and noise margin available dictate that analog circuit analysis techniques be used for MESFET-based integrated circuit designs.

The main loops in the design cycle involve tools to automate or facilitate the capture of a design and the use of simulation or technology-specific analysis tools to verify or critique some portion of a design against either a specification or a set of rules for that technology. From the designer's point of view, the most important attribute of a CAD system is *closure*—completeness and tool-to-tool compatibility. The designer should not have to write and debug programs in order to accomplish an integrated circuit design. The CAD system should support the design activity from the behavioral specification phase through foundry interface. Once fabrication has been completed, the same CAD data base should support characterization and testing.

An abbreviated tour of a typical integrated circuit design cycle follows. The boxes in Fig. 9-1 represent CAD programs and the ellipses represent data. Integrated circuit design is an iterative process and, as such, there are many loops within the development cycle. For simplicity the loops are not depicted in Fig. 9-1. In general there are arcs from the output side of each CAD program that lead back to all earlier stages. It is by traversal of these arcs that minor errors are caught and corrected and through which the design is detailed, iterated, and verified.

The earliest stage of development of an integrated circuit is the formulation of an external behavioral specification—a clear statement of *what* the part is intended to do from an external point of view. Ideally this specification addresses purely behavior and not implementation. It is typical that later stages of development (which *do* consider implementation) may influence behavior, but this normally takes on the character of fine-tuning, adding detail rather than grossly changing the specified behavior.

CAD support for the behavioral specification phase is primarily in the form of languages for precise specification and in tools and models to support very high-level simulation. The output from this phase is a definition of the intended behavior of the part at its I/O boundaries. This external specification is the primary input to the architecture and organization phase.

The focus of the architecture and organization phase is the *internal* subsystem-level definition, arrangement, and partitioning of function within the chip. This phase includes many subactivities and can benefit greatly from the availability of cell libraries and a data base of experience from past chip designs. There are several tightly interacting entities and activities whose interplay dominates the architecture and organization planning:

1. External behavioral specification. This determines *what* the part under development must do. It does not legislate *how* the intended behavior must be realized. Required timing and acceptable bounds for critical electrical parameters are typical and appropriate parts of this specification.

2. Cell library. The availability of a set of fully characterized and documented cells with which the designers are familiar and from which subsystems can be composed can strongly influence not only overall floor planning but also subsystem partitioning and basic algorithm selection. With a minimal cell library the tendency will be for most designs to be built up from logic gates and flip-flops using a gate-array-like structure.
3. Subsystem definition, partitioning, and floor planning. Constrained on one side by what functions must be performed and on another by a (possibly small) set of candidate approaches, the choice of subsystems, their partitioning, and their global interrelationships within the chip (floor planning and tiling) must occur together or with close iteration.
4. Data flow modeling and simulation. Specific or statistically realistic exercises of subsystems are often performed via simulation. This activity gives the designers information that aids in tradeoff decisions relating to bussing and interconnect alternatives within the part.

Once an architecture has been developed that is capable of meeting the external behavioral specification, the job of designing, detailing, and interconnecting subsystems begins. Within this phase the tools take on a more physical character since they are dealing with layouts, electrical parameters, and point-to-point intracell routing. The result of the subsystem implementation phase is a hierarchy of cell layouts with the entire chip layout represented by the topmost level.

Once a full chip layout is available, an army of analysis tools can be applied. Besides the design rule checking task, which acts directly on a layout, most analysis tools deal with an abstraction of the chip network which is derived from the layout geometry via circuit extraction. The extracted circuit is usually composed of device model primitives, lumped parasitic elements, and interconnections. Using this information, a variety of checks can be performed:

1. Schematic comparison. A determination of whether the implemented circuit, as represented by the extracted network, matches the intended schematic.
2. Electrical rules check. A static check for ratios, threshold drops, fan-in/fan-out limits, correct formation of gates and level-shifters, and so forth.
3. Propagation check. A gross node-by-node check to verify that each node is capable of being driven to logic '0' and logic '1', and that it can be affected by at least one input and can affect at least one output. By this technique floating or misconnected nodes can be discovered and suitable warnings can be generated.
4. Timing analysis. Reasonably accurate timing models based on lumped element approximations or on characterized cell library members can be manipulated to estimate and validate the timing relationships of a subsystem or an entire chip.
5. Switch-level simulation. Use of grossly oversimplified relay-like device models allows the logical simulation of large networks. For digital networks following conservative design rules and using a somewhat restricted class of circuits, switch-level simulation can be an effective tool for validating the functionality of an extracted network. Normally, this type of simulation is used only to verify logical

behavior without regard for timing. With suitable extraction of estimated parasitics, however, some switch-level simulators are capable of generating estimates of timing behavior.

6. Detailed analog simulation. Because of computational requirements, only relatively small subcircuits (with at most around a hundred devices) can be simulated by SPICE [5] [8]. Nothing approaching full-chip complexity can be considered. The role of detailed analog simulation lies at the cell library or small subsystem design level. Because of its accuracy as well as the flexibility and visibility it provides, a general-purpose circuit simulator such as SPICE or its equivalent is an essential tool for GaAs integrated circuit design.

The final stage of the prefabrication portion of the integrated circuit design cycle involves the foundry interface. The primary activity is mask generation. A wide variety of tools are available to perform the necessary steps in a completely automated fashion. Basic operations of union, intersection, bloat, shrink,[1] and the like are applied to the designer-specified layers. Alignment marks are added. For full-wafer composition, streets and wafer test structures are added. Finally, mask images are fractured into individual commands suitable for driving a specific mask-writing machine. These commands are written to magnetic tape or to other media suitable for interfacing with the mask-making equipment.

The primary emphasis of this chapter is computer-aided design tools and techniques for creation and analysis of GaAs IC layouts. There are a wide variety of systems available to support integrated circuit design. Many industrial CAD systems are table-driven in an effort to make them flexible, changeable, and technology-independent. Such systems can be retargetted to support new technologies merely by changing tables—by providing parameters that define the mask layers of the technology, their colors, their design rule interrelationships, and so forth.

9.2 THE MAGIC VLSI DESIGN SYSTEM

In the university VLSI community, there is wide acceptance and use of a computer-aided design package called MAGIC which was created at UC-Berkeley in 1984 [1]. MAGIC incorporates hierarchy management, a geometry editor, interactive design rule checks, a circuit extractor, and a foundry interface—all within one interactive tool. Because it has an easy-to-learn interface and runs on a wide variety of UNIX-based hosts, MAGIC has become popular and installations are widespread. Its weaknesses are the lack of any sort of schematic capture facility, a limitation to purely Manhattan layouts, and substantial memory-intensive processing requirements that ultimately limit performance on large integrated circuit designs.

---

[1] It is typical to find support for mask manipulation in the form of a simple set of well-defined parameterized operations that can be combined sequentially to provide a degree of programmability and generality. This approach attempts to facilitate, in a flexible way, automation of the translation from geometric design data to fabrication-compatible masks.

TABLE 9.1
Magic subsystems and technology file sections

MAGIC subsystem	Technology file sections
Data base	planes, tiles
Graphics and man/machine interface	styles
Foundry interface (CIF/CALMA)	cifinput, cifoutput
Design-rule check	drc
Router	router
Compactor	planes, tiles, contacts, compose
Plotting	plot
Extractor	planes, tiles, contacts, extract

The MAGIC program is technology-independent. Appendixes A and B give technology files describing two popular GaAs processes.[2] These two files are sufficient to configure MAGIC for supporting the design of GaAs chips. A perusal of the technology files reveals the inner organization of MAGIC—a set of subsystems, each supporting a specialized area of the design or analysis activities. Table 9.1 lists the MAGIC subsystems and matches them with one or more sections of the technology file. The sections that follow give an abbreviated tour of MAGIC. These sections are intended to provide the reader with sufficient background to understand and modify MAGIC technology files. An excellent tutorial on the use of MAGIC is available [2].

9.2.1 MAGIC Data Base

At the heart of MAGIC is a data base composed of a small set of *planes*, each containing rectangular *tiles*. A novel data structure known as *corner-stitching* [3] facilitates efficient manipulation and searching of the tile data base. The corner-stitched data structure provides a form of two-dimensional sorting. All of the facilities of MAGIC rest on the data base and are written in terms of tile manipulations. The technology independence of MAGIC stems in large part from this property. MAGIC learns about a new technology by reading the technology file at start-up time. This file declares the planes and tile types to be used with the given technology. It describes their appearance on-screen and in hardcopy plots, their interrelationships, and the design rules.

An understanding of the capabilities as well as the weaknesses of MAGIC requires knowledge of planes and tiles. *Planes* are logically independent levels within a technology. Each plane is made up of *tiles* covering all representable two-

---

[2] Such technology files reflect a set of design rules supported by a particular foundry or set of foundries. These rules are changed periodically as processes are further developed and improved. Compatibility with prior rule sets is not guaranteed, however, so careful management and maintenance of the technology file is required.

dimensional space. In addition to the tile types defined in the technology file, there are a few built-in types that are part of *all* technologies. One such built-in is the *space* tile. Each plane represents all space, appropriately divided into rectangular regions (implemented as tiles) and inter-linked by corner-stitching. Only one tile type can occupy any given location within a plane. Refer to Fig. 9-2 for an example of corner-stitching. In this figure, *space* tiles are indicated by dotted outlines.

The data base is always maintained in a canonical form; i.e., there is one and only one data base representation for a given geometry regardless of the operations or their order of application during build-up of the given geometry. Within a plane, the electrical connectivity of neighboring tiles is defined in the *connect* section of the technology file. The choice of tile types and planes is influenced by the external display appearance desired, the physical and electrical behavior of the various layers, the design rules to be checked, and the contacts available in the technology.

The *contact* section of the technology file defines a set of tile types which interlink one or more planes. The contact tiles have an image in each plane to which they connect in the real technology. MAGIC "knows" about interplane electrical continuity only through the definition and use of contact tiles.

Manipulation of tiles is most efficient when the subject tiles share a common plane. Hence, it is desirable to place tightly interacting layers together with their corresponding tile types sharing the same plane. In the GaAs selective-implant depletion

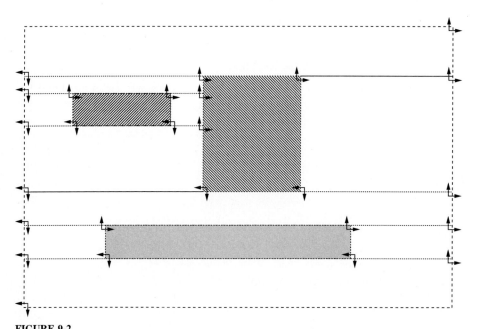

FIGURE 9-2
Corner-stitched data base—three nonspace tiles in a plane. The arrows represent pointers within the plane data base. *Space* tiles are dotted. Edge tiles extend to infinity. (*From Ousterhout et al.* [*1*]. *Used with permission.*)

mode technology, the *schottky-metal, n-plus*, and *n-minus* tile types are all defined to be on the same plane—*active*. A section of the MAGIC technology file where the planes and primitive tiles for this GaAs process are defined is given below:

```
planes
    active              /*contains substrate-level features plus schottky-metal*/
    metal2              /*contains second-metal features*/
    fab                 /*assorted fabrication-related tiles*/
end
types
    active   n_plus,np          /*note multiple names for a tile type*/
    active   n_minus,nm
    active   schottky_metal,sm
    metal2   second_metal,m2
    fab      glass,ov
    :        :                  /*other (non-primitive) tile types follow*/
end
```

In addition to these primitive tiles, tile types for composite structures such as *mesfet, logic-diode*, and *level-shift-diode* can be defined. The *mesfet* tile type occurs whenever both *n-minus* and *schottky-metal* coexist. Because of the one-tile-type-per-plane-per-point organization of the data base, it is impossible for both *n-minus* and *schottky-metal* to actually coexist; thus, the *mesfet* tile type is substituted to record this design feature. With this organization, it is easy for the MAGIC program to discover and handle this case.

In MAGIC, all composite structures are defined in the *compose* section. The *compose, decompose, paint,* and *erase* operators are MAGIC's built-in support for composite structures. From these operations one can describe, for example, how the addition of *schottky-metal* across an existing *n-minus* region produces a MESFET (with corresponding *mesfet* tile type) at the intersection, as shown in Fig. 9-3.

Most composite relationships are symmetric with respect to the operation of *compose* and *decompose*. All *compose* operations are also order-independent, i.e., they are commutative. For example,

$$\text{schottky-metal} + \text{n-minus} = \text{n-minus} + \text{schottky-metal} = \text{mesfet}$$

In the technology file, *compose* operations have the syntax below.

> compose   result   component-type   component-type

Thus, the composition rule for *mesfet* is written as

> compose   mesfet   n-minus   schottky-metal

Decompose operations, on the other hand, are order-dependent since they involve a starting configuration and an attempt to take away material of some type(s) from that initial configuration, e.g.,

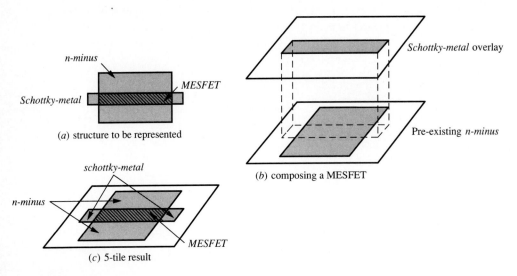

FIGURE 9-3
MESFET tile formed by composition rule (n-minus plus schottky-metal).

$$mesfet - n\text{-}minus = schottky\text{-}metal$$
$$mesfet - schottky\text{-}metal = n\text{-}minus$$

A *decompose* specification has the syntax

    decompose *initial-type component-type-A component-type-B*

The meaning of such a specification is that erasing *component-type-A* from *initial-type* yields *component-type-B* and erasing *component-type-B* from *initial-type* yields *component-type-A*.

    For more complex compositional relationships, the lower-level *paint* and *erase* built-ins can be used. These are needed to tell MAGIC exactly how to deal with *asymmetric* situations involving composite structures. As an example of such asymmetry, consider what should happen when *metal-2* is to be removed from a *via2* in the GaAs E/D technology. Since a *via2* connects *metal-1* to *metal-2*, the removal of *metal-2* should leave *metal-1* remaining. A *via2* is not a simple composite of *metal-1* and *metal-2*, however. Thus, there exists the asymmetric relationship

$$via2 - metal\text{-}2 = metal\text{-}1 \quad \text{and} \quad via2 - metal\text{-}1 = metal\text{-}2$$
$$\text{but} \quad metal\text{-}1 + metal\text{-}2 \neq via2$$

The syntax of the *paint* and *erase* descriptors is given below:

    paint *initial-type painted-type result*
    erase *initial-type erased-type result*

The *erase* built-in can be used to describe the *via2* asymmetry mentioned above as

$$\text{erase} \quad \text{via2} \quad \text{metal-1} \quad \text{metal-2}$$

$$\text{erase} \quad \text{via2} \quad \text{metal-2} \quad \text{metal-1}$$

A safer way to handle this particular asymmetry is to use erase rules to completely remove the *via2* whenever any layer of it is erased, thereby avoiding missed contacts due to hard-to-see partial-layer erasures which can happen inadvertently. The technology file entries to implement these rules are below:

$$\text{erase} \quad \text{via2} \quad \text{metal-1} \quad \text{space}$$

$$\text{erase} \quad \text{via2} \quad \text{metal-2} \quad \text{space}$$

This conservative approach is followed in the technology files supplied in Apps. A and B.

9.2.2 MAGIC Display and Plotting Styles

The appearance of tiles on color graphics screens is specified in the *styles* section of the technology file. For hardcopy plots, the stippling patterns and other attributes such as outlining and corner-to-corner "×"-ing are specified in the *plot* section. Each tile type has an externally visible name and zero or more aliases. The user can refer to the tile type by any of its symbolic forms.

For each tile type the screen appearance is chosen so as to give a recognizably distinct, yet globally meaningful, color. Screen appearance is established by combining one or more display styles, each composed of low-level graphics details such as bit plane mask, color index, solid vs. stipple-filled, outlined, crossed, transparent vs. opaque color. The palette of display styles is distinct from the technology file. By this separation, sets of mutually compatible display styles can be developed and reused across multiple technologies. The GaAs technology files described and included in this text do not require customized display styles: the widely available MAGIC "mos7bit" display styles are used. This set of styles is compatible with displays supporting seven or more bit planes, and is sufficient to support all MOS and GaAs technologies. The reader interested in lowest-level detail regarding display styles is referred to Ref. [4].

MAGIC supports hardcopy plotting on a variety of raster and dot-matrix printers. The interface to such printers is quite general, providing direct user control of the number of dots per inch, the number of dots per raster line, the swath height, and conversion/spooling program via run-time changeable plotting parameters. This type of interface abstracts and isolates the MAGIC data base access and rasterization process from the details of printer/plotter code conversion and I/O device drive. Final code conversion and printer spooling is left to the end user. Many different types of printers have been used with MAGIC, including electrostatic, laser, and dot-matrix types.

The appearance of each tile type is specified by detailing a 16-by-16 bit stippling pattern. Each such pattern is carefully chosen so as to provide a distinct appearance,

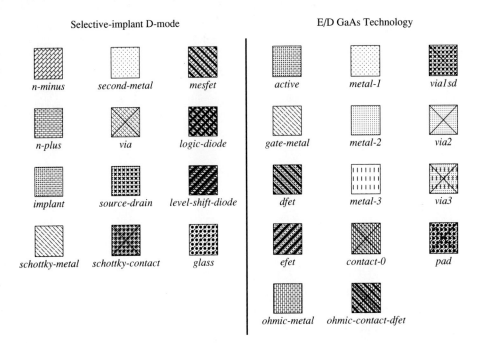

FIGURE 9-4
Stipple patterns for GaAs.

enabling the user to readily determine the layer(s) present at any given point of interest in a hardcopy plot. Of particular concern are the interactions between individual patterns when the layers they represent are overlaid—causing the corresponding stipple patterns to be ORed together. Density of stipple patterns and their interrelationships are the primary considerations in designing tile hardcopy appearance. Figure 9-4 shows the hardcopy stipple patterns defined for the two GaAs processes used in this text.

9.2.3 Design Rule Checks

The full specification of a technology involves more than just display and hardcopy output styles, data base formats with proper electrical continuity, contacts, and composition rules. Proper checking for adherence to geometric design rules requires considerably more information. In addition to simple minimum width and separation rules, the design rule checker of MAGIC supports very general *edge*-based checking. The corner-stitched data base facilitates efficient enforcement of overlaps, extensions, and other special cases as well as the more straightforward checking of minimum widths and spacings.

The edge-based model underlying MAGIC's design rule checker is shown in Fig. 9-5. Because they represent a significant fraction of all design rule checks, minimum width and spacing checks are specified directly as such (rather than in edge-based form into which they are internally converted) by giving a list of tile types, a

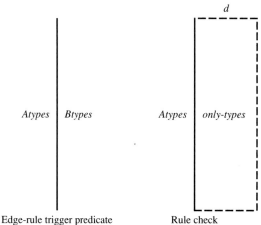

FIGURE 9-5
Model of edge-based DRC.

(Left) Edge-rule trigger predicate — A tile from set *Atypes* exists on one side of an edge and a tile from set *Btypes* shares the edge with it.

(Right) Rule check — Only tiles from the set *only-types* can exist within a box defined by the edge and distance d.

minimum width or spacing amount, and an error feedback message. The syntax of width and spacing check specifications is

> width   *tiletype[,tiletype]*. . .   *width-amount*   "*error feedback message*"
>
> spacing   *tiletype[,tiletype]*. . .   *tiletype[,tiletype]*. . .   *separation-amount* \
>   *touch-rule*   "*error feedback message*"

The width rule requires that all regions containing tiles of the specified set must be wider than *width-amount* in both dimensions. When violated, the area of the screen containing the specific violation is highlighted. The user can invoke the "drc why" command to receive the specific design rule error feedback message corresponding to the violation.

The *spacing* rule specifies two sets of tile types, a minimum separation distance, and a touching rule. The checker requires that any tile in the first set must be separated from any tile in the second set by the Manhattan distance given. The touching rule is one of the keywords **touching_ok** or **touching_illegal**. These allow (or disallow) tiles from the two sets being immediately adjacent. An example from the selective-implant D-mode GaAs technology file is

> spacing   mesfet,logic_diode,level_shift_diode \
>   source_drain,schottky_contact 2 \
>   touching_illegal \
>   "schottky-metal/implant separation must be at least 1 micron (2 lambda)"

This example spans multiple lines, so all but the last line ends with a special continuation escape character ("\") to indicate that the next line belongs to the same spacing rule. This rule checks that 2λ or more separation exists between devices (*mesfet*, *logic-diode*, or *level-shift-diode* tiles) and *schottky-metal* or *source-drain* regions (areas that contain Schottky metal) and that it is illegal for any two tiles from these classes to directly touch.

Each general *edge*-based rule is triggered by the occurrence in the data base of an edge satisfying the enabling predicate—expressed as two sets of tile types, one on each side of an edge. Once triggered, the rule causes checking in a neighboring region wherein only tiles from a given set of types are allowed. Though specified in an orientation-independent manner, each edge rule is checked only on the top or right edges of a tile whenever the firing predicate is satisfied. This top/right restriction can be overridden by specifying the rule as a "4way" rule. The reasons for two-way checking are improved speed and elimination of duplicate error messages. Double messages result, for example, from two objects occurring too close to one another, with the separations being checked from the perspectives of both tiles independently. Each tile sees the other one as being too close and two design rule errors are flagged for what is in actuality only a single spacing violation.

By default, corners are not included in the edge-based design rule checks. Special corner checks can be enabled when needed. Figure 9-6 illustrates how unchecked corners can lead to missed design rule violations when using the simplest form of edge-only checks.

The fully general (two-way) edge rule has the format below:

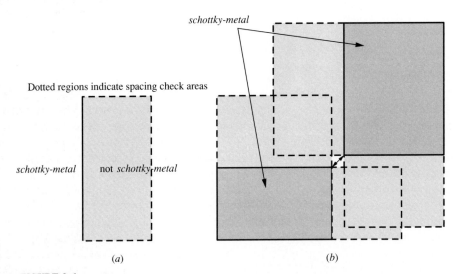

FIGURE 9-6
Ignoring corners can lead to missed violations (dotted regions are spacing check areas). (*a*) Simple edge rule for schottky-metal separation. (*b*) Corner-to-corner spacing violation missed.

edge *Atypes Btypes d OKtypes corner_types corner_distance error_message [plane]*

where

    Atypes ≡ list of types on "inside" edge

    Btypes ≡ list of types on "outside" edge

    d ≡ edge-check distance in λ

    OKtypes ≡ list of types allowed to occur within distance *d* on "outside" of edge

    corner_types ≡ list of types enabling corner checking

    corner_distance ≡ distance defining second dimension of corner check area

    error_message ≡ user feedback message

    plane ≡ plane within which to check constraint region (optional)

Figure 9-7 depicts the general edge rule schematically. The check for *corner_types* occurs in a 1λ × 1λ region on the *Atypes* side of the edge. If material of a type found in the set *corner_types* occurs in the 1λ × 1λ corner region, then the larger corner constraint area defined by *corner_distance* and *d* is checked. Only a type from *OKtypes* can be found in the corner region. An example edge rule from the selective-implant D-mode technology file is below:

    edge mesfet,level_shift_diode,logic_diode space 1 0 0 0 \
        "mesfet/diode overhang is missing"

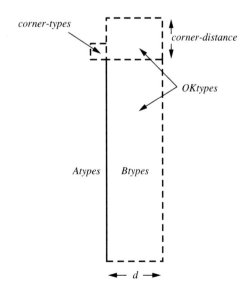

FIGURE 9-7
General edge rule, including corner check.

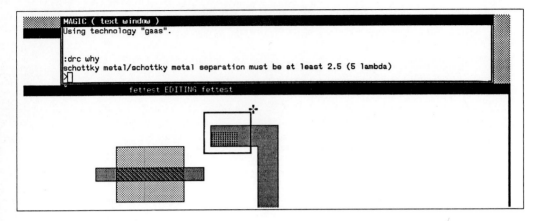

FIGURE 9-8
Partial screen image of MAGIC showing a DRC error and feedback message.

This rule checks for any edge with an active device on one side (any tile from the set *mesfet, level_shift_diode,* or *logic_diode*) and *space* on the other. Such a tile directly represents an error since there is a required overhang of *n-minus, n-plus,* or *implant* on such devices. Notice that the set of *OKtypes* in the example rule is null—specified as "0". This guarantees that the check will be violated, and thus the enabling condition alone causes the design rule error to be signaled.

Design rule violations are highlighted by placing error tiles with logically attached prose-form feedback messages into the data base. These tile types are normally given a bright, easy-to-see display style. In addition, there are subcommands within MAGIC to find design rule violations, center them within the display region, and print the appropriate feedback messages. Figure 9-8 is a portion of a MAGIC screen image showing a design rule violation and its feedback message.

The *drc* (design rule checker) subsystem of MAGIC runs in background mode, analyzing data base changes between regular user commands. This feature is quite useful when initially learning a new technology, since design rule violations appear boldly on the screen at the very moment when the rules are violated during user interaction. Interactive checking greatly facilitates rapid learning of geometric design rules. The *drc* subsystem can be deactivated with the "drc off" command. When this is done, regions requiring checking are recorded but the actual checking is not performed. When the *drc* system is subsequently restarted with a "drc on" or "drc recheck" command, MAGIC checks only the areas that were changed since the checker last ran.

9.2.4 Circuit Extraction

Circuit extraction is the process of deriving from the layout geometry a list of devices and their interconnections. The circuit extraction process often involves a "flattened" or fully instantiated (nonhierarchical) representation of the circuit. Such a representation can be extremely large; hence long running times and large memory requirements are typical.

Because of the corner-stitched data base and the use of hierarchy in MAGIC, circuit extraction can occur at the cell level[3] and can be performed both directly and rapidly. A traversal of the cell's tile data base yields both connectivity and circuit element size estimates. Transistor and diode sizes as well as parasitic resistances and capacitances are estimated by examining the areas and perimeters of various tiles representing circuit elements. The electrical characteristics of each layer type can be specified in the *extract* section of the technology file. This information allows MAGIC to compute reasonably accurate estimates for parasitic resistance and capacitance in the circuit.

The result of circuit extraction is an *extract* file. Such files are in a one-to-one relationship with corresponding layout files, matching the cells of a hierarchy directly. Layout files use the file extension ".mag" while extract files use ".ext". Figure 9-9 shows a simple cell "x" together with its layout ("x.mag") and extract ("x.ext") files. Note that both file types use an ASCII format. This aids understanding and facilitates file transfer between foreign machine types as well as simple interfaces to other tools.

The hierarchy is implemented as a tree structure (inverted) with the topmost layer represented as the root of the tree and the lowest-level detailed layout cells represented as leaves of the tree. The use of hierarchy considerably complicates the extraction process at nonleaf levels within the hierarchy. The principal problems stem from cells that overlap or abut. Since each cell uses a separate set of corner-stitched planes, hierarchies of cells must be carefully combined in order to know what connects to what. The design of MAGIC allows (but restricts) cell overlaps. The only overlaps that are allowed are those that do not form any new composite or contact structures. Figure 9-10 illustrates several hierarchical abutments and overlaps. Notice that a circuit extractor could easily handle these cases using a *flattened* representation; indeed, MAGIC can directly handle a flattened version as well. The cost of hierarchical circuit extraction is paid in terms of restrictions on overlaps and abutments between cells. The payoff is rapid processing—typically seconds or minutes for a hierarchical circuit extraction versus hours for extractions from a flattened version.

Nonleaf cells within a hierarchical design contain references (calls) to all next-level child cells along with placement and rotation/mirroring details. Nonleaf cells may also contain tiles specifying devices and interconnect. When extracted, a nonleaf cell is transformed into its corresponding ".ext" file which contains references to all of the child cells' ".ext" files, the extracted interconnectivity information, and *merge*

---

[3] We use the term *cell* to refer to a single "chunk" of layout—a set of tiles and references to subordinate cells that are treated at higher levels as one entity. In MAGIC, a cell corresponds one-to-one with a layout file and comprises the simplest independently manageable non-primitive part of the design hierarchy.

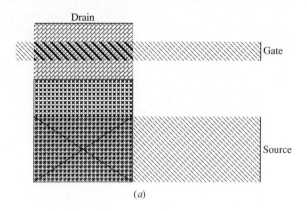

(a)

```
magic                              timestamp 581724944
tech gaas                          version 4.0
timestamp 581724944                tech gaas
<<n-minus>>                        scale 1000 1000 50
rect 39 13 50 15                   resistclasses
rect 39  9 50 11                   node "Source" 0 0 39 –2
<<schottky-metal>>                 node "Gate" 0 0 37 10
rect 37 11 39 13                   node "Drain" 0 0 39 12
rect 50 11 64 13                   fet n 39 10 40 11 22 26 !GND "Gate" 4 0
rect 50 –2 64  5                     "Drain" 11 0 "Source" 11 0
<<source-drain>>                                (c)
rect 39  5 50  9
<<schottky-contact>>
rect 39 –2 50  5
<<mesfet>>
rect 39 11 50 13
<<labels>>
rlabel schottky-metal 64 –2 64  5 3 Source
rlabel schottky-metal 64 11 64 13 3 Gate
rlabel n-minus 39 15 50 15 1 Drain
<<end>>
```

(b)

FIGURE 9-9
A simple cell layout and its corresponding .mag and .ext files. (a) Layout of cell "x". (b) MAGIC layout file (x.mag). (c) Extract file (x.ext).

directives.[4] It is at this level that cells of the hierarchy are correctly registered and signal equivalences are discovered and recorded. After the hierarchical data base has

---

[4] These extract file directives are instructions for building a flat netlist. Each *merge* mentions two signal nets that have been found to be attached and hence must be merged into one net when the hierarchy is flattened. Lumped parasitic values associated with each of the nodes to be merged are also carried along to facilitate more accurate estimation of parasitics in the final flattened circuit representation.

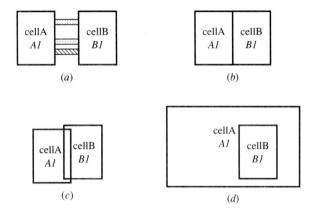

FIGURE 9-10
Several cell overlap and abutment styles. (*a*) Two non-overlapping cells. Interconnections are performed in parent cells. No DRC interaction is possible if separation exceeds DRC "radius." (*b*) Two abutted cells. No extra interconnect is required. DRC and extract issues may arise. (*c*) Partly-overlapping cells. DRC and extract issues arise. (*d*) Complete containment. DRC and extract issues arise.

been processed, a final pass over all ".ext" files is required in order to flatten the hierarchy and instantiate all circuit elements for simulation.

In order to provide an easy way to refer to nodes within a layout, *labels* are created. These are text strings attached during layout to tiles of a cell. Most labels are node names which are needed by routers, simulators, and timing analyzers. Some labels serve to attach *attributes* to tiles rather than to give them names. Attributes are provided to allow a designer to give additional information to circuit analysis tools.

Labels are an important part of the layout-extract-simulate cycle because they allow meaningful signal names to be used. Each label has a rectangular region and a tile type associated with it. The label is logically attached to that tile. If the tile is moved or copied, the label will be moved or copied with it. If the tile is erased or subdivided, the label is "disconnected" and subsequently reattached to some other tile type—one that covers the entire rectangular label space. Figure 9-11 depicts several labels, including point, line, and box styles.

When two cells overlap or abut, the placement of labels is significant to the circuit extractor. As hierarchical cell extraction proceeds, individual cells are extracted

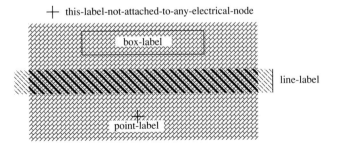

FIGURE 9-11
Label styles available in MAGIC.

with the results placed into the corresponding ".ext" files. Intermediate- or top-level cells containing multiple abutting or overlapping child cells will have *merge* commands in their extract files. These commands specify that when the final flattening pass is performed, the listed nodes are to be merged—considered as a single electrical node. In order to correctly generate and propagate user-defined node names efficiently, labels must be found at or very near the edge of a cell. It is good practice to label all signals that cross the boundaries of a cell. Since this practice defines a clean interface to/from each cell, it is also an excellent aid in situations involving multidesigner cooperation and documentation. Always remember to select short, meaningful text for labels and to attach each label to an unambiguous region near the cell edge. It is not good practice to place a label over several independent layers, since it would not be clear from a hardcopy plot to which layer the label was attached.

There are two types of labels: *local* and *global*. Any label which has text ending with "!" is considered to be global in scope. All nodes labeled with the same global label are assumed to be electrically connected. Even if the nodes are distinct, they will be considered to be the same node if labeled globally. This is often a useful feature since it allows early simulation of collections of subcircuit blocks even before they have been completely hooked up. The use of global labels comes with a strong caution, however. The layout determines what is and what is not hooked up. Global labeling does not add any physical interconnections—it only makes the simulators *pretend* that things are hooked up. For the final stages of chip extraction and verification, it is best not to rely on global labels.

The most common global labels are the power rails—"Vdd!", "Vss!", and "GND!". The capitalization and spelling of these particular labels are "built-ins" of several simulation and analysis programs and hence should be used exactly as given. Other global labels may of course be used, but the power rails are treated by analysis programs in a special way.

Any nonglobal label is considered to be *local*. Local node names refer to a particular node within a cell. They are unique within the cell; it is illegal to use the same local label on electrically distinct (unconnected) nodes within a cell. It is legal—even typical—to use the same local node name in different cells, however. In the completely flattened simulation file, all unique nets must have globally unique node names. Local names are made globally unique by prepending the cell instance name followed by a "/" to the local node name. If a cell "A" contains sub-cell "B" with local label "X," the unique node name for "X" in instance n of "A" would be "A_n/B_0/X". For deeply-nested hierarchies, some signal names can become rather long, but they will always be unique and easy to find and identify.

9.2.5 Ext2sim

The final conversion from hierarchical extract file to fully flattened simulation format is not performed within MAGIC. Once hierarchical extraction has been completed, the user exits from MAGIC and the "extract-to-simulation" conversion program *ext2sim* is used. Figure 9-12 illustrates two inverters in abutted cells. The figure shows layouts with labels, as well as the ".mag," ".ext," and the final simulation (".sim") file. Unlike the hierarchical extract files which exist for each cell in a design, there is a single ".sim" file for the entire circuit. *Ext2sim* reads the extract files produced by MAGIC's hierarchical extractor, building an internal representation of the signal hierarchy as it goes. Using this information it then traverses the signal

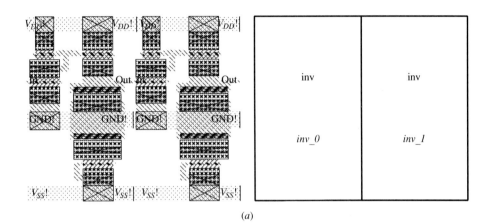

(a)

```
magic
tech gaas
timestamp 582833361
<<n-minus>>
rect 8  33  20  35
rect 39  33  59  35
rect 8  29  20  31
rect 39  29  59  31
rect 4  15  24  17
rect 4  11  24  13
rect 39  -38  59  -36
rect 39  -42  59  -40
<<schottky-metal>>
rect 22  33  37  35
rect 6  31  8  33
rect 20  31  39  33
rect 59  31  61  33
rect 26  25  30  31
rect 24  21  30  25
         :
<<end>>
```

(b)

FIGURE 9-12
A hierarchy of two inverters with .mag, .ext, and .sim files. (a) Expanded and unexpanded views of dual inverter cell "inv2." (b) MAGIC layout file excerpt (inv.mag).

```
timestamp 582833361
version 4.0
tech gaas
scale 1000 1000 50
resistclasses
node "Vss!" 0 0 33 -50
equiv "Vss!" "Vss!"
node "3_66_73#" 0 0 33 -36
node "3_62_47#" 0 0 31 -23
node "GND!" 0 0 4 -17
equiv "GND!" "GND!"
node "In" 0 0 2 13
node "Out" 0 0 31 7
node "3_8_30#" 0 0 4 15
node "Vdd!" 0 0 8 33
equiv "Vdd!" "Vdd!"
fet n 39 -40 40 -39 40 44 !GND "Vss!" 4 0 "3_66_73#" 20 0 "VSS!" 20 0
fet s 33 -23 34 -22 128 72 !GND "3_62_47#" 40 0 "3_66_73#" 32 0
fet s 33 7 34 8 128 72 !GND "Out" 40 0 "3_62_47#" 32 0
fet n 4 13 5 14 40 44 !GND "In" 4 0 "3_8_30#" 20 0 "GND!" 20 0
fet n 39 31 40 32 40 44 !GND "3_8_30#" 4 0 "Vdd!" 20 0 "Out" 20 0
fet n 8 31 9 32 24 28 !GND "3_8_30#" 4 0 "Vdd!" 12 0 "3_8_30#" 12 0
```

(c)

```
| units: 50  tech: gaas
n inv_1/3_8_30 Vdd inv_1/3_8_30 2 12 77 94
n inv_1/3_8_30 Vdd inv_1/Out 2 20 108 94
n inv_1/In inv_1/3_8_30 GND 2 20 73 76
s inv_1/Out inv_1/3_62_47 inv_1/3_62_47 4 32 102 70
s inv_1/3_62_47 inv_1/3_66_73 inv_1/3_66_73 4 32 102 40
n Vss inv_1/3_66_73 Vss 2 20 108 23
n inv_0/3_8_30 Vdd inv_0/3_8_30 2 12 6 94
n inv_0/3_8_30 Vdd inv_1/In 2 20 37 94
n inv_0/In inv_0/3_8_30 GND 2 20 2 76
s inv_1/In inv_0/3_62_47 inv_0/3_62_47 4 32 31 70
s inv_0/3_62_47 inv_0/3_66_73 inv_0/3_66_73 4 32 31 40
n Vss inv_0/3_66_73 Vss 2 20 37 23
```

(d) Simulation file (inv2.sim)

FIGURE 9-12 (*continued*)

(c) Extract file (inv.ext). (d) Simulation file (inv2.sim).

and cell hierarchy, producing a flat ".sim" file for use by analysis and simulation programs.

The ".sim" file contains a series of lines in ASCII, each beginning with a key letter. The key indicates the interpretation to be used for the rest of the line. The key letters and their meanings follow:

| *any-string-of-characters*

The "|" introduces a comment line, which is ignored by most analysis tools. There is one special-purpose comment, however, which, if present, must be the first line of the ".sim" file (see below).

| **units:** *u* **tech:** *tech*

This special comment line identifies the technology (*tech*) from which the ".sim" file was derived. The units parameter (*u*) gives the scale factor by which all linear dimensions in the file are multiplied to obtain centimicrons.

type gate source drain l w [x y]

The line above represents a device of type *type*. For depletion mode GaAs, the set of types includes **'d'** or **'n'** (for D-mode MESFETs), **'l'** (for logic diodes), and **'s'** (for level-shift diodes). For E/D GaAs, the same set of types is used with the addition of **'e'** (for enhancement mode MESFETs).[5] Following the *type* code are three node names representing the *gate, source,* and *drain* connections. Diodes are treated as three-terminal devices with the gate terminal as anode and source shorted to drain as cathode. Occasionally one of the source or drain terminals of a diode may be left unconnected. This practice may cause warning messages to be generated by some analysis tools.

Following the gate, source, and drain node names are the length (*l*) and width (*w*) of the active part of the device. Finally, an optional location (*x,y*) of the device may be specified. If present, (*x,y*) represents the coordinates of a point within the gate region of the device. Note that *l, w, x,* and *y* are specified in λ units.

C *node1 node2 cap*

This line defines a lumped capacitance of *cap* femtofarads between *node1* and *node2*.

= *node1 node2*

The " = " indicates an alias or rename, where *node2* becomes an alternate name for *node1*. There is no declarative construct in ".sim" files. Node names become defined by appearing in device definition lines. All nodes have a single name but may have one or more aliases.

A few other line types exist in the ".sim" format. Since they are not used in GaAs circuits, these miscellaneous line types are not described here. The ".sim" format is sufficient to represent the circuit as an interconnected network of MESFETs, diodes, and capacitors. With such a level of detail, analysis tools and switch-level

---

[5] For NMOS the types are **'e'** and **'d'** for enhancement and depletion mode MOSFETs respectively. For CMOS, the device type codes are **'p'** and **'n'**, representing P-channel and N-channel FETs respectively.

simulators can provide useful information to the designer. Section 9.3.1 describes *statG*—a static electrical rules check analysis tool. Section 9.3.2 discusses *esim*—a simple switch-level event simulator. The *sim2spice* utility—a program to derive SPICE "decks" automatically from a ".sim" file—is described in Sec. 9.3.3.

9.2.6 Foundry Interface

The final result of an integrated circuit designer's activities is the specification of a set of masks which are used during wafer processing. Each fabrication process has its own needs at the detailed mask level. Thus, most technology-independent CAD toolsets provide a set of layer manipulation primitives and rules for applying the primitives.

In MAGIC, the *cifoutput* section of the technology file specifies the translation of the design data base into a mask-level foundry interface. Although named for its Caltech Intermediate Format output, the mask specifications can be written either in CIF [6] or in CALMA GDS II stream format [9].

The mask manipulation primitives provided in the MAGIC foundry interface are listed below. These primitives are divided into three types: boolean operators, sizing operators, and others.

MAGIC cifout Primitives

Boolean	Sizing	Other
or	bloat-or	squares
and	grow	
and-not	shrink	

Boolean primitives combine a specified list of data base tile types with the foundry mask layer currently under construction according to a specified boolean operator. The sizing operations (*bloat-or, grow*, and *shrink*) allow selective bloating or shrinking with electrical connectivity maintained. Figure 9-13 illustrates how *or* and *shrink* operations are applied to the *source-drain* and *schottky-contact* layers in the depletion mode process to obtain the ohmic-metal mask layer GC.

9.2.7 MAGIC's User Interface

The user interface to MAGIC is via keyboard and mouse. Overall, the system is window-based and interactive. There may be many windows on the user's screen during a MAGIC session, but only one of them can be a MAGIC *text* window. It is via this window that user commands are entered and textual feedback messages appear.

MAGIC has approximately 80 built-in commands. These form the primitives from which more customized user interaction may be built. Built-in commands are invoked by typing the command name prefixed by ":" and followed by the command arguments. Since each and every command begins with ":", the colons are not printed in the examples within this chapter. A colon is used in order to distinguish the built-in commands from user-defined single-key shorthand macros. If a user types a noncolon

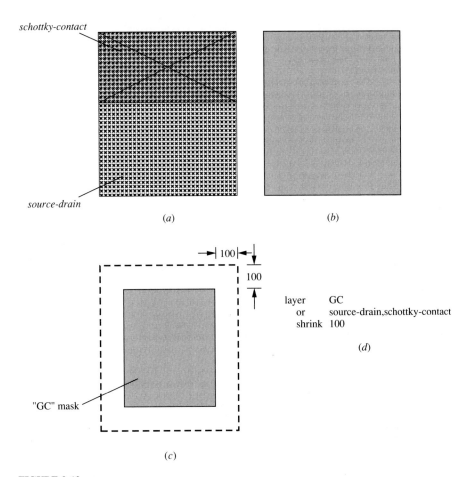

FIGURE 9-13
"GC" CIF layer made from *source-drain* and *schottky-contact* tiles. (*a*) Two adjoining tiles in plane database. (*b*) Tiles ORed together. (*c*) Result of shrink operation. (*d*) "cifout" commands in MAGIC technology file.

key when a command is expected, the key is considered a single-key macro and, if defined, the corresponding built-in command(s) will be executed. Users can define their own sets of single-key macros via the *macro* built-in command:

   macro   single-key   command-list

A typical set of predefined single-key macros is given in Table 9.2. Though described within the MAGIC documentation, these macros are not truly built into MAGIC. Rather, they are defined at the beginning of each user session. The mechanism for macro definition is the automatic inclusion and processing (known as *source*-ing) of a user-specific (or system-wide default) start-up script which has the filename ".magic". An individual user may define custom macros by putting *macro* definition commands in a private ".magic" start-up file located in the user's home directory.

All interactions with MAGIC can be accomplished by issuing commands from a keyboard, though greatly enhanced capabilities are available when a mouse is used.

TABLE 9.2
Default macros (as defined in ".magic" file)

macro	s	"select"	macro	W	"stretch down 1"
macro	S	"select more"	macro	E	"stretch up 1"
macro	a	"select area"	macro	R	"stretch right 1"
macro	A	"select more area"	macro	g	"grid"
macro	f	"select cell"	macro	G	"grid 2"
macro	C	"select clear"	macro	u	"undo"
macro	d	"delete"	macro	U	"redo"
macro	↑D	"erase $"	macro	v	"view"
macro	t	"move"	macro	z	"findbox zoom"
macro	T	"stretch"	macro	Z	"zoom 2"
macro	c	"copy"	macro	b	"box"
macro	↑X	"expand toggle"	macro	B	"findbox"
macro	x	"expand"	macro	,	"center"
macro	X	"unexpand"	macro	y	"drc why"
macro	q	"move left 1"	macro	↑L	"redraw"
macro	w	"move down 1"	macro	?	"help"
macro	e	"move up 1"	macro	o	"openwindow"
macro	r	"move right 1"	macro	O	"closewindow"
macro	Q	"stretch left 1"	macro	"*space*"	"tool"

The mouse serves as a pointing device for selection from menus and for establishing the location and size of the *box* and the *cursor*. Normally, a three-button mouse is used. All geometric manipulations involve the box, the cursor, and/or some selected tiles.

The *cursor* is a mark that directly tracks the mouse. The cursor can have one of several distinct appearances based on the current mode of MAGIC. The window containing the cursor is the one to which any subsequent user command applies.[6] The appearance of the cursor and box can be seen in the sample design rule check screen image of Fig. 9-8. The cursor is the plus-like target near the upper right-hand corner of the box.

The *box* is a rectangle that can be placed and sized by the user. Its purpose is to define a rectangular shape and location for use in subsequent commands. The ease of moving and shaping the *box* makes this method of specifying location and shape information very effective. Clicking the leftmost mouse button sets the fixed corner of the box (normally the lower-left corner though this can be changed under user control) at the position of the cursor, retaining the previous box size and shape.

---

[6] This can sometimes be confusing since the command keystrokes are echoed in the *text* window while most user commands are applied to layouts or other windows.

Clicking the rightmost mouse button sets the size and shape of the box by establishing the location of the nonfixed box corner (normally the upper right). Note that the box may degenerate to a line or even to a point (which is indicated by a thick "plus" mark) if one or both of the box's height or width is zero. Both line and point "boxes" have zero area but are useful for attaching point, line, or box labels to layout geometry.

The box can be manipulated without using the mouse via the commands below. In these commands, n is a distance in λ:

box height n

box width n

box *direction* n where *direction* is one of "up", "down", "right", "left"

box x_{ll} y_{ll} x_{ur} y_{ur}

Commands that reference or manipulate the box are only meaningful within a layout window.

The primary actions necessary for editing the tile data base are *paint, erase, select, move, stretch,* and *copy*. The *paint* command fills the area specified by the location and shape of the current box with a specified combination of tile types (known as the *paint*). Painting actions can be abbreviated with the mouse by clicking the middle mouse button while the cursor is positioned over the desired type of *paint*. In this mode, the cursor is thought to be as a paint brush dipping into the material to be painted.

In a similar (but opposite) way the *erase* command can remove a specified type of paint from the region under the box. Complete erasure can be abbreviated via the middle mouse button (while the cursor is positioned over *space*) but selective erasure can only be performed via the command

erase   typelist

The *typelist* argument is optional. When omitted, "*,labels" (all layout layers, plus the labels) is implied. In general, *typelist* is a comma-separated list of types. The special character "$" represents the type(s) of material currently present under the cursor. Thus, the command "erase $" can be used to selectively erase a particular type or set of types (whatever is found under the cursor) from within the region defined by the current box location. As an example, to erase all *second-metal* within the region covered by the box, issue the command "erase m2." The command "erase $," if invoked while the cursor is over a *second-metal* tile, is equivalent. If "erase" is invoked via the mouse (by clicking the middle mouse button with the cursor over *space*), everything within the boxed region will be erased.

Though paint and erase are sufficient to create or modify any desired layout, a richer set of layout editing commands is provided. In order to perform *move, copy,* or *stretch* commands, some existing tile geometry must first be selected. There are several methods provided to accomplish such selections. Single tiles can be selected (or added to the current selection) by placing the cursor over the desired tile and invoking the *select* (or *select more*) command. Single or multi-tile selections can be extended to include all electrically interconnected tiles by issuing a second *select* (or *select more*) without moving the cursor. This feature is convenient for investigating the connectivity within a complex network.[7] Selection can also be performed or augmented by containment within the box—use the *select [more] area [typelist]* command.

Once a set of tiles have been selected, the selection can be moved, copied, or stretched. *Move* and *copy* are straightforward operations using the fixed box corner and the current cursor location as from/to reference points. A special form of copy is the *array* operation which generates a one- or two-dimensional array of the selection using the box shape as an iteration offset.

The *stretch* operation is a useful variation of *move*. When the selection is moved via the stretch command, any paint crossing the trailing edge of the box is stretched to fill the gap. Material crossing the box's leading edge is consumed. Stretching is extremely useful for manipulating chunks of a layout without breaking electrical connectivity. Note that, unlike moving, stretching may involve motion in only a single Manhattan direction at a time.

In order to facilitate the use of meaningful node and signal names in simulations, MAGIC provides the capability to attach *labels* to tiles of a layout. The command to add a label is

label *label-string* [*position* [*layer*]]

where *position*, if present, is one of "top", "bottom", "left", "right", "center" or their abbreviations. The *label* command places a user-specified label at the position of the box. The user's *label-string* is displayed at the specified position relative to the box. Normally, the box is collapsed to either a point or a line, and the region under the box is occupied by a single-tile type. If *position* and *layer* are not specified, the label is logically attached to the layer under the box or to the *space* layer if no layer covers the entire area of the box. If tiles are moved or copied, the attached labels are also moved/copied, provided the move/copy always manipulates the entire boxed region of the label. When material underneath a label is subdivided, the label may get reattached to another tile type if one exists that occupies the entire label box. If no such tile type exists, the label will be attached to *space*. Whenever a label is detached or reattached, the user is notified via a text-window message.

All tiles of a layout are kept in a disk file with the filename "*cellname*.mag". In order to write the current edit cell to disk, the *save* command is issued. To load a cell from disk into MAGIC for viewing or editing, the command "load *cellname*" is given.

MAGIC supports multiple windows. Each window has a type from the set: *layout, netlist, colormap, text*. When entered, MAGIC creates two windows, one for layout editing and one for textual interaction. New windows can be created via the command "openwindow *type* [*x y*]." If unspecified, the location (*x,y*) comes

---

[7] Since selected tiles are highlighted on the graphics screen, this provides a quick and easy way to see detailed electrical continuity properties within a layout.

from the current cursor position. Windows can be closed via the *closewindow* command.

It is often desirable to open multiple windows so as to get more than one view of the chip or cell being edited. No matter how many windows are open, there can be only one box and one cursor. Of course, it is possible that the box may appear in part or in whole in several layout windows if their fields of view overlap. The cursor, however, serves to identify which window the mouse or keyboard command affects and thus has only one appearance on the screen.

The material on MAGIC presented in this chapter is provided in order to facilitate learning and getting results. It is neither complete nor detailed. The final word on MAGIC can be found in Refs. 4 and 7.

9.3 ANALYSIS TOOLS AND TECHNIQUES

9.3.1 Electrical Rules Check—statG

The files produced by MAGIC's extractor and the hierarchy flattener *ext2sim* contain a wealth of information. In fact, it is possible to analyze this information and critically evaluate the overall circuit. The *statG* program performs a static electrical rules check using the information found in the ".sim" file. The idea for such a program—originally for the NMOS technology—came from Baker and Terman in 1980 [10]. The GaAs electrical rules check program is a rewritten version of Baker's original *stat* program for NMOS. Many changes were necessary to support GaAs but the underlying data structures and analysis algorithms are largely the same as in *stat*.

StatG performs 12 checks on the network given as input. These checks involve some heuristics and so may not give meaningful results for all circuits. The operations and analyses performed by *statG* are:

Device classification and summary report
Location of network inputs (heuristic)
Location of pull-ups and pulled-up nodes
Location of network outputs (heuristic)
Location of pull-downs and pulled-down nodes
Location of pass transistors
Location of level-shift networks (heuristic)
Location and classification of logic gates
Location of superbuffers (heuristic)
Propagation analysis
Reporting of "strange" nodes, which are very likely to be errors
Rewriting of the network for switch-level simulation

These whole-chip analyses are vital sanity checks that can locate missed or accidental connections, incorrectly implemented gates, misconnected or disjoint subsystems, and floating nodes. Because the program checks for the presence of a small set of primitive structures and uses the results in its categorizations, it may not be

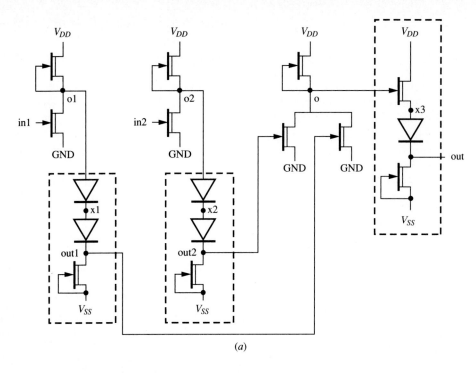

FIGURE 9-14
A 'statG' example. (*a*) Simple invert/NOR circuit schematic (identified level-shifter circuits are indicated in dashes). (*b*) Sim file (input to "statG"). (*c*) Modified "Sim2" file (output from "statG").

```
***** Static Electrical Rules Check Program - V3.5 *****

2 input node (explicitly declared)
1 output nodes (explicitly declared)
14 nodes, 0 enhancement, 11 depletion, 0 duplicates, 5 level-shift diodes
    3 d vss A vss
    1 d A B gnd
    3 d A A vdd
    5 s A B B
3 nodes simply pulled up, 1 unknown
7 pulldown transistors
0 pass transistors
2 inverters, 1 nor gates, 0 complex gates
1 level shift networks with 1 diode
2 level shift networks with 2 diodes
Propagation check OK
```

(d)

FIGURE 9-14 (*continued*)
(d) "StatG" printed output.

effective on new or experimental circuits, particularly those with primarily analog character. Future versions of the program may check fan-out rules, ratios, and threshold drops.

Figure 9-14a illustrates a simple circuit with two FET logic inverters driving a BFL NOR gate. A ".sim" file describing this circuit is shown in Fig. 9-14b. Though this file was prepared manually it has the same format as the ".sim" files extracted from MAGIC layouts after conversion by the *ext2sim* program. There is one addition to the standard ".sim" format, however—the declaration of input and output nodes (the last three lines of Fig. 9-14b). This information is necessary for *statG*'s propagation analysis phase. If these declarations had not been given, the program would have used heuristics to locate possible inputs and outputs.

For this example, *statG* correctly identified three level-shift networks, two inverters, and one NOR gate. These statistics are part of the printed output from *statG*, which is shown in Fig. 9-14d. The statistics produced by *statG* are an important output from the program. The designer can use them to quickly check whether the implemented circuit has the expected count of devices of each type as well as the expected number of gates, level-shifters, and the like. The first two lines of output following the title give an accounting of the number of inputs and outputs, including the method used in their determination—either heuristic or explicit user declarations. Following this, the circuit is summarized, giving the number of nodes, and all devices are categorized and counted by type. The device types are self-explanatory (except for the category "duplicates," which counts the number of FETs that are found to be connected in parallel—gate to gate, source to source, and drain to drain). Each such structure is considered to be one single composite FET with a width that is the sum of the widths of the paralleled component FETs.

Following the network summary is an indented usage summary showing counts of various usages of devices found within the network. The format of each line is

This section is

count type gate source drain

where *gate*, *source*, and *drain* are chosen from the set of symbols { **A, B, C, vdd, vss, gnd** }. When two like symbols appear on a usage report line (e.g., **3 d vss A vss**), it means that the two terminals—in this case the gate and drain terminals—are connected to the *same node*. Note that, like many of the other VLSI CAD tools, *statG* does not make any distinction between source and drain; in fact, it always considers a connection to a fixed voltage reference to be a connection to the drain terminal. Thus the example "**3 d vss A vss**" accounts for the three current sources that occur within the level-shifters of the Fig. 9-14*a* example. These depletion mode devices each have two terminals—their gate and source—connected to the **Vss** rail with the third terminal connected to a unique node. As another example, the line

1 d A B vdd

specifies that there is one D-mode MESFET with unique gate and source connections but with its drain connected to **Vdd**. This matches the source follower FET in the NOR circuit's level shifter.

Following the device usage summary, the *statG* program prints statistics on classifications of gates and other structures. It categorizes nodes with a single "d A A vdd" device as "simply pulled up" and uses such nodes and associated pull-down structures to recognize and classify gates as inverters, NORs, or others (which are lumped together under the heading of "complex" gates). Associated with the gate classification is a heuristic identification of level-shift networks and a report summarizing the findings. The structures enclosed within dashed boxes in Fig. 9-14*a* are correctly classified as level-shift networks. These are removed from the circuit for switch-level simulation purposes (diodes have no switch-level model and, in this case, their function is understood by heuristic-based recognition). The removal occurs by creation of an alternate simulation file—the so-called ".sim2" file. This file (as seen in Fig. 9-14*c*) is in the same format as the input ".sim" file and thus can be used directly with MOS switch-level simulators.

Because pull-ups and level shifters are heuristically identified, they can be changed or removed to make an equivalent switch-simulatable network. Devices removed by this action are changed into comment lines. Pull-up devices are changed into depletion loads and devices serving pull-down roles are changed into enhancement type. The *statG* program can be used to transform FL, BFL, and various ED circuits into "equivalent" switch-level networks that simple NMOS/CMOS simulators can handle. The use of NMOS-like networks facilitates the simplest form of switch simulation, where fights between a depletion load and an enhancement mode pull-down network are resolved by considering the enhancement as the stronger (less resistive) of the two types of elements. The device widths (though available) are not used by some switch-level simulators. One such simulator, *esim*, is described in detail in Sec. 9.3.2.

In the Fig. 9-14 example, each of the 14 nodes is labeled and the nodes names are shown within the schematic. It is a simple matter to match the ".sim" file description with the devices and their connectivity as shown in the schematic.

9.3.2 Switch-Level Simulation

A variety of levels of simulation are used during the design and analysis phases of an integrated circuit development. At the lowest level, an analog circuit simulator such as SPICE is used. With sufficiently small circuits and accurate device models, detailed analog waveforms can be simulated with relatively good agreement to measured data. Accuracies to within 10 percent are typical. Unfortunately, analog simulation is extremely computation-intensive and falls considerably short of handling the entire integrated circuit.

At the opposite end of the simulation spectrum lies behavioral simulation, a very high-level block data flow analysis that is used to prove correct problem decomposition as well as highest-level functionality and completeness. In between this and full analog circuit simulation lie a number of intermediate levels, including register-transfer, gate-, and switch-level simulations. Because high-level simulation ignores details at or below the circuit level, there is no substantive difference in simulation of GaAs systems. The highest level of simulation that can handle large circuits and at the same time work at a level low enough to depend on layout, device size, and circuit connectivity details is *switch-level* simulation.

At switch level, a gross simplification of switch (relay-like) behavior is made. Though this leads to an overall behavior quite unlike what really happens, the approximation is useful. The switch-level abstraction is only appropriate for *digital* integrated circuits. For systems relying on novel analog behavior or for more general analog circuit modeling, SPICE or its equivalent must be used. Because of its simplicity, switch-level simulation can handle extremely *large* circuits. It is not unusual to execute complete pad-to-pad tests, running test vectors though an entire integrated circuit design with tens of thousands of devices. In order to be able to deal with such a large number of devices, each device is modeled as a relay, as shown in Fig. 9-15. Specifically, the source and drain are considered to be the contact terminals of a relay controlled by the gate terminal. For MOS circuits, this approximation is not a bad

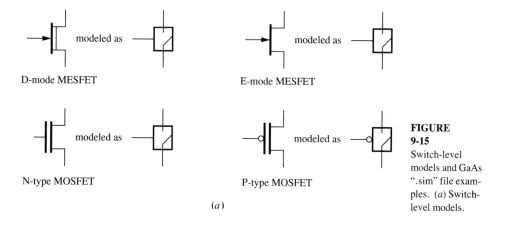

FIGURE 9-15
Switch-level models and GaAs ".sim" file examples. (*a*) Switch-level models.

446 GALLIUM ARSENIDE DIGITAL INTEGRATED CIRCUIT DESIGN

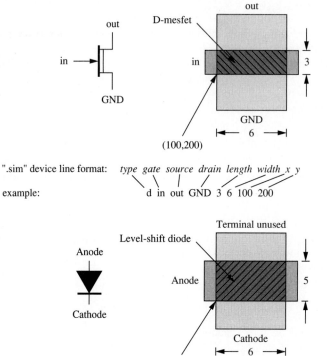

".sim" device line format: *type gate source drain length width x y*

example: d in out GND 3 6 100 200

".sim" device line format: *type gate source drain length width x y*

example: s anode cathode unused 5 6 1450 2100

FIGURE 9-15 (*continued*)

(*b*) (*b*) ".sim" file examples.

one. N-type MOSFETS "make" the relay contact circuit when the gate is high and break it when the gate is low. For P-type MOSFETs, low gate voltages (logic '0's) make the circuit and high gate voltages (logic '1's) break it.

The switch model for GaAs MESFETs is a very poor approximation of reality. Still, the model is useful for checking full-chip functionality. Because an entire chip can be simulated, connectivity and full functionality at the digital logic level can be verified. If the switch-level circuit is one that has been extracted from the geometric layout, then this type of simulation provides a valuable cross-check, comparing intended with implemented functionality.

There are several simulators with switch-level capability available today. A partial list includes *esim* [11], *rsim* [12], *mossim* [13], *cosmos* [14], *SILOS* [15], and *MagiCAD* [16]. Of the ones that directly accept ".sim" files as input, *esim* is perhaps the simplest to learn and use. Because of its particularly simplistic switch-level model which disregards device sizes, *esim* cannot correctly simulate some circuits. For circuits whose correct operation depends on device width ratios, *rsim, cosmos,* or another simulator based on a more sophisticated model must be used.

The syntax and semantics of *esim* commands follows. The description is broken into two parts: network description commands and simulation commands.

NETWORK DESCRIPTION COMMANDS OF esim

| *comment string* . . .

    Comment line—ignored by the simulator.

e *gate source drain* [*l w x y*]

    Define an enhancement device. Device length (l), width (w), and location (x, y) are ignored.

d *gate source drain* [*l w x y*]

    Define a depletion device. Length (l), width (w), and location (x, y) are ignored.

n *gate source drain* [*l w x y*]

    Define an N-channel MOSFET device. Length (l), width (w), and location (x, y) are ignored.

p *gate source drain* [*l w x y*]

    Define a P-channel MOSFET device. Length (l), width (w), and location (x, y) are ignored.

C *node1 node2 cap*

    Define lumped capacitor between *node1* and *node2*—ignored.

= *node1 node2*

    Define *node2* as alias for *node1*.

SIMULATION COMMANDS OF esim

h *node1* [*node2* . . .]

    Force the named nodes to the "high" logic state.

l *node1* [*node2* . . .]

    Force the named nodes to the "low" logic state.

x *node1* [*node2* . . .]

    Remove the forced state from named nodes.

I

    Initialize the network heuristically in an effort to get valid digital levels on all nodes. This should be used multiple times, until the number of events required to stabilize the network is reported as zero. This command is heuristic and does not always bring the network to a fully digital state.

w *node1* [*node2* ...]

Watch all specified nodes. At the end of each simulation step the value (state) of all watched nodes is printed. Node names prefixed with "−" are *removed* from the list of watched nodes.

W *vectorname node1* [*node2* ...]

Watch an ordered set of named nodes, displaying them as a bit vector.

v

View—print state of all watched nodes and vectors.

? *node1* [*node2* ...]

Print a detailed report of each named node showing its current state and the state of all devices affecting it.

! *node1* [*node2* ...]

Print a list of all devices controlled by each of the specified nodes.

V *node bit-sequence*

Define a sequence of input values for a node. On successive clock cycles (see the '**K**' command for clock cycle definition) the node is driven to the values given in *bit-sequence*.

K *node1 vector1* [*node2 vector2* ...]

Define the system clock cycle. A cycle consists of each of the named nodes running through the corresponding sequence vector with a simulation step ('**s**') command in between.

c

Cycle once through the system clock, as defined by the '**K**' command.

N

New—clear all vectors defined by the '**V**' command.

G [*n*]

Go—run the simulator through *n* clock cycles. If *n* is not specified it defaults to the length of the longest currently defined vector.

P *any text* ...

Print arbitrary constant text to the output. This is useful in annotating output generated by long-running simulations or complex test "scripts."

@ *file*

Include the contents of file *file* as if typed interactively. This allows a "script" of *esim* commands to be "played" and repeatably replayed to create a given set of precise test conditions.

s

Perform one simulation step. Propagate new values for inputs through the network until it has settled. If it never settles due to oscillatory behavior, the simulation step will run forever!

Example. A complete session with *esim* follows. The circuit consists of a single FET-logic inverter with level-shifter. After processing by *statG*, the ".sim2" file looks exactly like an NMOS inverter: one D-mode pull-up load and one E-mode pull-down, with four circuit nodes (V_{dd}, GND, *in*, and *out*). The two-line ".sim2" file is simply

e in out GND

d out out Vdd

The invocation line for running *esim* specifies the ".sim2" file. *Esim* output is shown in boldface type; user commands are shown in italics.

esim example.sim2

2 transistor, 4 nodes (1 pulled up)

sim> *V in 01*

sim> *w in out*

sim> *G 50*

0101010101010101010101010101010101010101010101010101>in

1010101010101010101010101010101010101010101010101010>out

sim> *q*

Notice that the **G** (go) command requested 50 clock cycles. Since the vector defined by the **V** command was shorter in length then 50, its sequence was simply recycled (starting at the beginning on the next cycle after the last specified value was used). This behavior is convenient for defining periodic stimuli.

A more substantive subsystem simulation is illustrated in Fig. 9-16. The subsystem presented is a 10-stage shift register (as shown in Fig. 9-16*a*, color insert). The *statG* output of Fig. 9-16*b* reveals 30 propagation errors. These errors demonstrate a quirk of many CAD tools, particularly simulators. In this case, the static analysis program correctly discovers and reports a circular dependence condition. A report line of the form

Propagate (· · I ·): *nodename*

***** Static Electrical Rules Check Program - V3.5 *****

1 input nodes (explicitly declared)
10 output nodes (explicitly declared)
236 nodes, 0 enhancement, 280 depletion, 0 duplicates, 96 level-shift diodes
　　40 d A B C
　　48 d vss A vss
　　96 d A B gnd
　　48 d A B vdd
　　48 d A A vdd
　　96 s A B B
48 nodes simply pulled up, 48 unknown
184 pulldown transistors
0 pass transistors
4 inverters, 4 nor gates, 40 complex gates
48 level shift networks with 2 diodes

30 nodes cannot be affected by the inputs
Propagate (I): ff_0[1]/3_312_38
Propagate (I): ff_1[3]/3_312_38
Propagate (I): ff_1[3]/3_312_205
Propagate (I): ff_0[2]/3_312_38
Propagate (I): ff_1[4]/3_312_38
Propagate (I): ff_0[0]/3_312_205
Propagate (I): ff_1[0]/3_312_38
Propagate (I): ff_0[3]/3_318_333
Propagate (I): ff_1[1]/3_312_38
Propagate (I): ff_0[2]/3_312_205
Propagate (I): ff_1[0]/3_318_333
Propagate (I): ff_1[2]/3_312_38
Propagate (I): ff_1[3]/3_318_333
Propagate (I): ff_0[0]/3_318_333
Propagate (I): ff_1[2]/3_312_205
Propagate (I): ff_0[2]/3_318_333
Propagate (I): ff_1[4]/3_312_205
Propagate (I): ff_0[1]/3_312_205
Propagate (I): ff_0[4]/3_312_205
Propagate (I): ff_1[2]/3_318_333
Propagate (I): ff_1[1]/3_312_205
Propagate (I): ff_1[4]/3_318_333
Propagate (I): ff_0[1]/3_318_333
Propagate (I): ff_0[4]/3_318_333
Propagate (I): ff_0[3]/3_312_38
Propagate (I): ff_0[3]/3_312_205
Propagate (I): ff_0[4]/3_312_38
Propagate (I): ff_1[1]/3_318_333
Propagate (I): ff_1[0]/3_312_205
Propagate (I): ff_0[0]/3_312_38

(b)

FIGURE 9-16
A session with statG as esim. (b) *statG* analysis of 10-stage shift register.

| Attempt to initialize the network
|

I
I

| Establish a valid initial condition
| (CLOCK low, Phi19 high)
|

l clock
s
h Phi19
s

| Re-attempt initialization
|

I
I

| Release the forced state of Phi19
|
x Phi19
s

| Define a periodic wave train on node 'CLOCK'
|

V CLOCK 0 1

| Watch all input and output nodes
|

w CLOCK Phi11 Phi12 Phi13 Phi14 Phi15 Phi16 Phi17 Phi18 Phi19 Phi20

| Execute the simulation for 70 clock cycles
|

G 70

| Quit
|
q

(c)

FIGURE 9-16
(*continued*)
(*c*) Command input file.

```
136 transistors, 91 nodes (48 pulled up)
initialization took 212 steps
initialization took 0 steps
step took 81 events
step took 23 events
initialization took 2 steps
initialization took 0 steps
step took 45 events
> 0 1 0 1 0 1 0 1 0 1 0 1 0 1 0 1 0 1 0 1 0 1 0 1 0 1 0 1 0 1 0 1 0 1 0 1 0 1 0 1 0 1 0 1 0 1 0 1 0 1 0 1 0 1 0 1 0 1 0 1 0 1 0 1 0 1 0 1 0 1 0 1 0 1 0 1 0 1 0 1 0 1 :CLOCK
> X 0 0 0 0 0 0 0 0 0 0 0 0 0 0 0 0 0 1 1 0 0 0 0 0 0 0 0 0 0 0 0 0 0 0 0 1 1 0 0 0 0 0 0 0 0 0 0 0 0 0 0 0 0 0 0 1 1 0 0 0 0 0 0 0 0 0 0 0 0 0 0 0 0 1 1 0 0 0 0 0 0 0 0 :Phi1
> X X X 0 0 0 0 0 0 0 0 0 0 0 0 0 0 0 0 0 1 1 0 0 0 0 0 0 0 0 0 0 0 0 0 0 0 0 0 1 1 0 0 0 0 0 0 0 0 0 0 0 0 0 0 0 0 0 1 1 0 0 0 0 0 0 0 0 0 0 0 0 0 0 0 1 1 0 0 0 0 0 0 :Phi12
> X X X X X 0 0 0 0 0 0 0 0 0 0 0 0 0 0 0 0 0 1 1 0 0 0 0 0 0 0 0 0 0 0 0 0 0 0 0 0 1 1 0 0 0 0 0 0 0 0 0 0 0 0 0 0 0 0 0 1 1 0 0 0 0 0 0 0 0 0 0 0 0 0 0 0 1 1 0 0 0 0 :Phi13
> X X X X X X X 0 0 0 0 0 0 0 0 0 0 0 0 0 0 0 0 0 1 1 0 0 0 0 0 0 0 0 0 0 0 0 0 0 0 0 0 1 1 0 0 0 0 0 0 0 0 0 0 0 0 0 0 0 0 0 1 1 0 0 0 0 0 0 0 0 0 0 0 0 0 0 0 1 1 0 0 0 :Phi14
> X X X X X X X X X 0 0 0 0 0 0 0 0 0 0 0 0 0 0 0 0 0 1 1 0 0 0 0 0 0 0 0 0 0 0 0 0 0 0 0 0 1 1 0 0 0 0 0 0 0 0 0 0 0 0 0 0 0 0 0 1 1 0 0 0 0 0 0 0 0 0 0 0 0 0 0 1 1 0 :Phi15
> X X X X X X X X X X X 0 0 0 0 0 0 0 0 0 0 0 0 0 0 0 0 0 1 1 0 0 0 0 0 0 0 0 0 0 0 0 0 0 0 0 0 1 1 0 0 0 0 0 0 0 0 0 0 0 0 0 0 0 0 0 1 1 0 0 0 0 0 0 0 0 0 0 0 0 0 0 1 :Phi16
> X X X X X X X X X X X X X 0 0 0 0 0 0 0 0 0 0 0 0 0 0 0 0 0 1 1 0 0 0 0 0 0 0 0 0 0 0 0 0 0 0 0 0 1 1 0 0 0 0 0 0 0 0 0 0 0 0 0 0 0 1 1 0 0 0 0 0 0 0 0 0 0 0 0 0 0 :Phi17
> X X X X X X X X X X X X X X X 0 0 0 0 0 0 0 0 0 0 0 0 0 0 0 0 0 1 1 0 0 0 0 0 0 0 0 0 0 0 0 0 0 0 1 1 0 0 0 0 0 0 0 0 0 0 0 0 0 0 0 1 1 0 0 0 0 0 0 0 0 0 0 0 0 0 0 :Phi18
> 1 X X X X X X X X X X X X X X X X 0 0 0 0 0 0 0 0 0 0 0 0 0 0 0 0 0 1 1 0 0 0 0 0 0 0 0 0 0 0 0 0 0 0 1 1 0 0 0 0 0 0 0 0 0 0 0 0 0 0 0 1 1 0 0 0 0 0 0 0 0 0 0 0 0 0 :Phi19
> X 1 1 X X X X X X X X X X X X X X X X X 0 0 0 0 0 0 0 0 0 0 0 0 0 0 0 0 0 1 1 0 0 0 0 0 0 0 0 0 0 0 0 0 0 0 1 1 0 0 0 0 0 0 0 0 0 0 0 0 0 0 0 1 1 0 0 0 0 0 0 0 0 0 0 0 :Phi20
```

(d)

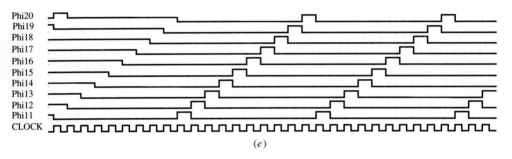

(e)

FIGURE 9-16 (*continued*)
(*d*) *Esim*. (*e*) Logic analyzer-style plot of *esim* output.

warns that the indicated node (*nodename*) does not depend on any input signal. Input propagation errors are denoted with the character "I". Other propagation errors can also be reported—in this case none have occurred. For a node that cannot be pulled low (indicated by the character "0") or high (indicated by "1"), the report lines will be

Propagate (0 · · ·): *nodename*

Propagate (· 1 · ·): *nodename*

When a node that can have absolutely no effect on any output signal is found, it is reported with a line of the form

Propagate (· · ·O): *nodename*

If a node has more than one of the above propagation errors, the reports are combined with the (**01IO**) field indicating the types of propagation anomaly that exist.

In the case of Fig. 9-16*b* a set of nodes are marked with "I" errors. These are due to circular dependencies. A circular dependency exists when there is a node, call it *A*, that can be affected by an input provided that some other node *B* can also be affected, but *B*'s propagation condition similarly depends on *A*. In the network of

Fig. 9-16, there are 10 flip-flops, each with three internal nodes that are difficult to initialize. The flip-flops rely on cross-coupled complementary signals to resolve at power-on into digital levels and latch in one of two valid states. *StatG*'s warning is useful because it points to potential problem areas where changes to the design might be made, resulting in a circuit that is more cleanly resettable, simpler to simulate, and generally more robust.

A simulation session with the *esim* simulator is shown in Fig. 9-16. The input file is presented in Fig. 9-16*c* and the simulator output is shown in Fig. 9-16*d*. Note the considerable workaround[8] that is necessary in order to get the network initialized. Since the network is a 10-stage circular shift register, once any one of the flip-flops exhibits digital behavior the simulation shows that the entire shift register gets "on track" within 10 clocks.

For more complex integrated systems the problem of creating test vectors exists. If the design has followed a structured design-for-testability strategy such as one of the types discussed in Chap. 6, then computer-aided testing tools and test pattern generation programs can be used to partially automate this task. Without a test strategy, the designer is left with *ad hoc* approaches which are generally inadequate in test coverage and are ultimately doomed to failure as circuit complexities rise.

9.3.3 Converting Netlists to SPICE Decks—sim2spice

It is desirable to close the loop from a layout to obtain detailed (simulated) performance information. The same file used for switch-level simulation can be used, after conversion, as input to SPICE. Of course, one cannot attempt to perform detailed analog simulation of an entire integrated circuit. However, if a general-purpose analog simulator existed that was fast and robust enough to handle such a large system, the capability does exist to generate a SPICE-format input deck. In practice, SPICE verification is performed on extracted cells and even on small subsystems, to formally verify layout and to model the timing and waveform effects due to interconnect and other parasitics.

The program *sim2spice* reads a file in ".sim" format and creates a new file in SPICE format. The conversion is guided by a *definitions* file, which maps certain node names found in the ".sim" file to special predefined SPICE node numbers. The definition files for the selective-implant depletion mode GaAs process and the self-aligned enhancement/depletion process are given in Table 9.3.

Sim2spice produces a SPICE deck containing a network of transistors, diodes, and capacitors. The user must add the SPICE model and simulation control lines to make the input file complete. The SPICE input is produced in a file with

---

[8] After several heuristic-based initializations ('**I**' commands) are performed, it is still necessary to force a digital state into one of the flip-flops (node **Phi19** is held high while the **CLOCK** node is low). While these nodes are held, another heuristic-based initialization is requested. Finally, the forced nodes are released via the '**x**' command. Note that each time a node is explicitly changed by the user, the simulator must be "stepped" ('**s**' command) so that the implications of the change can take effect.

TABLE 9.3
Sim2spice default definitions

(a) GaAs selective-implant D-mode process

set	GND	0	gaas
set	GND!	0	gaas
set	Vdd!	1	gaas
set	VDD	1	gaas
set	VDD!	1	gaas
set	Vdd	1	gaas
set	Vss!	2	gaas
set	VSS	2	gaas
set	VSS!	2	gaas

(b) GaAs self-aligned E/D process

set	GND	0	edgaas
set	GND!	0	edgaas
set	Vdd!	1	edgaas
set	VDD	1	edgaas
set	VDD!	1	edgaas
set	Vdd	1	edgaas
set	Vss!	2	edgaas
set	VSS	2	edgaas
set	VSS!	2	edgaas

the extension ".spice". *Sim2spice* also produces a file with the extension ".names" containing the mapping between ".sim" file node names and SPICE node numbers.

9.3.4 SPICE

Since its creation in the early 1970s, the SPICE program has been the dominant general-purpose analog circuit simulation tool used by electrical engineers. In the intervening years the world of electronics has grown and changed dramatically. Though it has continually undergone improvement, translation, and ports to many different computer systems, the basic structure, philosophy, and methods used within SPICE remain largely the same as the original program.

For GaAs integrated circuit design, accurate analog circuit simulation is vital. Unfortunately, many of the CAD programs available today to perform this function are deceptively named the same—SPICE or some variant—yet they sometimes have radically differing underlying device models. To make things even more confusing, some versions of SPICE have device models that use parameters with identical names but with *differing* meanings. The reader is cautioned that it is imperative to use a known version of SPICE—one that has been calibrated against the same fabrication process that is to be used in producing the design. The fabricator will often provide a version of SPICE that is compatible with their process line.

9.4 SYNTHESIS TOOLS AND EXAMPLES

As the base technologies for GaAs evolve and mature, they gradually become better understood. This evolution may occur over a span of years. One approach to design which enhances the probability of obtaining working circuits during process evolution is to insulate the designer to the greatest extent possible from process-dependent details. This approach pays the cost of relatively low density and lower overall speed but it facilitates the early use of the GaAs technology.

The isolation of function from process involves an intermediate level of expression of the design. This is nothing new—such forms as logic gates, state machine transition graphs, and boolean equations can express the functionality of a system without including circuits or process details. It is precisely at this level that a host of automated synthesis tools exist.

9.4.1 PLA Generation Tools

The circuits and layouts for simple GaAs programmable logic arrays (PLAs) are discussed in Sec. 8.5.1. In this section the focus is on automation of the conversion from intermediate functional form to PLA personality matrices. Once a personality matrix exists, a PLA can be generated merely by stacking together the appropriate component blocks per the characters in the matrix. Figure 9-17 (color insert) shows such a matrix and the resulting PLA.

The UC-Berkeley PLA tools [18] are useful for this purpose. This suite of computer-aided design programs treats PLAs in a technology-independent way. Such treatment makes the creation of PLAs for a new technology relatively straightforward. As discussed in Sec. 8.5.1, GaAs layout rules and fan-out limitations restrict the creation of densely packed MOS-like PLAs. Nevertheless, PLAs can be made and, because of the ease of doing so and the inherent correct-by-construction property, their use is appropriate for control circuits or non-time-critical portions of a GaAs integrated circuit. PLAs can directly implement boolean combinational logic as well as small finite state machines (once latches are added for holding the state variables).

All PLAs are described in terms of their so-called "personality matrices." The format of such a description is simply a file with control information (lines beginning with "."), comments, and information about the AND and OR planes. PLA personality matrix format is also known as truth table (or "tt") format.

#anything		*comment*
.i	*n*	*specifies number of PLA inputs*
.o	*n*	*specifies number of PLA outputs*
.ilb	*label . . .*	*specifies input label(s)*
.ob	*label . . .*	*specifies output label(s)*
.p	*n*	*specifies number of product terms*

.na *symbol*   specifies the name of resulting PLA
.e   end of file

Following the "**.p**" line is a description of the AND and OR planes of the PLA with one line per product term. Double rail inputs are assumed to be available within any PLA. For personalizations, connections in the AND plane are represented with a "1" for a connection to the noninverted line and a "0" for connection to the inverted line. No connection to a given input is indicated by one of the characters "x", "X", or "−". Connections in the OR plane are represented by a "1". No connection in the OR plane is indicated by "x", "X", "0", or "−". Note that a "0" in the OR plane specification does NOT mean that the function has the value "0". Instead it means that the product term represented by the current line is *not to be connected* to the given function output line in the OR plane.

For combinational logic, a PLA personality matrix—a truth table—can be obtained from boolean equations. The equation-to-truth table conversion utility *eqntott* [17] performs this transformation. The input file given to *eqntott* contains a description of the required function(s) written in statements from a simple specification language.

The *eqntott* specification file begins with the line "NAME = *name*;" which gives a name to the PLA. This is followed by declarations, identifying the symbols used for the circuit's inputs and outputs. Inputs (outputs) are declared on a line with the keyword INORDER (OUTORDER) and the format

INORDER = *name* . . . ;

OUTORDER = *name* . . . ;

As the inputs and outputs are declared, their sequence determines the order of appearance along the periphery of the PLA to be generated. Once the inputs and outputs have been declared, the rest of the specification consists of equations—one or more for each output—of the form

OUTname = expression;

Multiple assignments to the same output name are equivalent in meaning to a single assignment of the logical disjunction (OR) of the original expressions. Operators allowed within expressions are "|" (OR), "&" (AND), "!" (NOT). By default, AND operators associate left to right and have the same precedence as NOT operators. OR operators also associate left to right, but have a lower precedence than AND. This precedence makes traditional sum-of-products expressions easy to write. Parentheses may be used to override the default precedences.

Figure 9-18 illustrates the conversion process starting from equations and producing a personality matrix for a completely combinational example PLA implementing a single-bit full-adder circuit. Once the conversion from equations is complete the truth table file can be used to generate a PLA. For large functions, it may first be desirable to perform boolean minimization, however. Several programs are available that can perform such optimizations on a PLA in "tt" form. The programs *espresso* [19] and *presto* take a "tt" format file as input and produce a minimized "tt" format file as output. These minimization tools are part of the UC-Berkeley VLSI CAD suite, along with MAGIC, *ext2sim, esim,* and many other CAD programs.

```
NAME = adder;                /* assigns the name "adder" to the PLA*/
INORDER = A B Cin;           /*declaration of inputs and their order*/
OUTORDER = Cout Sum;         /*declaration of outputs and their order*/

Sum = A & !B & !Cin | !A & B & !Cin | !A & !B & Cin | A & B & Cin;
Cout = A & B | A & Cin | B & Cin;
```

(a)

```
%eqntott -1 adder.eqn
  .i 3                truth table has three inputs
  .o 2                truth table has two outputs
  .na adder           specifies the name "adder"
  .ilb A B Cin        specifies input labels and order
  .ob Cout Sum        specifies output labels and order
  .p 7                seven product terms follow
  001 0 1
  010 0 1
  –11 1 0
  100 0 1
  1–1 1 0
  11– 1 0
  111 0 1
  .e                  marks end of truth table file
```

(b)

FIGURE 9-18
Full-adder PLA personality matrix generated from equations. (a) *eqntott* input specification file ("adder.eqn"). (b) Program invocation and output.

PLA generation progresses by running *mpla*. This program uses the supplied personality matrix and a PLA *style* to specify how to stack predefined circuit layout blocks together to form the final PLA. *Mpla* is technology-independent. Because GaAs PLAs are inefficient and rather difficult to lay out, there are relatively few styles. The invocation of *mpla* is

$$\text{mpla } [\text{ options }] -\text{s } style -\text{o } output\text{-}file\ input\text{-}file$$

where *style* is one of the keywords **BFLcis** or **EDcis**. Figure 9-17 (color insert) shows the PLA resulting from running *mpla* with *style* of **EDcis** using the personality matrix of the full-adder example (Fig. 9-17*a*).

HOMEWORK PROBLEMS

9.1. Learn your CAD system by reading the available tutorial material. If your system is MAGIC, execute the MAGIC tutorials 1-6. Prove your proficiency by performing each of the following on your CAD system:
(*a*) Make a palette cell showing all available layers with names and colors.
(*b*) Produce a hardcopy plot of the palette cell of part (*a*).

(c) Enter the circuit layout of Fig. 9-12a and save as cell "INV."
(d) Stack four "INV" cells together in a row. Do this using the ":get" and ":array" commands.

9.2. Use MAGIC's *drc* subsystem to check a layout. Insert some design rule errors and observe the discovery of them. Use the ":drc why" command to get prose-form feedback on each error.

9.3. Design the following simple cells and do a layout for each. Verify their functionality from the extracted layout using switch-level models and the *esim* (or equivalent) simulator.
(a) NOR gate
(b) Full adder
(c) Master/slave flip-flop
(d) 4-bit binary up-counter

9.4. Use a PLA generator (*mpla* or equivalent) to make a PLA implementing an 8-bit priority encoder. The truth table is given below.

Inputs	Active	S2	S1	S0
0 0 0 0 0 0 0 0	0	0	0	0
1 - - - - - - -	1	0	0	0
0 1 - - - - - -	1	0	0	1
0 0 1 - - - - -	1	0	1	0
0 0 0 1 - - - -	1	0	1	1
0 0 0 0 1 - - -	1	1	0	0
0 0 0 0 0 1 - -	1	1	0	1
0 0 0 0 0 0 1 -	1	1	1	0
0 0 0 0 0 0 0 1	1	1	1	1

(a) Extract the PLA and convert it to ".sim" format.
(b) Exercise the simulated circuit with a switch-level simulator to verify its correct functionality.
(c) Make a SPICE deck from the ".sim" file and use SPICE to estimate the timing of one of the output functions.

REFERENCES

1. Ousterhout, J. K., Hamachi, G. T., and Mayo, R. N.: "The Magic VLSI Layout System," *IEEE Design and Test,* pp. 19–30, February 1985.
2. "Magic Tutorials," in *Berkeley VLSI Tools: More Works by the Original Artists,* Technical report UCB/CSD/85-225, University of California, Berkeley, Calif., February 1985.
3. Ousterhout, J. K.: "Corner Stitching: A Data Structuring Technique for VLSI Layout Tools," *IEEE Trans. on CAD,* vol. CAD-3, no. 1, pp. 87–99, January 1984.
4. "MAGIC Technology File Maintainer's Manual," in *Berkeley VLSI Tools: More Works by the Original Artists,* Technical report UCB/CSD/85-225, University of California, Berkeley, Calif., February 1985.
5. Nagel, L. W.: "SPICE2: A Computer Program to Simulate Semiconductor Circuits," Memorandum ERL-M520, Electronics Research Lab, University of California, Berkeley, Calif., May 1975.
6. Hon, R., and Sequin, C.: "A Guide to LSI Implementation," 2d ed., Technical memo SSL-79-7, XEROX Palo Alto Research Center, January 1980.

7. "MAGIC—VLSI Layout Editor," Manual pages from the UC-Berkeley 1985 VLSI Tools Distribution, Report UCD/CSD 85/225, University of California, Berkeley, Calif., February 1985.
8. Quarles, T., et al.: "SPICE 3B1 User's Guide," Department of Electrical Engineering and Computer Science, University of California, Berkeley, CA 94720.
9. CALMA Corporation, *GDS-II Stream Format,* July 1984.
10. Baker, C., and Terman, C.: "Tools for Verifying Integrated Circuit Designs," *Lambda*, pp. 22–30, fourth quarter 1980.
11. "ESIM—Event Driven Switch-Level Simulator," Manual pages from the UC-Berkeley 1985 VLSI Tools Distribution, Report UCB/CSD 85/225, University of California, Berkeley, Calif., February 1985.
12. Terman, C.: "RSIM—A Logic-Level Timing Simulator," *Proceedings of the IEEE International Conference on Computer Design,* pp. 437–440, October 1983.
13. Bryant, R.: "A Switch-Level Model and Simulator for MOS Digital Systems," *IEEE Trans. Computers,* vol. C-33, no. 2, pp. 160–177, February 1984.
14. Bryant, R., Beatty, D., Brace, K., Cho, K., and Sheffler, T.: "COSMOS: A Compiled Simulator for MOS Circuits," *Proceedings of the 24th ACM/IEEE Design Automation Conference,* pp. 9–16, 1987.
15. "SILOS—Logic and Switch-Level Simulator," SimuCad, Inc., 1985.
16. "MagiCAD User's Guide," Mayo Clinic, 1988.
17. "EQNTOTT—Generate Truth Table from Boolean Equations," Manual pages from the UC-Berkeley 1985 VLSI Tools Distribution, Report UCB/CSD 85/225, University of California, Berkeley, Calif., February 1985.
18. "MPLA—Technology-Independent PLA Generator," Manual pages from the UC-Berkeley 1985 VLSI Tools Distribution, Report UCB/CSD 85/225, University of California, Berkeley, Calif., February 1985.
19. Brayton, R., Hachtel, G., McMullen, C., and Sangiovanni-Vincentelli, A.: *Logic Minimization Algorithms for VLSI Synthesis,* Kluwer Academic Publishers, 1984.

APPENDIX A

MAGIC TECHNOLOGY FILE FOR SELECTIVE-IMPLANT D-MODE PROCESS

```
/*******************************************************
********************************************************
**
**    gaas.tech — Technology file for MAGIC
**
**
**    Defines the selective-implant depletion mode technology.
**
**    Suitable for multiple foundries
**
**    Developed by the University of California-Santa Barbara 1988
**
********************************************************
*******************************************************/
tech
    gaas
end
```

```
planes
    active
    metal2
    fab
end

types

/*  Primary active and interconnection layers  */

    active      n_plus,np
    active      n_minus,nm
    active      schottky_metal,sm
    metal2      second_metal,m2
    fab         glass,ov

/*  Derived layers  */

    active      implant,im

/*  Contacts (actually derived tile types) between layers  */

    active      via,v
    active      ohmic_contact,oc
    active      schottky_contact,sc

/*  Active devices  */

    active      mesfet,mf
    active      level_shift_diode,ls
    active      logic_diode,log
end

/*  On-screen appearance of each tile type. Uses standard MOS styles  */

styles
    styletype           mos

    schottky_metal      1

    n_minus             2

    n_plus              4

    implant             3

    second_metal        20

    ohmic_contact       2
    ohmic_contact       33

    schottky_contact    1
    schottky_contact    33

    via                 1
    via                 20
    via                 32

    ls                  6
    ls                  10
```

	mesfet	6		
	mesfet	7		
	log	6		
	log	11		
	glass	20		
	glass	34		
	error_s	42		
	error_p	42		
	error_ps	42		

end

/* Contacts (connections between planes) */

contact
    via   schottky_metal   second_metal
end

/* Composition and decomposition rules */

compose

/* The following rules allow for the composition of mesfets,
 * level shift diodes and logic diodes from schottky metal
 * and implant (n+ and n−)
 */

compose	mesfet	schottky_metal	nm
compose	implant	np	nm
compose	logic_diode	schottky_metal	np
compose	level_shift_diode	schottky_metal	implant
compose	schottky_contact	schottky_metal	ohmic_contact
erase	level_shift_diode	np	mesfet
erase	level_shift_diode	nm	logic_diode
paint	mesfet	np	level_shift_diode
paint	mesfet	im	level_shift_diode
paint	logic_diode	nm	level_shift_diode
paint	logic_diode	im	level_shift_diode
erase	ohmic_contact	nm	space
erase	ohmic_contact	np	space
erase	ohmic_contact	im	space
erase	schottky_contact	nm	schottky_metal
erase	schottky_contact	np	schottky_metal
erase	schottky_contact	im	schottky_metal
erase	glass	second_metal	space

end

/* Specify the adjoining tile types that propagate continuity */

connect
    schottky_metal   oc,sc,via,ls,mf,log

```
    n_minus          np,im,oc,sc
    n_plus           im,oc,sc
    im               oc,sc
    second_metal     via
    ohmic_contact    sc
    schottky_contact via,ls,mf,log
end
```

/* The cifoutput section controls CIF and CALMA generation.
 * Only one output style is given, the lambda=0.5 micron style.
 */

```
cifoutput
    style       dmode
    scalefactor 50

    layer GN nm,im,oc,sc,mf,ls
        calma 2 1

    layer GI np,im,oc,sc,ls,log
        calma 3 1

    layer GC oc,sc
        shrink 100
        calma 4 1

    layer GS sm,sc,via,ls,mf,log
        calma 6 1

    layer GV via
        shrink 100
        calma 7 1

    layer GM m2,via
        calma 8 1

    layer GO glass
        calma 10 1
end
```

/* The cifinput section is defined for files generated using this
 * technology file only. Files generated by 'external' means should
 * be hand-checked for missing or misinterpreted features.
 */

```
cifinput
    style       dmode
    scalefactor 50

    layer n_minus        GN

    layer n_plus         GI

    layer ohmic_contact  GC
        grow             100
```

```
                and             GN
                and             GI
        layer   via             GV
                grow            100
                and             GS
                and             GM
        layer   schottky_metal  GS
        layer   second_metal    GM
        layer   glass           GO
        calma   GN      2 *
        calma   GI      3 *
        calma   GC      4 *
        calma   GS      6 *
        calma   GV      7 *
        calma   GM      8 *
end

/*  Design rules  */

drc
#define allim         nm,np,im,oc,sc,mf,log,ls
#define allschottky   sm,sc,via,mf,log,ls
#define allmetal      m2,via
#define allohmic      oc,sc
#define allactive     mf,ls,log

/*****Width rules*****/

        width allim 5 "Implant width must be at least 2.5 microns (5 lambda)"

        width allmetal 6 "Second metal width must be at least 3 microns (6 lambda)"

        width via 10 "Vias must be at least 5 microns square (10 lambda)"

        width allohmic 11 \
        "Ohmic contacts must be at least 5.5 microns square (11 lambda)"

        width allactive 2 \
        "Mesfet/diode length must be at least 1 micron (2 lambda)"

        width allschottky 2 \
        "Schottky metal width must be at least 1 micron (2 lambda)"

/******Spacing rules*****/

        spacing allim allim 7 touching_ok \
             "Implant separation must be at least 3.5 microns (7 lambda)"

        spacing allschottky allschottky 5 touching_ok \
             "Schottky metal/schottky metal separation must be at least 2.5 (5 lambda)"
```

spacing allschottky allim 2 touching_ok \
  "Schottky metal/implant separation must be at least 1 micron (2 lambda)"

spacing allmetal allmetal 6 touching_ok \
  "Second metal/second metal separation must be at least 3 microns (6 lambda)"

spacing second_metal schottky_metal 0 touching_illegal \
  "Second metal/schottky metal edges cannot be coincident"

/*****Composition rules*****/

edge4way sc oc 4 allohmic 0 0 \
  "Malformed schottky metal/ohmic contact connection"

edge4way oc sc 4 allohmic 0 0 \
  "Malformed schottky metal/ohmic contact connection"

edge allactive space 1 0 0 0 \
  "Mesfet/diode overhang is missing"

edge space allactive 1 0 0 0 \
  "Mesfet/diode overhang is missing"

edge4way allactive sm 2 sm 0 0 \
  "Schottky metal must overhang mesfet/diode by at least 1 micron (2 lambda)"

edge4way allactive allim 3 allim 0 0 \
  "Implant must overhang mesfet/diode by at least 1.5 microns (3 lambda)"

end

/* Auto-router section (untested) */

router

layer1	second_metal	6	allmetal	6
layer2	schottky_metal	4	allschottky	5
contacts	via	10		
gridspacing		10		

end

/* Circuit extractor parameters */

extract

```
    style    default
    lambda 50
    step     50
    resist   nm             2000000
    resist   np,im          500000
    resist   sm,mf,ls,log,sc,oc 80
    resist   m2,ov,via         40

    fet      mesfet    nm 2    n    !GND 0 0
    fet      ls        im 1    s    !GND 0 0
    fet      log       np 1    g    !GND 0 0
```

end

/* Layers for use with Magic's wiring tool */

wiring

/* It is assumed that all wiring will be done in either schottky or second metal.
 * For this reason, the ohmic contact/schottky contact connection is not
 * included in this section.
 */

    contact   via   10   m2   0   sm   0

end

/* Parameters to control Magic's plow command */

plowing

    fixed   mf,ls,log,oc,glass
    covered   mf,ls,log,oc
    drag   mf,ls,log,oc

end

/* Stipple patterns for raster plotting */

plot

    style versatec   /* Stipple fill patterns for the various layers */

    /* Same as Gremlin stipple 9: */
    ls,log \
    07c0 0f80 1f00 3e00 \
    7c00 f800 f001 e003 \
    c007 800f 001f 003e \
    007c 00f8 01f0 03e0

    /* Same as Gremlin stipple 10: */
    mesfet,log \
    1f00 0f80 07c0 03e0 \
    01f0 00f8 007c 003e \
    001f 800f c007 e003 \
    f001 f800 7c00 3e00

    /* Same as Gremlin stipple 11: */
    oc,sc \
    c3c3 c3c3 0000 0000 \
    0000 0000 c3c3 c3c3 \
    c3c3 c3c3 0000 0000 \
    0000 0000 c3c3 c3c3

    /* Same as Gremlin stipple 12: */
    glass \
    0040 0080 0100 0200 \
    0400 0800 1000 2000 \
    4000 8000 0001 0002 \
    0004 0008 0010 0020

    /* Same as Gremlin stipple 17: */
    im,ls,oc,sc \

```
    0000  4242  6666  0000 \
    0000  2424  6666  0000 \
    0000  4242  6666  0000 \
    0000  2424  6666  0000

/*   Same as Gremlin stipple 19:   */
sm,via/active,mesfet,ls,log,sc \
    0808  0400  0202  0101 \
    8080  4000  2020  1010 \
    0808  0004  0202  0101 \
    8080  0040  2020  1010

/*   Same as Gremlin stipple 22:   */
m2,via/metal2,glass \
    8080  0000  0000  0000 \
    0808  0000  0000  0000 \
    8080  0000  0000  0000 \
    0808  0000  0000  0000

/*   Same as Gremlin stipple 23:   */
glass \
    0000  0000  1c1c  3e3e \
    3636  3e3e  1c1c  0000 \
    0000  0000  1c1c  3e3e \
    3636  3e3e  1c1c  0000

/*   Same as Gremlin stipple 29:   */
np,ls,log \
    0000  0808  5555  8080 \
    0000  8080  5555  0808 \
    0000  0808  5555  8080 \
    0000  8080  5555  0808

/*   Same as Gremlin stipple 31:   */
nm,ls,mesfet \
    0808  1010  2020  4040 \
    8080  4141  2222  1414 \
    0808  1010  2020  4040 \
    8080  4141  2222  1414
    via                    X
    oc,sc                  B

style gremlin
    ls,log                        9
    mesfet,log                   10
    oc                           11
    glass                        12
    im,mesfet,ls,log,oc          17
    sm,via/active,mesfet,ls,log,oc  19
    m2,via,glass                 22
    via                           X
    oc                            B
end
```

APPENDIX B

MAGIC TECHNOLOGY FILE FOR GaAs SELF-ALIGNED E/D-MODE PROCESS

```
/***************************************************************
 *
 * edgaas.tech
 *
 * Defines the GaAs E/D process including third metal.
 *
 * Derived from Vitesse's Foundry Design Guide, Mar 1989
 * (used with permission).
 *
 * (Lambda = 0.4 micron)
 *
 * Version 0.11 — August 1, 1989
 *
 ***************************************************************
 */
tech
        edgaas
end
```

planes
    substrate
    metal1
    metal2
    metal3
end

types
    /*   Primary substrate and interconnection layers   */
    substrate   active_area,act
    substrate   gate_metal,gm
    substrate   ohmic_metal,om,source_drain_metal,sdm
    metal1   metal_1,m1
    metal2   metal_2,m2
    metal3   metal_3,m3
    /*   Contacts between layers and planes   */
    substrate   contact_0,ohmic_contact,oc
    substrate   ohmic_contact_dfet,ocd
    metal1   via1g,vg
    metal1   via1sd,vsd
    metal2   via2,v2
    metal3   via3,v3
    metal3   pad
    /*   Active devices   */
    substrate   efet,ef
    substrate   dfet,df
end

styles
    styletype   mos
    gate_metal   1
    source_drain_metal   2
    source_drain_metal   33
    contact_0   1
    contact_0   33
    ocd   1
    ocd   7
    ocd   33
    active_area   2
    metal_1   20
    metal_2   21
    metal_3   5
    via1g   1
    via1g   20

```
        via1g             32
        via1sd            2
        via1sd            20
        via1sd            32

        via2              20
        via2              21
        via2              32

        via3              5
        via3              21
        via3              32

        efet              6
        efet              11

        dfet              6
        dfet              7

        pad               5
        pad               21
        pad               32
        pad               33

        error_s           42
        error_p           42
        error_ps          42
end

contact
        via1g    gate_metal        metal_1
        via1sd   active_area       metal_1
        via2     metal_1           metal_2
        via3     metal_2           metal_3
        pad      metal_2           metal_3
end

compose

/*  The following rules allow for the composition of
 *  enhancement and depletion mode mesfets. Diodes are
 *  constructed as depletion mode mesfets with source and
 *  drain shorted.
 *  Also included is the composition of the ohmic/gate metal
 *  contact for the modified Vitesse process.
 */
        compose     efet        gate_metal    act
        compose     contact_0   gate_metal    om
        compose     ocd         dfet          om
        decompose   dfet        gate_metal    act
        paint       dfet        act           dfet
        paint       dfet        gate_metal    dfet
```

paint	pad	via3	pad
paint	pad	metal_2	pad
paint	pad	metal_3	pad
erase	pad	metal_2	space
erase	pad	metal_3	space

end

connect

active_area	ohmic_metal,ohmic_contact,ocd,via1sd
gate_metal	ohmic_contact,ocd,via1g,efet,dfet
ohmic_metal	ohmic_contact,ocd,via1sd
metal_1	via1g,via1sd,via2
metal_2	via2,pad,via3
metal_3	via3,pad
ohmic_contact	via1g,dfet,efet
ohmic_contact_dfet	via1g,dfet
via1g	via2,via1sd
via1sd	via2
via2	via3

end

cifoutput

    style   vitesse0.4
    scalefactor   40

    layer GAC   active_area,via1sd,efet,dfet,oc,ocd,om
        calma 1   1

    layer GDI   dfet,ocd
        grow   100
        calma 2   1

    layer GGM   gm,via1g,efet,dfet,oc,ocd
        calma 3   1

    layer GOM   via1sd/substrate,om
        shrink   20
        bloat-or oc,ocd * 0 via1sd/substrate,om 20
        grow   40
        shrink   40
        calma 4   1

    layer GV1   via1sd
        shrink   25
        or   via1g
        shrink   50
        calma 5   1

    layer GM1   m1,vg,vsd,v2
        calma 6   1

    layer GV2   via2
        shrink   60
        calma 7   1

```
        layer   GM2     m2,v2,v3,pad
                calma 8  1

        layer   GV3     pad
                shrink  100
                or      via3
                shrink  150
                calma 9  1

        layer   GM3     pad
                shrink  100
                or      m3,v3
                calma 10 1

        layer   GOG     pad
                shrink  750
                calma 11 1

end

cifinput
        style                           vitesse0.4
        scalefactor                     40

        layer       active_area         GAC

        layer       dfet                GDI
                    shrink              100
                    and                 GAC
                    and                 GGM

        layer       om                  GOM

        layer       gate_metal          GGM

        layer       via1sd              GV1
                    grow                50
                    and                 GOM
                    and                 GAC
                    and                 GM1

        layer       via1g               GV1
                    grow                50
                    and                 GGM
                    and                 GM1

        layer       metal_1             GM1

        layer       via2                GV2
                    grow                60
                    and                 GM1
                    and                 GM2

        layer       metal_2             GM2

        layer       via3                GV3
                    grow                150
                    and                 GM2
```

	and		GM3
layer	metal_3		GM3
layer	pad		GOG
	grow		700
calma	GAC	1	*
calma	GDI	2	*
calma	GGM	3	*
calma	GOM	4	*
calma	GV1	5	*
calma	GM1	6	*
calma	GV2	7	*
calma	GM2	8	*
calma	GV3	9	*
calma	GM3	10	*
calma	GOG	11	*

end

drc

```
#define    allactive      act,efet,dfet,vsd,oc,ocd,om
#define    justactive     act,vsd,om
#define    allgate        gm,vg,efet,dfet,oc,ocd
#define    close_ohmic    om,vsd
#define    far_ohmic      oc,ocd
#define    allohmic       oc,om,ocd,vsd
#define    allm1          m1,vsd,vg,v2
#define    allvia1        vg,vsd
#define    allm2          m2,v2,v3/metal2,pad/metal2
#define    allm3          m3,v3,pad
#define    allfet         efet,dfet,oc,ocd
```

/* Level 1 – Active Area Definition */

edge4way gm dfet,ocd 5 dfet,ocd 0 0 \
"11W1(T): active area of transistor must be at least 2.0 microns (5 lambda)"

edge4way gm efet 5 efet 0 0 \
"11W1(T): active area of transistor must be at least 2.0 microns (5 lambda)"

edge4way allfet allactive 5 allactive 0 0 \
"11W1(T): active area of transistor must be at least 2.0 microns (5 lambda)"

width allactive 4 \
"11LW1(R): active as resistor must be at least 1.6 microns (4 lambda)"

spacing allactive allactive 5 touching_ok \
"11SP1: active area separation must be at least 2.0 microns (5 lambda)"

/* Level 2 – Depletion Implant Definition   */

   spacing dfet dfet 10 touching_ok \
   "22SP1: dfet/dfet separation must be at least 4.0 microns (10 lambda)"

   spacing efet dfet 5 touching_illegal \
   "21SW1: efet/dfet separation must be at least 2.0 microns (5 lambda)"

/* Level 3 – Gate Metal Definition   */

   width allgate 3 \
   "33LW1: gate metal width must be at least 1.2 microns (3 lambda)"

   spacing allgate allgate 4 touching_ok \
   "33SP1: gate metal/gate metal separation must be at least 1.6 microns (4 lambda)"

   edge4way allfet gm 2 gm 0 0 \
   "31OL1: gate metal must overhang transistor by at least 0.8 microns (2 lambda)"

   edge4way allfet space 1 0 0 0 \
   "31OL1: gate metal must overhang transistor by at least 0.8 microns (2 lambda)"

   edge4way space allfet 1 0 0 0 \
   "31OL1: gate metal must overhang transistor by at least 0.8 microns (2 lambda)"

   spacing allgate close_ohmic 2 touching_ok \
   "34SP1: gate metal/ohmic metal separation must be at least 0.8 microns (2 lambda)"

   spacing allgate far_ohmic 3 touching_ok \
   "34SP1: gate metal/ohmic metal separation must be at least 1.2 microns (3 lambda)"

   spacing allactive allgate 2 touching_ok \
   "31SP2: active area/gate metal separation must be at least 0.8 microns (2 lambda)"

   edge4way justactive gm,vg 2 space 0 0 \
   "31SP2: active area/gate metal separation must be at least 0.8 microns (2 lambda)"

   edge4way gm allfet 3 allfet 0 0 \
   "31OL2: gate metal edge must be at least 1.2 microns (3 lambda) from parallel active area edge"

   edge4way justactive allfet 3 allfet 0 0 \
   "31OL2: gate metal edge must be at least 1.2 microns (3 lambda) from parallel active area edge"

   edge4way allfet justactive 3 justactive 0 0 \
   "31OL2: gate metal edge must be at least 1.2 microns (3 lambda) from parallel active area edge"

/* Level 4 – Ohmic Metal Definition   */

   width om,far_ohmic 5 \
   "44LW1: ohmic metal width must be at least 2.0 microns (5 lambda)"

   width om,via1sd 6 \
   "44LW1+ : ohmic metal width must be at least 2.4 microns (6 lambda)"

   spacing allohmic allohmic 4 touching_ok \
   "44SP1: ohmic metal/ohmic metal separation must be at least 1.6 microns (4 lambda)"

   width ohmic_contact 4 \
   "43OL1: ohmic contact width must be 1.6 microns (4 lambda)"

edge4way om oc,ocd 4 oc,ocd 0 0 \
"43OL2: ohmic contact must extend at least 1.6 microns (4 lambda) beyond edge"

edge4way oc,ocd om 5 om,vsd/substrate 0 0 \
"43OL2: ohmic metal must extend at least 2.0 microns (5 lambda) beyond edge"

edge4way oc,ocd allgate 3 allgate 0 0 \
"43OL3: gate metal must extend at least 1.2 microns (3 lambda) beyond ohmic contact"

/* Level 5 – Via 1 Definition   */

width via1g 6 \
"55LW1: via1g width must be at least 2.4 microns (6 lambda)"

width via1sd 7 \
"55LW1: via1sd width must be at least 2.8 microns (7 lambda)"

spacing vg vsd 2 touching_illegal \
"55SP1: via1g/via1sd separation must be at least 0.8 microns (2 lambda)"

spacing vg vg 2 touching_ok \
"55SP1: via1g/via1g separation must be at least 0.8 microns (2 lambda)"

spacing allgate via1sd 1 touching_illegal \
"54OL1: gate metal/via1sd separation must be at least 0.4 microns (1 lambda)"

/* Level 6 – Metal 1 Interconnect Definition   */

width allm1 5 \
"66LW1: metal 1 width must be at least 2.0 microns (5 lambda)"

spacing allm1 allm1 5 touching_ok \
"66SP1: metal 1/metal 1 separation must be at least 2.0 microns (5 lambda)"

/* Level 7 – Via 2 Definition   */

width via2 8 \
"77LW1: via 2 width must be at least 3.2 microns (8 lambda)"

spacing via2 via2 2 touching_ok \
"77SP1: via2/via2 separation must be at least 0.8 microns (2 lambda)"

/* Level 8 – Metal 2 Interconnect Definition   */

width allm2 6 \
"88LW1: metal 2 width must be at least 2.4 microns (6 lambda)"

spacing allm2 allm2 7 touching_ok \
"88SP1: metal 2/metal 2 separation must be at least 2.8 microns (7 lambda)"

/* Level 9 – Via 3 Definition   */

width via3 15 \
"99LW1: via3 width must be at least 6.0 microns (15 lambda)"

/* Level 10 – Metal 3 Definition */

    width allm3 15 \
    "1010LW1: metal 3 width must be at least 6.0 microns (15 lambda)"

    spacing allm3 allm3 23 touching_ok \
    "1010SP1: metal 3/metal 3 separation must be at least 9.2 microns (23 lambda)"

/* Level 11 – Passivation Contact Definition */

    width pad 275 \
    "BP4: pads must be at least 110 microns (275 lambda)"

    spacing pad pad 100 touching_illegal \
    "BP4: pad/pad separation must be at least 40 microns (100 lambda)"

/* Bonding Pads */

    spacing pad allm3 23 touching_ok \
    "1010SP1: pad/metal 3 separation must be at least 9.2 microns (23 lambda)"

    spacing pad/metal2 allm2 7 touching_ok \
    "88SP1: pad/metal 2 separation must be at least 2.8 microns (7 lambda)"

    edge4way space pad 275 space 0 0 \
    "BP-6: Bonding pad not allowed over circuitry" substrate

end

router
end

/* - - - - - Extractor section — NOTE(!) - - - - - - - - - - - -

Because of the requirement that all capacitances be specified as an integral number of attoFarads per square lambda, the extremely small actual capacitance values, and the need for accuracy in SPICE simulations, we have specified the values in this section 1000 TIMES LARGER THAN ACTUAL. It is therefore necessary to use the UCSB-modified 'sim2spice' program in order that the values get properly re-scaled when SPICE decks are generated.

- - - - - - - - - - - - - - - - - - - NOTE(!) - - - - - - - - - - - -
*/

extract
    style   default
    lambda   40

    resist   gm   1500   /* milliohms per square */
    resist   om   10000
    resist   m1   70
    resist   m2   50
    resist   m3   25

    overlap   m3   m1   8640   /* attoFarads (X1000) / square lambda */
    overlap   m3   m2   5600

overlap	m3	gm	4640				
overlap	m3	om	4640				
areacap		m3	3680	/* attoFarads (X1000) / square lambda */
perimc	m3	m2	1920	/* attoFarads (X1000) / lambda */
perimc	m3	m1	1400				
perimc	m3	gm	1200				
perimc	m3	om	1200				
perimc	m3	space	1400				
overlap	m2	m1	12160				
overlap	m2	gm	8480				
overlap	m2	om	8480				
areacap		m2	5600				
perimc	m2	m1	1960				
perimc	m2	gm	1800				
perimc	m2	om	1800				
perimc	m2	space	1680				
overlap	m1	gm	21440				
overlap	m1	om	21440				
areacap		m1	9120				
perimc	m1	gm	2040				
perimc	m1	om	2040				
perimc	m1	space	1760				
areacap		gm	12640				
perimc	m1	space	1800				
fet	efet	active		2	e	!GND	0	0
fet	dfet	active,om,ocd,oc		1	d	!GND	0	0

end

wiring

contact	via1sd	7	act	0	m1	0
contact	via1g	6	gm	0	m1	0
contact	via2	8	m1	0	m2	0
contact	via3	15	m2	0	m3	0

end

plowing
end

plot

    style versatec

        /* Same as Gremlin stipple 9: */
        efet \
            07c0   0f80   1f00   3e00 \

MAGIC TECHNOLOGY FILE FOR GaAs SELF-ALIGNED E/D-MODE PROCESS

```
    7c00    f800    f001    e003 \
    c007    800f    001f    003e \
    007c    00f8    01f0    03e0
/*  Same as Gremlin Stipple 10:  */
dfet,ocd \
    1f00    0f80    07c0    03e0 \
    01f0    00f8    007c    003e \
    001f    800f    c007    e003 \
    f001    f800    7c00    3e00
/*  Same as Gremlin stipple 11:  */
via1sd \
    c3c3    c3c3    0000    0000 \
    0000    0000    c3c3    c3c3 \
    c3c3    c3c3    0000    0000 \
    0000    0000    c3c3    c3c3
/*  Same as Gremlin stipple 12:  */
pad \
    0040    0080    0100    0200 \
    0400    0800    1000    2000 \
    4000    8000    0001    0002 \
    0004    0008    0010    0020
/*  Same as Gremlin stipple 17:  */
act,via1sd/substrate,efet,dfet \
    0000    4242    6666    0000 \
    0000    2424    6666    0000 \
    0000    4242    6666    0000 \
    0000    2424    6666    0000
/*  Same as Gremlin stipple 19:  */
gm,via1g/substrate,efet,dfet,oc,ocd \
    0808    0400    0202    0101 \
    8080    4000    2020    1010 \
    0808    0004    0202    0101 \
    8080    0040    2020    1010
/*  Same as Gremlin stipple 30:  */
oc,ocd,om \
    1414    2222    0000    2222 \
    4141    2222    0000    2222 \
    1414    2222    0000    2222 \
    4141    2222    0000    2222
/*  Same as Gremlin stipple 22:  */
m1,via1g/metal1,via1sd/metal1,pad \
    8080    0000    0000    0000 \
    0808    0000    0000    0000 \
    8080    0000    0000    0000 \
    0808    0000    0000    0000
```

```
            /*  Same as Gremlin stipple 23:    */
            pad \
                0000    0000    1c1c    3e3e \
                3636    3e3e    1c1c    0000 \
                0000    0000    1c1c    3e3e \
                3636    3e3e    1c1c    0000

            /*  Same as Gremlin stipple 28:    */
            m2,v2,pad \
                0000    1111    0000    0000 \
                0000    1111    0000    0000 \
                0000    1111    0000    0000 \
                0000    1111    0000    0000

            /*  Same as Gremlin stipple 15:    */
            m3,v3,pad \
                2020    2020    2020    2020 \
                2020    2020    2020    2020 \
                0000    0000    0000    0000 \
                0000    0000    0000    0000

            vg,vsd,v2,v3         X

style gremlin
        efet                             9
        dfet,ocd                        10
        pad                             12
        act,vsd/substrate,efet,dfet,ocd 17
        gm,vg/substrate,efet,dfet,ocd   19
        m1,vsd,vg,v2/metal1,pad         22
        oc,ocd,om                       30
        pad                             23
        m2,v2,pad                       28
        m3,v3,pad                       15
        vg,vsd,v2,v3                     X
end
```

INDEX

A

AND function, 234
AND/OR/INVERT function, 225, 351–352
AOI, 351–352
Arbitration delay, 363
Asynchronous handshake, 356

B

Backgating effects, 46, 111
Barrier height, 21
Boston geometry, 386
Branch trailer, 367–368
Bus throughput, 362–363

C

Caltech Interchange Format, 386
    C primitive, 387
    DF primitive, 387
    DS primitive, 387
    L primitive, 387
    symbol mirroring, 388
    symbol rotation, 388
    symbol translation, 388
Cell library, 417
Characteristic impedance, 248
CIF format. *See* Caltech Interchange Format
Circuit extraction, 428–432

Clock
    distribution, 354
    loading, 355
    phasing, 355–356
    skew, 354
Common mode rejection ratio, 217
Control/decode circuits, 412
Coplanar lines, 255, 257
Crosstalk, 287, 350
Current crowding, 25

D

Depletion layer depth, 27
Depletion mode logic circuits, 196
    buffered FET logic (BFL), 203
    capacitive feedforward logic circuits (CDFL), 198
    enhancement/depletion BFL, 223
    pass transistor logic, 225
    unbuffered FET logic (FL), 151, 196
Design cycle, 415–418
    architecture and organization phase, 416
    behavioral specification phase, 416
    subsystem implementation phase, 417
Design "logs," 389
Design rules, 384
    electrical, 384, 395, 401
    geometric, 384
    coincident edges, 395, 400
    d-mode active devices, 390–393
    d-mode contacts, 393–394

d-mode summary, 391
d-mode width/separation, 390
E/D active devices, 398
E/D width/separation, 397–398
maximum width MESFET, 396, 400
Design styles, 384
    lambda-based, 384
    register transfer, 410–411
    scalable, 384
Diametrical inversion, 378–379
Dielectric constant, 248
Differential amplifier, 217
Diode layout style, 391–392
Diode logic, 209
Double-rail encoding, 376
Drift velocity, 13,15
Dynamic logic circuits, 231
    asynchronous dynamic circuit, 238
    domino logic, 233
    minimum clock frequency, 231
    single-phase dynamic circuits, 233
    two-phase dynamic circuits, 237
Dynamic memory, 409
Dynamic pull-down, 205

E

E/D self-aligned process, 55, 396–401
    contact-0, 399
    design rules, 397–401
    ohmic-contact, 399
    pad, 400
    primary layers, 397
    source-drain-metal, 399
    via1g, 398
    via1sd, 398
    via2, 399
    via3, 399
Electrical rules check, 441–444, 449–450, 452
Electromigration, 262, 395–396
Energy band diagram, 12, 61
Enhancement/depletion mode logic circuits, 210
    complementary JFET logic, 230
    direct coupled FET logic (DCFL), 167, 211
    e/d BFL, 223
    source coupled FET logic, 216
    superbuffer FET logic, 215, 369
Eqntott, 456
Equipotential region, 376
Esim, 447–449, 451–452

F

Fabrication, 53
    air bridge, 55
    directly aligned gate fabrication, 56
    lift off, 55
    self-aligned gate fabrication, 55
Fetch/execute cycle, 364–366
Finite state machine
    implemented using PLA, 407–409
    single-phase clocking requirements, 408
Flow control, 355
Foundry interface, 384, 418, 435–436
Function blocks, 410

G

Gallium arsenide
    material properties, 13
    semi-insulating, 17

H

Handshake
    2-cycle, 358
    4-cycle, 357–358
Heterojunction, 58
    energy band diagrams, 61
    pseudomorphic, 60
    quantum well, 60
Heterojunction FET, 58
    HIGFET, 69
    MODFET, 64
    p-channel, 68
    SISFET, 67
    transconductance, 64
Heterostructure bipolar transistor model, 120–134
    base resistance, 126
    basewidth modulation, 132
    charge storage, 126
    collector-base capacitance, 131
    diffusion capacitances, 127
    Early voltage, 133
    Ebers-Moll model, 122
    high current region, 131
    junction capacitances, 127
    low current region, 129
    model parameters, 134
    parasitic resistances, 43, 124

Heterostructure bipolar transistors (HBT), 69, 120–134
  current gain, 70
  switching delay, 71
Hierarchical design, 429
Hot spot contention, 363
Hybrid semiconductor module, 352–353, 368

I

ICN, 361–363
Instruction set
  branch trailer, 367–368
  reduced, 347–348
Interconnection network, 361–363
  throughput, 362–363
Interconnections
  attenuation, 278, 307
  bus-type, 309, 356–361, 364
  crossover capacitance, 270
  crosstalk, 287, 350
  delay estimation, 269, 280
  effective source resistance of driver, 271
  even and odd modes, 289
  impedance-matched, 350
  maximum length, 278
  parallel, 349
  point-to-point, 310, 363–364
  serial, 349
  skin effect, 276, 307, 350
  termination resistor, 306
  unterminated, 306
  wiring, 253
Inverter, 1, 143, 149

J

Junction capacitance, 30

L

Labels, 430, 440
  global, 432
  local, 432
Lambda-based design, 385
Layout examples
  2-input NOR (E/D), 402
  3-input NOR (SDFL), 402
  4-input NOR (d-mode), 401
  GaAs-to-GaAs pad driver, 403
  latch (FL), 402–403
Level-shift circuit, 24, 204
Level-shift diode, 391
Logic circuit design
  channel resistance, 4
  dc design, 1, 149
  fall time, 172
  fan-in, fan-out effects, 170, 188
  forward conduction limit, 4
  graphical analysis, 150
  input capacitance of logic gate, 175
  inverter, 2
  large signal equivalent capacitance, 175
  level shift diodes, 155
  maximum clock frequency, 189, 329
  Miller effect, 176
  noise margin. *See* Noise
  power dissipation, 162
  power-delay product, 187
  propagation delay, 172, 326
  rise time, 172
  threshold voltage, 154
  transient response, 5, 170–186
  width ratio, 153
  wiring capacitance, 253, 270
  yield, 164–170
Logic voltage levels, 143–148

M

MAGIC, 418–441
  box and cursor, 437–439
  circuit extraction, 428–432
  corner-stitching, 419–420
  design-rule check, 424
  display styles, 423, 462–463, 470–471
  edge-based design rule check, 424–428
  planes, 419, 462, 470
  plotting styles, 423–424, 467–468, 479–480
  technology files, 419, 461–480
  tiles, 419, 462, 470
Manhattan geometry, 386
MESFET, Gallium Arsenide
  backgating, 46, 111
  current gain-bandwidth product (f_T), 42, 171
  cutoff, 33

MESFET, Gallium Arsenide (cont.)
  depletion mode, 28
  doping profile, 28
  drain conductance, 37, 116
  enhancement mode, 28
  fabrication methods, 53–56
  frequency dependent output conductance, 49, 116
  gate conduction, 44, 92
  gate length dependence of drain current, 35
  gradual channel model, 34
  lag effect, 49, 116
  ohmic or linear region, 33
  output conductance, 116
  parasitic capacitance, 43, 108
  p-channel, 16
  pinch-off voltage, 26
  regions of operation, 32
  saturation region, 34
  short-channel effect, 56
  sidegating, 46, 111
  source resistance, 6, 43
  structure, 11
  substrate/channel interface, 46
  subthreshold current, 50, 114
  temperature dependence, 50, 120
  threshold voltage, 26
  transconductance, 39
  velocity saturation, 35
Mobility, 13
MODFET. See Heterojunction FET
MOSFET, silicon, 3, 140
Muller signaling, 377

N

NAND function, 107, 160
Node names, 430, 440
Noise
  dynamic noise margin, 190
  factors influencing noise margin, 148
  intrinsic noise margin, 144
  maximum square noise margin, 147
  maximum width noise margin, 146
  slope = −1 definition, 147
  sources, 142
Non-restoring logic, 226
NOR function, 159, 204, 207, 223
NRZ signaling, 377

O

OAI, 351–352
Ohmic contact, 21
OR/AND/Invert, 351–352
OR/NAND function, 208
OR/NAND/wired AND function, 208

P

Package, hybrid, 352–353, 368
Packaging, 301
  package trace impedance, 303
Pad driver
  distributed NOR, 372–373
  GaAs-to-ECL, 370
  GaAs-to-GaAs, 368–369, 403
Pad receiver
  ECL-to-GaAs, 372
  ECL-to-GaAs with threshold adjust, 374–375
  GaAs-to-GaAs, 371
  GaAs-to-GaAs bus-type, 373–374
Peak load current–static current ratio, 203, 214
Phase velocity, 246
Pipeline interlock, 367
Pipelining, 364, 366
PLA generation, 455–457
  mpla, 457
  personality matrix, 456
  truth table form, 456
Position-time diagram, 252
Power dissipation, 162
Power supply and ground design, 260, 311
  bypass capacitor, 311
  electromigration, 262, 395
  inductance, 263
  resistance, 261
Printed circuit boards, 301
Probe. See Testing, wafer probing
Producer/consumer transactions, 364, 367
Propagation delay test structures, 326
  asynchronous test structure, 326
  delayed Johnson counter, 328
  D-type flip flop, 222, 327
  frequency divider, 327
  ring oscillator, 326

INDEX **485**

Q

Quasi-complementary buffer, 215

R

Ratioed logic, 231
Reflections
    reflection coefficient, 250
    reflection diagram, 252
    transmission coefficient, 251
Register transfer design, 410–411
RISC, 347–348, 366–368
RTZ signaling, 377

S

Saturated drift velocity, 13
Scalable design, 385
Schottky barrier diode, 18, 80
    barrier height, 21
    current crowding, 25, 80
    depletion layer depth, 27
    junction capacitance, 30, 81
    series resistance, 25, 80
    SPICE diode model parameters, 82
    temperature dependence, 53
Schottky diode FET logic, 206
Selective-implant process, 56, 388
    composite layers, 389–390
    primary mask layers, 388–389
Semi-insulating GaAs, 17
Signal names, 430, 440
Signaling
    2-cycle, 358, 377
    4-cycle, 357, 377
    delay-insensitive, 376
    NRZ, 358, 377
    RTZ, 358, 377
    self-timed, 376–377
Sim2spice, 453–454
Skin effect, 276, 307, 350
Source follower, 200
SPICE, 183
    backgating model, 111–113
    channel-length modulation parameter, 88
    drain current transient effects, 116
    dual gate MESFET model, 106–108
    edge fringing capacitance, 109
    frequency dependent output conductance, 49
    gate capacitance, 93, 104
    gate current model, 92
    geometric fringing capacitance, 110
    hyperbolic tangent (Curtice) model, 95–100
    input generation from extracted circuit, 453–454
    lag effect, 48
    MESFET models, 86–110
    model parameters, 94, 97, 105
    parameter extraction, 90, 102
    parameter optimization, 96
    parasitic capacitances, 43, 108
    space-charge-limited current, 115
    SPICE JFET model, 88–95
    SPICE3—Raytheon model, 100–105
    substrate leakage current, 111, 114
    subthreshold current model, 114
    temperature effects, 50, 120
    transconductance parameter, 51, 88
StatG, 441–444, 449–450, 452
Steering logic, 409–410
Superbuffer, 214–215, 369
Switch-level simulation, 444–449
Synchronous design, 354

T

Ternary encoding, 378
Ternary inversion, 378–379
Ternary NAND, 378
Ternary NOR, 378
Testing, high speed
    built-in test, 325, 330
    electrooptical methods, 341
    parallel multiplier, 331
    propagation delay, 326
    pseudo-random testing, 337
    scan-path techniques, 334
    selectable feedback path, 331
Testing, wafer probing, 321
    clock feedthrough, 324
    common mode crosstalk, 324
    coplanar probes, 325
Threshold adjust, 374–375
Transconductance, 6, 39, 64

Transmission lines
    backward coupling, 294, 297
    characteristic impedance, 248
    coplanar stripline, 257
    coplanar waveguide, 255
    crosstalk, 287, 350
    effective dielectric constant, 248, 254, 291
    elliptic integral, 256
    even and odd modes, 289
    forward coupling, 295, 297
    impedance discontinuities, 251, 303
    microstrip, 257
    RC approximation, 299
    reflection coefficient, 250
    relative dielectric constant, 248
    source termination, 309, 322
    stripline, 258
    superposition of modes, 291
    theory, 245

U

Unbuffered FET logic, 151, 196

V

Velocity of propagation, 246
Von Neumann computer, 348, 364–365

Y

Yield analysis, 164–170